Special Functions for Applied Scientists

A.M. Mathai • Hans J. Haubold

Special Functions for Applied Scientists

 Springer

A.M. Mathai
McGill University
Montreal, QC
Canada
and

Centre for Mathematical Sciences
Pala Campus, Kerla
India
mathai@math.mcgill.ca

Hans J. Haubold
Office for Outer Space Affairs
UN, Vienna
Austria
hans.haubold@unvienna.org

ISBN 978-1-4419-2610-4   e-ISBN 978-0-387-75894-7
DOI: 10.1007/978-0-387-75894-7

Preface

The first S.E.R.C. (Science and Engineering Research Council of the Department of Science and Technology, Government of India, New Delhi) School on Special Functions was sponsored by DST (Department of Science and Technology), New Delhi, and conducted by CMS (Centre for Mathematical Sciences) for six weeks in 1995. The second S.E.R.C. School on Special Functions and Functions of Matrix Argument was sponsored by DST, Delhi, and conducted by CMS for five weeks during the period 29th May to 30th June 2000. In the second School, the main lectures were given by Dr. H.L. Manocha of Delhi, India, Dr. S. Bhargava of Mysore, India, Dr. K. Srinivasa Rao and Dr. R. Jagannathan of Chennai, India and Dr. A.M.Mathai of Montreal, Canada. Supplementary lectures were given and problem sessions were supervised by Dr. K.S.S. Nambooripad and Dr. S.R. Chettiyar of CMS, Dr. R.N. Pillai (retired), Dr. T.S.K. Moothathu and Dr. Yageen Thomas of the University of Kerala and Dr. E. Krishnan of the University College Trivandrum. Lecture Notes were brought out as Publication No.31 of CMS soon after the School was completed. The first two Schools were conducted in Trivandrum (Thiruvananthapuram) area. Dr. A.M. Mathai was the Director and Dr. K.S.S. Nambooripad was the Co-Director of these two Schools.

The third S.E.R.C. School on Special Functions and Functions of Matrix Argument: Recent Developments and Recent Applications in Statistics and Astrophysics, sponsored by DST, Delhi, was conducted for five weeks from 14th March to 15th April 2005 by CMS at its Pala Campus, Kerala, India. This time DST, wanted the lecture notes to be collected from the main lecturers in advance, compiled and distributed prior to the start of the School. The main lectures for the 3rd School were given by Dr. Hans J. Haubold of the Office of Outer Space Affairs of the United Nations, Dr. Serge B. Provost, Professor of Actuarial Sciences and Statistics of the University of Western Ontario, Canada, Dr. R.K. Saxena of Jodhpur, India, Dr. S. Bhargava of Mysore, India and Dr. A.M. Mathai of Montreal, Canada. Supplementary lectures were given by Dr. A. Sukumaran Nair (Chairman, CMS), Dr. K.K. Jose (Director-in-Charge, CMS Pala Campus), Dr. R.N. Pillai, Dr. Yageen Thomas, Dr. V. Seetha Lekshmi, Dr. Alice Thomas, Dr. E. Sandhya, Dr. S. Satheesh and

Dr. K. Jayakumar. Problem sessions were supervised by Dr. Sebastian George and other lecturers. Extra training in the use of statistical packages, LATEX and Mathematica/Maple were given by Joy Jacob, Seemon Thomas and K.M. Kurian of the Department of Statistics of St. Thomas College, Pala. A special lecture sequence on Matlab was conducted by Alexander Haubold of Columbia University, U.S.A.

The 4th S.E.R.C. School on Special Functions and Functions of Matrix Argument: Recent Advances and Applications to Stochastic Processes, Statistics and Astrophysics, sponsored by DST, Delhi, was conducted by CMS for five weeks from 6th March to 7th April 2006 at its Pala Campus. The main lectures were scheduled to be given by Dr. Hans Joachim Haubold of the Office of Outer Space Affairs of the United Nations, Dr. P.N. Rathie of the University of Brasilia, Brazil, Dr. S. Bhargava of Mysore, India, Dr. R.K. Saxena of Jodhpur, India, Dr. K.K. Jose of Pala, India and Dr. A.M. Mathai of Montreal, Canada. Supplementary lectures were given by Dr. A. Sukumaran Nair (Chairman, CMS), Dr. K.S.S. Nambooripad (Director-in-Charge, CMS Trivandrum Campus), Dr. R. Y Denis of Gorakhpur, India, Dr. N. Unnikrishnan Nair (former Vice-Chancellor of Cochin University of Science and Technology, Kerala, India), Dr. Yageen Thomas and Dr. K. Jayakumar. One week was devoted to Stochastic Processes and recent advances in this area in the 4th School. Problem sessions were supervised by Dr. Joy Jacob, Dr. Sebastian George and other lecturers.

The 5th S.E.R.C. School in this sequence, titled "Special Functions and Functions of Matrix Argument: Recent Advances and Applications in Stochastic Processes, Statistics, Wavelet Analysis and Astrophysics" was conducted by CMS at its Pala Campus from 23rd April to 25th May 2007. The lecture notes for the 5th School were assembled by November 2006 and were printed and distributed as Publication No.34 in the Publications Series of CMS. The main lectures were scheduled to be given by Dr. Hans J. Haubold (Office of Outer Space Affairs of the United Nations, Austria), Dr. H.M. Srivastava (University of Victoria, Canada), Dr. R.Y. Denis (Gorakhpur University, India), Dr.R.K. Saxena (Jodhpur University, India), Dr. S. Bhargava (Mysore University, India), Dr. D.V. Pai (IIT Bombay, India), Dr. Yageen Thomas (University of Kerala, India), Dr. K.K. Jose (Mahatma Gandhi University, India), Dr. J.J. Xu (China/Canada), Dr. K. Jayakumar (Calicut University, India) and Dr. A.M. Mathai (Canada/India). But Dr. H.M. Srivastava, Dr. S. Bhargava and Dr. Xu could not reach the venue on time due to unexpected emergencies. Supplementary lectures were given by Dr. A. Sukumaran Nair (Chairman,CMS), Dr. R.N. Pillai (former Head, Department of Statistics, University of Kerala, India), Dr. N. Mukunda (IISc, Bangalore, India) and the problem sessions were supervised by Dr. Sebastian George, Dr. Joy Jacob, Dr. Seemon Thomas and the lecturers. For the 2005, 2006 and 2007 Schools, Dr. A.M. Mathai was the Director and Dr. K.K. Jose was the Co-Director of the Schools.

The participants for the S.E.R.C. Schools are selected on all-India basis. All the expenses of the selected candidates, total number of seats is 30, including travel, accommodation, food, lecture materials, local travels etc are met by the DST, Delhi, Government of India. Foreign participation is allowed under the conditions that the participants must come with their own return air tickets and must attend all the lectures, and problem sessions and take all the examinations and tests from the beginning till the end. All their local expenses are met and lecture materials are provided by the Schools. There is no fee for attending the Schools. The Schools are mainly research orientation courses aimed at young faculty members in colleges and universities across India, below 35 years of age, and fresh graduates with M.Sc, M.Phil, Ph.D degrees below 30 years of age, in Mathematics/Statistics/Theoretical Physics. In most of the Indian universities rigid compartmentalization is the order and as a result a M.Sc graduate in mathematics may not have even taken a very basic course in probability and statistics. Even though basic differential and integral calculus and matrix theory are required subjects for statistics and physics students, these students may have forgotten these subjects because they teach the compartmentalized topics in their own areas and do not usually do research even in their narrow areas of their own fields. For maintaining their jobs and getting regular increments in their salaries and all other monetary and other benefits, research work and further reading and learning process are not required of them. As a result, the quality of teaching and the information passed on to the students go down from year to year. In order to remedy this situation a little bit, S.E.R.C. Schools in various areas were started by DST, Delhi. Dr. A.M. Mathai was asked to run S.E.R.C. Schools in mathematics. For taking up a challenging research problem in any applied area a good background in basic mathematics, probability and statistics is required. In order to give the basic ideas in probability and statistics and to bring a number of topics in the area of Special Functions and Functions of Matrix Argument and their applications to the current research level, these S.E.R.C. Schools on Special Functions were established. The current sequence of five Schools really achieved the aim and almost all the participants in the first four Schools have become research oriented towards a career in research and teaching, and the participants of the 5th School are also expected to follow the same footsteps of their predecessors.

Chapter 1 introduces elementary classical special functions. Gamma, beta, psi, zeta functions, hypergeometric functions and the associated special functions, generalizations to Meijer's G and Fox's H-functions are also examined here. Discussion is confined to basic properties and some applications. Introduction to statistical distribution theory is given here. Some recent extensions of Dirichlet integrals and Dirichlet densities are also given. A glimpse into multivariable special functions such as Appell's functions and Lauricella functions is also given. Special functions as solutions of differential equations are also examined here.

Chapter 2 is devoted to fractional calculus. Fractional integrals and fractional derivatives of various kinds are discussed here. Then their applications to reaction-diffusion problems in physics, input-output analysis and Mittag-Leffler stochastic processes are examined here. Chapter 3 deals with q-hypergeometric or basic hypergeometric functions and Chapter 4 goes into basic hypergeometric functions and Ramanujan's work on elliptic and theta functions. Chapter 5 examines the topic of Special Functions and Lie Groups.

Chapters 6 to 9 are devoted to applications of Special Functions to various areas. Applications to stochastic processes, geometric infinite divisibility of random variables, Mittag-Leffler processes, α-Laplace processes, density estimation, order statistics and various astrophysics problems, are dealt with in these chapters.

Chapter 10 is devoted to Wavelet Analysis. An introduction to wavelet analysis is given here. Chapter 11 deals with the Jacobians of matrix transformations. Various types of matrix transformations and the associated Jacobians are given here. Chapter 12 is devoted to the discussion of functions of matrix argument. Only the real case is considered here. Functions of matrix argument and the pathway models, recently introduced by Mathai (2005), are also discussed here, along with their applications to various areas.

In all the S.E.R.C. Schools conducted under the Directorship of Dr. A. M. Mathai a serious effort is made so that the participants absorb the materials covered in the Schools. The classes started at 8.30 am and went until 6.00 pm. The first lecture of 08.30 to 10.30 was followed by a problem session from 10.30 to 13.00 hrs on the materials covered in the first lecture. The second lecture of the day was from 14.00 to 16.00 hrs followed by problem session from 16.00 to 18.00 hrs. At the end of every week a written examination was conducted, followed by a personal interview of each participant by the lecturer of that week in the form of an oral examination. Cumulative grades of such weekly examinations appeared in the final certificates distributed to them. The main aim was to inculcate in them a habit of long and sustained hard work, which would help them in their careers whatever they may be. During the first week the participants, especially the teachers from colleges and universities, found it difficult to adjust to the routine of long hours of hard work but starting from the second week, in all the Schools, the participants started enjoying, especially the problem sessions, because for the first time, they started understanding and appreciating the meanings and significance of theorems that they learnt or memorized when they were students.

The lecture notes are written up in a style for self-study. Each topic is developed from first principles with lots of worked examples and exercises. Hence the material in this book can be used for self-study and will help anyone to understand the basic ideas in the area of Special Functions and Functions of Matrix Argument and they will be able to make use of these results in their own problems in applied areas, especially in Statistics, Physics and Engineering problems. Applications in various areas are illustrated in this book. Insights into recent developments in the applications of fractional calculus, in the developments in various other topics are also given in the book so that the readers who are interested in any of the topics discussed in the book can directly go into a research problem in the topics.

Several people have contributed enormously for the success of the S.E.R.C. Schools and in making the publications of four Lecture Notes and this final publication of summarized lecture notes possible. Dr. B.D. Acharya, Advisor to Government of India and Dr. Ashok K. Singh of the Mathematical Sciences Division of DST, New Delhi, are the driving force behind the re-energized mathematical activities in India now. They were kind enough to pursue the matter and get the funds released for running the Schools as well as for the preparation of various publications, including this one. Since the basic materials for this publication are supplied by various lecturers in the form of their lecture notes, there will be some overlaps. Very obvious inconsistencies are removed but some overlapping materials are left there to make the discussions self-contained. Dr. R.K. Saxena, Dr. S. Bhargava, Dr. H.J. Haubold, Dr. P.N. Rathie, Dr. K.K. Jose, Dr. K. Jayakumar, Dr. H.L. Manocha, Dr. R. Jagannathan, Dr. K. Srinivasa Rao, Dr. K.S.S. Nambooripad, Dr. Serge B. Provost, Dr. Yageen Thomas, Dr. D.V. Pai, and Dr. A.M. Mathai are thanked for making their notes available in advance for the S.E.R.C. Schools. Most of the material was typeset at CMS office by Miss K.H. Soby, Dr. Joy Jacob, Seemon Thomas, Dr. Sebastian George, Dr. K.K. Jose and Dr. A.M. Mathai. Part of the material was typeset by Barbara Haubold of the United Nations, Vienna Office, fully free of charge as a voluntary service to the Schools. Notes and programs for a series of lectures on Matlab were supplied to CMS by Alexander Haubold of Columbia University, USA. Those notes are not included in this book to keep the materials within the focus of the book. Dr. Sebastian George, Dr. Joy Jacob, Dr. Seemon Thomas, K.M. Kurian, Jaisymol Thomas and Ashly P. Jose, who spent a lot of time in running problem sessions, in running separate sessions on the use of statistical packages and Maple program, use of LATEX etc and for checking the typed materials, are thanked. CMS would like to thank each and every one who helped to make S.E.R.C. Schools on Special Functions a grand success and who helped to make this publication possible.

The authors would like to express their sincere thanks to the Department of Science and Technology, Government of India, New Delhi, India, for the financial assistance under the project No. SR/S4/MS:287/05 titled "Building up a Core Group of Researchers/Faculty and Facilities at CMS, Trivandrum and Pala Campuses", which enabled the collaboration on this book project possible. We express our sincere gratitude to the Management and especially to Dr. Mathew John K. (Principal), St. Thomas College Pala, Kerala, India, for providing all facilities in the College during the preparation of this book.

Pala, Kerala, India   A.M. Mathai

1st February, 2007   H.J. Haubold

Glossary of Symbols

$J_{x,\infty}^{\alpha,\beta,\eta}(\cdot)$	fractional integral / derivative of the first kind	Definition 2.7.8	117
$T_n(Z),\, u_N(Z)$	Chebyshev polynomials	Exercises 3.1.12	138
C_n^m	Gegenbauer (ultraspherical) polynomial	Exercise 3.1.14	138
$(a;q)_n$	q-shifted factorial	Section 3.2	139
$[\alpha]_q$	$[1-q^\alpha]/[1-q]$	Notation 3.2.1	139
$_r\phi_s(a_1,...,a_r;$ $b_1,...,b_s;q,z)$	q-hypergeometric function	Definition 3.2.2	139
$(a;q)_\infty$	q-shifted factorial	Notation 3.2.3.	140
$e_q(z),\, E_q(z)$	q-exponential functions	Definition 3.2.4	145
$\begin{bmatrix} n \\ k \end{bmatrix}_q$	q-binomial coefficient	Definition 3.2.5	148
D_q	q-derivative operator	Definition 3.3.1	149
$\int_a^b f(z)d_qz$	q-integral	Definition 3.3.2	150
$\bar{\Gamma}_q(x)$	q-gamma function	Definition 3.4.1	151
$\bar{B}_q(x,y)$	q-beta function	Definition 3.4.2	152
$\sin_q(x),\cos_q(x)$	q-trigonometric functions	Exercise 3.5.4	154
$\mathrm{Sin}_q(x),\mathrm{Cos}_q(x)$	q-trigonometric functions	Exercise 3.5.4	154
$\theta_3(q,z)$	Jacobi theta function	Remark 4.1.1	159
$(a)_n = (a;q)_n,$ $(a)_\infty = \lim_{n\to\infty}(a;q)_n$	q-shifted factorials	Section 4.2	164
$a(q),b(q),c(q)$	cubic theta functions	Section 4.6	177
$a(q,z),b(q,z)$	cubic theta functions	Section 4.8.1.	193
$c(q,z),a'(q,z)$	cubic theta functions	Section 4.8.1.	194
$a(q,\tau,z),b(q,\tau,z)$	cubic theta functions	Section 4.9.1.	201
$c(q,\tau,z),a'(q,\tau,z)$	cubic theta functions	Section 4.9.1.	201
$\mathrm{Ker}f$	Kernel of f	Definition 5.1.9	213
$L_n^{(\alpha)}(x)$	Laguerre polynomial	Definition 5.6.1	231
$\{x(t,w);t \in T; w \in \Omega\}$	stochastic process	Section 6.1	248
$ML(\alpha)$	Mittag-Leffler distribution	Section 6.2.7	266
$SML(\alpha)$	semi-Mittag-Leffler distribution	Section 6.2.7	267
$AR(p)$	autoregressive model of order p	Section 6.3.2	270
$ARMA(p,q)$	autoregressive moving average models	Section 6.3.2	270
$TEAR(1)$	tractable exponential autoregressive process	Section 6.4	276
$TMLAR(1)$	tractable Mittag-Leffler autoregressive process	Section 6.4.1	276
$TSMLAR(1)$	tractable semi-Mittag-Leffler autoregressive process	Section 6.4.1	276
$NEAR(1)$	new exponential autoregressive process	Section 6.4.2	277
$EAR(1)$	exponential autoregressive process	Section 6.4.2	278

$NMLAR(1)$	new Mittag-Leffler autoregressive process	Section 6.4.3	278
$NSMLAR(1)$	new semi-Mittag-Leffler autoregressive process	Section 6.4.4	280
$MLTAR(1)$	Mittag-Leffler tailed autoregressive process	Section 6.5.2	283
$SMLTAR(1)$	semi-Mittag-Leffler tailed autoregressive process	Section 6.5.2	284
$MOSW(1)$	Marshall-Olkin Semi-Weibull process	Section 6.6.2	287
$MOSP(1)$	Marshall-Olkin semi-Pareto process	Section 6.6.2	288
$MOGW(1)$	Marshall-Olkin generalized Weibull process	Section 6.6.4	291
$AR(1)$	first order autoregressive process	Section 6.6.5	292
$Gamma(\alpha,\beta)$	gamma random variable	Section 7.5.1	306
$JacobiP(\cdot)$	Jacobi polynomial	Section 7.5.3	307
$HermiteH(\cdot)$	Hermite polynomial	Section 7.5.4	307
$x_{r:n}$	r^{th} order statistic from a sample of n observations	Section 8.1.1	313
$F_{r:n}(x)$	distribution function	Section 8.1.1	313
$f_{r:n}(x)$	density, r-th order statistic	Section 8.1.2	314
iid	independently and identically distributed	Section 8.1.2	314
$f_{r,s:n}(x)$	joint density, order statistics	Section 8.1.4	315
$\mu_{r:n}^{(k)}$	moments, order statistics	Section 8.1.4	315
$BLUE$	best linear unbiased estimator	Section 8.4.2	328
$R_\odot$	solar radius	Section 9.2.1	346
$M(r)$	solar mass at r	Section 9.2.1	346
$M_\odot$	solar mass	Section 9.2.1	346
$\rho(r)$	solar density at r	Section 9.2.1	346
$P(r)$	pressure at r	Section 9.2.1	347
$T(r)$	temperature at r	Section 9.2.1	347
$<\cdot>$	expected value	Section 9.2.3	349
$L_\odot$	solar luminosity	Section 9.2.3	350
BG	Boltzmann-Gibbs	Section 9.6	364
(f,g)	inner product	Section 10.1	390
$\|f\|_2$	norm	Section 10.1	390
$\langle f,g \rangle$	inner product	Section 10.1	390
$\psi_{a,b}(t)$	basic wavelet	Definition 10.1.3	393
$(W_\psi f)$	wavelet transform	Definition 10.1.3	393
$\psi^H(t)$	Haar wavelet	Definition 10.1.3	393
$tr(X)$	trace of X	Section 11.0	409
X'	transpose of X	Section 11.0	409

$dx \wedge dy$	wedge product	Section 11.0	409
$|(.)|$	determinant of $(\cdot)$	Section 11.1	409
J	Jacobian	Section 11.1	410
$\Gamma_p(\alpha)$	real matrix-variate gamma	Definition 11.2.1	418
$V_{p,n}$	Stiefel manifold	Corollary 11.3.1	424
$B_p(\alpha, \beta)$	real matrix-variate beta	Definition 12.1.1	431
$L_f(t_1, ... t_k)$	Laplace transform	Section 12.2	435
$L_f(T^*)$	Laplace transform	Definition 12.2.1	435
$C_K(Z)$	zonal polynomial	Section 12.3.2	441
$\Gamma_p(\alpha, \kappa)$	generalized gamma	Section 12.3.2	442
$M_\alpha(f)$	M transform	Section 12.3.3	443

Contents

Chapter 1
Basic Ideas of Special Functions and Statistical Distributions

[This chapter is based on the lectures of Professor A.M. Mathai of McGill University, Canada (Director of the SERC Schools).]

1.0 Introduction

Some preliminaries of special functions and statistical distributions are given here. Details are available from the following sources, which are accessible to the participants of the SERC Schools:

1. *Notes of the 2nd SERC School.* (Publication No 31 of the Centre for Mathematical Sciences (CMS)), 2000.
2. *Notes of the 3rd SERC School.* (Publication No 32 of CMS), 2005.
3. *Notes of the 4th SERC School.* (Publications No 33,33A of CMS), 2006.
4. Mathai, A.M. (1993). *"A Handbook of Generalized Special Functions for Statistical and Physical Sciences"*, Oxford University Press, Oxford, U.K.
5. Mathai, A.M. and Saxena, R.K. (1978). *"The H-Function with Applications in Statistics and Other Disciplines"*, Wiley Halsted, New York.
6. Mathai, A.M. and Saxena (1973). *"Generalized Hypergeometric Functions with Applications in Statistics and Physical Sciences"*, Lecture Notes No 348, Springer-Verlag, Heidelberg, Germany.

Notation 1.0.1. Pochhammer symbol

$$(b)_r = b(b+1)\cdots(b+r-1), \quad (b)_0 = 1, \ b \neq 0. \tag{1.0.1}$$

For example,

$$\left(-\frac{1}{4}\right)_2 = \left(-\frac{1}{4}\right)\left(-\frac{1}{4}+1\right) = -\frac{3}{16}; \quad (-2)_3 = (-2)(-1)(0) = 0;$$

$$\left(\frac{1}{3}\right)_4 = \left(\frac{1}{3}\right)\left(\frac{1}{3}+1\right)\left(\frac{1}{3}+2\right)\left(\frac{1}{3}+3\right) = \frac{280}{81}; (7)_0 = 1; (0)_5 = \text{not defined.}$$

1

The following general property holds for m, n non-negative integers

$$(a)_{m+n} = (a)_m(a+m)_n = (a)_n(a+n)_m. \tag{1.0.2}$$

Notation 1.0.2.   **Factorial n or n factorial**

$$n! = (1)(2)\cdots(n), \ 0! = 1 \ (\text{convention }). \tag{1.0.3}$$

For example,

$$\frac{1}{3}! = \text{not defined}; (-5)! = \text{not defined}; 1! = 1; 0! = 1 \ (\text{convention});$$

$$3! = (1)(2)(3) = 6; \quad 4! = (1)(2)(3)(4) = 24.$$

Notation 1.0.3.   **Number of combinations of n taken r at a time**

$$\binom{n}{r} = \text{number of subsets of } r \text{ distinct objects from a set of } n \text{ distinct objects}$$

$$= \frac{n(n-1)\cdots(n-(r-1))}{r!} = \frac{n(n-1)\cdots(n-r+1)}{r!} = \frac{n!}{r!(n-r)!}, \ 0 \le r \le n. \tag{1.0.4}$$

For example,

$$\binom{4}{1} = \frac{4}{1!} = 4; \ \binom{4}{0} = \frac{4!}{0!(4-0)!} = \frac{4!}{1\,(4!)} = 1; \ \binom{4}{4} = \frac{(4)(3)(2)(1)}{4!} = 1;$$

$$\binom{n}{1} = \frac{n}{1!} = \binom{n}{n-1} = \frac{n(n-1)\cdots(n-(n-1))}{(n-1)!} = n; \ \binom{n}{r} = \binom{n}{n-r},$$

$$r = 0, 1, \cdots, n;$$

$$\binom{n}{0} = \binom{n}{n} = 1; \ \binom{n}{r} = \binom{n-1}{r} + \binom{n-1}{r-1};$$

$$\binom{1/4}{1} = \text{not defined as a combination};$$

$$\binom{-3}{2} = \text{not defined as a combination}; \ \binom{0}{2} = \text{not defined as a combination};$$

$$\binom{n}{r} = \frac{n(n-1)\cdots(n-r+1)}{r!} = \frac{(-1)^r(-n)(-n+1)\cdots(-n+r-1)}{r!}$$

$$= \frac{(-1)^r(-n)_r}{r!}. \tag{1.0.5}$$

If $\binom{n}{r}$ is interpreted not as the number of combinations but as in equation (1.0.5) then one can give interpretations when n is not a positive integer. For example

$$\binom{-1/3}{2} = \frac{(-1)^2 \left(\frac{1}{3}\right)_2}{2!} = \frac{(-1)^2}{2!} \left(\frac{1}{3}\right)\left(\frac{1}{3}+1\right) = \frac{1}{2!}\frac{4}{9} = \frac{2}{9};$$

$$\binom{1/2}{2} = \frac{(-1)^2 \left(-\frac{1}{2}\right)_2}{2!} = \frac{(-1)^2}{2!} \left(-\frac{1}{2}\right)\left(-\frac{1}{2}+1\right) = \frac{1}{2!}\frac{(-1)}{4} = -\frac{1}{8}.$$

1.1 Gamma Function

Notation 1.1.1. $\Gamma(z) = \mathbf{gamma}\ z$

A gamma function $\Gamma(z)$ can be defined in many ways. $\Gamma(z)$ exists for all values of z, negative, positive and complex values of z, except at $z = 0, -1, -2, \cdots$. Also $\Gamma(z)$ has an integral representation for $\Re(z) > 0$ where $\Re(\cdot)$ means the real part of $(\cdot)$. Thus we may note for example that $\Gamma(5)$ exists; $\Gamma(-\frac{1}{3})$ exists; $\Gamma(\frac{2}{5})$ exists; $\Gamma(0)$ does not exist; $\Gamma(-3)$ does not exist; $\Gamma(-\frac{7}{2})$ exists. Some of the definitions of $\Gamma(z)$ are the following:

Definition 1.1.1.

$$\Gamma(z) = \lim_{n \to \infty} \frac{n!\, n^z}{z(z+1)\cdots(z+n)}, \quad z \neq 0, -1, -2, \cdots \tag{1.1.1}$$

Definition 1.1.2.

$$\Gamma(z) = z^{-1} \prod_{n=1}^{\infty} \left(1 + \frac{1}{n}\right)^z \left(1 + \frac{z}{n}\right)^{-1}. \tag{1.1.2}$$

Definition 1.1.3.

$$\frac{1}{\Gamma(z)} = z \lim_{n \to \infty} \left\{ n^{-z} \prod_{k=1}^{n} \left(1 + \frac{z}{k}\right) \right\}. \tag{1.1.3}$$

Definition 1.1.4.

$$\frac{1}{\Gamma(z)} = z e^{\gamma z} \prod_{n=1}^{\infty} \left[\left(1 + \frac{z}{n}\right) e^{-\frac{z}{n}} \right] \tag{1.1.4}$$

where γ is the Euler's constant, defined as follows:

Notation 1.1.2. Euler's constant γ

$$\gamma = \lim_{n \to \infty} \left\{ 1 + \frac{1}{2} + \frac{1}{3} + \cdots + \frac{1}{n} - \ln n \right\} \approx 0.5772156649015328606065512. \tag{1.1.5}$$

Definition 1.1.5.

$$\Gamma(z) = p^z \int_0^{\infty} t^{z-1} e^{-pt} dt, \Re(p) > 0, \Re(z) > 0. \tag{1.1.6}$$

Definition 1.1.6.

$$\frac{1}{\Gamma(z)} = \frac{1}{2\pi i} \int_{c-i\infty}^{c+i\infty} t^{-z} e^t dt, \ c > 0, \Re(z) > 0, \ i = \sqrt{-1} \qquad (1.1.7)$$

where π is the mathematical constant,

$$\pi \approx 3.141592653589793238462643. \qquad (1.1.8)$$

1.1.1 Some basic properties of gamma functions

From all the definitions of $\Gamma(z)$ it is not difficult to show that

$$\Gamma(z) = (z-1)\Gamma(z-1) \qquad (1.1.9)$$

whenever the gammas are defined. It is obvious from the integral representation in (1.1.6). Take $p = 1$ and integrate by parts by using the formula

$$\int u dv = uv - \int v du$$

and by taking $dv = e^{-t}$ and $u = t^{z-1}$. If the integral representation is used then we need the conditions $\Re(z) > 0, \Re(z-1) > 0$ which means $\Re(z) > 1$. This restriction is not needed for the Definitions 1.1.1 - 1.1.4. [Verification and derivation of the result in (1.1.9) by using the Definitions 1.1.1 - 1.1.4 are left to the reader]. Continuing the process in (1.1.9) we have the following:

$$\Gamma(z) = (z-1)(z-2)\cdots(z-r)\Gamma(z-r) \qquad (1.1.10)$$

whenever the gammas exist. As a consequence of (1.1.10) we may note that for $n = 1, 2, \cdots$

$$\Gamma(n) = (n-1)(n-2)\cdots 1\Gamma(1) = (n-1)(n-2)\cdots 1 = (n-1)! \qquad (1.1.11)$$

since $\Gamma(1) = 1$. Thus, $\Gamma(z)$ can be looked upon as a generalization of $(z-1)!$. Thus for example,

$$\Gamma(5) = 4! = 24; \Gamma(-2) = \text{not defined}; \ \Gamma\left(\frac{5}{2}\right) = \left(\frac{3}{2}\right)\left(\frac{1}{2}\right)\Gamma\left(\frac{1}{2}\right) = \frac{3}{4}\Gamma\left(\frac{1}{2}\right);$$

$$\Gamma\left(\frac{1}{2}\right) = \left(-\frac{1}{2}\right)\left(-\frac{3}{2}\right)\left(-\frac{5}{2}\right)\left(-\frac{7}{2}\right)\Gamma\left(-\frac{7}{2}\right) = \frac{105}{16}\Gamma\left(-\frac{7}{2}\right)$$

$$\Rightarrow \Gamma\left(-\frac{7}{2}\right) = \frac{16}{105}\Gamma\left(\frac{1}{2}\right).$$

Thus whenever $\Gamma(z)$ is defined we can write it in the form

$$\Gamma(z) = (\text{a few factors}) \, \Gamma(\alpha), \ 0 < \alpha \le 1. \tag{1.1.12}$$

But $\Gamma(\alpha)$ for $0 < \alpha \le 1$ is extensively tabulated. Hence for computational purposes we may use the formula in (1.1.10) and the extensive tables for $\Gamma(\alpha)$ for $0 < \alpha \le 1$.

Example 1.1.1. Evaluate $\Gamma\left(\frac{51}{2}\right) \Gamma\left(-\frac{27}{2}\right)$.

Solution 1.1.1: By using (1.1.10) we have the following:

$$\Gamma\left(\frac{1}{2}\right) = \left(-\frac{1}{2}\right)\left(-\frac{3}{2}\right)\cdots\left(-\frac{27}{2}\right)\Gamma\left(-\frac{27}{2}\right)$$

$$= \frac{(-1)^{14}(1)(3)\cdots(27)}{2^{14}} \Gamma\left(-\frac{27}{2}\right) \Rightarrow$$

$$\Gamma\left(-\frac{27}{2}\right) = \frac{2^{14}}{(1)(3)\cdots(27)} \Gamma\left(\frac{1}{2}\right)$$

$$\Gamma\left(\frac{51}{2}\right) = \left(\frac{49}{2}\right)\left(\frac{47}{2}\right)\cdots\left(\frac{1}{2}\right)\Gamma\left(\frac{1}{2}\right).$$

Hence

$$\Gamma\left(\frac{51}{2}\right)\Gamma\left(-\frac{27}{2}\right) = \frac{(49)(47)\cdots(27)(25)\cdots(1)\,\Gamma\left(\frac{1}{2}\right)}{2^{25}} \frac{2^{14}}{(1)(3)\cdots(27)}\Gamma\left(\frac{1}{2}\right)$$

$$= \frac{(49)(47)\cdots(29)}{2^{11}} \left[\Gamma\left(\frac{1}{2}\right)\right]^2.$$

Direct computation of this quantity will overflow in the computer. Hence take logarithms, simplify and then take antilogarithm to obtain the exact result. It can be shown that $\Gamma\left(\frac{1}{2}\right) = \sqrt{\pi}$ where π is the mathematical constant and hence one may use (1.1.8) while computing $\Gamma\left(\frac{1}{2}\right)$.

Example 1.1.2. Show that $\Gamma\left(\frac{1}{2}\right) = \sqrt{\pi}$.

Solution 1.1.2: A simple proof can be given with the help of the integral representation for gamma functions.

$$\left[\Gamma\left(\frac{1}{2}\right)\right]^2 = \Gamma\left(\frac{1}{2}\right)\Gamma\left(\frac{1}{2}\right) = \left[\int_0^\infty x^{\frac{1}{2}-1}e^{-x}dx\right]\left[\int_0^\infty y^{\frac{1}{2}-1}e^{-y}dy\right]$$

$$= \int_0^\infty\int_0^\infty x^{-\frac{1}{2}}y^{-\frac{1}{2}}e^{-(x+y)}dx \wedge dy.$$

Put $x = r\cos^2\theta$, $y = r\sin^2\theta$, $0 \le r < \infty$, $0 \le \theta \le \frac{\pi}{2}$. Then the Jacobian is $2r\sin\theta\cos\theta$. Then,

$$\left[\Gamma\left(\frac{1}{2}\right)\right]^2 = \int_{r=0}^{\infty}\int_{\theta=0}^{\frac{\pi}{2}}(r\cos^2\theta)^{-\frac{1}{2}}(r\sin^2\theta)^{-\frac{1}{2}}2r\cos\theta\sin\theta e^{-r}dr\wedge d\theta$$

$$= \left(2\int_0^{\frac{\pi}{2}}d\theta\right)\left(\int_0^{\infty}e^{-r}dr\right) = \pi \Rightarrow \Gamma\left(\frac{1}{2}\right) = \sqrt{\pi} \qquad (1.1.13)$$

where $dr \wedge d\theta$ denotes the wedge product or skew symmetric product of the differentials dr and $d\theta$.

Also one can give a representation of the Pochhammer symbol in terms of gamma functions .

$$(a)_n = \frac{\Gamma(a+n)}{\Gamma(a)} \qquad (1.1.14)$$

whenever the gammas exist.

1.1.2 Wedge product and Jacobians of transformations

Notation 1.1.3. $\wedge =$ **wedge product**, $dx \wedge dy =$ **wedge product of dx and dy.**

Definition 1.1.7.

$$dx \wedge dy = -dy \wedge dx \Rightarrow dx \wedge dx = -dx \wedge dx = 0.$$

Thus, a wedge product is a skew symmetric product. As a consequence of Definition 1.1.7 we can evaluate the Jacobians when transforming a set of variables to another set of variables. As an example, let us consider two scalar functions of two real scalar variables x_1 and x_2. Let

$$y_1 = f_1(x_1,x_2) \text{ and } y_2 = f_2(x_1,x_2).$$

Then

$$dy_1 = \frac{\partial f_1}{\partial x_1}dx_1 + \frac{\partial f_1}{\partial x_2}dx_2 \text{ and } dy_2 = \frac{\partial f_2}{\partial x_1}dx_1 + \frac{\partial f_2}{\partial x_2}dx_2$$

where $\frac{\partial}{\partial x}$ denotes the partial derivative operator. Then

$$dy_1 \wedge dy_2 = \left[\frac{\partial f_1}{\partial x_1}dx_1 + \frac{\partial f_1}{\partial x_2}dx_2\right] \wedge \left[\frac{\partial f_2}{\partial x_1}dx_1 + \frac{\partial f_2}{\partial x_2}dx_2\right]$$

$$= \frac{\partial f_1}{\partial x_1}\frac{\partial f_2}{\partial x_1}dx_1 \wedge dx_1 + \frac{\partial f_1}{\partial x_1}\frac{\partial f_2}{\partial x_2}dx_1 \wedge dx_2$$

$$+ \frac{\partial f_1}{\partial x_2} \frac{\partial f_2}{\partial x_1} dx_2 \wedge dx_1 + \frac{\partial f_1}{\partial x_2} \frac{\partial f_2}{\partial x_2} dx_2 \wedge dx_2$$

$$= 0 + \frac{\partial f_1}{\partial x_1} \frac{\partial f_2}{\partial x_2} dx_1 \wedge dx_2 + \frac{\partial f_1}{\partial x_2} \frac{\partial f_2}{\partial x_1} dx_2 \wedge dx_1 + 0$$

since $dx_1 \wedge dx_1 = 0$ and $dx_2 \wedge dx_2 = 0$

$$= \left[\frac{\partial f_1}{\partial x_1} \frac{\partial f_2}{\partial x_2} - \frac{\partial f_1}{\partial x_2} \frac{\partial f_2}{\partial x_1} \right] dx_1 \wedge dx_2 = Jdx_1 \wedge dx_2$$

since $dx_2 \wedge dx_1 = -dx_1 \wedge dx_2$

where J is the Jacobian given by the expression

$$J = \frac{\partial f_1}{\partial x_1} \frac{\partial f_2}{\partial x_2} - \frac{\partial f_1}{\partial x_2} \frac{\partial f_2}{\partial x_1} = \begin{vmatrix} \frac{\partial f_1}{\partial x_1} & \frac{\partial f_1}{\partial x_2} \\ \frac{\partial f_2}{\partial x_1} & \frac{\partial f_2}{\partial x_2} \end{vmatrix}. \tag{1.1.15}$$

Observe that a 2×2 determinant is evaluated as

$$\begin{vmatrix} a & b \\ c & d \end{vmatrix} = (a)(d) - (c)(b),$$

where $\|[.]\|$ denotes the determinant of the matrix $[.]$.

Example 1.1.3. Evaluate the Jacobian in the transformation $y_1 = x_1 + x_2$ and $y_2 = x_1$.

Solution 1.1.3: The Jacobian J is given by

$$J = \begin{vmatrix} \frac{\partial f_1}{\partial x_1} & \frac{\partial f_1}{\partial x_2} \\ \frac{\partial f_2}{\partial x_1} & \frac{\partial f_2}{\partial x_2} \end{vmatrix} = \begin{vmatrix} 1 & 1 \\ 1 & 0 \end{vmatrix} = (1)(0) - (1)(1) = -1$$

where

$$\frac{\partial f_1}{\partial x_1} = \frac{\partial}{\partial x_1}(x_1 + x_2) = 1, \frac{\partial}{\partial x_2}(x_1 + x_2) = 1, \frac{\partial}{\partial x_1}(x_1) = 1, \frac{\partial}{\partial x_2}(x_1) = 0.$$

Example 1.1.4. Evaluate the Jacobian in the transformation $x = r\cos^2\theta$, $y = r\sin^2\theta$.

Solution 1.1.4: The partial derivatives with respect to r and θ are the following:

$$\frac{\partial x}{\partial r} = \cos^2\theta, \frac{\partial x}{\partial \theta} = -2r\cos\theta\sin\theta, \ \frac{\partial y}{\partial r} = \sin^2\theta, \ \frac{\partial y}{\partial \theta} = 2r\sin\theta\cos\theta.$$

$$J = \begin{vmatrix} \frac{\partial x}{\partial r} & \frac{\partial x}{\partial \theta} \\ \frac{\partial y}{\partial r} & \frac{\partial y}{\partial \theta} \end{vmatrix} = \begin{vmatrix} \cos^2\theta & -2r\cos\theta\sin\theta \\ \sin^2\theta & 2r\cos\theta\sin\theta \end{vmatrix}$$

$$= \cos^2\theta[2r\cos\theta\sin\theta] + \sin^2\theta[2r\cos\theta\sin\theta]$$

$$= 2r\cos\theta\sin\theta[\cos^2\theta + \sin^2\theta] = 2r\cos\theta\sin\theta.$$

This means,

$$\mathrm{d}x \wedge \mathrm{d}y = J\,\mathrm{d}r \wedge \mathrm{d}\theta = 2r\cos\theta\sin\theta \ \mathrm{d}r \wedge \mathrm{d}\theta. \tag{1.1.16}$$

Note that the various steps in the solution of Example 1.1.2 are done with the help of (1.1.16).

1.1.3 Multiplication formula for gamma functions

$$\Gamma(mz) = (2\pi)^{\frac{1-m}{2}} m^{mz-\frac{1}{2}}\Gamma(z)\Gamma\left(z+\frac{1}{m}\right)\cdots\Gamma\left(z+\frac{m-1}{m}\right), \ m = 1,2,\cdots \tag{1.1.17}$$

For $m = 2$ we obtain the duplication formula for gamma functions, namely,

$$\Gamma(2z) = (2\pi)^{\frac{1-2}{2}} 2^{2z-\frac{1}{2}}\Gamma(z)\Gamma\left(z+\frac{1}{2}\right) = \pi^{-\frac{1}{2}}2^{2z-1}\Gamma(z)\Gamma\left(z+\frac{1}{2}\right). \tag{1.1.18}$$

We may simplify gamma products with the help of (1.1.17). For example,

$$1 = \Gamma(1) = \Gamma\left[2\left(\frac{1}{2}\right)\right] = \pi^{-\frac{1}{2}}2^{1-1}\Gamma\left(\frac{1}{2}\right)\Gamma\left(\frac{1}{2}+\frac{1}{2}\right) = \pi^{-\frac{1}{2}}\Gamma\left(\frac{1}{2}\right) \Rightarrow \Gamma\left(\frac{1}{2}\right) = \sqrt{\pi}.$$

$$1 = \Gamma(1) = \Gamma\left[3\left(\frac{1}{3}\right)\right] = (2\pi)^{\frac{1-3}{2}}3^{1-\frac{1}{2}}\Gamma\left(\frac{1}{3}\right)\Gamma\left(\frac{2}{3}\right)\Gamma(1) \Rightarrow \Gamma\left(\frac{1}{3}\right)\Gamma\left(\frac{2}{3}\right) = \frac{2\pi}{\sqrt{3}}.$$

By using the product formulae for trigonometric functions we can establish the following results:

$$\Gamma(z)\Gamma(1-z) = \pi \ \mathrm{cosec} \ \pi z \tag{1.1.19}$$

$$\Gamma(z)\Gamma(-z) = -\frac{\pi}{z} \ \mathrm{cosec} \ \pi z \tag{1.1.20}$$

$$\Gamma\left(\frac{1}{2}+z\right)\Gamma\left(\frac{1}{2}-z\right) = \pi \ \sec \pi z. \tag{1.1.21}$$

1.1.4 Asymptotic formula for a gamma function

For $|z| \to \infty$ and α a bounded quantity, it can be shown that

$$\ln \Gamma(z+\alpha) = \frac{1}{2}\ln(2\pi) + \left(z+\alpha-\frac{1}{2}\right)\ln z - z$$
$$+ \sum_{k=1}^{\infty} \frac{(-1)^{k+1}B_{k+1}(\alpha)}{k(k+1)z^k}, \quad |\arg(z+\alpha)| \leq \pi - \varepsilon, \; \varepsilon > 0 \quad (1.1.22)$$

where $B_{k+1}(\alpha)$ is a Bernoulli polynomial. A brief description is given here and for more details see Mathai (1993).

1.1.5 Bernoulli polynomials

Notation 1.1.4.

$$B_k^{(a)}(x) = \textbf{generalized Bernoulli polynomial of order } k$$
$$B_k^{(1)}(x) = B_k(x) = \textbf{Bernoulli polynomial of order } k$$
$$B_k(0) = B_k = \textbf{Bernoulli number of order } k$$

Definition 1.1.8.

$$\frac{t^{\alpha}e^{xt}}{(e^t-1)^{\alpha}} = \sum_{k=0}^{\infty} \frac{t^k}{k!}B_k^{(\alpha)}(x), \quad |t| < 2\pi; \quad (1.1.23)$$

$$\frac{te^{xt}}{e^t-1} = \sum_{k=0}^{\infty} \frac{t^k}{k!}B_k(x), \quad |t| < 2\pi; \quad (1.1.24)$$

$$\frac{t}{e^t-1} = \sum_{k=0}^{\infty} B_k, \quad |t| < 2\pi. \quad (1.1.25)$$

1.1.6 Some basic properties of generalized Bernoulli polynomials

$$B_k^{(0)}(x) = x^k; \quad (1.1.26)$$
$$B_0^{(\alpha)}(x) = 1; \quad (1.1.27)$$
$$B_k^{(\alpha)}(x) = \frac{d^k}{dt^k}\left\{ e^{xt}\left[\frac{t^{\alpha}}{e^t-1)^{\alpha}}\right]\right\} \text{ at } t = 0. \quad (1.1.28)$$

For computational purposes we need the first few Bernoulli polynomials. These will be listed here.

1.1.7 The first three generalized Bernoulli polynomials

$$B_0^{(\alpha)}(x) = 1; B_1^{(\alpha)}(x) = x - \frac{\alpha}{2}; B_2^{(\alpha)}(x) = x^2 - \alpha x + \frac{\alpha(3\alpha - 1)}{12}. \qquad (1.1.29)$$

From here one has the Bernoulli polynomials and Bernoulli numbers:

$$B_0(x) = 1, \ B_1(x) = x - \frac{1}{2}, \ B_2(x) = x^2 - x + \frac{1}{6} \qquad (1.1.30)$$

$$B_0 = 1, \ B_1 = -\frac{1}{2}, \ B_2 = \frac{1}{6}. \qquad (1.1.31)$$

The first part of (1.1.22) is known as *Stirling's approximation* for a gamma function, namely,

$$\Gamma(z + \alpha) \approx (2\pi)^{\frac{1}{2}} z^{z+\alpha-\frac{1}{2}} e^{-z} \qquad (1.1.32)$$

for $|z| \to \infty$ and α a bounded quantity.
For example, taking $z = 90$ and $\alpha = 0.5$ we have

$$\Gamma(90.5) \approx \sqrt{2\pi}(90)^{90} e^{-90}.$$

For α and β bounded and $|z| \to \infty$ we have

$$\frac{\Gamma(z+\alpha)}{\Gamma(z+\alpha+\beta)} \approx \frac{(2\pi)^{\frac{1}{2}} z^{z+\alpha-\frac{1}{2}} e^{-z}}{(2\pi)^{\frac{1}{2}} z^{z+\alpha+\beta} e^{-z}} = z^{-\beta}. \qquad (1.1.33)$$

Example 1.1.5. Evaluate the following integrals:

$$(1) \int_0^\infty x^4 e^{-x^8} dx; \quad (2) \int_{-\infty}^\infty e^{-2|x|} dx; \quad (3) \int_0^\infty x^3 e^{-2x^{\frac{1}{2}}} dx.$$

Solutions 1.1.5:

(1): Put $u = x^8 \Rightarrow x = u^{\frac{1}{8}} \Rightarrow dx = \frac{1}{8} u^{\frac{1}{8}-1} du, \ x^4 = u^{\frac{1}{2}}.$

$$\int_0^\infty x^4 e^{-x^8} dx = \frac{1}{8} \int_0^\infty u^{\frac{1}{2}+\frac{1}{8}-1} e^{-u} du = \frac{1}{8} \Gamma\left(\frac{5}{8}\right).$$

(2): Since the integrand is an even function, $f(x) = f(-x)$, we have

$$\int_{-\infty}^\infty e^{-2|x|} dx = 2 \int_0^\infty e^{-2x} dx = \int_0^\infty e^{-y} dy, (2x = y) = 1.$$

(3): Put $y = 2x^{\frac{1}{2}} \Rightarrow x = \frac{y^2}{4} \Rightarrow dx = \frac{2y}{4} dy = \frac{y}{2} dy.$

$$\int_0^\infty x^3 e^{-2x^{\frac{1}{2}}} dx = \frac{1}{2(4^3)} \int_0^\infty y^7 e^{-y} dy = \frac{1}{2(4^3)} \Gamma(8) = \frac{7!}{2(4^3)}$$

$$= \frac{315}{8}.$$

Exercises 1.1.

1.1.1. Evaluate the following whenever they exist:

(a) $\left(-\frac{2}{3}\right)_4$;   (b) $(-2)_3$;   (c) $(1)_n$;   (d) $(0)_3$.

1.1.2. Evaluate the following, interpreting as the number of combinations, whenever they exist:

(a) $\binom{1/3}{2}$;   (b) $\binom{-1}{2}$;   (c) $\binom{3}{5}$;   (d) $\binom{5}{2}$;   (e) $\binom{90}{4}$.

1.1.3. Evaluate the following in terms of $\Gamma(\alpha), 0 < \alpha \le 1$.

(a) $\Gamma(-\frac{7}{2})$;   (b) $\Gamma(-\frac{5}{4})$;   (c) $\Gamma(\frac{5}{2})$;   (d) $\Gamma(7)$.

1.1.4. Evaluate the following:

(a) $\Gamma(\frac{1}{4})\Gamma(\frac{3}{4})$;   (b) $\Gamma(\frac{1}{6})\Gamma(\frac{5}{6})$.

1.1.5. Prove that Definitions 1.1.3 and 1.1.4 are one and the same.

1.1.6. Prove that $z\Gamma(z) = \Gamma(z+1)$ by using Definitions 1.1.1 and 1.1.2.

1.1.7. Evaluate the following integrals:

(a) $\int_0^\infty x^{\frac{1}{2}} e^{-3x^5} dx$; (b) $\int_0^\infty x^{\alpha-1} e^{-ax^\delta} dx$ (state the conditions).

1.1.8. Evaluate the following integrals:

(a) $\int_{-\infty}^\infty e^{-3|x|} dx$; (b) $\int_{-\infty}^\infty |x|^{\alpha-1} e^{-a|x|^\delta} dx$ (state the conditions).

1.1.9. Show that for $\Re(\alpha) > 0$, $\Re(\beta) > 0$,

$$\Gamma(\alpha)\Gamma(\beta) = \Gamma(\alpha+\beta) \int_0^1 x^{\alpha-1}(1-x)^{\beta-1} dx = \Gamma(\alpha+\beta) \int_0^1 x^{\beta-1}(1-x)^{\alpha-1} dx.$$

1.1.10. Show that for $\Re(\alpha) > 0$, $\Re(\beta) > 0$,

$$\Gamma(\alpha)\Gamma(\beta) = \Gamma(\alpha+\beta) \int_0^\infty x^{\alpha-1}(1+x)^{-(\alpha+\beta)} dx = \Gamma(\alpha+\beta) \int_0^\infty x^{\beta-1}(1+x)^{-(\alpha+\beta)} dx.$$

1.2 The Psi and Zeta Functions

The logarithmic derivative of a gamma function is the psi function and successive derivatives give generalized zeta functions.

Notation 1.2.1. $\psi(z)$: **psi z**

Definition 1.2.1.

$$\psi(z) = \frac{d}{dz} \ln \Gamma(z) = \frac{1}{\Gamma(z)} \frac{d}{dz} \Gamma(z) \quad (1.2.1)$$

$$\ln \Gamma(z) = \int_1^z \psi(x) dx.$$

By taking logarithm and then differentiating one can obtain many properties for psi functions from the corresponding properties of gamma functions. For example, from (1.1.10.) we have

$$\psi(z) = \frac{1}{z-1} + \frac{1}{z-2} + \cdots + \frac{1}{z-r} + \psi(z-r). \quad (1.2.2)$$

The following are some further properties:

$$\psi(z) = -\gamma - \frac{1}{z} + z \sum_{k=1}^{\infty} \frac{1}{k(z+k)} \quad (1.2.3)$$

$$\psi(z) = -\gamma + (z-1) \sum_{k=0}^{\infty} \frac{1}{(k+1)(z+k)} \quad (1.2.4)$$

$$\psi(1) = -\gamma \quad (1.2.5)$$

$$\psi\left(\frac{1}{2}\right) = -\gamma - 2\ln 2 \quad (1.2.6)$$

$$\psi(z) - \psi(1-z) = -\pi \cot \pi z \quad (1.2.7)$$

$$\psi\left(\frac{1}{2}+z\right) - \psi\left(\frac{1}{2}-z\right) = \pi \tan \pi z \quad (1.2.8)$$

where γ is the Euler's constant.

1.2.1 Generalized zeta function

Notation 1.2.2.

$$\zeta(\rho, a): \textbf{ generalized zeta function}$$
$$\zeta(\rho): \textbf{ Riemann zeta function}$$

Definition 1.2.2.

$$\zeta(\rho,a) = \sum_{k=0}^{\infty} \frac{1}{(k+a)^\rho}, \ \Re(\rho) > 1, \ a \neq 0, -1, -2, \cdots \tag{1.2.9}$$

$$\zeta(\rho) = \sum_{k=1}^{\infty} \frac{1}{k^\rho}, \ \Re(\rho) > 1. \tag{1.2.10}$$

For $\rho \leq 1$ the series is divergent. Successive derivatives of (1.2.4) yield the following results:

$$\frac{d^2}{dz^2} \ln \Gamma(z) = \frac{d}{dz} \psi(z) = \sum_{k=0}^{\infty} \frac{1}{(z+k)^2} = \zeta(2,z) \tag{1.2.11}$$

$$\frac{d^r}{dz^r} \ln \Gamma(z) = \frac{d^{r-1}}{dz^{r-1}} \psi(z) = \begin{cases} \psi(z), & \text{for } r = 1 \\ (-1)^r (r-1)! \zeta(r,z), & \text{for } r \geq 2 \end{cases}$$

$$= (-1)^r (r-1)! \sum_{k=0}^{\infty} \frac{1}{(z+k)^r}. \tag{1.2.12}$$

Explicit evaluations can be done in a few cases.

$$\zeta(2) = \zeta(2,1) = \sum_{k=1}^{\infty} \frac{1}{k^2} = \frac{\pi^2}{6} \tag{1.2.13}$$

$$\zeta(4) = \zeta(4,1) = \sum_{k=1}^{\infty} \frac{1}{k^4} = \frac{\pi^4}{90} \tag{1.2.14}$$

$$\zeta(2n) = \zeta(2n,1) = \sum_{k=1}^{\infty} \frac{1}{k^{2r}} = \frac{(-1)^{r+1}(2\pi)^{2r}}{2(2r)!} B_{2r} \tag{1.2.15}$$

where B_{2r} is a Bernoulli number. For these and other results see Mathai (1993).

Exercises 1.2.

1.2.1. Prove formula (1.2.4) by using (1.1.10).

1.2.2. Prove formula (1.2.3).

1.2.3. Prove formula (1.2.6) by using the duplication formula for gamma functions.

1.2.4. Show that

$$\psi(1+n) = 1 + \frac{1}{2} + \cdots + \frac{1}{n} - \gamma.$$

1.2.5. Evaluate $\psi(-\frac{3}{2})$.

1.2.6. Evaluate $\psi(5)$.

1.2.7. If $\ln\Gamma(z+1) = a_0 + a_1 z + \cdots + a_n z^n + \cdots$ evaluate a_n, $n = 0,1,2,\cdots$

1.2.8. Show that $\zeta(k,\frac{1}{2}) = (2^k - 1)\zeta(k)$.

1.2.9. Show that $\zeta\left(k,-\frac{3}{2}\right) = (-1)^k\left(2^k\right)\left[1+\frac{1}{3^k}\right] + \zeta\left(k,\frac{1}{2}\right)$.

1.2.10. Show that

$$\zeta\left(k, z - \frac{2r+1}{2}\right) = \frac{1}{\left(z-\frac{1}{2}\right)^k} + \cdots + \frac{1}{\left(z-\frac{2r+1}{2}\right)^k}$$
$$+ \zeta\left(k, z + \frac{1}{2}\right), r = 0,1,\cdots, \quad k = 2,3,\cdots$$

1.3 Integral Transforms

Basic integral transforms are the Mellin transform, the Laplace transform and the Fourier transform. Once the transforms are given, the unique functions which are recovered from these transforms are known as the inverse transforms such as inverse Mellin transform, inverse Laplace transform and inverse Fourier transform respectively. Depending upon the kernel function in the integral transform we have many other transforms such as the Bessel transform, Whittaker transform, Hankel transform, Stieltjes transform, Laguerre transform, hypergeometric transform, K-transform, Y-transform, G-transform, H-transform etc.

1.3.1 Mellin transform

Notation 1.3.1. $M_f(s)$: **Mellin transform of $f(x)$ with parameter s**

Definition 1.3.1. The Mellin transform of a real scalar function $f(x)$ with parameter s is defined as

$$M_f(s) = \int_0^\infty x^{s-1} f(x) dx \qquad (1.3.1)$$

whenever $M_f(s)$ exists.

It is a function of the arbitrary parameter s. Existence conditions for the Mellin and inverse Mellin transforms are available from books on complex analysis. Some detailed conditions are given in Mathai (1993). Since the participants of the School are a mixed group it is unwise to go into the theory of integral transforms and the details of existence conditions. We will introduce the basic transforms and illustrate how to use these transforms to solve practical problems.

Example 1.3.1. Evaluate the Mellin transform of the function $f(x) = e^{-x}$ for $x > 0$.

Solution 1.3.1:

$$M_f(s) = \int_0^\infty x^{s-1} e^{-x} dx = \Gamma(s) \text{ for } \Re(s) > 0 \tag{1.3.2}$$

from the integral representation of the gamma function in (1.1.6).

Thus given $M_f(s) = \Gamma(s)$ what is that function which gives rise to this $M_f(s)$. We know that one such function, if there exists many functions, is e^{-x}. Under the conditions of uniqueness for the existence of the inverse, the inverse function is uniquely determined as e^{-x}. The formula for the inverse Mellin transform is the following:

$$f(x) = \frac{1}{2\pi i} \int_L M_f(s) x^{-s} ds, i = \sqrt{-1} \tag{1.3.3}$$

and L is a suitable contour, usually $L = \{c - i\infty, c + i\infty\}$ for some real c. Let us evaluate the inverse transform for $M_f(s)$ in Example 1.3.1.

Example 1.3.2. Given $M_f(s) = \Gamma(s)$ evaluate $f(x)$.

Solution 1.3.2: From equation (1.3.3), $f(x)$ is given by the formula

$$f(x) = \frac{1}{2\pi i} \int_{c-i\infty}^{c+i\infty} \Gamma(s) x^{-s} ds. \tag{1.3.4}$$

This is a contour integral or an integral in the complex domain. The poles of the integrand $\Gamma(s) x^{-s}$ are coming from the poles of $\Gamma(s)$, which are at the points $s = 0, -1, -2, \cdots$. (see Definition 1.1.1). By the residue theorem in complex analysis, $f(x)$ in (1.3.3) is available as the sum of the residues of the integrand at the poles $s = 0, -1, -2, \cdots$. The residue at $s = -v$, denoted by $\Re_v$ is given by

$$\Re_v = \lim_{s \to -v} (s+v)[\Gamma(s) x^{-s}].$$

Since direct substitution will give an indeterminate quantity we may seek help from the property of gamma function in (1.1.10). That is,

$$(s+v)\Gamma(s) = (s+v)\frac{(s+v-1)\cdots s\Gamma(s)}{(s+v-1)\cdots s} = \frac{\Gamma(s+v+1)}{(s+v-1)\cdots s}.$$

Now, direct substitution is possible and hence

$$\mathcal{R}_v = \lim_{s \to -v} \frac{\Gamma(s+v+1)x^{-s}}{(s+v-1)\cdots s} = \frac{\Gamma(1)x^v}{(-1)(-2)\cdots(-v)} = \frac{(-1)^v}{v!}x^v.$$

Hence the sum of the residues is given by

$$\sum_{v=0}^{\infty} \mathcal{R}(v) = \sum_{v=0}^{\infty} \frac{(-1)^v}{v!}x^v = e^{-x}$$

which is the inverse function recovered from $M_f(s) = \Gamma(s)$. We may also observe from (1.3.4) that the poles are at $s = 0, -1, -2, \cdots$ and hence if a straight line contour $c - i\infty$ to $c + i\infty$ is taken with any $c > 0$ then all the poles of the integrand in (1.3.4) lie to the left of the contour. Then an infinite semi-circle can enclose all these poles and the residue theorem applies immediately.

Definition 1.3.2. *Residue at $z = a$.* If the scalar function $\phi(z)$ in the complex domain has a pole of order m at $z = a$ then the residue at $z = a$, denoted by $R_{a,m}$ is given by the following:

$$R_{a,m} = \lim_{z \to a} \left\{ \frac{1}{(m-1)!} \left[\frac{d^{m-1}}{dz^{m-1}}(z-a)^m \phi(z) \right] \right\} \tag{1.3.5}$$

$$= \lim_{z \to a} [(z-a)\phi(z)] \text{ for } m = 1 \text{ or for a simple pole at } z = a.$$

For example,

$$\phi_1(z) = \frac{z^5}{(z-1)(z-3)}$$

has simple poles or poles of order 1 at $z = 1$ and at $z = 3$, whereas

$$\phi_2(z) = \frac{e^{-z}}{(z-2)^3(z+1)}$$

has a simple pole at $z = -1$ and a pole of order 3 at $z = 2$. The residue of $\phi_1(z)$ at $z = 1$ is then

$$R_1 = \lim_{z \to 1} \frac{(z-1)z^5}{(z-1)(z-3)} = \lim_{z \to 1} \frac{z^5}{z-3} = \frac{1^5}{1-3} = -\frac{1}{2}$$

and the residue of $\phi_1(z)$ at $z = 3$ is given by,

$$R_3 = \lim_{z \to 3} \frac{(z-3)z^5}{(z-1)(z-3)} = \lim_{z \to 3} \frac{z^5}{z-1} = \frac{3^5}{3-1} = \frac{3^5}{2}.$$

The residue at $z = 2$ in $\phi_2(z)$ is given by the following:

$$R_{2,3} = \lim_{z \to 2} \left\{ \frac{1}{2!} \left[\frac{d^2}{dz^2} (z-2)^3 \frac{e^{-z}}{(z-2)^3(z+1)} \right] \right\}$$

$$= \lim_{z \to 2} \left\{ \frac{1}{2} \left[\frac{d^2}{dz^2} \frac{e^{-z}}{z+1} \right] \right\}$$

$$= \lim_{z \to 2} \left\{ \frac{1}{2} \left[e^{-z} \left(\frac{1}{z+1} + \frac{2}{(z+1)^2} + \frac{2}{(z+1)^3} \right) \right] \right\}$$

$$= \frac{1}{2} e^{-2} \left(\frac{1}{3} + \frac{2}{9} + \frac{2}{27} \right).$$

1.3.2 Laplace transform

Notation 1.3.2. $L_f(t)$: **Laplace transform of f with parameter t.**

Definition 1.3.3. The Laplace transform of a real scalar function $f(x)$ of the real variable x, with parameter t, is defined as

$$L_f(t) = \int_0^\infty e^{-tx} f(x) dx \tag{1.3.6}$$

whenever $L_f(t)$ exists. The inverse Laplace transform is given by

$$f(x) = \frac{1}{2\pi i} \int_L L_f(t) e^{tx} dt, \ \ i = \sqrt{-1} \tag{1.3.7}$$

where L is a suitable contour. Thus (1.3.6) and (1.3.7) are known as the Laplace inverse Laplace pair.

Example 1.3.3. Evaluate the Laplace transform of the following

$$f(x) = \frac{x^{\alpha-1}}{\Gamma(\alpha)} e^{-x}, \ x > 0, \Re(\alpha) > 0 \text{ and } f(x) = 0 \text{ elsewhere.}$$

Solution 1.3.3:

$$L_f(t) = \int_0^\infty e^{-tx} \frac{x^{\alpha-1} e^{-x}}{\Gamma(\alpha)} dx = \int_0^\infty \frac{x^{\alpha-1} e^{-(1+t)x}}{\Gamma(\alpha)} dx$$

$$= (1+t)^{-\alpha} \text{ for } 1+t > 0.$$

The Laplace transform in (1.3.6) need not exist always. But if t is replaced by it $i = \sqrt{-1}$, then

$$e^{-itx} = \cos tx - i \sin tx \text{ and } |e^{-itx}| = |\cos tx - i \sin tx| = 1.$$

Hence

$$\left| \int_{-\infty}^{\infty} e^{-itx} f(x) dx \right| \le \int_{-\infty}^{\infty} |f(x)| dx.$$

Therefore, if $f(x)$ is an absolutely integrable function in the sense $\int_{-\infty}^{\infty} |f(x)| dx < \infty$ then $\int_{-\infty}^{\infty} e^{-itx} f(x) dx$ always exists. This is known as the Fourier transform of $f(x)$, denoted by $F_f(t)$. That is,

$$F_f(t) = \int_{-\infty}^{\infty} e^{-itx} f(x) dx, \quad i = \sqrt{-1}. \tag{1.3.8}$$

More aspects of the basic transforms will be discussed after introducing statistical densities.

Exercises 1.3.

1.3.1. Convolution property for Mellin transform. Let

$$g(u) = \int_0^{\infty} \frac{1}{v} f_1(v) f_2 \left(\frac{u}{v} \right) dv. \tag{1.3.9}$$

Then show that the Mellin transform of $g(u)$ with parameter s, denoted by $h(s)$, is the product of the Mellin transforms of $f_1(x)$ and $f_2(y)$ respectively. That is, $h(s) = h_1(s)h_2(s), \ h_1(s) = \int_0^{\infty} x^{s-1} f_1(x) dx, \ h_2(s) = \int_0^{\infty} y^{s-1} f_2(y) \, dy.$

1.3.2. Show that

$$\int_0^1 x^{\alpha-1}(1-x)^{\beta-1} dx = \int_0^1 y^{\beta-1}(1-y)^{\alpha-1} dy = \frac{\Gamma(\alpha)\Gamma(\beta)}{\Gamma(\alpha+\beta)}, \quad \Re(\alpha) > 0, \Re(\beta) > 0.$$

1.3.3. Show that

$$\int_0^{\infty} x^{\alpha-1}(1+x)^{-(\alpha+\beta)} dx = \int_0^{\infty} y^{\beta-1}(1+y)^{-(\alpha+\beta)} dy$$

$$= \frac{\Gamma(\alpha)\Gamma(\beta)}{\Gamma(\alpha+\beta)}, \quad \Re(\alpha) > 0, \Re(\beta) > 0.$$

1.3.4. By using Exercise 1.3.2., or otherwise, evaluate the Mellin transform of the function

$$f(x) = \frac{\Gamma(\alpha+\beta)}{\Gamma(\alpha)\Gamma(\beta)} x^{\alpha-1}(1-x)^{\beta-1}, \quad 0 \le x \le 1,$$

$\Re(\alpha) > 0, \Re(\beta) > 0$ and $f(x) = 0$ elsewhere.

1.3.5. By using Exercise 1.3.3., or otherwise, evaluate the Mellin transform of the function

$$f(x) = \frac{\Gamma(\alpha+\beta)}{\Gamma(\alpha)\Gamma(\beta)} x^{\alpha-1}(1+x)^{-(\alpha+\beta)},$$

$x > 0, \Re(\alpha) > 0, \Re(\beta) > 0$ and $f(x) = 0$ elsewhere.

1.3.6. Evaluate the Laplace transform of the function in Exercise 1.3.4. if it exists.

1.3.7. Evaluate the Laplace transform of the function in Exercise 1.3.5. if it exists.

1.3.8. Convolution property of Laplace transforms. Let

$$g(u) = \int_0^u f_1(u-x)f_2(x)dx. \tag{1.3.10}$$

Then the Laplace transform of $g(u)$ is the product of the Laplace transforms of $f_1(x)$ and $f_2(x)$ respectively, whenever the Laplace transforms exist.

1.3.9. Let $f(x)$ be a real scalar function of the real and positive variable x. Consider the transformation $y = -\ln x$. Then show that the Mellin transform of $f(x)$ with parameter s is the same as the Laplace transform of the corresponding function of y with parameter $s-1$ when the transforms exist.

1.3.10. Evaluate the inverse Laplace transform of

$$L_f(t) = \frac{1}{(1+2t)(1+3t)}.$$

1.4 Some Statistical Preliminaries

A *random experiment* is an experiment where the outcomes are *not deterministic*. If the purpose of an experiment is to see whether a gold coin will sink in water then the outcome is predetermined. The coin will sink in water. It is not a random experiment. Consider an experiment of throwing a coin. Call one side of the coin "head" and the other side "tail". If the aim is to see whether head or tail will turn up when the coin is thrown once then the outcome is not predetermined. There is a chance that the head may turn up. There is also a chance that the tail may be the one turning up. This is a random experiment. If we assume that the coin will not stand on its edge and that it will fall head (H) or tail (T) for sure then there are two possible outcomes in this random experiment. The outcome set, called *"sample space"*, is then $\{H,T\}$. If a die (a cube with the six faces marked $1,2,3,4,5,6$) is rolled once then either 1 may turn up or 2 may turn up, $\cdots$, or 6 may turn up. The sample space here is the set $\{1,2,3,4,5,6\}$.

An *event* is a subset of the sample space. In the case of the die the event of rolling a number between 3 and 5 (inclusive) is the set $\{3,4,5\}$. The event of rolling an even number is the set $\{2,4,6\}$. In the case of throwing a coin once the event of getting a head $A = \{H\}$ and the event of getting a tail $B = \{T\}$. The "chance" of the occurrence of the event A is called the probability of A, denoted by $P_r(A)$ or simply $P(A)$. Let us assign a number between 0 and 1 to measure $P(A)$. In the experiment of throwing a coin once the event of getting two heads is impossible or it is a *null set*, denoted by O. The event of getting either a head or tail when the coin is thrown once is sure to happen. It is a *sure event*, denoted by S. Let us assign the number zero to the *impossible event* O and the number one to the sure event S. Then any event $C \subset S$ (subset of S) has the probability $0 \le P(C) \le 1$. In the random experiment of throwing a coin once the events $A = \{H\}$ and $B = \{T\}$ are *mutually exclusive* because when A occurs B cannot occur or their intersection is null or $A \cap B = O$. In this experiment the union of these two events is the sure event itself, that is $A \cup B = S$. In the experiment of rolling a die once consider the following events: $A_1 = \{1\}, A_2 = \{3,5\}, A_3 = \{4\}, A_4 = \{2,6\}$. Then $A_i \cap A_j = O$ for all $i \ne j$ and $A_1 \cup A_2 \cup A_3 \cup A_4 \cup = S$. These events are then called *mutually exclusive* (intersections are null sets) and *totally exhaustive* (union is the sure event). The probability of an event will be defined by using the following postulates. For any sample space S of a random experiment let A be an event, $A \subset S$, with probability of A denoted by $P(A)$. Then $P(A)$ is assumed to satisfy the following postulates:

$$(i) \quad 0 \le P(A) \le 1$$
$$(ii) \quad P(S) = 1 \tag{1.4.1}$$
$$(iii) \quad P(A_1 \cup A_2 \cup \cdots) = P(A_1) + P(A_2) + \cdots \quad \text{whenever } A_1, A_2, \cdots$$

are mutually exclusive.

The above postulates will not help to evaluate the probability of an event in a given situation. What is the probability that it will rain at 12 noon tomorrow over this lecture hall? There are two possibilities: $A =$ event that it will rain, $B =$ event that it will not rain. These are mutually exclusive and totally exhaustive and hence

$$1 = P(S) = P(A \cup B) = P(A) + P(B)$$

from postulates (ii) and (iii) and we know that $0 \le P(A) \le 1$. Since there are only two possibilities A and B we cannot conclude that $P(A) = \frac{1}{2}$. These two events obviously do not have equal probabilities. A meteorologist will be able to give a good estimate for $P(A)$.

In the case of throwing a coin once what is the probability of getting a head? If the events are $A = \{H\}$ and $B = \{T\}$ then as before we can come to the equation

$$1 = P(A) + P(B) \quad \text{with } 0 \le P(A) \le 1.$$

We cannot say that $P(A) = \frac{1}{2}$ claiming that there are only two possibilities. In the case of rain we have seen that probabilities cannot be assigned by simply looking at the number of possibilities. If there is no way of preferring one event to another or if there is symmetry in the outcomes of the random experiment then one way is to assign equal probabilities. If there is symmetry with respect to all characteristics of the coin then we say that the coin is *balanced*. In this case we will assign equal probabilities $P(A) = \frac{1}{2}$ and $P(B) = \frac{1}{2}$.

Example 1.4.1. If a "balanced" coin is tossed twice the possibilities are $\{HH\}$, $\{HT\}$, $\{TH\}$, $\{TT\}$ where the first letter denotes the outcome in the first trial. Since we assumed symmetry in the outcomes we assign equal probabilities to the events $P(A_1) = \frac{1}{4}$, $P(A_2) = \frac{1}{4}$, $P(A_3) = \frac{1}{4}$, $P(A_4) = \frac{1}{4}$ where $A_1 = \{H,H\}$, $A_2 = \{H,T\}$, $A_3 = \{T,H\}$, $A_4 = \{T,T\}$. Let x denote the number of heads in the individual outcomes. Then x can take the values $2,1,0$. Thus x is a variable. Further, we can assign probabilities to the values x takes. Probability that $x = 2$ is the probability of the event A_1. But $x = 1$ means either A_2 or A_3 has occurred with probabilities $\frac{1}{4}$ each. Hence we have the following *probability function* for x, denoted by $f(x)$

$$f(x) = \begin{cases} \frac{1}{4}, & \text{for } x = 2 \\ \frac{2}{4}, & \text{for } x = 1 \\ \frac{1}{4}, & \text{for } x = 0 \\ 0, & \text{elsewhere.} \end{cases}$$

This is an example of a *discrete random variable*, discrete in the sense of taking individually distinct values, such as $2,1,0$, with nonzero probabilities. Observe also that we can define a function of the following type $F(y) =$ probability that x is less than or equal to y, or written as

$$F(y) = Pr\{x \le y\}, \text{cumulative probabilities up to } y$$

for all real values of y. In our example above, $F(y) = 0$ for all y such that $-\infty < y < 0$. But at $y = 0$ there is a probability $\frac{1}{4}$ and there is no probability for the interval $0 < y < 1$. But at $y = 1$ there is another probability of $\frac{1}{2}$ and no probability over the interval $1 < y < 2$ and then $\frac{1}{4}$ at $y = 2$. Thus $F(y)$ is a step function of the following form:

$$F(y) = \begin{cases} 0, & -\infty < y < 0 \\ \frac{1}{4}, & 0 \le y < 1 \\ \frac{3}{4}, & 1 \le y < 2 \\ 1, & 2 \le y < \infty. \end{cases}$$

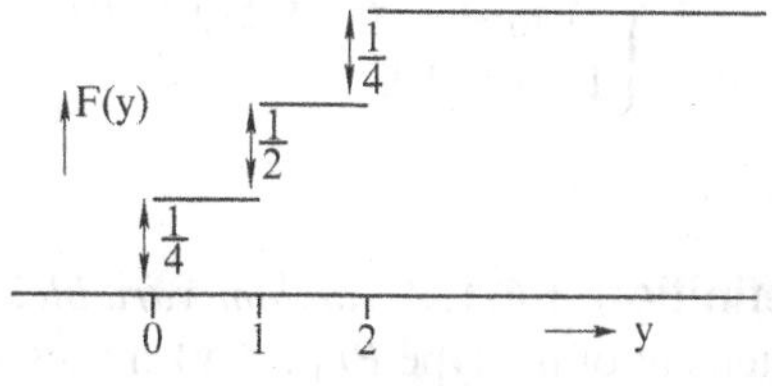

Fig. 1.4.1 A distribution function

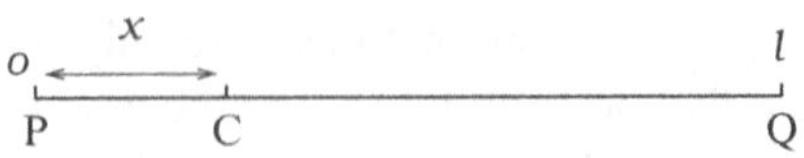

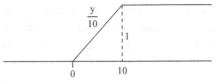

Fig. 1.4.2 Random cut

Example 1.4.2. *Random cut.* A child played with scissors and cut a string of length l into two pieces. Let one end of the uncut string be denoted by P, the other end by Q and the point of cut by C. Let the distance PC be denoted by x. Note that x is a variable and we can also make probability statements on x. A convenient way to assign probabilities to x is to assign "relative lengths" as probabilities. What is the probability that the cut C is between 8.2 and 9.4, when $l = 10$.

Then

$$Pr\{8.2 \le x \le 9.4\} = \frac{9.4 - 8.2}{10} = \frac{1.2}{10} = 0.12$$

is the probability that x falls between 8.2 and 9.4 or that C falls between 8.2 and 9.4. What is the probability that the cut is between 11 and 11.5. This, of course is zero because it is an impossible event. What is the probability that C is on an interval of length Δx over the closed interval $[0, 10]$? The answer is obviously $\frac{\Delta x}{10}$. Then we may associate a function $f(x)$ with this random variable x that

$$f(x) = \begin{cases} \frac{1}{10}, & 0 \le x \le 10 \\ 0, & elsewhere. \end{cases}$$

This x is defined on a continuum of points with nonzero probabilities and hence it is called a *continuous random variable* as opposed to a discrete random variable. In this case what is the probability that $x = 2.3$? The length is zero here and hence

$$Pr\{x = 2.3\} = \frac{2.3 - 2.3}{10} = 0.$$

In this example also one can look at the cumulative probability function. Let

$$F(y) = Pr\{x \le y\}$$
$$= \begin{cases} 0, & -\infty < y < 0 \\ \int_0^y \frac{1}{10} dx = \frac{y}{10}, & 0 \le y \le 10 \\ 1, & y \ge 10. \end{cases}$$

Fig. 1.4.3 A distribution function

Definition 1.4.1. *A random variable.* A real variable x for which a probability statement of the type $Pr\{x \le y\}$ makes sense for all real $y, -\infty < y < \infty$, is called a *real random variable*.

If x takes individually distinct values with nonzero probabilities (as in Example 1.4.1) then x is called a *discrete random variable* whereas if x takes a continuum of points with nonzero probabilities (as in Example 1.4.2) then x is called a *continuous random variable*.

Definition 1.4.2. The *distribution function* or *cumulative probability function*. For any random variable x, $F(y) = Pr\{x \leq y\}$ is called the *distribution function* associated with x.

If x is discrete then $F(y)$ will be a step function as in Figure 1.4.1. In general, whether x is discrete or continuous or mixed, $F(y)$ satisfies the following conditions:

$$\text{(i)} \quad F(-\infty) = 0$$
$$\text{(ii)} \quad F(\infty) = 1$$
$$\text{(iii)} \quad F(a) \leq F(b) \text{ for } a < b. \tag{1.4.2}$$

If $F(x)$ corresponds to a continuous random variable and if $F(x)$ is differentiable then

$$f(x) = \frac{\mathrm{d}}{\mathrm{d}x} F(x) \tag{1.4.3}$$

is the *density function* for the continuous random variable x. If $F(x)$ is a step function then the *probability function* $f(x)$ of the discrete random variable x is available by taking successive differences. A density or probability function satisfies the following conditions:

$$\text{(i)} \quad f(x) \geq 0 \text{ for all } x \tag{1.4.4}$$

$$\text{(ii)} \quad \int_{-\infty}^{\infty} f(x)\mathrm{d}x = 1 \text{ if } x \text{ is continuous, and } \sum_{-\infty}^{\infty} f(x) = 1 \tag{1.4.5}$$

if x is discrete, where Σ denotes a sum.

Example 1.4.3. Check whether the following are density functions corresponding to a continuous real random variable x.

(a) $f(x) = \lambda e^{-\lambda x}, x > 0, \lambda > 0$ and $f(x) = 0$ elsewhere.

(b) $f(x) = \dfrac{x^{\alpha-1} e^{-\frac{x}{\beta}}}{\beta^{\alpha} \Gamma(\alpha)}$, $x > 0, \alpha > 0, \beta > 0$ and $f(x) = 0$ elsewhere.

(c) $f(x) = \dfrac{1}{b-a}$, $a \leq x \leq b, a < b$ and $f(x) = 0$ elsewhere.

Solution 1.4.1:

(a) Is $f(x) \geq 0$ for all x? Obviously it is true. Is the total integral 1?

$$\int_{-\infty}^{\infty} f(x)\mathrm{d}x = 0 + \int_{0}^{\infty} \lambda e^{-\lambda x}\mathrm{d}x = \int_{0}^{\infty} e^{-y}\mathrm{d}y, \ (y = \lambda x)$$
$$= 1.$$

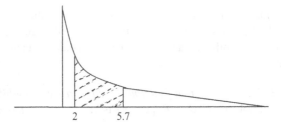

Fig. 1.4.4 Probability of an event

Hence it is a density function. This density is known as the exponential density. What is the probability that x takes values between 2 and 5.7 in this case?

$$Pr\{2 \leq x \leq 5.7\} = \int_2^{5.7} \lambda e^{-\lambda x} dx$$

$=$ area under the curve between the ordinates at $x = 2$ and $x = 5.7$.

In the continuous case, in general when a single real random variable is involved, the probabilities are the areas under the curve and the total area = total probability = 1.

(b) It is obvious that $f(x) \geq 0$ for all x. Is the total integral 1?

$$\int_{-\infty}^{\infty} f(x)dx = 0 + \int_0^{\infty} \frac{x^{\alpha-1}e^{-\frac{x}{\beta}}}{\beta^{\alpha}\Gamma(\alpha)} dx = \int_0^{\infty} \frac{y^{\alpha-1}e^{-y}}{\Gamma(\alpha)} dy, \quad \left(y = \frac{x}{\beta} \right)$$

$$= \frac{\Gamma(\alpha)}{\Gamma(\alpha)} = 1.$$

It is a density. This is called a *gamma density* or a *two parameter gamma density*. It is a density whatever be the values of $\alpha > 0$ and $\beta > 0$. Such unknowns in a density function, or in a probability function, are called *parameters*. Here there are two parameters and in the example (a) above there was one parameter λ.

(c) The first condition is obvious and hence we check the second condition.

$$\int_{-\infty}^{\infty} f(x)dx = 0 + \int_a^b \frac{1}{b-a} dx = \left[\frac{x}{b-a} \right]_a^b = \frac{b-a}{b-a} = 1.$$

This is called a *uniform density* in the sense that the probability is uniformly distributed over the closed interval $[a,b]$ in the sense that the probabilities, which are areas under the curve, over intervals of equal lengths are equal wherever be the intervals taken from $[a,b]$.

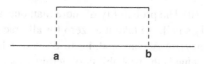

Fig. 1.4.5 Uniform density

Example 1.4.4. Check whether the following are probability functions corresponding to some discrete random variable x.

(a) $f(x) = \binom{n}{x} p^x (1-p)^{n-x}, x = 0, 1, 2, \cdots, n, 0 < p < 1$ and $f(x) = 0$ elsewhere.

(b) $f(x) = \dfrac{\lambda^x}{x!} e^{-\lambda}, x = 0, 1, 2, \cdots, \lambda > 0$ and $f(x) = 0$ elsewhere.

Solution 1.4.2:

(a) The first condition is obvious. Hence we check the second condition.

$$\sum_{-\infty}^{\infty} f(x) = 0 + \sum_{x=0}^{n} \binom{n}{x} p^x (1-p)^{n-x} = [p + (1-p)]^n = 1^n = 1.$$

Hence it is a probability function. It is called the *binomial probability function* and x here is called a *binomial random variable* because the probability function $f(x)$ is the general term in the binomial expansion $[a+b]^n$ where $a = p, b = 1 - p$.

(b) The first condition is obvious. Consider

$$\sum_{-\infty}^{\infty} f(x) = 0 + \sum_{x=0}^{\infty} \frac{\lambda^x}{x!} e^{-\lambda} = e^{-\lambda} \sum_{x=0}^{\infty} \frac{\lambda^x}{x!} = e^{-\lambda} e^{\lambda} = 1.$$

This is known as the *Poisson probability function*, named after its inventor S. Poisson, a French mathematician. Note that we have only probability masses at individual points $x = 0, x = 1, \cdots$. For example,

$$Pr\{x = 0\} = \frac{\lambda^0}{0!} e^{-\lambda} = e^{-\lambda}, \quad Pr\{x = 1\} = \frac{\lambda^1}{1!} e^{-\lambda} = \lambda e^{-\lambda},$$

and so on.

Example 1.4.5. *Poisson arrivals.* Consider an event taking place over time t, such as the arrival of telephone calls to a switchboard, arrival of customers at a checkout counter, arrival of cars for repair in a repair garage, occurrence of earth quakes at a particular locality, and so on. Let the arrivals be governed by the following conditions: (i) The probability of an arrival in time interval t to $t + \Delta t$ is proportional to its length, say $\lambda \Delta t$. (ii) The probability of more than one arrival in this interval of length Δt is negligibly small, we take it as zero for all practical purposes. (iii) Arrival or non-arrival in $[t, t + \Delta t]$ has nothing to do with what happened before. Under these conditions, what is the probability of getting exactly x arrivals in time t. Let us denote this probability of x arrivals in time t by $f(x,t)$. Then $f(x,t+\Delta t)$ is the probability of x arrivals in time $t + \Delta t$ or in the interval $[0, t + \Delta t]$. This can happen in two ways: exactly x arrivals in $[0,t]$ and no arrivals in $[t,t+\Delta t]$ or exactly $x-1$ arrivals in $[0,t]$ and one arrival in $[t,t+\Delta t]$. These are mutually exclusive events also. Hence,

$$f(x,t+\Delta t) = f(x,t)[1 - \lambda \Delta t] + f(x-1,t)\lambda \Delta t. \tag{1.4.6}$$

That is,

$$\frac{f(x,t+\Delta t) - f(x,t)}{\Delta t} = \lambda[f(x-1,t) - f(x,t)].$$

Taking the limit as $\Delta t \to 0$ we have the difference-differential equation

$$\frac{\partial}{\partial t} f(x,t) = -\lambda[f(x,t) - f(x-1,t)]. \tag{1.4.7}$$

This can be solved successively by noting that at $x = 0, f(x-1,t) = 0$ since the number of arrivals has to be zero or more. Thus,

$$\frac{\partial}{\partial t} f(0,t) = -\lambda f(0,t) \Rightarrow f(0,t) = e^{-\lambda t}.$$

Solving successively (the reader may also verify) we have,

$$f(x,t) = \frac{(\lambda t)^x}{x!} e^{-\lambda t}, \lambda > 0, t > 0, x = 0,1,2,\cdots. \tag{1.4.8}$$

and zero elsewhere, or we have a Poisson probability law with parameter λt.

Exercises 1.4.

1.4.1. Bernoulli probability law. Show that $f(x) = p^x(1-p)^{1-x}, x = 0,1, 0 < p < 1$ and $f(x) = 0$ elsewhere is a probability function.

1.4.2. Discrete hypergeometric law. Show that

$$f(x) = \binom{a}{x}\binom{b}{n-x} / \binom{a+b}{n}, x = 0,1,\cdots,n \text{ or } a;$$

b, a positive integers, is a probability function.

1.4.3. Bose-Einstein density. Show that

$$f(x) = \frac{1}{c[-1+\exp(\alpha+\beta x)]}, 0 < x < \infty, \beta > 0$$

is a density where c is the *normalizing constant* (in the sense that c will make the total integral unity). Evaluate c also.

1.4.4. Cauchy density. Show that

$$f(x) = \frac{c}{\Delta^2 + (x-\mu)^2}, -\infty < x < \infty, \Delta > 0$$

is a density where c is the normalizing constant. Evaluate c also.

1.4.5. Fermi-Dirac density. Show that

$$f(x) = \frac{1}{c[1+\exp(\alpha+\beta x)]}, 0 < x < \infty, \alpha \neq 0, \beta > 0,$$

is a density where c is the normalizing constant. Evaluate c also.

1.4.6. Generalized gamma (gamma, chisquare, exponential, Weibull, Rayleigh etc special cases). Show that

$$f(x) = cx^{\alpha-1}e^{-ax^\delta}, \ \alpha > 0, a > 0, \delta > 0, \ 0 < x < \infty.$$

is a density function where c is the normalizing constant. Evaluate c also.

1.4.7. Helley's density. Show that

$$f(x) = \left(\frac{mg}{KT}\right)e^{-(mgx)/(KT)}, \ x > 0, m > 0, g > 0, T > 0$$

is a density function.

1.4.8. Helmert density. Show that

$$f(x) = \frac{n^{\frac{(n-1)}{2}}\left(\frac{x}{\sigma}\right)^{n-2}e^{-\left(\frac{nx^2}{2\sigma^2}\right)}}{\sigma 2^{\frac{(n-3)}{2}}\Gamma(\frac{n-1}{2})}, \ 0 < x < \infty, \sigma > 0,$$

n a positive integer, is a density function.

1.4.9. Normal or Gaussian density. Show that

$$f(x) = c\,e^{-\frac{1}{2}(\frac{x-\mu}{\sigma})^2}, -\infty < x < \infty, -\infty < \mu < \infty, \ \sigma > 0$$

is a density. Evaluate the normalizing constant c.

1.4.10. Maxwell-Boltzmann density. Show that

$$f(x) = \frac{4}{\sqrt{\pi}}\beta^{\frac{3}{2}}x^2 e^{-\beta x^2}, 0 < x < \infty, \beta > 0$$

is a density.

1.5 Some Properties of Random Variables

A few essential properties of random variables will be given here so that the relevance of special functions can be appreciated.

Notation 1.5.1. $E[\psi(x)]$: **Expected value of** $\psi(x)$

Definition 1.5.1.

$$E[\psi(x)] = \int_{-\infty}^{\infty} \psi(x) f(x) \mathrm{d}x \tag{1.5.1}$$

when x is continuous with density function $f(x)$

$$= \sum_{-\infty}^{\infty} \psi(x) f(x) \tag{1.5.2}$$

when x is discrete with probability function $f(x)$.

Example 1.5.1. Evaluate the expected value of x^r when x has an exponential density

$$f(x) = \lambda e^{-\lambda x}, \ x > 0, \lambda > 0 \text{ and } f(x) = 0 \text{ elsewhere.}$$

Solution 1.5.1:

$$E(x^r) = 0 + \int_0^{\infty} x^r \lambda e^{-\lambda x} \mathrm{d}x = \lambda^{-r} \int_0^{\infty} y^{r+1-1} e^{-y} \mathrm{d}y, y = \lambda x$$
$$= \lambda^{-r} \Gamma(r+1) = r! \lambda^{-r}$$

by evaluating with the help of a gamma function. For example, the expected values of x in this case is

$$E(x) = \frac{1}{\lambda}.$$

Example 1.5.2. Evaluate the expected value of a Poisson random variable x with parameter λ.

Solution 1.5.2:

$$E(x) = 0 + \sum_{x=0}^{\infty} x \frac{\lambda^x}{x!} e^{-\lambda} = e^{-\lambda} \sum_{x=1}^{\infty} x \frac{\lambda^x}{x!}$$
$$= \lambda e^{-\lambda} \sum_{x=1}^{\infty} \frac{\lambda^{x-1}}{(x-1)!} = \lambda e^{-\lambda} e^{\lambda} = \lambda.$$

The expected value of the Poisson random variable is the parameter itself.

Definition 1.5.2. $\mu_r' = E(x^r)$ is called the rth moment of x and $E(x - E(x))^r = \mu_r$ is called the rth central moment of x.

$$\mu_1' = E(x)$$

is also called the *mean value* of x or the *centre of gravity* in x. When x is discrete, taking values $x_1, \cdots, x_k$ with the corresponding probabilities $p_1, \cdots, p_k$, $p_i \geq 0$, $i = 1, \cdots, k$, $p_1 + \cdots + p_k = 1$ then

$$E(x) = \sum_{i=1}^{k} p_i x_i = \frac{\sum_{i=1}^{k} p_i x_i}{\sum_{i=1}^{k} p_i}. \tag{1.5.3}$$

This can be considered to be a physical system with weights $p_1, \cdots, p_k$ at $x_1, \cdots, x_k$ then $E(x)$ is the center of gravity of the system.

Fig. 1.5.1 Centre of gravity

$$\mu_2 = E[x - E(x)]^2 = E[x^2 - xE(x) + [E(x)]^2]$$
$$= E(x^2) - [E(x)]^2 = \text{Var}(x) \tag{1.5.4}$$

is called the *variance of* x and the positive square root $\sigma = \sqrt{\text{Var}(x)}$ is called the *standard deviation* of x. Observe that σ can measure the spread or *dispersion* in x from the point $E(x)$. In a physical system μ_2 can also represent the moment of inertia of the system. Observe that from the definition of expected value it follows that

$$E[a\psi(x) + b] = aE[\psi(x)] + b, \tag{1.5.5}$$

where a and b are constants.

$$E[e^{tx}] = M(t); \quad M(-t) = L_f(t), \tag{1.5.6}$$

is called the moment *generating function* of x. Observe that when t is replaced by $-t$ and when the variable is continuous with density function $f(x)$ for a positive random variable x then $M(-t)$ is the Laplace transform of $f(x)$. When t is replaced by $it, i = \sqrt{-1}$ we obtain the *characteristic function* of x and when it is replaced by $-it$ we obtain the Fourier transform of the density of x.

Example 1.5.3. Evaluate the variance of the random variable x having the density function

$$f(x) = \begin{cases} x, & 0 \leq x \leq 1 \\ 2 - x, & 1 \leq x \leq 2 \\ 0, & \text{elsewhere.} \end{cases}$$

Solution 1.5.3:

$$E(x) = 0 + \int_0^1 x(x)dx + \int_1^2 x(2-x)dx$$

$$= \left[\frac{x^3}{3}\right]_0^1 + \left[\frac{2x^2}{2} - \frac{x^3}{3}\right]_1^2 = 1.$$

$$E(x^2) = 0 + \int_0^1 x^2(x)dx + \int_1^2 x^2(2-x)dx = \frac{7}{6}.$$

$$\text{Var}(x) = E(x^2) - [E(x)]^2 = \frac{7}{6} - 1^2 = \frac{1}{6}.$$

Example 1.5.4. Evaluate the moment generating function for the Gaussian or normal density

$$f(x) = \frac{1}{\sigma\sqrt{2\pi}} e^{-\frac{1}{2}(\frac{x-\mu}{\sigma})^2}, \quad -\infty < x < \infty, \quad -\infty < \mu < \infty, \sigma > 0.$$

Solution 1.5.4:

$$M(t) = E(e^{tx}) = e^{t\mu} E[e^{t(x-\mu)}] \text{ since } e^{t\mu} \text{ is a constant}$$

$$= e^{t\mu} \int_{-\infty}^{\infty} e^{t(x-\mu)} \frac{1}{\sigma\sqrt{2\pi}} e^{-\frac{1}{2}(\frac{x-\mu}{\sigma})^2} dx.$$

Put $y = \frac{x-\mu}{\sigma} \Rightarrow dx = \sigma dy$. Then for $z = (y - t\sigma)/\sqrt{2}$

$$M(t) = e^{t\mu} \int_{-\infty}^{\infty} \frac{e^{t\sigma y - \frac{1}{2}y^2}}{\sqrt{2\pi}} dy = e^{t\mu + \frac{t^2\sigma^2}{2}} \int_{-\infty}^{\infty} \frac{e^{-z^2}}{\sqrt{\pi}} dz,$$

$$= e^{t\mu + \frac{t^2\sigma^2}{2}}.$$

The last part is evaluated with the help of a gamma function.

$$\int_{-\infty}^{\infty} \frac{e^{-z^2}}{\sqrt{\pi}} dz = 2 \int_0^{\infty} \frac{e^{-z^2}}{\sqrt{\pi}} dz \text{ due to evenness}$$

$$= \int_0^{\infty} \frac{w^{\frac{1}{2}-1} e^{-w}}{\sqrt{\pi}} dw, w = z^2$$

$$= \frac{\Gamma(\frac{1}{2})}{\sqrt{\pi}} = \frac{\sqrt{\pi}}{\sqrt{\pi}} = 1.$$

1.5.1 Multivariate analogues

A function $f(x_1, \cdots, x_k)$ of k real variables $(x_1, \cdots, x_k)$ is a probability function or a density function if it satisfies the following conditions:

(i) $f(x_1, \cdots, x_k) \geq 0$ for all $x_1, \cdots, x_k$

(ii) $\int_{-\infty}^{\infty} \cdots \int_{-\infty}^{\infty} f(x_1, \cdots, x_k) dx_1 \wedge \cdots \wedge dx_k = 1$ if $x_1, \cdots, x_k$ are continuous and

$$\sum_{-\infty}^{\infty} \cdots \sum_{-\infty}^{\infty} f(x_1, \cdots, x_k) = 1 \text{ if } x_1, \cdots, x_k \text{ are discrete.} \qquad (1.5.7)$$

For mixed cases when some variables are discrete and others continuous sum up the discrete ones and integrate the continuous ones.

One popular multivariate (many variables case) discrete situation is the multinomial probability law given by the probability function

$$f(x_1, \cdots, x_k) = \frac{n!}{x_1! \cdots x_k!} p_1^{x_1} \cdots p_k^{x_k}, p_i > 0, \ i = 1, \cdots, k,$$

$$p_1 + \cdots + p_k = 1, \ x_i = 0, 1, \cdots, n, \ i = 1, \cdots k, \ x_1 + \cdots + x_k = n.$$

This is a $k-1$ variate probability function. Since $x_1 + \cdots + x_k = n$ there are only $k-1$ free variables.

For any multivariate probability density function we can define expected values, product moments, joint moment generating function, joint characteristic function etc.

$$M(t_1, \cdots, t_k) = E[e^{t_1 x_1 + \cdots + t_k x_k}]$$

$$= \int_{-\infty}^{\infty} \cdots \int_{-\infty}^{\infty} e^{t_1 x_1 + \cdots + t_k x_k} f(x_1, \cdots, x_k) dx_1 \wedge \cdots \wedge dx_k \qquad (1.5.8)$$

if $x_1, \cdots, x_k$ continuous with density $f(x_1, \cdots, x_k)$

$$= \sum_{-\infty}^{\infty} \cdots \sum_{-\infty}^{\infty} e^{t_1 x_1 + \cdots + t_k x_k} f(x_1, \cdots, x_k) \qquad (1.5.9)$$

if $x_1, \cdots, x_k$ are discrete, is the joint moment generating function of $x_1, \cdots, x_k$. If t_i is replaced by $-t_i$ for $i = 1, \cdots, k$ then we obtain the Laplace transform of the density $f(x_1, \cdots, x_k)$ for $x_i > 0$, $i = 1, \cdots, k$. If t_j is replaced by $-it_j, i = \sqrt{-1}$ for $j = 1, \cdots, k$ we have the Fourier transform of $f(x_1, \cdots, x_k)$ for x_j continuous for $j = 1, \cdots, k$.

1.5.2 Marginal and conditional densities

If $f(x_1, \cdots, x_k)$ is the joint density of the random variables $x_1, \cdots, x_k$ then if we integrate out a few of the variables we obtain the joint *marginal density* of the remaining variables. For example

$$f_1(x_1) = \int_{-\infty}^{\infty} \int_{-\infty}^{\infty} f(x_1, x_2, x_3) dx_2 \wedge dx_3 \tag{1.5.10}$$

is the marginal density of x_1 when $f(x_1, x_2, x_3)$ is the joint density of x_1, x_2 and x_3.

If $x_1, \cdots, x_k$ have the joint density $f(x_1, \cdots, x_k)$ and if $x_1, \cdots, x_r$ and $x_{r+1}, \cdots, x_k$ have the joint marginal densities $g_1(x_1, \cdots, x_r)$ and $g_2(x_{r+1}, \cdots, x_k)$ respectively then the conditional density of $x_1, \cdots, x_r$ given $x_{r+1} = a_{r+1}, \cdots, x_k = a_k$, where $a_{r+1} \cdots, a_k$ are given numbers, is given by $h(x_1, \cdots, x_r | x_{r+1} = a_{r+1} \cdots, x_k = a_k)$

$$= \frac{f(x_1, \cdots, x_k)}{g_2(x_{r+1}, \cdots, x_k)} \text{ at } x_{r+1} = a_{r+1}, \cdots, x_k = a_k \tag{1.5.11}$$

provided $g_2(a_{k+1}, \cdots, a_k) \neq 0$.

Example 1.5.5. Evaluate the marginal densities of x_1 and x_2 and the conditional density of x_1 given $x_2 = \frac{1}{3}$ from the function

$$f(x_1, x_2) = x_1 + x_2, 0 \le x_1 \le 1, \ 0 \le x_2 \le 1 \text{ and } f(x_1, x_2) = 0 \text{ elsewhere},$$

provided it is a joint density function. Check whether it is a joint density.

Solution 1.5.5: Let the marginal densities be denoted by $f_1(x_1)$ and $f_2(x_2)$ respectively.

$$f_1(x_1) = \int_{x_2} f(x_1, x_2) dx_2 = \int_0^1 (x_1 + x_2) dx_2 = x_1 + \frac{1}{2}$$

$$f_2(x_2) = \int_{x_1} f(x_1, x_2) dx_1 = \int_0^1 (x_1 + x_2) dx_1 = x_2 + \frac{1}{2}$$

$f(x_1, x_2) \ge 0$ for all x_1 and x_2. Further,

$$\int_{-\infty}^{\infty} \int_{-\infty}^{\infty} f(x_1, x_2) dx_1 \wedge dx_2 = 0 + \int_0^1 \int_0^1 (x_1 + x_2) dx_1 \wedge dx_2 = \int_0^1 \left(x_1 + \frac{1}{2}\right) dx_1 = 1.$$

Hence $f(x_1, x_2)$ is a joint density and the marginal densities are as given above. The conditional density of x_1 given $x_2 = \frac{1}{3}$ is given by

$$h\left(x_1 | x_2 = \frac{1}{3}\right) = \frac{f(x_1, x_2)}{f_2(x_2)} \text{ at } x_2 = \frac{1}{3}$$

$$= \frac{x_1 + x_2}{x_2 + \frac{1}{2}}\Big|_{\frac{1}{3}} = \frac{x_1 + \frac{1}{3}}{\frac{1}{3} + \frac{1}{2}} = \frac{6}{5}\left(x_1 + \frac{1}{3}\right), 0 \le x_1 \le 1$$

and $h(x_1 | x_2 = \frac{1}{3}) = 0$ elsewhere.

Observe that the notation for $x_1 | x_2$ (x_1 given x_2) is a vertical bar after the first set of variables, and not a division symbol.

Exercises 1.5.

1.5.1. Evaluate the conditional density of y given x and (1) the conditional expectation of y given x, denoted by $E(y|x)$, (2) the conditional variance of y given x, denoted by $\text{Var}(y|x)$ if the following is a joint density function of x and y. Verify that it is a joint density.

$$f(x,y) = \frac{1}{x^2\sigma\sqrt{2\pi}}e^{-\frac{1}{2}(\frac{y-2-3x}{\sigma})^2}, \quad -\infty < y < \infty, \ 1 \le x < \infty, \sigma > 0$$

and $f(x,y) = 0$ elsewhere.

1.5.2. If the joint density is the product of the marginal densities then the random variables are said to be *independent* or *independently distributed*. Show that in (i) below the variables are independently distributed whereas in (ii) the variables are not independent.

(i) $f(x_1,x_2,x_3) = 6e^{-x_1-2x_2-3x_3}, 0 \le x_1 < \infty, 0 \le x_2 < \infty, 0 \le x_3 < \infty$, and $f(x_1,x_2,x_3) = 0$ elsewhere.

(ii) $f(x,y) = x+y, \ 0 \le x \le 1, \ 0 \le y \le 1$ and zero elsewhere.

1.5.3. Evaluate the conditional density of x_1 given x_2 and x_3, denoted by $g_1(x_1|x_2,x_3)$ in (i) of Exercise 1.5.2 and the conditional density of x given y, denoted by $g_2(x|y)$ in (ii) of Exercise 1.5.2. Evaluate the conditional expectations $E(x_1|x_2 = 5, x_3 = 10), E(x|y = \frac{2}{3})$ and show that $E(x_1|x_2,x_3) = E(x_1)$ and $E(x|y) \ne E(x)$. [When the variables are independently distributed the conditional expectation is the same as the marginal expectation or it is free of the conditions imposed on the conditioned variables].

1.5.4. Let x_j have the gamma density

$$f_j(x_j) = \frac{x_j^{\alpha_j-1}}{\beta_j^{\alpha_j}\Gamma(\alpha_j)}, \ e^{-\frac{x_j}{\beta_j}} \ x_j \ge 0, \ \alpha_j > 0, \ \beta_j = \beta > 0$$

and $f_j(x_j) = 0$ elsewhere, for $j = 1,2$. Assume that x_1 and x_2 are independently distributed [the joint density is the product of marginal densities when independent]. Consider $u = x_1 + x_2, v = \frac{x_1}{x_1+x_2}$ and $w = \frac{x_1}{x_2}$. Show that (i) u and v are independently distributed [Hint: Consider the transformation $x_1 = r\cos^2\theta, x_2 = r\sin^2\theta$], (ii) u is gamma distributed (u has a gamma density), and (iii) evaluate the densities of v and w.

1.5.5. Evaluate the joint moment generating function in Exercise 1.5.2 (i) and show that it is a product of the marginal (individual) moment generating functions due to independence of the variables.

1.5.6.   Evaluate the joint moment generating function in Exercise 1.5.2 (ii) and show that it is not the product of the marginal (individual) moment generating functions.

1.5.7.   The covariance between two random variables x and y, denoted by $Cov(x,y)$, is defined for non-degenerate random variables $(Var(x) \neq 0, Var(y) \neq 0)$. It is a measure of joint variation in (x,y) and it is defined as

$$Cov(x,y) = E[x - E(x)][y - E(y)]$$

and the *linear correlation* in (x,y) is defined as $\rho = \dfrac{Cov(x,y)}{\sqrt{Var(x)Var(y)}} = \rho(x,y)$. Show that whatever be the non-degenerate random variables x and $y, -1 \leq \rho \leq 1$ and $\rho = \pm 1$ if and only if $y = a + bx, b \neq 0, a, b$ constants.

1.5.8.   Show that

(i) $Cov(x,y) = E(xy) - E(x)E(y)$

(ii) $\rho(x,y) = \rho(ax + b, cy + d), \ a > 0, c > 0, a, b, c, d$ constants.

1.5.9.   Evaluate $Cov(x,y)$ and $\rho(x,y)$ in the joint density in Exercise 1.5.2 (ii)

1.5.10.   Show that the following is a joint density of x and y:

$$f(x,y) = 2, 0 \leq x \leq y \leq 1 \ \text{ and zero elsewhere.}$$

For this joint density evaluate $Cov(x,y)$ and $\rho(x,y)$.

1.6 Beta and Related Functions

Notation 1.6.1. $B(\alpha, \beta)$: **Beta function**

Definition 1.6.1.

$$B(\alpha, \beta) = \frac{\Gamma(\alpha)\Gamma(\beta)}{\Gamma(\alpha + \beta)}, \Re(\alpha) > 0, \Re(\beta) > 0. \tag{1.6.1}$$

One can give several types of integral representations for the beta function.

$$B(\alpha, \beta) = \int_0^1 x^{\alpha-1}(1-x)^{\beta-1}dx, \ 0 \leq x \leq 1, \ \Re(\alpha) > 0, \Re(\beta) > 0, \tag{1.6.2}$$
$$= \int_0^1 y^{\beta-1}(1-y)^{\alpha-1}dy.$$

These are known as type-1 integral representations of a beta function. We can also show that

$$B(\alpha,\beta) = \int_0^\infty x^{\alpha-1}(1+x)^{-(\alpha+\beta)}dx$$

$$= \int_0^\infty y^{\beta-1}(1+y)^{-(\alpha+\beta)}dy, \quad \Re(\alpha) > 0, \Re(\beta) > 0. \tag{1.6.3}$$

These are known as type-2 integral representations of a beta function. The derivations of these integral representations can be done starting from the definition of a gamma function. Consider the integral representation

$$\Gamma(\alpha)\Gamma(\beta) = \int_0^\infty x^{\alpha-1}e^{-x}dx \int_0^\infty y^{\beta-1}e^{-y}dy = \int_0^\infty \int_0^\infty x^{\alpha-1}y^{\beta-1}e^{-(x+y)}dx \wedge dy$$

for $\Re(\alpha) > 0, \Re(\beta) > 0$. Make the transformation

$$x = r\cos^2\theta, y = r\sin^2\theta \Rightarrow dx \wedge dy = 2r\sin\theta\cos\theta\, dr \wedge d\theta.$$

Then

$$\Gamma(\alpha)\Gamma(\beta) = \int_{r=0}^\infty \int_{\theta=0}^{\pi/2} r^{\alpha+\beta-1}e^{-r}(\cos^2\theta)^{\alpha-1}(\sin^2\theta)^{\beta-1}2\sin\theta\cos\theta\, dr \wedge d\theta$$

$$= \int_{r=0}^\infty r^{\alpha+\beta-1}e^{-r}dr \int_{\theta=0}^{\pi/2}(\cos^2\theta)^{\alpha-1}(\sin^2\theta)^{\beta-1}2\sin\theta\cos\theta\, d\theta.$$

$$= \Gamma(\alpha+\beta)\int_0^1 u^{\alpha-1}(1-u)^{\beta-1}du, \quad [u=\cos^2\theta \Rightarrow du = -2\cos\theta\sin\theta d\theta].$$

Hence

$$\int_0^1 u^{\alpha-1}(1-u)^{\beta-1}du = \frac{\Gamma(\alpha)\Gamma(\beta)}{\Gamma(\alpha+\beta)}, \quad \Re(\alpha) > 0, \Re(\beta) > 0. \tag{1.6.4}$$

$$= \int_0^1 v^{\beta-1}(1-v)^{\alpha-1}dv, [v = 1 - u].$$

Put

$$w = \frac{u}{1-u} \Rightarrow \frac{1}{(1+w)^2}dw = du, \quad u = \frac{w}{1+w}.$$

Then

$$\int_0^1 u^{\alpha-1}(1-u)^{\beta-1}du = \int_0^\infty w^{\alpha-1}(1+w)^{-(\alpha+\beta)}dw, \quad \Re(\alpha) > 0, \Re(\beta) > 0$$

$$= \int_0^\infty t^{\beta-1}(1+t)^{-(\alpha+\beta)}dt, [t = \frac{1}{w}]. \tag{1.6.5}$$

With the help of type-1 and type-2 beta functions we can define the corresponding beta densities.

Definition 1.6.2. Type-1 beta density

$$f_1(x) = \frac{1}{B(\alpha,\beta)}x^{\alpha-1}(1-x)^{\beta-1}, \ 0 \le x \le 1, \ \Re(\alpha) > 0, \Re(\beta) > 0 \tag{1.6.6}$$

and $f_1(x) = 0$ elsewhere.

Definition 1.6.3. Type-2 beta density

$$f_2(x) = \frac{1}{B(\alpha,\beta)}x^{\alpha-1}(1+x)^{-(\alpha+\beta)}, \quad \Re(\alpha) > 0, \Re(\beta) > 0, x > 0 \qquad (1.6.7)$$

and $f_2(x) = 0$ elsewhere.

Note that both $f_1(x)$ and $f_2(x)$ satisfy non-negativity with the total integral being unity.

Example 1.6.1. Evaluate the h-th moment of x if x has

(i) a gamma distribution with density

$$f_1(x) = \frac{x^{\alpha-1}e^{-x}}{\Gamma(\alpha)}, x \geq 0, \Re(\alpha) > 0, \Re(\beta) > 0, \text{ and } f_1(x) = 0 \text{ elsewhere};$$

(ii) a type-1 beta distribution with density

$$f_2(x) = \frac{\Gamma(\alpha+\beta)}{\Gamma(\alpha)\Gamma(\beta)}x^{\alpha-1}(1-x)^{\beta-1}, 0 \leq x \leq 1, \Re(\alpha) > 0, \text{ and } f_2(x) = 0 \text{ elsewhere};$$

(iii) a type-2 beta density

$$f_3(x) = \frac{\Gamma(\alpha+\beta)}{\Gamma(\alpha)\Gamma(\beta)}x^{\alpha-1}(1+x)^{-(\alpha+\beta)}, \ 0 \leq x < \infty, \Re(\alpha) > 0, \Re(\beta) > 0$$

and $f_3(x) = 0$ elsewhere.

Solution 1.6.1:

(i)

$$E(x^h) = \int_0^\infty x^h \frac{x^{\alpha-1}e^{-x}}{\Gamma(\alpha)}dx = \frac{1}{\Gamma(\alpha)}\int_0^\infty x^{\alpha+h-1}e^{-x}dx$$

$$= \frac{\Gamma(\alpha+h)}{\Gamma(\alpha)} \text{ for } \Re(\alpha+h) > 0. \qquad (1.6.8)$$

Thus the h-th moment exists for negative values of h also provided $\alpha + h > 0$ if α and h are real. In statistical problems usually the parameters are all real. For $h = s - 1$ one has the Mellin transform of $f_1(x)$.

(ii)

$$E(x^h) = \int_0^1 x^h \frac{\Gamma(\alpha+\beta)}{\Gamma(\alpha)\Gamma(\beta)}x^{\alpha-1}(1-x)^{\beta-1}dx$$

$$= \frac{\Gamma(\alpha+\beta)}{\Gamma(\alpha)\Gamma(\beta)}\int_0^1 x^{\alpha+h-1}(1-x)^{\beta-1}dx$$

$$= \frac{\Gamma(\alpha+\beta)}{\Gamma(\alpha)\Gamma(\beta)}\frac{\Gamma(\alpha+h)\Gamma(\beta)}{\Gamma(\alpha+\beta+h)} \text{ for } \Re(\alpha+h) > 0$$

$$= \frac{\Gamma(\alpha+h)}{\Gamma(\alpha)}\frac{\Gamma(\alpha+\beta)}{\Gamma(\alpha+\beta+h)}. \qquad (1.6.9)$$

For $h = s - 1$ one has the Mellin transform of $f_2(x)$. Thus, as an inverse Mellin transform, $f_2(x)$, is available from (1.6.9). That is,

$$f_2(x) = \frac{1}{2\pi i} \int_{c-i\infty}^{c+i\infty} \frac{\Gamma(\alpha+s-1)}{\Gamma(\alpha)} \frac{\Gamma(\alpha+\beta)}{\Gamma(\alpha+\beta+s-1)} x^{-s} ds, \qquad (1.6.10)$$

$i = \sqrt{-1}$. Evaluating this contour integral as the sum of the residues at the poles of $\Gamma(\alpha+s-1)$ one obtains $f_2(x)$ as given in the example.

(iii)

$$E(x^h) = \int_0^\infty x^h \frac{\Gamma(\alpha+\beta)}{\Gamma(\alpha)\Gamma(\beta)} x^{\alpha-1} (1+x)^{-(\alpha+\beta)} dx$$

$$= \frac{\Gamma(\alpha+\beta)}{\Gamma(\alpha)\Gamma(\beta)} \int_0^\infty x^{\alpha+h-1} (1+x)^{-[(\alpha+h)+(\beta-h)]} dx$$

$$= \frac{\Gamma(\alpha+\beta)}{\Gamma(\alpha)\Gamma(\beta)} \frac{\Gamma(\alpha+h)\Gamma(\beta-h)}{\Gamma(\alpha+\beta)} \quad \text{for } \Re(\alpha+h) > 0, \Re(\beta-h) > 0$$

$$= \frac{\Gamma(\alpha+h)}{\Gamma(\alpha)} \frac{\Gamma(\beta-h)}{\Gamma(\beta)} \quad \text{for } -\Re(\alpha) < \Re(h) < \Re(\beta). \qquad (1.6.11)$$

Thus, only a few moments satisfying the condition $-\alpha < h < \beta$ can exist when α and β are real.

1.6.1 Dirichlet integrals and Dirichlet densities

A multivariate integral, which is a generalization of a beta integral, is the Dirichlet integral. We looked at type-1 and type-2 beta integrals. Here we consider type-1 and type-2 Dirichlet integrals and their generalizations. Analogously we will also define the corresponding statistical densities.

Notation 1.6.2. Dirichlet function: $D(\alpha_1, \cdots, \alpha_k; \alpha_{k+1})$ **(real scalar case)**

Definition 1.6.4.

$$D(\alpha_1, \cdots, \alpha_k; \alpha_{k+1}) = \frac{\Gamma(\alpha_1)\Gamma(\alpha_2)\cdots\Gamma(\alpha_{k+1})}{\Gamma(\alpha_1 + \cdots + \alpha_{k+1})} \quad \text{for } \Re(\alpha_j) > 0, j = 1, \cdots, k+1.$$
$$(1.6.12)$$

Note that for $k = 1$ we have the beta function in the real scalar case. Consider the following integral:

$$D_1 = \int_\Omega \cdots \int x_1^{\alpha_1-1} \cdots x_k^{\alpha_k-1} (1 - x_1 - \cdots - x_k)^{\alpha_{k+1}-1} dx_1 \wedge \cdots \wedge dx_k \qquad (1.6.13)$$

where $\Omega = \{(x_1, \cdots, x_k) | 0 \le x_i \le 1, i = 1, \cdots, k, 0 \le x_1 + \cdots + x_k \le 1\}$. Since $1 - x_1 - \cdots - x_k \ge 0$ we have $0 \le x_1 \le 1 - x_2 - \cdots - x_k$. Integration over x_1 yields the following:

$$\int_{x_1=0}^{1-x_2-\cdots-x_k} x_1^{\alpha_1-1}(1-x_1-x_2-\cdots-x_k)^{\alpha_{k+1}-1} dx_1 = (1-x_2-\cdots-x_k)^{\alpha_{k+1}-1}$$

$$\times \int_{x_1=0}^{1-x_2-\cdots-x_k} x_1^{\alpha_1-1}\left[1-\frac{x_1}{1-x_2-\cdots-x_k}\right]^{\alpha_{k+1}-1} dx_1.$$

Put, for fixed $x_2, \cdots, x_k$,

$$y_1 = \frac{x_1}{1-x_2-\cdots-x_k} \Rightarrow dx_1 = (1-x_2-\cdots-x_k)dy_1.$$

Then the integral over x_1 yields,

$$(1-x_2-\cdots-x_k)^{\alpha_1+\alpha_{k+1}-1}\int_0^1 y_1^{\alpha_1-1}(1-y_1)^{\alpha_{k+1}-1}dy_1$$

$$= (1-x_2-\cdots-x_k)^{\alpha_1+\alpha_{k+1}-1}\frac{\Gamma(\alpha_1)\Gamma(\alpha_{k+1})}{\Gamma(\alpha_1+\alpha_{k+1})}$$

for $\Re(\alpha_1) > 0$, $\Re(\alpha_{k+1}) > 0$. Integral over x_2 yields,

$$\frac{\Gamma(\alpha_1)\Gamma(\alpha_{k+1})}{\Gamma(\alpha_1+\alpha_{k+1})}\frac{\Gamma(\alpha_2)\Gamma(\alpha_1+\alpha_{k+1})}{\Gamma(\alpha_1+\alpha_2+\alpha_{k+1})} = \frac{\Gamma(\alpha_1)\Gamma(\alpha_2)\Gamma(\alpha_{k+1})}{\Gamma(\alpha_1+\alpha_2+\alpha_{k+1})}.$$

Proceeding like this, we have the final result:

$$D_1 = D(\alpha_1, \cdots, \alpha_k; \alpha_{k+1}) = \frac{\Gamma(\alpha_1)\Gamma(\alpha_2)\cdots\Gamma(\alpha_{k+1})}{\Gamma(\alpha_1+\cdots+\alpha_{k+1})}, \quad \Re(\alpha_j) > 0, j = 1, \cdots, k+1.$$

$$(1.6.14)$$

Here, (1.6.13) is the type-1 Dirichlet integral. Hence by normalizing the integrand in (1.6.13) we have the type-1 Dirichlet density.

Definition 1.6.5. Type-1 Dirichlet density $f_1(x_1, \cdots, x_k)$.

$$f_1(x_1, \cdots, x_k) = \frac{1}{D(\alpha_1, \cdots, \alpha_k; \alpha_{k+1})}x_1^{\alpha_1-1}\cdots x_k^{\alpha_k-1}(1-x_1-\cdots-x_k)^{\alpha_{k+1}-1},$$

$$0 \le x_j \le 1, j = 1, \cdots, k, \ 0 \le x_1 + \cdots + x_k \le 1, \ \Re(\alpha_j) > 0, \quad (1.6.15)$$

$$j = 1, \cdots, k+1, \text{ and } f_1(x_1, \cdots, x_k) = 0 \text{ elsewhere.}$$

Consider the type-2 Dirichlet integral

$$D_2 = \int_0^\infty \cdots \int_0^\infty x_1^{\alpha_1-1}\cdots x_k^{\alpha_k-1}(1+x_1+\cdots+x_k)^{-(\alpha_1+\cdots+\alpha_{k+1})}dx_1 \wedge \cdots \wedge dx_k.$$

$$(1.6.16)$$

This can be integrated by writing

$$(1+x_1+\cdots+x_k) = (1+x_2+\cdots+x_k)\left[1+\frac{x_1}{1+x_2+\cdots+x_k}\right]$$

and then integrating out with the help of type-2 beta integrals. The final result will agree with the Dirichlet function

$$D_2 = D(\alpha_1,\cdots\alpha_k;\alpha_{k+1}). \tag{1.6.17}$$

Thus, we can define a type-2 Dirichlet density.

Definition 1.6.6. Type-2 Dirichlet density.

$$f_2(x_1,\cdots,x_k) = \frac{1}{D(\alpha_1,\cdots,\alpha_k;\alpha_{k+1})} x_1^{\alpha_1-1}\cdots x_k^{\alpha_k-1}(1+x_1+\cdots+x_k)^{-(\alpha_1+\cdots+\alpha_{k+1})},$$

$$0 \le x_j < \infty, \; j=1,\cdots,k, \; \Re(\alpha_j) > 0, \; j=1,\cdots,k+1, \tag{1.6.18}$$

and $f_2(x_1,\cdots,x_k) = 0$ elsewhere.

It is easy to observe that if $(x_1,\cdots,x_k)$ has a k-variate type-1 Dirichlet density then any subset of r of the variables have a r-variate type-1 Dirichlet density for $r = 1,\cdots,k$. Similarly if $(x_1,\cdots,x_k)$ have a type-2 Dirichlet density then any subset of them will have a type-2 Dirichlet density.

Example 1.6.2. Evaluate the marginal densities from the following bivariate density:

$$f(x_1,x_2) = \frac{\Gamma(\alpha_1+\alpha_2+\alpha_3)}{\Gamma(\alpha_1)\Gamma(\alpha_2)\Gamma(\alpha_3)} x_1^{\alpha_1-1} x_2^{\alpha_2-1}(1-x_1-x_2)^{\alpha_3-1}, 0 \le x_j \le 1, j=1,2,3,$$

$$0 \le x_1+x_2+x_3 \le 1, \; \Re(\alpha_j) > 0, j=1,2,3, \text{ and } f(x_1,x_2) = 0 \text{ elsewhere.}$$

Solution 1.6.2: Let the marginal densities be denoted by $f_1(x_1)$ and $f_2(x_2)$ respectively.

$$f_1(x_1) = \int_{x_2} f(x_1,x_2)dx_2 = \frac{\Gamma(\alpha_1+\alpha_2+\alpha_3)}{\Gamma(\alpha_1)\Gamma(\alpha_2)\Gamma(\alpha_3)} x_1^{\alpha_1-1}$$

$$\times \int_{x_2=0}^{1-x_1} x_2^{\alpha_2-1}(1-x_1-x_2)^{\alpha_3-1}dx_2$$

$$= \frac{\Gamma(\alpha_1+\alpha_2+\alpha_3)}{\Gamma(\alpha_1)\Gamma(\alpha_2)\Gamma(\alpha_3)} x_1^{\alpha_1-1}(1-x_1)^{\alpha_3-1}\int_{x_2=0}^{1-x_1} x_2^{\alpha_2-1}\left[1-\frac{x_2}{1-x_1}\right]^{\alpha_3-1}dx_2.$$

Put, for fixed x_1,

$$y_2 = \frac{x_2}{1-x_1} \Rightarrow dx_2 = (1-x_1)dy_2.$$

$$f_1(x_1) = \frac{\Gamma(\alpha_1 + \alpha_2 + \alpha_3)}{\Gamma(\alpha_1)\Gamma(\alpha_2)\Gamma(\alpha_3)} x_1^{\alpha_1 - 1}(1 - x_1)^{\alpha_2 + \alpha_3 - 1} \int_0^1 y_2^{\alpha_2 - 1}(1 - y_2)^{\alpha_3 - 1} dy_2.$$

Evaluating the y_2-integral with the help of a type-1 beta integral we obtain $\frac{\Gamma(\alpha_2)\Gamma(\alpha_3)}{\Gamma(\alpha_2 + \alpha_3)}$. Hence,

$$f_1(x_1) = \frac{\Gamma(\alpha_1 + \alpha_2 + \alpha_3)}{\Gamma(\alpha_1)\Gamma(\alpha_2 + \alpha_3)} x_1^{\alpha_1 - 1}(1 - x_1)^{\alpha_2 + \alpha_3 - 1}, 0 \le x_1 \le 1,$$

and zero elsewhere. From symmetry,

$$f_2(x_2) = \frac{\Gamma(\alpha_1 + \alpha_2 + \alpha_3)}{\Gamma(\alpha_2)\Gamma(\alpha_1 + \alpha_3)} x_2^{\alpha_2 - 1}(1 - x_2)^{\alpha_1 + \alpha_3 - 1} \, 0 \le x_2 \le 1,$$

and zero elsewhere. Thus, the marginal densities of x_1 and x_2 are type-1 beta densities.

Example 1.6.3. Evaluate the normalizing constant c if the following is a density function:

$$f(x_1, x_2) = cx_1^{\alpha_1 - 1}(1 - x_1)^{\beta_1} x_2^{\alpha_2 - 1}(1 - x_1 - x_2)^{\alpha_3 - 1}, 0 \le x_j \le 1, \quad (1.6.19)$$

$0 \le x_1 + x_2 \le 1, j = 1, 2, \; \Re(\alpha_j) > 0, \; j = 1, 2, 3$ and $f(x_1, x_2) = 0$ elsewhere.

Solution 1.6.3: Let us integrate out x_2 first.

$$\int_{x_2=0}^{1-x_1} x_2^{\alpha_2 - 1}(1 - x_1 - x_2)^{\alpha_3 - 1} dx_2$$

$$= (1 - x_1)^{\alpha_2 + \alpha_3 - 1} \int_0^1 y_2^{\alpha_2 - 1}(1 - y_2)^{\alpha_3 - 1} dy_2, y_2 = \frac{x_2}{1 - x_1}$$

$$= (1 - x_1)^{\alpha_2 + \alpha_3 - 1} \frac{\Gamma(\alpha_2)\Gamma(\alpha_3)}{\Gamma(\alpha_2 + \alpha_3)}, \; \Re(\alpha_2) > 0, \; \Re(\alpha_3) > 0.$$

Now, integrating out x_1 we have,

$$\int_0^1 x_1^{\alpha_1 - 1}(1 - x_1)^{\beta_1 + \alpha_2 + \alpha_3 - 1} dx_1 = \frac{\Gamma(\alpha_1)\Gamma(\beta_1 + \alpha_2 + \alpha_3)}{\Gamma(\alpha_1 + \alpha_2 + \alpha_3 + \beta_1)}$$

$\Re(\alpha_1) > 0, \; \Re(\beta_1 + \alpha_2 + \alpha_3) > 0$. Hence

$$c = \frac{\Gamma(\alpha_2 + \alpha_3)\Gamma(\alpha_1 + \alpha_2 + \alpha_3 + \beta_1)}{\Gamma(\alpha_1)\Gamma(\alpha_2)\Gamma(\alpha_3)\Gamma(\alpha_2 + \alpha_3 + \beta_1)}, \; \Re(\alpha_j) > 0, j = 1, 2, 3, \Re(\alpha_2 + \alpha_3 + \beta_1) > 0.$$

A generalization to k variable case is one of the generalizations of type-1 Dirichlet density and the corresponding type-1 Dirichlet function.

Example 1.6.4. Evaluate the normalizing constant if the following is a density function:

$$f(x_1,x_2,x_3) = cx_1^{\alpha_1-1}(x_1+x_2)^{\beta_2} x_2^{\alpha_2-1}(x_1+x_2+x_3)^{\beta_3} x_3^{\alpha_3-1}(1-x_1-x_2-x_3)^{\alpha_4-1},$$

$$0 \leq x_1 + \cdots + x_j \leq 1, \; j = 1,2,3,4, \; \Re(\alpha_j) > 0, \; j = 1,2,3,4,$$

$$\Re(\alpha_1 + \cdots + \alpha_j + \beta_2 + \cdots + \beta_j) > 0, j = 1,2,3,4$$

$$(1.6.20)$$

and $f(x_1,x_2,x_3) = 0$ elsewhere.

Solution 1.6.4: Let $u_1 = x_1$, $u_2 = x_1 + x_2$, $u_3 = x_1 + x_2 + x_3$ and let the joint density of u_1, u_2, u_3 be denoted by $g(u_1, u_2, u_3)$. Then

$$g(u_1, u_2, u_3) = c\, u_1^{\alpha_1-1} u_2^{\beta_2}(u_2-u_1)^{\alpha_2-1} u_3^{\beta_3}(u_3-u_2)^{\alpha_3-1}(1-u_3)^{\alpha_4-1},$$

$$0 \leq u_1 \leq u_2 \leq u_3 \leq 1.$$

Note that $0 \leq u_1 \leq u_2$. Integration over u_1 yields the following:

$$\int_{u_1=0}^{u_2} u_1^{\alpha_1-1}(u_2-u_1)^{\alpha_2-1} \mathrm{d}u_1 = u_2^{\alpha_2-1} \int_{u_1=0}^{u_2} u_1^{\alpha_1-1}\left(1-\frac{u_1}{u_2}\right)^{\alpha_2-1} \mathrm{d}u_1$$

$$= u_2^{\alpha_1+\alpha_2-1} \int_0^1 y_1^{\alpha_1-1}(1-y_1)^{\alpha_2-1} \mathrm{d}y_1, \; y_1 = \frac{u_1}{u_2}$$

$$= u_2^{\alpha_1+\alpha_2-1} \frac{\Gamma(\alpha_1)\Gamma(\alpha_2)}{\Gamma(\alpha_1+\alpha_2)}, \; \Re(\alpha_1) > 0, \Re(\alpha_2) > 0.$$

Integration over u_2 yields the following:

$$\int_{u_2=0}^{u_3} u_2^{\alpha_1+\alpha_2+\beta_2-1}(u_3-u_2)^{\alpha_3-1} \mathrm{d}u_2 = u_3^{\alpha_1+\alpha_2+\alpha_3+\beta_2-1} \frac{\Gamma(\alpha_3)\Gamma(\alpha_1+\alpha_2+\beta_2)}{\Gamma(\alpha_1+\alpha_2+\alpha_3+\beta_2)}$$

$$\text{for } \Re(\alpha_3) > 0, \Re(\alpha_1+\alpha_2+\beta_2) > 0.$$

Finally, integral over u_3 yields the following:

$$\int_{u_3=0}^1 u_3^{\alpha_1+\alpha_2+\alpha_3+\beta_2+\beta_3-1}(1-u_3)^{\alpha_4-1} \mathrm{d}u_3 = \frac{\Gamma(\alpha_4)\Gamma(\alpha_1+\alpha_2+\alpha_3+\beta_2+\beta_3)}{\Gamma(\alpha_1+\cdots+\alpha_4+\beta_2+\beta_3)},$$

$$\Re(\alpha_4) > 0, \Re(\alpha_1+\alpha_2+\alpha_3+\beta_2+\beta_3) > 0.$$

Hence

$$c^{-1} = \Gamma(\alpha_1)\Gamma(\alpha_2)\Gamma(\alpha_3)\Gamma(\alpha_4)\frac{\Gamma(\alpha_1+\alpha_2+\beta_2)}{\Gamma(\alpha_1+\alpha_2+\alpha_3+\beta_2)}$$

$$\times \frac{\Gamma(\alpha_1+\alpha_2+\alpha_3+\beta_2+\beta_3)}{\Gamma(\alpha_1+\alpha_2+\alpha_3+\alpha_4+\beta_2+\beta_3)}$$

for $\Re(\alpha_j) > 0, j = 1,2,3,4, \Re(\alpha_1+\cdots+\alpha_j+\beta_2+\cdots+\beta_j) > 0, \; j = 2,3.$

Note that one can generalize the function in (1.6.20) to a k- variables situation. This will produce another generalization of the type-1 Dirichlet function as well as the type-1 Dirichlet density. Corresponding situations in the type-2 case will provide generalizations of the type-2 Dirichlet integral and density.

Exercises 1.6.

1.6.1. Let $f(x_1,x_2,x_3) = cx_1^{\alpha_1-1}(1+x_1)^{-(\alpha_1+\beta_1)}x_2^{\alpha_2-1}(1+x_1+x_2)^{-(\alpha_2+\beta_2)}x_3^{\alpha_3-1}$ $(1+x_1+x_2+x_3)^{-(\alpha_3+\beta_3)}$, $0 \le x_j < \infty$, $j = 1,2,3$ and $f(x_1,x_2,x_3) = 0$ elsewhere. If $f(x_1,x_2,x_3)$ is a density function then evaluate c and write down the conditions on the parameters.

1.6.2. Generalize the density in Exercises 1.6.1 to k-variables case, evaluate the corresponding c and write down the conditions.

1.6.3. Write down the k-variables situation in Example 1.6.3 and evaluate the normalizing constant, and give the conditions on the parameters.

1.6.4. Write down the general density corresponding to Example 1.6.4 and evaluate the normalizing constant, and give the conditions on the parameters.

1.6.5. By using the gamma structure in the normalizing constant in Exercise 1.6.4 show that the joint density in Exercise 1.6.4 can also be obtained as the joint density of k mutually independently distributed real scalar type-1 beta random variables, and identify the parameters in these independent type-1 beta random variables.

1.7 Hypergeometric Series

A general hypergeometric series with p upper or numerator parameters and q lower or denominator parameters is denoted and defined as follows:

Notation 1.7.1.

$$_pF_q(a_1,\cdots,a_p;b_1,\cdots,b_q;z) = {}_pF_q((a_p);(b_q);z) = {}_pF_q(z)$$

Definition 1.7.1.

$$_pF_q(z) = \sum_{r=0}^{\infty} \frac{(a_1)_r \cdots (a_p)_r}{(b_1)_r \cdots (b_q)_r} \frac{z^r}{r!} \tag{1.7.1}$$

where $(a_j)_r$ and $(b_j)_r$ are the Pochhammer symbols of (1.0.1). The series in (1.7.1) is defined when none of the b_j 's, $j = 1,\cdots,q$, is a negative integer or zero. If a b_j is a negative integer or zero then $(b_j)_r$ will be zero for some r. A b_j can be zero provided there is a numerator parameter a_k such that $(a_k)_r$ becomes zero first before

$(b_j)_r$ becomes zero. If any numerator parameter a_j is a negative integer or zero then (1.7.1) terminates and becomes a polynomial in z. From the ratio test it is evident that the series in (1.7.1) is convergent for all z if $q \geq p$, it is convergent for $|z| < 1$ if $p = q+1$ and divergent if $p > q+1$. When $p = q+1$ and $|z| = 1$ the series can converge in some cases. Let

$$\beta = \sum_{j=1}^{p} a_j - \sum_{j=1}^{q} b_j.$$

It can be shown that when $p = q+1$ the series is absolutely convergent for $|z| = 1$ if $\Re(\beta) < 0$, conditionally convergent for $z = -1$ if $0 \leq \Re(\beta) < 1$ and divergent for $|z| = 1$ if $1 \leq \Re(\beta)$.

Some special cases of a $_pF_q$ are the following: When there are no upper or lower parameters we have,

$$_0F_0(;;\pm z) = \sum_{r=0}^{\infty} \frac{(\pm z)^r}{r!} = e^{\pm z}. \tag{1.7.2}$$

Thus $_0F_0(.)$ is an exponential series.

$$_1F_0(\alpha;;z) = \sum_{r=0}^{\infty} (\alpha)_r \frac{z^r}{r!} = (1-z)^{-\alpha} \text{ for } |z| < 1. \tag{1.7.3}$$

This is the binomial series. $_1F_1(.)$ is known as confluent hypergeometric series or *Kummers's hypergeometric series and* $_2F_1(.)$ is known as *Gauss' hypergeometric series* .

Example 1.7.1. *Incomplete gamma function.* Evaluate the incomplete gamma function

$$\gamma(\alpha,b) = \int_0^b x^{\alpha-1}e^{-x}dx, \ b < \infty$$

and write it in terms of a Kummer's hypergeometric function.

Solution 1.7.1: Since b is finite we may expand the exponential part and integrate term by term.

$$\gamma(\alpha,b) = \int_0^b x^{\alpha-1} \left\{ \sum_{r=0}^{\infty} \frac{(-1)^r}{r!} x^r \right\} dx = \sum_{r=0}^{\infty} \frac{(-1)^r}{r!} \int_0^b x^{\alpha+r-1}dx$$

$$= \sum_{r=0}^{\infty} \frac{(-1)^r}{r!} \frac{b^{\alpha+r}}{\alpha+r} = \frac{b^\alpha}{\alpha} \sum_{r=0}^{\infty} \frac{(-1)^r}{r!} \frac{(\alpha)_r}{(\alpha+1)_r} b^r$$

$$= \frac{b^\alpha}{\alpha} \, _1F_1(\alpha;\alpha+1;-b). \tag{1.7.4}$$

Hence the upper part

$$\Gamma(\alpha,b) = \int_b^{\infty} x^{\alpha-1}e^{-x}dx = \Gamma(\alpha) - \gamma(\alpha,b). \tag{1.7.5}$$

Example 1.7.2. *Incomplete beta function.* Evaluate the incomplete beta function

$$b(\alpha, \beta; t) = \int_0^t x^{\alpha-1}(1-x)^{\beta-1} dx, t < 1$$

and write it in terms of a Gauss' hypergeometric function.

Solution 1.7.2: Note that since $0 < x < 1$,

$$(1-x)^{\beta-1} = (1-x)^{-(1-\beta)} = \sum_{r=0}^{\infty} \frac{(1-\beta)_r}{r!} x^r.$$

Hence,

$$b(\alpha, \beta; t) = \sum_{r=0}^{\infty} \frac{(1-\beta)_r}{r!} \int_0^t x^{\alpha+r-1} dx = \sum_{r=0}^{\infty} \frac{(1-\beta)_r}{r!} \frac{t^{\alpha+r}}{\alpha+r}$$

$$= \frac{t^{\alpha}}{\alpha} \sum_{r=0}^{\infty} \frac{(1-\beta)_r(\alpha)_r}{(\alpha+1)_r} \frac{t^r}{r!} = \frac{t^{\alpha}}{\alpha} \,_2F_1(1-\beta, \alpha; \alpha+1; t). \qquad (1.7.6)$$

Hence the upper part,

$$B(\alpha, \beta; t) = \int_t^1 x^{\alpha-1}(1-x)^{\beta-1} dx = \frac{\Gamma(\alpha)\Gamma(\beta)}{\Gamma(\alpha+\beta)} - b(\alpha, \beta; t). \qquad (1.7.7)$$

Example 1.7.3. Obtain an integral representation for a $_2F_1$.

Solution 1.7.3: Consider the integral,

$$\int_0^1 x^{a-1}(1-x)^{c-a-1}(1-zx)^{-b} dx, \text{ for } |z| < 1$$

$$= \sum_{r=0}^{\infty} \frac{z^r}{r!} (b)_r \int_0^1 x^{a+r-1}(1-x)^{c-a-1} dx, \text{ (expanding } (1-zx)^{-b}$$

by binomial expansion)

$$= \sum_{r=0}^{\infty} \frac{z^r}{r!} (b)_r \frac{\Gamma(a+r)\Gamma(c-a)}{\Gamma(c+r)} \text{ (by using a type-1 beta integral)}$$

$$= \frac{\Gamma(a)\Gamma(c-a)}{\Gamma(c)} \sum_{r=0}^{\infty} \frac{z^r}{r!} \frac{(b)_r(a)_r}{(c)_r} \text{ (by writing } \Gamma(a+r) = (a)_r\Gamma(a))$$

$$= \frac{\Gamma(a)\Gamma(c-a)}{\Gamma(c)} \,_2F_1(a, b; c; z).$$

That is,

$$_2F_1(a, b; c; z) = \frac{\Gamma(c)}{\Gamma(a)\Gamma(c-a)} \int_0^1 x^{a-1}(1-x)^{c-a-1}(1-zx)^{-b} dx \qquad (1.7.8)$$

$$\text{for } \Re(a) > 0, \ \Re(c-a) > 0 \ |z| < 1.$$

This is the famous integral representation for $_2F_1$. From the integral representation note that when $z = 1$ one can evaluate the integral with the help of a type-1 beta integral. That is,

$$
\begin{aligned}
_2F_1\,(a,b;c;1) &= \frac{\Gamma(c)}{\Gamma(a)\Gamma(c-a)} \int_0^1 x^{a-1}(1-x)^{c-a-b-1}dx \\
&= \frac{\Gamma(c)}{\Gamma(a)\Gamma(c-a)}\,\frac{\Gamma(a)\Gamma(c-a-b)}{\Gamma(c-b)}. \\
&= \frac{\Gamma(c)\Gamma(c-a-b)}{\Gamma(c-a)\Gamma(c-b)}
\end{aligned}
\tag{1.7.9}
$$

when the arguments of all the gammas are positive. This is the famous summation formula for a $_2F_1$ series.

1.7.1 Evaluation of some contour integrals

Since the technique of Mellin and inverse Mellin transforms is frequently used for solving some problems in applied areas we will look into the evaluation of some contour integrals with the help of residue theorem. We will not go into the theory of analytic functions and residue calculus. We will need to know only how to apply the residue theorem for evaluating some integrals where the integrands contain gamma functions. In order to illustrate the technique let us redo a known result.

Example 1.7.4. Evaluate the contour integral, which is also an inverse Mellin transform,

$$
f(x) = \frac{1}{2\pi i} \int_{c-i\infty}^{c+i\infty} \frac{\Gamma(a_1-s)\cdots\Gamma(a_p-s)}{\Gamma(b_1-s)\cdots\Gamma(b_q-s)}\Gamma(s)(-z)^{-s}ds
\tag{1.7.10}
$$

as the sum of the residues at the pole of $\Gamma(s)$.

Solution 1.7.4: The poles are at $s = -v, v = 0, 1, \cdots$. The residue at $s = -v$ is given by the following:

$$
\mathfrak{R}_v = \lim_{s \to -v} \left\{ (s+v)\Gamma(s)\frac{\Gamma(a_1-s)\cdots\Gamma(a_p-s)}{\Gamma(b_1-s)\cdots\Gamma(b_q-s)}(-z)^{-s} \right\}.
$$

By using the process in Example 1.3.2 we have,

$$
\begin{aligned}
\mathfrak{R}_v &= \frac{(-1)^v}{v!}\,\frac{\Gamma(a_1+v)\cdots\Gamma(a_p+v)}{\Gamma(b_1+v)\cdots\Gamma(b_q+v)}(-z)^v \\
&= \left\{ \frac{\prod_{j=1}^p \Gamma(a_j)}{\prod_{j=1}^q \Gamma(b_j)} \right\} \frac{(a_1)_v\cdots(a_p)_v}{(b_1)_v\cdots(b_q)_v}\frac{z^v}{v!}.
\end{aligned}
$$

Hence the sum of the residues is given by,

$$\sum_{v=0}^{\infty} \Re_v = K \sum_{v=0}^{\infty} \frac{(a_1)_v \cdots (a_p)_v}{(b_1)_v \cdots (b_q)_v} \frac{z^v}{v!} = K \; {}_pF_q(a_1, \cdots, a_p; b_1, \cdots, b_q; z)$$

where K is the constant

$$K = \frac{\Pi_{j=1}^{p} \Gamma(a_j)}{\Pi_{j=1}^{q} \Gamma(b_j)}.$$

Thus $\frac{1}{K}$ times the right side in (1.7.10) is the Mellin-Barnes representation for a general hypergeometric function. If poles of higher orders are involved then one may use the general formula. If $\phi(z)$ has a pole of order m at $z = a$ then the residue at $z = a$, denoted by $\Re_{a,m}$, is given by the following formula:

$$\Re_{a,m} = \lim_{z \to a} \left\{ \frac{1}{(m-1)!} \left[\frac{d^{m-1}}{dz^{m-1}} (z-a)^m \phi(z) \right] \right\}. \tag{1.7.11}$$

Some illustrations of this formula will be given when we solve some problems in astrophysics later on.

1.7.2 Residues when several gammas are involved

Let

$$\phi(z) = \Gamma(b_1 + z) \cdots \Gamma(b_m + z) x^{-z}$$
$$= h(z) x^{-z} \quad \text{with}$$
$$h(z) = \Gamma(b_1 + z) \cdots \Gamma(b_m + z).$$

Depending upon the values of $b_1, \cdots, b_m$ one can expect poles of orders $1, 2, ..., m$ if the b_j's differ by integers. Let $z = a$ be a pole of order k for $\phi(z)$. Then the residue of $\phi(z)$ at $z = a$ is given by the following:

$$R_{a,k} = \lim_{z \to a} \left\{ \frac{1}{(k-1)!} \frac{\partial^{k-1}}{\partial z^{k-1}} [(z-a)^k \phi(z)] \right\}$$
$$= \lim_{z \to a} \left\{ \frac{1}{(k-1)!} \frac{\partial^{k-1}}{\partial z^{k-1}} [(z-a)^k h(z) x^{-z}] \right\}.$$

Note that a convenient operator can be used to take x^{-z} outside. Consider the operator

$$\left[\frac{\partial}{\partial z} + (-\ln z) \right]^{k-1} = \sum_{r=0}^{k-1} \binom{k-1}{r} (-\ln x)^{k-1-r} \frac{\partial^r}{\partial z^r}. \tag{1.7.12}$$

Then

$$\frac{\partial^{k-1}}{\partial z^{k-1}}[(z-a)^k h(z)x^{-z}]$$

$$= x^{-z}\left[\frac{\partial}{\partial z} + (-\ln x)\right]^{k-1}[(z-a)^k h(z)]$$

$$= x^{-z}\sum_{r=0}^{k-1}\binom{k-1}{r}(-\ln x)^{k-1-r}\frac{\partial^r}{\partial z^r}[(z-a)^k h(z)].$$

Let

$$B(z) = (z-a)^k h(z) \text{ and } A(z) = \frac{\partial}{\partial z}\ln B(z).$$

Then

$$\frac{\partial^r}{\partial z^r}B(z) = \frac{\partial^{r-1}}{\partial z^{r-1}}\left[\frac{\partial}{\partial z}B(z)\right]$$

$$= \frac{\partial^{r-1}}{\partial z^{r-1}}[B(z)A(z)]$$

$$= \sum_{r_1=0}^{r-1}\binom{r-1}{r_1}A^{(r-1-r_1)}(z)B^{(r_1)}(z) \tag{1.7.13}$$

where, for example,

$$A^{(m)}(z) = \frac{\partial^m}{\partial z^m}A(z).$$

The above recurrence relation can be used when computing the residues. Thus

$$R_{a,k} = \frac{x^{-a}}{(k-1)!}\sum_{r=0}^{k-1}\binom{k-1}{r}(-\ln x)^{k-1-r}\{\sum_{r_1=0}^{r-1}\binom{r-1}{r_1}A_0^{(r-1-r_1)}$$

$$\times \sum_{r_2=0}^{r_1-1}\binom{r_1-1}{r_2}A_0^{(r_1-1-r_2)}\cdots\}B_0 \tag{1.7.14}$$

where

$$B_0 = \lim_{z\to a}B(z) \text{ and } A_0^{(m)} = \lim_{z\to a}A^{(m)}(z). \tag{1.7.15}$$

For convenience of computations the first few terms of the differential operator

$$\left\{\frac{\partial}{\partial z} + (-\ln x)\right\}^v B(z) = H_v(z)B(z) \tag{1.7.16}$$

will be listed here explicitly, where

$$A^{(0)} = A, \quad A^r = [A(z)]^r, \quad A^{(m)}$$

is the m-th derivative of $A(z), A(z) = \frac{d}{dz}\ln B(z)$.

$$
\begin{aligned}
H_0 &= 1 \\
H_1 &= (-\ln x) + A \\
H_2 &= (-\ln x)^2 + 2(-\ln x)A + A^{(1)} + A^2 \\
H_3 &= (-\ln x)^3 + 3(-\ln x)^2 A + 3(-\ln x)(A^{(1)} + A^2) \\
&\quad + (A^{(2)} + 3A^{(1)}A + A^3).
\end{aligned}
\tag{1.7.17}
$$

Exercises 1.7.

1.7.1. For a Gauss' hypergeometric function ${}_2F_1$ derive the following relationships:

$$
\begin{aligned}
{}_2F_1(a,b;c;z) &= (1-z)^{-b}\, {}_2F_1\left(c-a,b;c;-\frac{z}{1-z}\right), z \neq 1 \\
&= (1-z)^{-a}\, {}_2F_1\left(a,c-b;c;-\frac{z}{1-z}\right), z \neq 1 \\
&= (1-z)^{c-a-b}\, {}_2F_1(c-a,c-b;c;z).
\end{aligned}
$$

1.7.2. Let x_1 and x_2 be independently distributed real scalar gamma random variables with the parameters $(\alpha_1, 1)$ and $(\alpha_2, 1)$ respectively. Let $u = x_1 x_2$. Evaluate the density of u by using Mellin transformation technique when α_1 and α_2 do not differ by integers or zero.

1.7.3. Let x_1 and x_2 be independently distributed real type-1 beta random variables with the parameters (α_1, β_1) and (α_2, β_2) respectively. Let $u = x_1 x_2$. Evaluate the density of u by using Mellin transform technique if α_1 and α_2 do not differ by integers or zero.

1.7.4. Repeat the problem in Exercise 1.7.3 if x_1 and x_2 are type-2 beta distributed, where $\alpha_1 - \alpha_2 \neq \pm \lambda, \lambda = 0, 1, \cdots \beta_1 - \beta_2 \neq \pm v, v = 0, 1, 2, \cdots$.

1.7.5. Let $f(x) = \frac{1}{2\pi i}\int_{c-i\infty}^{c+i\infty} \Gamma(\alpha - s)\Gamma(s)x^{-s}ds$. Evaluate $f(x)$ as the sum of residues at the poles of $\Gamma(s)$. Then evaluate it again at the poles of $\Gamma(\alpha - s)$. Then compare the two results. In the first case we get the function for $|x| < 1$ and in the second case for $|x| > 1$.

1.7.6. Evaluate the integral

$$
f(x) = \frac{1}{2\pi i}\int_{c-i\infty}^{c+i\infty} \phi(s)x^{-s}ds
$$

where

$$\phi(s) = \frac{\Gamma\left(\frac{3}{2}+s-1\right)}{\Gamma\left(\frac{3}{2}\right)} \frac{\Gamma\left(\frac{7}{2}\right)}{\Gamma\left(\frac{7}{2}+s-1\right)} \frac{\Gamma(2+s-1)}{\Gamma(2)} \frac{\Gamma(4)}{\Gamma(4+s-1)}$$

for $\Re\left(\frac{1}{2}+s\right) > 0$.

1.7.7. Evaluate the integral

$$f(x) = \frac{1}{2\pi i} \int_{c-i\infty}^{c+i\infty} \frac{\Gamma\left(\frac{1}{3}+s\right)\Gamma\left(\frac{5}{6}+s\right)}{\Gamma\left(\frac{7}{3}+s\right)\Gamma\left(\frac{17}{6}+s\right)} x^{-s} ds$$

for $c > -\frac{1}{3}, i = \sqrt{-1}, 0 < x < 1$.

1.7.8. Evaluate

$$f(x) = \frac{1}{3\sqrt{\pi}} \frac{1}{2\pi i} \int_{c-i\infty}^{c+i\infty} \Gamma(3+s)\Gamma\left(\frac{1}{2}+s\right) x^{-s} ds,$$

for $x > 0, c > -\frac{1}{2}$.

1.7.9. Evaluate

$$f(x) = \frac{1}{144} \frac{1}{2\pi i} \int_{c-i\infty}^{c+i\infty} \Gamma(3+s)\Gamma(4+s) x^{-s} ds, x > 0, c > -3.$$

1.7.10. Prove that

$$\frac{1}{2\pi i} \int_{c-i\infty}^{c+i\infty} \frac{\Gamma(\alpha+s-1)\Gamma(\alpha+s-\frac{1}{2})}{\Gamma(\alpha+\beta+s-1)\Gamma(\alpha+\beta+s-\frac{1}{2})} x^{-s} ds$$

$$= \frac{2^{2\beta-1}}{\Gamma(2\beta)} x^{\alpha-1} (1-x^{\frac{1}{2}})^{2\beta-1},$$

$$0 < x < 1, \Re(\alpha) > 0, \Re(\beta) > 0, c > -\Re(\alpha-1).$$

1.8 Meijer's G-function

A generalization of the hypergeometric function in the real scalar case is Meijer's G-function. It is defined in terms of a Mellin-Barnes integral.

Notation 1.8.1.

$$G_{p,q}^{m,n}\left[z\Big|_{b_1,\ldots,b_q}^{a_1,\ldots,a_p}\right] = G_{p,q}^{m,n}\left[z\Big|_{(b_q)}^{(a_p)}\right] = G_{p,q}^{m,n}(z) = G(z).$$

Definition 1.8.1. G-function.

$$G_{p,q}^{m,n}[z|_{b_1,\ldots,b_q}^{a_1,\ldots,a_p}] = \frac{1}{2\pi i}\int_L \frac{\left\{\prod_{j=1}^m \Gamma(b_j+s)\right\}\left\{\prod_{j=1}^n \Gamma(1-a_j-s)\right\}z^{-s}ds}{\left\{\prod_{j=m+1}^q \Gamma(1-b_j-s)\right\}\left\{\prod_{j=n+1}^p \Gamma(a_j+s)\right\}} \quad (1.8.1)$$

where L is a contour separating the poles of $\Gamma(b_j+s)$, $j=1,\cdots,m$ from those of $\Gamma(1-a_j-s)$, $j=1,\cdots,n$. Three types of contours are described and the conditions of existence for the G-function are discussed in Mathai (1993). The simplified conditions are the following: $G(z)$ exists for the following situations:

$$\begin{aligned}
&\text{(i)} \quad q \geq 1, q > p, \text{ for all } z, z \neq 0\\
&\text{(ii)} \quad q \geq 1, q = p, \text{ for } |z| < 1\\
&\text{(iii)} \quad p \geq 1, p > q, \text{ for all } z, z \neq 0\\
&\text{(iv)} \quad p \geq 1, p = q, \text{ for } |z| > 1.
\end{aligned} \quad (1.8.2)$$

Example 1.8.1. Evaluate

$$f(x) = G_{1,1}^{1,0}\left[x|_{\alpha}^{\alpha+\beta+1}\right].$$

Solution 1.8.1:   As per our notation, $m=1, n=0, p=1, q=1$.

$$G_{1,1}^{1,0}\left[x|_{\alpha}^{\alpha+\beta+1}\right] = \frac{1}{2\pi i}\int_L \frac{\Gamma(\alpha+s)}{\Gamma(\alpha+\beta+1+s)}x^{-s}ds.$$

As per situation (ii) above we should obtain a convergent function for $|x| < 1$ if we evaluate the integral as the sum of the residues at the poles of $\Gamma(\alpha+s)$. The poles are at $s = -\alpha - v, v = 0, 1, \cdots$ and the sum of the residues

$$\sum_{v=0}^{\infty} \Re_v = \sum_{v=0}^{\infty} \frac{(-1)^v}{v!} \frac{x^{v+\alpha}}{\Gamma(\beta+1-v)}; \Gamma(\beta+1-v) = \frac{(-1)^v\Gamma(\beta+1)}{(-\beta)_v}.$$

$$G_{1,1}^{1,0}[x|_{\alpha}^{\alpha+\beta+1}] = \frac{x^{\alpha}}{\Gamma(\beta+1)}\sum_{v=0}^{\infty}\frac{(-\beta)_v x^v}{v!} = \frac{x^{\alpha}}{\Gamma(\beta+1)}(1-x)^{\beta}, |x| < 1 \quad (1.8.3)$$

for $\Re(\beta+1) > 0$.

Example 1.8.2.   Let $u = x_1 x_2 \cdots x_p$ where $x_1, \cdots, x_p$ are independently distributed real random variables with (1) : x_j gamma distributed with parameters $(\alpha_j, 1), j = 1, \cdots, p$; (2) : x_j type-1 beta distributed with parameters $(\alpha_j, \beta_j), j = 1, \cdots, p$; (3) : x_j is type-2 beta distributed with parameters $(\alpha_j, \beta_j), j = 1, \cdots, p$. Evaluate the density of u in (1),(2) and (3).

Solution 1.8.2: Taking the $(s-1)^{th}$ moment of u or the Mellin transform of the density of u we have the following:

$$E(u^{s-1}) = E(x_1 \cdots x_p)^{s-1} = E(x_1^{s-1} \cdots x_p^{s-1}) = E(x_1^{s-1}) \cdots E(x_p^{s-1})$$

due to independence

$$= \prod_{j=1}^{p} E(x_j^{s-1}) = \prod_{j=1}^{p} \frac{\Gamma(\alpha_j + s - 1)}{\Gamma(\alpha_j)}, \Re(\alpha_j + s - 1) > 0, j = 1, \cdots, p$$

for case (1)

$$= \prod_{j=1}^{p} \frac{\Gamma(\alpha_j + s - 1)}{\Gamma(\alpha_j)} \frac{\Gamma(\alpha_j + \beta_j)}{\Gamma(\alpha_j + \beta_j + s - 1)}, \Re(\alpha_j + s - 1) > 0, j = 1, \cdots, p$$

for case (2)

$$= \prod_{j=1}^{p} \frac{\Gamma(\alpha_j + s - 1)}{\Gamma(\alpha_j)} \frac{\Gamma(\beta_j - s + 1)}{\Gamma(\beta_j)}, \Re(\alpha_j + s - 1) > 0, \Re(\beta_j - s + 1) > 0,$$

$j = 1, \cdots, p$ for case (3).

Let the densities be denoted by $g_1(u), g_2(u)$ and $g_3(u)$ respectively. They are available from the respective inverse Mellin transforms which can be written as G-functions as follows:

$$g_1(u) = \frac{1}{2\pi i} \int_{c-i\infty}^{c+i\infty} \left\{ \prod_{j=1}^{p} \frac{\Gamma(\alpha_j - 1 + s)}{\Gamma(\alpha_j)} \right\} u^{-s} ds$$

$$= \frac{1}{\left\{ \prod_{j=1}^{p} \Gamma(\alpha_j) \right\}} G_{0,p}^{p,0}[u|_{\alpha_j - 1, j=1,\cdots,p}], \text{ for } u > 0, \Re(\alpha_j) > 0, j = 1, \cdots, p$$

$$(1.8.4)$$

and zero elsewhere.

$$g_2(u) = \frac{1}{2\pi i} \int_{c-i\infty}^{c+i\infty} \left\{ \prod_{j=1}^{p} \frac{\Gamma(\alpha_j + s - 1)}{\Gamma(\alpha_j)} \frac{\Gamma(\alpha_j + \beta_j)}{\Gamma(\alpha_j + \beta_j + s - 1)} \right\} u^{-s} ds$$

$$= \left\{ \prod_{j=1}^{p} \frac{\Gamma(\alpha_j + \beta_j)}{\Gamma(\alpha_j)} \right\} G_{p,p}^{p,0}[u|_{\alpha_j - 1, j=1,\cdots,p}^{\alpha_j + \beta_j - 1, j=1,\cdots,p}], 0 < u < 1, \quad (1.8.5)$$

$\Re(\alpha_j) > 0, \Re(\beta_j) > 0, j = 1, \cdots, p$ and zero elsewhere .

$$g_3(u) = \frac{1}{2\pi i} \int_{c-i\infty}^{c+i\infty} \left\{ \prod_{j=1}^{p} \frac{\Gamma(\alpha_j + s - 1)}{\Gamma(\alpha_j)} \frac{\Gamma(\beta_j - s + 1)}{\Gamma(\beta_j)} \right\} u^{-s} ds \quad (1.8.6)$$

$$= \frac{1}{\left\{ \prod_{j=1}^{p} \Gamma(\alpha_j) \Gamma(\beta_j) \right\}} G_{p,p}^{p,p}[u|_{\alpha_j - 1, j=1,\cdots,p}^{-\beta_j, j=1,\cdots,p}], u > 0,$$

$\Re(\alpha_j) > 0, \Re(\beta_j) > 0, j = 1, \cdots, p$ and zero elsewhere.

Example 1.8.3. Evaluate the following integral, a particular case of which is the reaction rate integral in astrophysics.

$$I(\alpha,a,b,\rho) = \int_0^\infty x^{\alpha-1}e^{-ax-bx^{-\rho}}dx, a > 0, b > 0, \rho > 0. \qquad (1.8.7)$$

Solution 1.8.3:   Since the integrand can be taken as a product of positive integrable functions we can apply statistical distribution theory to evaluate this integral or such similar integrals. The procedure to be discussed here is suitable for a wide variety of problems. Let x_1 and x_2 be two real scalar random variables with density functions $f_1(x_1)$ and $f_2(x_2)$. Let $u = x_1x_2$ and let x_1 and x_2 be independently distributed. Then the joint density of x_1 and x_2, denoted by $f(x_1,x_2)$, is the product of the marginal densities due to statistical independence of x_1 and x_2. That is,

$$f(x_1,x_2) = f_1(x_1)f_2(x_2).$$

Consider the transformation $u = x_1x_2$ and $v = x_1 \Rightarrow dx_1 \wedge dx_2 = \frac{1}{v}du \wedge dv$. Hence the joint density of u and v, denoted by $g(u,v)$, is available as,

$$g(u,v) = \frac{1}{v}f_1(v)f_2\left(\frac{u}{v}\right). \qquad (1.8.8)$$

Then the density of u denoted by $g_1(u)$, is available by integrating out v from $g(u,v)$. That is,

$$g_1(u) = \int_v \frac{1}{v}f_1(v)f_2\left(\frac{u}{v}\right)dv. \qquad (1.8.9)$$

Here (1.8.8) and (1.8.9) are general results and the method described here is called the *method of transformation of variables* for obtaining the density of $u = x_1x_2$. Now, consider (1.8.7). Let

$$f_1(x_1) = c_1 x_1^\alpha e^{-ax_1} \text{ and } f_2(x_2) = c_2 e^{-zx_2^\rho}, 0 \le x_1 < \infty, 0 \le x_2 < \infty \qquad (1.8.10)$$

$a > 0, z > 0$, where c_1 and c_2 are the normalizing constants. These normalizing constants can be evaluated by using the property.

$$1 = \int_0^\infty f_1(x_1)dx_1 \text{ and } 1 = \int_0^\infty f_2(x_2)dx_2.$$

Since we do not need the explicit forms of c_1 and c_2 we will not evaluate them here. With the f_1 and f_2 in (1.8.10) let us evaluate (1.8.9). We have

$$g_1(u) = c_1c_2 \int_{v=0}^\infty \frac{1}{v}v^\alpha e^{-av}e^{-z\left(\frac{u}{v}\right)^\rho}dv = c_1c_2 \int_{v=0}^\infty v^{\alpha-1}e^{-av}e^{-(zu^\rho)v^{-\rho}}dv.$$
$$(1.8.11)$$

Note that with $b = zu^\rho$, (1.8.11) is (1.8.7) multiplied by c_1 and c_2. Thus, we have identified the integral to be evaluated as a constant multiple of the density of u. This density of u is unique. Let us evaluate the density through Mellin and inverse Mellin transform technique.

$$E(u^{s-1}) = E(x_1^{s-1})E(x_2^{s-1})$$

due to statistical independence of x_1 and x_2. But

$$E(x_1^{s-1}) = c_1 \int_0^\infty x_1^{\alpha+s-1} e^{-ax_1} dx_1 = c_1 a^{-(\alpha+s)} \Gamma(\alpha+s), \Re(\alpha+s) > 0 \quad (1.8.12)$$

and

$$E(x_2^{s-1}) = c_2 \int_0^\infty x_2^{s-1} e^{-zx_2^\rho} dx_2 = \frac{c_2}{\rho z^{s/\rho}} \int_0^\infty y^{\frac{s}{\rho}-1} e^{-y} dy = \frac{c_2}{\rho z^{s/\rho}} \Gamma\left(\frac{s}{\rho}\right), \Re(s) > 0. \quad (1.8.13)$$

Hence

$$E(u^{s-1}) = c_1 c_2 \frac{a^{-\alpha}}{\rho} \left(az^{\frac{1}{\rho}}\right)^{-s} \Gamma(\alpha+s)\Gamma\left(\frac{s}{\rho}\right). \quad (1.8.14)$$

Therefore, the density of u, denoted by $g_1(u)$, is available from the inverse Mellin transform.

$$g_1(u) = c_1 c_2 \frac{a^{-\alpha}}{\rho} \frac{1}{2\pi i} \int_{c-i\infty}^{c+i\infty} \Gamma(\alpha+s)\Gamma\left(\frac{s}{\rho}\right) \left(az^{\frac{1}{\rho}}u\right)^{-s} ds. \quad (1.8.15)$$

Now, compare (1.8.15) with (1.8.11) to obtain the following:

$$\int_0^\infty v^{\alpha-1} e^{-av} e^{-(zu^\rho)v^{-\rho}} dv = \frac{a^{-\alpha}}{\rho} \frac{1}{2\pi i} \int_{c-i\infty}^{c+i\infty} \Gamma(\alpha+s)\Gamma\left(\frac{s}{\rho}\right) \left(az^{\frac{1}{\rho}}u\right)^{-s} ds. \quad (1.8.16)$$

On the right side in (1.8.15) the coefficient of s in $\Gamma(\frac{s}{\rho})$ is $\frac{1}{\rho} \neq 1$. Hence (1.8.15) is not a G-function but it can be written as an H-function, which will be considered next. In reaction rate theory in physics, $\rho = \frac{1}{2}$ and then

$$\Gamma\left(\frac{s}{\rho}\right) = \Gamma(2s) = \pi^{-\frac{1}{2}} 2^{2s-1} \Gamma(s)\Gamma\left(s+\frac{1}{2}\right)$$

by using the duplication formula for gamma functions. Then the right side of (1.8.16) reduces to

$$\frac{1}{2\rho a^\alpha \pi^{\frac{1}{2}}} \frac{1}{2\pi i} \int_{c-i\infty}^{c+i\infty} \Gamma(\alpha+s)\Gamma(s)\Gamma\left(s+\frac{1}{2}\right) \left(\frac{auz^{1/\rho}}{4}\right)^{-\rho} ds$$

$$= \frac{1}{2\rho a^\alpha \pi^{\frac{1}{2}}} G_{0,3}^{3,0}\left[\frac{auz^{1/\rho}}{4} \Big|_{\alpha,0,\frac{1}{2}}\right], u > 0.$$

But

$$b = zu^\rho \Rightarrow \frac{auz^{1/\rho}}{4} = \frac{ab^{1/\rho}}{4}.$$

Hence, for $\rho = \frac{1}{2}$,

$$\int_0^\infty v^{\alpha-1} e^{-av-bv^{-\rho}} dv = \frac{1}{2\rho a^\alpha \pi^{\frac{1}{2}}} G_{0,3}^{3,0}\left[\frac{ab^{1/\rho}}{4} \Big|_{\alpha,0,\frac{1}{2}}\right] \text{ for } \rho = \frac{1}{2}$$

$$= \frac{1}{a^\alpha \pi^{\frac{1}{2}}} G_{0,3}^{3,0}\left[\frac{ab^2}{4} \Big|_{\alpha,0,\frac{1}{2}}\right], u > 0. \quad (1.8.17)$$

Exercises 1.8.

Write down the Mellin-Barnes representations in Exercises 1.8.1 - 1.8.5 where the series forms are given. Here is an illustration.

$$_1F_0(\alpha;;x) = \sum_{r=0}^{\infty} (\alpha)_r \frac{x^r}{r!} = \frac{1}{\Gamma(\alpha)} \sum_{r=0}^{\infty} \Gamma(\alpha+r) \frac{x^r}{r!}$$

$$= \frac{1}{\Gamma(\alpha)} \frac{1}{2\pi i} \int_{c-i\infty}^{c+i\infty} \Gamma(\alpha-s)\Gamma(s)(-x^{-s})ds.$$

The last expression is the Mellin-Barnes representation for the series form $_1F_0(\alpha;;x)$.

1.8.1. $_0F_0(;;-z) = e^{-z} = \sum_{r=0}^{\infty} \frac{(-z)^r}{r!}$ (Exponential series)

1.8.2. $_2F_1(a,b;c;z) = \sum_{r=0}^{\infty} \frac{(a)_r(b)_r}{(c)_r} \frac{z^r}{r!}$ (Gauss' hypergeometric series)

1.8.3. $_1F_1(a;b;z) = \sum_{r=0}^{\infty} \frac{(a)_r}{(b)_r} \frac{z^r}{r!}$ (Confluent hypergeometric series)

1.8.4. $\sum_{r=0}^{\infty} \frac{(-1)^r}{r!} \frac{(z/2)^{v+2r}}{\Gamma(v+r+1)}$ (Bessel function $J_v(z)$)

1.8.5. $\sum_{r=0}^{\infty} \frac{(z/2)^{v+2r}}{r!\Gamma(v+r+1)}$ (Bessel function $I_v(z)$).

Write the series form from the Mellin-Barnes representation in Exercise 1.8.6 and list the conditions for convergence and existence also.

1.8.6. $\frac{\Gamma(1+2v)}{\Gamma(\frac{1}{2}+v-\mu)} e^{-z/2} z^{v+\frac{1}{2}} \frac{1}{2\pi i} \int_{c-i\infty}^{c+i\infty} \frac{\Gamma(s)\Gamma(\frac{1}{2}+v-\mu-s)}{\Gamma(1+2v-s)} (-z)^{-s} ds$ (Whittaker function $M_{\mu,v}(z)$)

Represent the following in Exercises 1.8.7 to 1.8.10 as G-functions and write down the conditions.

1.8.7. $z^\beta (1+az^\alpha)^{-1}$

1.8.8. $z^\beta (1+az^\alpha)^{-\gamma}$

1.8.9. (a) $\sin z$; (b) $\cos z$; (c) $\sinh z$; (d) $\cosh z$

1.8.10. (a) $\ln(1\pm z)$; (b) $\ln\left(\frac{1+z}{1-z}\right)$.

1.9 The H-function

This function is a generalization of the G-function. This was available in the literature as a Mellin-Barnes integral but Charles Fox made a detailed study of it in 1960's and hence the function is called Fox's H-function. The Mellin-Barnes representation is the following:

Notation 1.9.1. H-function

$$H_{p,q}^{m,n}\left[z\Big|_{(b_1,\beta_1),\cdots,(b_q,\beta_q)}^{(a_1,\alpha_1),\cdots,(a_p,\alpha_p)}\right] = H_{p,q}^{m,n}\left[z\Big|_{[(b_q,\beta_q)]}^{[(a_p,\alpha_p)]}\right] = H_{p,q}^{m,n}(z) = H(z).$$

Definition 1.9.1.

$$H_{p,q}^{m,n}\left[z\Big|_{(b_1,\beta_1),\cdots,(b_q,\beta_q)}^{(a_1,\alpha_1),\cdots,(a_p,\alpha_p)}\right] = \frac{1}{2\pi i}\int_L \phi(s)z^{-s}ds,$$

$$\phi(s) = \frac{\left\{\prod_{j=1}^{m}\Gamma(b_j+\beta_j s)\right\}\left\{\prod_{j=1}^{n}\Gamma(1-a_j-\alpha_j s)\right\}}{\left\{\prod_{j=m+1}^{q}\Gamma(1-b_j-\beta_j s)\right\}\left\{\prod_{j=n+1}^{p}\Gamma(a_j+\alpha_j s)\right\}},$$

(1.9.1)

where $\alpha_1,\cdots,\alpha_p,\beta_1,\cdots,\beta_q$ are real positive numbers (integers, rationals or irrationals), a_j's and b_j's are, in general, complex quantities, $i=\sqrt{-1}$ and the contour L separates the poles of $\Gamma(b_j+\beta_j s), j=1,\cdots,m$ from those of $\Gamma(1-a_j-\alpha_j s), j=1,\cdots,n$. Three paths L, similar to the ones for a G-function, can be given for the H-function also. Details of the existence conditions, various properties and applications may be seen from Mathai and Saxena (1978) and Mathai (1993). A simplified set of existence conditions is the following: Let

$$\mu = \sum_{j=1}^{q}\beta_j - \sum_{j=1}^{p}\alpha_j \text{ and } \beta = \left\{\prod_{j=1}^{q}\alpha_j^{\alpha_j}\right\}\left\{\prod_{j=1}^{q}\beta_j^{-\beta_j}\right\}.$$

(1.9.2)

The H-function exists for the following cases:

(i) $q \geq 1, \mu > 0$, for all $z, z \neq 0$

(ii) $q \geq 1, \mu = 0$, for $|z| < \beta^{-1}$

(iii) $p \geq 1, \mu < 0$, for all $z, z \neq 0$

(iv) $p \geq 1, \mu = 0$, for $|z|, z > \beta^{-1}$.

(1.9.3)

Two special cases, which follow from the definition itself, may be noted. When $\alpha_1 = 1 = \cdots = \alpha_p = \beta_1 = 1 = \cdots = \beta_q$ then the H-function reduces to a G-function. When all the α_j's and β_j's are rational numbers, that is ratios of two positive integers since by definition the α_j's and β_j's are positive real numbers, we may make a transformation $\frac{s}{u} = s_1$ where u is the common denominator for all the $\alpha_j, j = 1,\cdots,p$ and $\beta_j, j = 1,\cdots,q$. Under this transformation each coefficient of s_1 in each gamma in (1.9.1) becomes a positive integer. Then we may expand all the gammas by using the multiplication formula for gamma functions. Then the coefficients of s_1 in every gamma becomes ± 1 and then the H-function becomes a G-function. An illustration of this aspect was seen in Example 1.8.3.

Example 1.9.1. Evaluate the following reaction rate integral in physics and write it as an H-function.

$$I(\alpha, a, b, \rho) = \int_0^\infty x^{\alpha-1} e^{-ax-bx^{-\rho}} dx.$$

Solution 1.9.1:   From (1.8.16) in Example 1.8.3 we have

$$\int_0^\infty x^{\alpha-1} e^{-ax-bx^{-\rho}} dx = \frac{a^{-\alpha}}{\rho} \frac{1}{2\pi i} \int_{c-i\infty}^{c+i\infty} \Gamma(\alpha+s) \Gamma\left(\frac{s}{\rho}\right) \left(ab^{1/\rho}\right)^{-s} ds. \quad (1.9.4)$$

Writing the right side with the help of (1.9.1) we have the following:

$$\int_0^\infty x^{\alpha-1} e^{-ax-bx^{-\rho}} dx = \frac{1}{\rho a^\alpha} H_{0,2}^{2,0}\left[ab^{\frac{1}{\rho}} \Big|_{(\alpha,1),\left(0,\frac{1}{\rho}\right)}\right]. \quad (1.9.5)$$

Example 1.9.2. Let $x_1, \cdots, x_k$ be independently distributed real scalar gamma random variables with the parameters $(\alpha_j, 1), j = 1, \cdots, k$. Let $\gamma_1, \cdots, \gamma_k$ be real constants.
Let

$$u = x_1^{\gamma_1} x_2^{\gamma_2} \cdots x_k^{\gamma_k}.$$

Evaluate the density of u.

Solution 1.9.2:   Let us take the $(s-1)^{th}$ moment of u or the Mellin transform of the density of u.

$$E(u^{s-1}) = E\left[x_1^{\gamma_1} \cdots x_k^{\gamma_k}\right]^{s-1} = E\left(x_1^{\gamma_1(s-1)}\right) \cdots E\left[x_k^{\gamma_k(s-1)}\right]$$

due to independence. But for a real gamma random variable, with parameters $(\alpha_j, 1)$, the $[\gamma_j(s-1)]^{th}$ moment is the following:

$$E\left[x_j^{\gamma_j(s-1)}\right] = \frac{\Gamma(\alpha_j + \gamma_j(s-1))}{\Gamma(\alpha_j)} = \frac{\Gamma(\alpha_j - \gamma_j + \gamma_j s)}{\Gamma(\alpha_j)} \text{ for } \Re(\alpha_j + \gamma_j(s-1)) > 0.$$

$$(1.9.6)$$

Then

$$E(u^{s-1}) = \prod_{j=1}^k \frac{\Gamma(\alpha_j - \gamma_j + \gamma_j s)}{\Gamma(\alpha_j)}.$$

The density of u, denoted by $g(u)$, is available from the inverse Mellin transform. That is,

$$g(u) = \frac{1}{\left\{\prod_{j=1}^k \Gamma(\alpha_j)\right\}} \frac{1}{2\pi i} \int_{c-i\infty}^{c+i\infty} \left\{\prod_{j=1}^k \Gamma(\alpha_j - \gamma_j + \gamma_j s)\right\} u^{-s} ds$$

$$= \begin{cases} \frac{1}{\left\{\prod_{j=1}^k \Gamma(\alpha_j)\right\}} H_{0,k}^{k,0}\left[u \Big|_{(\alpha_j-\gamma_j,\gamma_j), j=1,\cdots,k}\right], u > 0, \\ 0, \quad \text{elsewhere.} \end{cases}$$

This is the density function for the product of arbitrary powers of independently distributed real scalar gamma random variables.

By using similar procedures one can obtain the densities of products of arbitrary powers of real scalar type-1 beta and type-2 beta random variables or arbitrary powers of products and ratios of real scalar gamma, type-1, type-2 beta or other such positive variables and write in terms of H-functions. Some details may be seen from Mathai (1993) and Mathai and Saxena (1978). The existence conditions may be seen from the existence condtions of the corresponding $(s-1)$th moments, as seen from (1.9.6).

Exercises 1.9.

1.9.1. Prove that

$$H_{p,q}^{m,n}\left[z\Big|\begin{matrix}(a_1,\alpha_1),\cdots,(a_p,\alpha_p)\\(b_1,\beta_1),\cdots,(b_q,\beta_q)\end{matrix}\right] = H_{q,p}^{n,m}\left[\frac{1}{z}\Big|\begin{matrix}(1-b_1,\beta_1),\cdots,(1-b_q,\beta_q)\\(1-a_1,\alpha_1),\cdots,(1-a_p,\alpha_p)\end{matrix}\right].$$

1.9.2. Prove that

$$z^\sigma H_{p,q}^{m,n}\left[z\Big|\begin{matrix}(a_1,\alpha_1),\cdots,(a_p,\alpha_p)\\(b_1,\beta_1),\cdots,(b_q,\beta_q)\end{matrix}\right] = H_{p,q}^{m,n}\left[z\Big|\begin{matrix}(a_1+\sigma\alpha_1,\alpha_1),\cdots,(a_p+\sigma\alpha_p,\alpha_p)\\(b_1+\sigma\beta_1,\beta_1),\cdots,(b_q+\sigma\beta_q,\beta_q)\end{matrix}\right].$$

1.9.3. Evaluate the Mellin-Barnes integral

$$E_\alpha(z) = \frac{1}{2\pi i}\int_{c-i\infty}^{c+i\infty}\frac{\Gamma(s)\Gamma(1-s)}{\Gamma(1-\alpha s)}(-z)^{-s}ds \tag{1.9.7}$$

and show that $E_\alpha(z)$ is the Mittag-Leffler series

$$E_\alpha(z) = \sum_{r=0}^\infty \frac{z^r}{\Gamma(\alpha r+1)}. \tag{1.9.8}$$

1.9.4. Evaluate the Laplace transform of $E_\alpha(z^\alpha)$ of Exercise 1.9.3, in (1.9.8), with parameter p.

1.9.5. A generalization of Mittag-Leffler function is $E_{\alpha,\beta}(z) = \sum_{r=0}^\infty \frac{z^r}{\Gamma(\alpha r+\beta)}$. Evaluate the Laplace transform of $t^{\beta-1}E_{\alpha,\beta}(z^\alpha)$.

1.9.6. If $\alpha = m, m = 1, 2, \cdots$ in (1.9.8) show that

$$E_m(z) = (2\pi)^{\frac{m-1}{2}} m^{-\frac{1}{2}} {}_0F_{m-1}\left(;\frac{1}{m},\frac{2}{m},\cdots,\frac{m-1}{m};\frac{z}{m^m}\right)\frac{1}{\Gamma(\frac{1}{m})\Gamma(\frac{2}{m})\cdots\Gamma(\frac{m-1}{m})}.$$

1.9.7. Write $E_\alpha(z)$ as an H-function.

1.9.8. If $\alpha = m, m = 1, 2, \cdots$ write down $E_\alpha(z)$ as a G-function.

1.9.9. Let x_1 and x_2 be independently distributed real gamma random variables with the parameters $(\alpha, 1), (\alpha + \frac{1}{2}, 1)$ respectively. Let $u = x_1 x_2$. Evaluate the density of u and show that the density of u, denoted by $g(u)$, is given by the following:

$$g(u) = \frac{2^{2\alpha-1}}{\Gamma(2\alpha)} u^{\alpha-1} e^{-2u^{\frac{1}{2}}}, u \geq 0 \text{ and zero elsewhere.}$$

1.9.10. Let x_1, x_2, x_3 be independently distributed real gamma random variables with the parameters $(\alpha, 1), (\alpha + \frac{1}{3}, 1), (\alpha + \frac{2}{3}, 1)$ respectively. Let $u = x_1 x_2 x_3$. Evaluate the density of u and show that it can be written as an H-function of the following type, where $g(u)$ denotes the density of u.

$$g(u) = \frac{27}{\Gamma(3\alpha)} H_{0,1}^{1,0}[27u \big|_{(3\alpha-3,3)}], u \geq 0 \text{ and zero elsewhere.}$$

1.10 Lauricella Functions and Appell's Functions

Another set of multivariable functions in frequent use in applied areas is the set of Lauricalla functions, and special cases of those are the Appell's functions. Lauricella functions f_A, f_B, f_C, and f_D are the following:

Definition 1.10.1. Lauricella function f_A

$$f_A(a, b_1, \cdots, b_n; c_1, \cdots, c_n; x_1, \cdots, x_n)$$
$$= \sum_{m_1=0}^{\infty} \cdots \sum_{m_n=0}^{\infty} \frac{(a)_{m_1+\cdots+m_n}(b_1)_{m_1} \cdots (b_n)_{m_n}}{(c_1)_{m_1} \cdots (c_n)_{m_n}} \frac{x_1^{m_1} \cdots x_n^{m_n}}{m_1! \cdots m_n!} \qquad (1.10.1)$$

for $|x_1| + \cdots + |x_n| < 1$.

Definition 1.10.2. Lauricella function f_B

$$f_B(a_1, \cdots, a_n, b_1, \cdots, b_n; c; x_1, \cdots, x_n)$$
$$= \sum_{m_1=0}^{\infty} \cdots \sum_{m_n=0}^{\infty} \frac{(a_1)_{m_1} \cdots (a_n)_{m_n}(b_1)_{m_1} \cdots (b_n)_{m_n}}{(c)_{m_1+\cdots+m_n}} \frac{x_1^{m_1} \cdots x_n^{m_n}}{m_1! \cdots m_n!} \qquad (1.10.2)$$

for $|x_1| < 1, |x_2| < 1, \cdots, |x_n| < 1$.

Definition 1.10.3. Lauricella function f_C

$$f_C(a, b; c_1, \cdots, c_n; x_1, \cdots, x_n) = \sum_{m_1=0}^{\infty} \cdots \sum_{m_n=0}^{\infty} \frac{(a)_{m_1+\cdots+m_n}(b)_{m_1+\cdots+m_n}}{(c_1)_{m_1} \cdots (c_n)_{m_n}} \frac{x_1^{m_1} \cdots x_n^{m_n}}{m_1! \cdots m_n!}$$
$$\qquad (1.10.3)$$

for $|\sqrt{x_1}| + \cdots + |\sqrt{x_n}| < 1$.

Definition 1.10.4. Lauricella function f_D

$$f_D(a,b_1,\cdots,b_n;c;x_1,\cdots,x_n)$$

$$= \sum_{m_1=0}^{\infty} \cdots \sum_{m_n=0}^{\infty} \frac{(a)_{m_1+\cdots+m_n}(b_1)_{m_1}\cdots\cdots(b_n)_{m_n}}{(c)_{m_1+\cdots+m_n}} \frac{x_1^{m_1}\cdots x_n^{m_n}}{m_1!\cdots m_n!} \qquad (1.10.4)$$

for $|x_1| < 1, |x_2| < 1, \cdots, |x_n| < 1$.

When $n = 2$ we have Appell's functions F_1, F_2, F_3, F_4. Also when $n = 1$ all these functions reduce to a Gauss' hypergeometric function ${}_2F_1$. We will list some of the basic properties of Lauricella functions.

1.10.1 Some properties of f_A

$$\int_0^1 \cdots \int_0^1 u_1^{b_1-1}\cdots u_n^{b_n-1}(1-u_1)^{c_1-b_1-1}\cdots(1-u_n)^{c_n-b_n-1}$$

$$\times (1-u_1x_1-\cdots-u_nx_n)^{-a}\,du_1\wedge\cdots\wedge du_n$$

$$= \left\{\prod_{j=1}^n \frac{\Gamma(b_j)\Gamma(c_j-b_j)}{\Gamma(c_j)}\right\} f_A(a,b_1,\cdots,b_n;c_1,\cdots,c_n;x_1,\cdots,x_n),$$

$$(1.10.5)$$

for $\Re(b_j) > 0$, $\Re(c_j-b_j) > 0, j = 1,\cdots,n$.

The result can be easily established by expanding the factor $(1-u_1x_1-\cdots-u_nx_n)^{-a}$ by using a multinomial expansion and then integrating out $u_j, j = 1,\cdots,n$ with the help of type-1 beta integrals.

$$\int_0^\infty e^{-t}t^{a-1}\,{}_1F_1(b_1;c_1;x_1t)\,{}_1F_1(b_2;c_2;x_2t)\cdots{}_1F_1(b_n;c_n;x_nt)dt \qquad (1.10.6)$$

$$= \Gamma(a)f_A(a,b_1,\cdots,b_n;c_1,\cdots,c_n;x_1,\cdots,x_n), \text{ for } \Re(a) > 0.$$

This can be established by taking the series forms for ${}_1F_1$'s and then integrating out t.

$$\frac{1}{(2\pi i)^n}\int\cdots\int \frac{\Gamma(a+t_1+\cdots+t_n)\Gamma(b_1+t_1)\cdots\Gamma(b_n+t_n)}{\Gamma(c_1+t_1)\cdots\Gamma(c_n+t_n)}$$

$$\times\Gamma(-t_1)\cdots\Gamma(-t_n)(-x_1)^{t_1}\cdots(-x_n)^{t_n}\,dt_1\wedge\cdots\wedge dt_n \qquad (1.10.7)$$

$$= \Gamma(a)\frac{\Gamma(b_1)\cdots\Gamma(b_n)}{\Gamma(c_1)\cdots\Gamma(c_n)}f_A(a,b_1,\cdots,b_n;c_1,\cdots,c_n;x_1,\cdots,x_n), i = \sqrt{-1}.$$

This can be established by evaluating the integrand as the sum of the residues at the poles of $\Gamma(-t_1),\cdots,\Gamma(-t_n)$, one by one.

1.10.2 Some properties of f_B

$$\int \cdots \int t_1^{a_1-1} \cdots t_n^{a_n-1} (1-t_1-\cdots-t_n)^{c-a_1-\cdots-a_n-1} \tag{1.10.8}$$

$$\times (1-t_1 x_1)^{-b_1} \cdots (1-t_n x_n)^{-b_n} \, dt_1 \wedge \cdots \wedge dt_n$$

$$= \frac{\Gamma(a_1)\cdots\Gamma(a_n)\Gamma(c-a_1-\cdots-a_n)}{\Gamma(c)} f_B(a_1,\cdots,a_n,b_1,\cdots,b_n;c;x_1,\cdots,x_n),$$

for $\Re(a_j) > 0, j = 1,\cdots,n, \Re(c-a_1-\cdots-a_n) > 0, t_j > 0, j = 1,\cdots,n$, and $1-t_1-\cdots-t_n > 0$.

This result can be established by opening up $(1-t_j x_j)^{-b_j}, j = 1,\cdots,n$ by using binomial expansions and then integrating out $t_1,\cdots,t_n$ with the help of a type-1 Dirichlet integral of Section 1.6.

$$\int_0^\infty \cdots \int_0^\infty s_1^{a_1-1} \cdots s_n^{a_n-1} t_1^{b_1-1} \cdots t_n^{b_n-1} e^{-s_1-\cdots-s_n-t_1-\cdots-t_n} \tag{1.10.9}$$

$$\times {}_0F_1\,(;c;s_1 t_1 x_1 + \cdots + s_n t_n x_n) \, ds_1 \wedge \cdots \wedge ds_n \wedge dt_1 \wedge \cdots \wedge dt_n$$

$$= \left\{ \prod_{j=1}^{n} \Gamma(a_j)\Gamma(b_j) \right\} f_B(a_1\cdots,a_n,b_1,\cdots,b_n;c;x_1,\cdots,x_n),$$

for $\Re(a_j) > 0, \; \Re(b_j) > 0, j = 1,\cdots,n$.

First, open up the ${}_0F_1$ as a power series in $(s_1 t_1 x_1 + \cdots + s_n t_n x_n)^k$. Since k is a positive integer open up by using a multinomial expansion. Then integrate out $s_1,\cdots,s_k$ and $t_1,\cdots,t_k$ by using gamma functions, to see the result.

$$\frac{1}{(2\pi i)^n} \int \cdots \int \frac{\Gamma(a_1+t_1)\cdots\Gamma(a_n+t_n)\Gamma(b_1+t_1)\cdots\Gamma(b_n+t_n)}{\Gamma(c+t_1+\cdots+t_n)} \tag{1.10.10}$$

$$\times \Gamma(-t_1)\cdots\Gamma(-t_n)(-x_1)^{t_1}\cdots(-x_n)^{t_n} \, dt_1 \wedge \cdots \wedge dt_n, \; i = \sqrt{-1}$$

$$= \left\{ \prod_{j=1}^{n} \frac{\Gamma(a_j)\Gamma(b_j)}{\Gamma(c)} \right\} f_B(a_1,\cdots,a_n,b_1,\cdots,b_n;c;x_1,\cdots,x_n).$$

Assume that $a_j - b_j \neq \pm v, v = 0,1,\cdots$. Then evaluate the integrand as the sum of the residues at the poles of $\Gamma(-t_1),\cdots,\Gamma(-t_n)$, one by one, to obtain the result.

1.10.3 Some properties of f_C

$$\int_0^\infty \cdots \int_0^\infty s^{a-1} t^{b-1} e^{-s-t} \, {}_0F_1\,(;c_1;x_1 st) \cdots {}_0F_1\,(;c_n;x_n st) ds \wedge dt \tag{1.10.11}$$

$$= \Gamma(a)\Gamma(b) f_C(a,b;c_1,\cdots,c_n;x_1,\cdots,x_n), \; \text{for } \Re(a) > 0, \Re(b) > 0.$$

Open up the ${}_0F_1$'s, then integrate out t and s with the help of gamma integrals to see the result.

$$\frac{1}{(2\pi i)^n} \int \cdots \int \frac{\Gamma(a+t_1+\cdots+t_n)\Gamma(b+t_1+\cdots+t_n)}{\Gamma(c_1+t_1)\cdots\Gamma(c_n+t_n)}$$
$$\times \Gamma(-t_1)\cdots\Gamma(-t_n)(-x_1)^{t_1}\cdots(-x_n)^{t_n}dt_1 \wedge \cdots \wedge dt_n$$
$$= \frac{\Gamma(a)\Gamma(b)}{\Gamma(c_1)\cdots\Gamma(c_n)} f_C(a,b;c_1,\cdots,c_n;x_1,\cdots,x_n), i = \sqrt{-1}.$$

$$(1.10.12)$$

Evaluate the integrand as the sum of the residues at the poles of $\Gamma(-t_1),\cdots,\Gamma(-t_n)$, one by one, to obtain the result.

1.10.4 Some properties of f_D

$$\int \cdots \int u_1^{b_1-1} \cdots u_n^{b_n-1}(1-u_1-\cdots-u_n)^{c-b_1-\cdots-b_n-1} \qquad (1.10.13)$$
$$\times (1-u_1x_1-\cdots-u_nx_n)^{-a}du_1 \wedge \cdots \wedge du_n$$
$$= \frac{\Gamma(b_1)\cdots\Gamma(b_n)\Gamma(c-b_1-\cdots-b_n)}{\Gamma(c)} f_D(a,b_1,\cdots,b_n;c;x_1,\cdots,x_n), \text{ for }$$
$$0 < u_j < 1, \ j=1,\cdots,n, \ 0 < u_1+\cdots+u_n < 1, \ 0 < x_1u_1+\cdots+x_nu_n < 1,$$
$$\Re(b_j) > 0, \ j=1,\cdots,n, \ \Re(c-b_1-\cdots-b_n) > 0.$$

Open up $(1-u_1x_1-\cdots-u_nx_n)^{-a}$ by using a multinomial expansion and then integrate out $u_1,\cdots,u_n$ by using a type-1 Dirichlet integral of Subsection 1.6.1

$$\int_0^1 u^{a-1}(1-u)^{c-a-1}(1-ux_1)^{-b_1}\cdots(1-ux_n)^{-b_n}du \qquad (1.10.14)$$
$$= \frac{\Gamma(a)\Gamma(c-a)}{\Gamma(c)} f_D(a,b_1,\cdots,b_n;c;x_1,\cdots,x_n) \text{ for } \Re(a) > 0, \ \Re(c-a) > 0.$$

Expand $(1-ux_j)^{-b_j}, j=1,\cdots,n$ by using binomial expansions and then integrate out u by using a type-1 beta integral to see the result.

$$\int_0^\infty \cdots \int_0^\infty t_1^{b_1-1}\cdots t_n^{b_n-1}e^{-t_1-\cdots-t_n} {}_1F_1(a;c;x_1t_1+\cdots+x_nt_n)dt_1 \wedge \cdots \wedge dt_n$$

$$(1.10.15)$$
$$= \Gamma(b_1)\cdots\Gamma(b_n)f_D(a,b_1,\cdots,b_n;c;x_1,\cdots,x_n), \text{ for } \Re(b_j) > 0, j=1,\cdots,n.$$

Expand ${}_1F_1$ as a series, then open up the general term with the help of a multinomial expansion for positive integral exponent, then integrate out $t_1,\cdots,t_n$ to see the result.

$$\frac{1}{(2\pi i)^n} \int \cdots \int \frac{\Gamma(a+t_1+\cdots+t_n)\Gamma(b_1+t_1)\cdots\Gamma(b_n+t_n)}{\Gamma(c+t_1+\cdots+t_n)} \qquad (1.10.16)$$
$$\times \Gamma(-t_1)\cdots\Gamma(-t_n)(-x_1)^{t_1}\cdots(-x_n)^{t_n}dt_1, \wedge \cdots \wedge dt_n$$
$$= \frac{\Gamma(a)\Gamma(b_1)\cdots\Gamma(b_n)}{\Gamma(c)} f_D(a,b_1,\cdots,b_n;c;x_1,\cdots,x_n), i = \sqrt{-1}.$$

Follow through the same method of evaluation of the contour integrals as in f_A, f_B and f_C to see the result.

$$f_D(a, b_1, \cdots, b_n; c; x, \cdots, x) = {}_2F_1(a, b_1 + \cdots + b_n; c; x). \tag{1.10.17}$$

Use the integral representation in (1.10.14) and put $x_1 = \cdots = x_n = x$ to see the result.

$$f_D(a, b_1, \cdots, b_n; c; 1, 1 \cdots, 1) = \frac{\Gamma(c)\Gamma(c - a - b_1 - \cdots - b_n)}{\Gamma(c - a)\Gamma(c - b_1 - \cdots - b_n)}. \tag{1.10.18}$$

Evaluate (1.10.17) at $x = 1$ to see the result.

There are other functions in the category of multivariable hypergeometric functions known as Humbert's functions, Kampé de Fériet functions and so on. These will not be discussed here. For a brief description of these, along with some of their properties, see for example Mathai (1993, 1997) and Srivastava and Karlsson (1985).

Example 1.10.1. Show that

$$f_A(a, b_1, \cdots, b_n; c_1, \cdots, c_n; x_1, \cdots, x_n) \tag{1.10.19}$$

$$= \sum_{m_1=0}^{\infty} \cdots \sum_{m_{n-1}=0}^{\infty} \frac{(a)_{m_1 + \cdots + m_{n-1}} (b)_{m_1} \cdots (b_n)_{m_n}}{(a)_{m_1} \cdots (c_n)_{m_n}}$$

$$\times \frac{x_1^{m_1} \cdots x_{n-1}^{m_{n-1}}}{m_1! \cdots m_{n-1}!} {}_2F_1(a + m_1 + \cdots + m_{n-1}, b_n; c_n; x_n), |x_1| + \cdots + |x_n| < 1.$$

Solution 1.10.1: This can be seen by summing up with respect to m_n by observing that $(a)_{m_1 + \cdots + m_n} = (a + m_1 + \cdots + m_{n-1})_{m_n}$. Then the sum is the following:

$$\sum_{m_n=0}^{\infty} \frac{(a + m_1 + \cdots + m_{n-1})_{m_n} (b_n)_{m_n}}{(c_n)_{m_n}} \frac{x^{m_n}}{m_n!} = {}_2F_1(a + m_1 + \cdots + m_{n-1}, b_n; c_n; x_n).$$

Example 1.10.2. Show that

$$\Gamma(a) f_C\left(\frac{a}{2}, \frac{a+1}{2}; c_1, \cdots, c_n; x_1, \cdots, x_n\right) \tag{1.10.20}$$

$$= \int_0^{\infty} t^{a-1} e^{-t} {}_0F_1\left(; c_1; \frac{t^2 x_1}{4}\right) \cdots {}_0F_1\left(; c_n; \frac{t^2 x_n}{4}\right) dt.$$

Solution 1.10.2: Expand the ${}_0F_1$'s. Then the right side becomes,

$$\sum_{m_1=0}^{\infty} \cdots \sum_{m_n=0}^{\infty} \frac{\int_0^{\infty} t^{a + 2m_1 + \cdots + 2m_n - 1}}{(c_1)_{m_1} \cdots (c_n)_{m_n}} e^{-t} \frac{x_1^{m_1}}{4^{m_1}} \cdots \frac{x_n^{m_n}}{4^{m_n}} \frac{1}{m_1! \cdots m_n!} dt.$$

Integral over t yields

$$\int_0^\infty t^{a+2m_1+\cdots+2m_n-1}e^{-t}dt = \Gamma(a+2m_1+\cdots+2m_n).$$

Expanding $\Gamma(a+2m_1+\cdots+2m_n) = \Gamma[2(\frac{a}{2}+m_1+\cdots+m_n)]$ by using the duplication formula, we have,

$$\Gamma\left[2\left(\frac{a}{2}+m_1+\cdots+m_n\right)\right] = \pi^{-\frac{1}{2}}2^{a+2m_1+\cdots+2m_n-1}\Gamma\left(\frac{a}{2}+m_1+\cdots+m_n\right)$$
$$\times \Gamma\left(\frac{a+1}{2}+m_1+\cdots+m_n\right)$$
$$= \pi^{-\frac{1}{2}}2^{a-1}\Gamma\left(\frac{a}{2}\right)\Gamma\left(\frac{a+1}{2}\right)$$
$$\times \left(\frac{a}{2}\right)_{m_1+\cdots+m_n}\left(\frac{a+1}{2}\right)_{m_1+\cdots+m_n}(4)^{m_1+\cdots+m_n}$$
$$= \Gamma(a)\left(\frac{a}{2}\right)_{m_1+\cdots+m_n}\left(\frac{a+1}{2}\right)_{m_1+\cdots+m_n}(4)^{m_1+\cdots+m_n}$$

(duplication formula is again applied on $\Gamma(a) = \Gamma[2(\frac{a}{2})]$). Now, substituting and interpreting as a f_C the result follows.

Example 1.10.3. Show that $f_B(a_1,\cdots,a_n,b_1,\cdots,b_n;c;x_1,\cdots,x_n)$

$$= \frac{\Gamma(c)}{\Gamma(d_1)\cdots\Gamma(d_n)\Gamma(c-d_1-\cdots-d_n)}\int\cdots\int u_1^{d_1-1}\cdots u_n^{d_n-1} \qquad (1.10.21)$$
$$\times (1-u_1-\cdots-u_n)^{c-d_1-\cdots-d_n-1}$$
$$\times {}_2F_1(a_1,b_1;d_1;u_1x_1)\cdots{}_2F_1(a_n,b_n;d_n;u_nx_n)du_1\wedge\cdots\wedge du_n$$

for $\Re(d_j) > 0, j = 1,\cdots,n$, $\Re(c-d_1-\cdots-d_n) > 0, |x_j| < 1, j = 1,\cdots,n$.

Solution 1.10.3: Expand the product of ${}_2F_1$'s first.

$${}_2F_1(a_1,b_1;d_1;u_1x_1)\cdots{}_2F_1(a_n,b_n;d_n;u_nx_n) \qquad (1.10.22)$$
$$= \sum_{m_1=0}^\infty\cdots\sum_{m_n=0}^\infty \frac{(a_1)_{m_1}\cdots(a_n)_{m_n}(b_1)_{m_1}\cdots(b_n)_{m_n}}{(d_1)_{m_1}\cdots(d_n)_{m_n}}\frac{(u_1x_1)^{m_1}\cdots(u_nx_n)^{m_n}}{m_1!\cdots m_n!}.$$

Now, evaluate the integral over $u_1,\cdots,u_n$ by using a type-1 Dirichlet integral.

$$\int\cdots\int u_1^{d_1+m_1-1}\cdots u_n^{d_n+m_n-1}(1-u_1-\cdots-u_n)^{c-d_1-\cdots-d_n-1}du_1\wedge\cdots\wedge du_n$$
$$= \frac{\Gamma(d_1+m_1)\cdots\Gamma(d_n+m_n)\Gamma(c-d_1-\cdots-d_n)}{\Gamma(c+m_1+\cdots+m_n)}$$
$$= \frac{\Gamma(c-d_1-\cdots-d_n)\Gamma(d_1)\cdots\Gamma(d_n)}{\Gamma(c)}\frac{(d_1)_{m_1}\cdots(d_n)_{m_n}}{(c)_{m_1+\cdots+m_n}} \qquad (1.10.23)$$

for $\Re(d_j) > 0, j = 1,\cdots,n, \Re(c-d_1-\cdots-d_n) > 0$. Now, substituting (1.10.23) and (1.10.22) on the right side of (1.10.21) the result follows.

Exercises 1.10.

1.10.1. Establish the result in (1.10.5)

1.10.2. Establish the result in (1.10.6)

1.10.3. Establish the result in (1.10.7)

1.10.4. Establish the result in (1.10.8)

1.10.5. Establish the result in (1.10.9)

1.10.6. Establish the result in (1.10.10)

1.10.7. Establish the result in (1.10.11)

1.10.8. Establish the result in (1.10.12)

1.10.9. Establish the result in (1.10.13)

1.10.10. Establish the result in (1.10.14)

1.11 Special Functions as Solutions of Differential Equations and Applications

[*This section is based on the lectures of Professor P. N. Rathie of the Department of Statistics, University of Brasília, Brazil.*]

1.11.0 Introduction

Certain special functions occur often in fields like physics and engineering. We study these functions (exponential to Mejer's G-functions) as solutions of differential equations because the behavior of a physical system is generally represented by a differential equation. A powerful method for solving differential equations is to assume a power series solution.

1.11.1 Sine, cosine and exponential functions

The sine, cosine and the exponential functions are the most elementary special functions. The following example illustrates how sine and cosine functions are obtained as solutions of a differential equation in a mathematical physics problem.

Example 1.11.1. *Motion of an elastically bound particle in one dimension.* The position x of a particle of mass m that moves under the influence of a force

$$F = -mw^2x \qquad (1.11.1)$$

is given at time t by

$$x = b\sin(wt + \phi) \tag{1.11.2}$$

where b and ϕ are constants. By Newton's second law of motion,

$$F = m\frac{d^2x}{dt^2}. \tag{1.11.3}$$

If the particle obeys *Hooke's law*, then (1.11.1) is valid with w as a constant. Thus (1.11.3) and (1.11.1) yield

$$\frac{d^2x}{dt^2} + w^2x = 0. \tag{1.11.4}$$

Assuming the power series solution

$$x(t) = \sum_{n=0}^{\infty} a_n t^{s+n} \tag{1.11.5}$$

where the a_n are constant coefficients to be determined, we obtain

$$s(s-1)a_0 = 0 \tag{1.11.6}$$
$$(s+1)sa_1 = 0 \tag{1.11.7}$$
$$(s+n+2)(s+n+1)a_{n+2} + w^2 a_n = 0. \tag{1.11.8}$$

Solution 1.11.1: (1.11.6) and (1.11.7) are satisfied for $s = 0$. With this choice (1.11.8) gives

$$a_n = \frac{(-1)^{n/2}w^n}{n!} a_0, \text{ for } n \text{ even} \tag{1.11.9}$$

$$a_n = \frac{(-1)^{(n-1)/2}w^{n-1}}{n!} a_1 \text{ for } n \text{ odd.} \tag{1.11.10}$$

Thus, the general solution of (1.11.4) is given by

$$x(t) = a_0 \sum_{\substack{n=0 \\ n=\text{even}}}^{\infty} \frac{(-1)^{n/2}w^n t^n}{n!} + a_1 \sum_{\substack{n=1 \\ n=\text{odd}}}^{\infty} \frac{(-1)^{(n-1)/2}w^{n-1}t^n}{n!} \tag{1.11.11}$$

$$= a_0 \sum_{k=0}^{\infty} \frac{(-1)^k(wt)^{2k}}{(2k)!} + \frac{a_1}{w} \sum_{k=0}^{\infty} \frac{(-1)^k(wt)^{2k+1}}{(2k+1)!}$$

$$= a_0 \cos(wt) + \frac{a_1}{w} \sin(wt).$$

Thus the general solution of the second-order differential equation (1.11.4) is the sum of two linearly independent solutions. The constants a_0 and a_1 are determined from the initial conditions. For example, for $t = 0$, $x(0) = x_0$ and $\frac{dx}{dt} = v = v_0$ then $a_0 = x_0$ and $a_1 = v_0$. Thus

$$x(t) = x_0 \cos(wt) + \frac{v_0}{w} \sin(wt) \tag{1.11.12}$$
$$= b \sin(wt + \phi)$$

with $b = \sqrt{x_0^2 + (v_0/w)^2}$ and $\phi = \tan^{-1}(x_0 w/v_0)$.

Exercises 1.11.

1.11.1. Find the solution of the differential equation

$$\frac{dy(x)}{dx} - y(x) = 0.$$

1.11.2. Obtain the solution of Eq. (1.11.4) for the following cases

(1) $s = 1$,
(2) $s = -1$.

1.11.2 Linear second order differential equations

Any linear, second-order, homogeneous differential equation can be written in the form

$$\frac{d^2}{dz^2} u(z) + P(z)\frac{d}{dz} u(z) + Q(z)u(z) = 0. \tag{1.11.13}$$

Assuming $u(z)$ and $\frac{d}{dz} u(z)$ at $z = z_0$ and successive differentiations, we can get the Taylor series for $u(z)$ as

$$u(z) = \sum_{n=0}^{\infty} \frac{\frac{d^n}{dz^n} u(z_0)}{n!} (z - z_0)^n. \tag{1.11.14}$$

If the series in (1.11.14) has a nonzero radius of convergence, then the solution exists. If $u(z)$ and $\frac{d}{dz} u(z)$ can be assigned arbitrary values at $z = z_0$, then we say that the point z_0 is an *ordinary point* of the differential equation (1.11.13), otherwise it is a *singular point*. Consider the differential equation,

$$z^2 \frac{d^2}{dz^2} u(z) + az \frac{d}{dz} u(z) + bu(z) = 0$$

where a and b are constants. For $z = 0$ we see that if $u(0)$ has any value other than zero, either $\frac{d}{dz} u(0)$ or $\frac{d^2}{dz^2} u(0)$ must be infinity and the Taylor series for $u(z)$ cannot be obtained around $z = 0$. If $P(z)$ or $Q(z)$ has a singularity (not a branch point) at $z = z_0$ so that $\frac{d^2}{dz^2}u(z_0)$ cannot be obtained to construct the Taylor series of $u(z)$, then the differential equation has a *regular singularity* if and only if both $(z - z_0)P(z)$ and $(z - z_0)^2 Q(z)$ are analytic at z_0. Otherwise, the singularity is *irregular*.

1.11.3 Hypergeometric function

The Gauss hypergeometric differential equation is given by

$$z(1-z)\frac{d^2u}{dz^2} + [c - (a+b+1)z]\frac{du}{dz} - abu = 0 \qquad (1.11.15)$$

or

$$[\delta(\delta+c-1) - z(\delta+a)(\delta+b)]u = 0, \quad \text{where } \delta = z\frac{d}{dz}. \qquad (1.11.16)$$

Assuming the solution

$$u(z) = \sum_{n=0}^{\infty} a_n z^{n+s} \qquad (1.11.17)$$

we get from (1.11.15), the following relations,

$$s(s+c-1)a_0 = 0 \qquad (1.11.18)$$

and

$$a_{n+1} = \frac{(n+s)(n+s+a+b)+ab}{(n+s+1)(n+s+c)}a_n. \qquad (1.11.19)$$

The trivial solution $u(z) = 0$ is obtained if we assume $a_0 = 0$. For the nontrivial solution we assume $a_0 \neq 0$. Equation (1.11.18) yields $s = 0$ or $s = 1 - c$. For $s = 0$, the solution of (1.11.15) is given by

$$u(z) = a_0 \sum_{n=0}^{\infty} \frac{(a)_n (b)_n}{n!(c)_n}z^n = a_0\,{}_2F_1(a,b;c;z), |z| < 1 \qquad (1.11.20)$$

where $(a)_n$ stands for the Pochhammer symbol.

Example 1.11.2. From (1.11.20), it is easy to see that

$$_2F_1(1,b;b;z) = \sum_{n=0}^{\infty} z^n, |z| < 1 \quad \text{(geometric series)}$$

and

$$_2F_1(-s,b;b;z) = (1-z)^s, |z| < 1 \quad \text{(binomial expansion)}.$$

The general solution (for $c \neq$ an integer) consists of two linearly independent solutions,

$$u(z) = c_1 u_1(z) + c_2 u_2(z) \tag{1.11.21}$$

where $u_1(z)$ is given by (1.11.20) and $u_2(z)$ corresponding to $s = -c+1$ is given by

$$u_2(z) = z^{1-c}{}_2F_1(1+a-c, 1+b-c; 2-c; z). \tag{1.11.22}$$

The arbitrary constants c_1 and c_2 are to be determined by the boundary conditions.

We may observe the following:

(1) The Gauss hypergeometric equation (1.11.15) has three singularities at 0, 1 and ∞.

(2) If c is an integer, $u_2(z)$ is not a new solution. For example, $c = 1$, $u_1(z) = u_2(z)$. If neither a nor b is zero or a negative integer, two linearly independent solutions are

$${}_2F_1(a, b; 1; z) \tag{1.11.23}$$

and (logarithmic solution)

$${}_2F_1(a, b; 1; z) = \ln(z) + \sum_{n=1}^{\infty} \frac{(a)_n (b)_n z^n}{(n!)^2} \left[\sum_{i=1}^{n} \frac{1}{a+i-1} + \sum_{i=1}^{n} \frac{1}{b+i-1} - \sum_{i=1}^{n} \frac{2}{i} \right]. \tag{1.11.24}$$

Example 1.11.3. *Simple pendulum*

A simple pendulum consists of a point mass m attached to one end of a massless cord of length l, and the other end fixed at a point such that the system can swing freely under gravity. Let T_1 be the tension in the cord when it is inclined at an angle θ with the vertical. Then by Newton's second law, we have

$$T_1 - mg\cos\theta = \frac{mv^2}{l} \tag{1.11.25}$$

$$-mg\sin\theta = m\frac{d^2}{dt^2}(l\theta). \tag{1.11.26}$$

The equation (1.11.25), takes the following form

$$\frac{d^2\theta}{dt^2} + \frac{g}{l}\sin\theta = 0. \tag{1.11.27}$$

Multiplying (1.11.26) by $\dfrac{d\theta}{dt}$ gives us

$$\frac{d}{dt}\left[\frac{1}{2}\left(\frac{d\theta}{dt}\right)^2 - \frac{g}{l}\cos\theta\right] = 0.$$

which implies that

$$\frac{1}{2}\left(\frac{d\theta}{dt}\right)^2 - \frac{g}{l}\cos\theta$$

is a constant. Let $\theta = \theta_0$ when $\dfrac{d\theta}{dt} = 0$ (pendulum at rest). This gives the value of constant as $\dfrac{g}{l}\cos\theta_0$ so that

$$\frac{d\theta}{dt} = \sqrt{\frac{2g}{l}(\cos\theta - \cos\theta_0)}.$$

Integration gives the period of oscillation as

$$T = 2\pi\sqrt{\frac{l}{g}}\,{}_2F_1\left(\frac{1}{2},\frac{1}{2};1;\sin^2\left(\frac{\theta_0}{2}\right)\right). \tag{1.11.28}$$

Thus, we see that the period of oscillation of a simple pendulum depends on the amplitude of the oscillation. Note that if the amplitude θ_0 is small, then the period is given by

$$T = 2\pi\sqrt{\frac{l}{g}}$$

Exercises 1.11.

1.11.3. Write down the general solution of the differential equation

$$z(1-z)\frac{d^2}{dz^2}u(z) + \left(\frac{5}{4}-2z\right)\frac{d}{dz}u(z) + \frac{3}{4}u(z) = 0.$$

1.11.4. Find the general solution of the differential equation

$$x\frac{d^2}{dx^2}y(x) + \mu\frac{d}{dx}y(x) + \lambda y(x) = 0$$

where λ and μ are constants. Discuss the conditions to be imposed on λ and μ.

1.11.5. Solve the differential equation

$$x\frac{d^2}{dx^2}f(x) + 2\frac{d}{dx}f(x) + xf(x) = 0$$

by series method. Is it possible to write the solution in terms of elementary special functions?

1.11.6. Find the general solution of the differential equation

$$z(1-z)\frac{d^2}{dz^2}u(z) + \mu(1-z)\frac{d}{dz}u(z) + \mu u(z) = 0,$$

where μ is not an integer. Is one of the solutions a polynomial?

1.11.7. Show that

$$y(z) = z^{-\alpha}e^{-f(z)}{}_2F_1(a,b;c;h(z))$$

satisfies the second order differential equation:

$$\frac{h(h-1)}{(h')^2}y'' + \left\{\frac{h(h-1)}{(h')^3}\left(\frac{2\alpha h'}{z} + 2f'h' - h''\right) + \frac{(a+b+1)h-c}{h'}\right\}y'$$
$$+ \left\{\left(\frac{\alpha}{z} + f'\right)\left(\frac{(a+b+1)h-c}{h'}\right) + \frac{h(h-1)}{(h')^3}\left[\frac{\alpha(\alpha-1)h'}{z^2} + \frac{2\alpha f'h'}{z}\right.\right.$$
$$\left.\left. + f''h' + (f')^2h' - \frac{\alpha h''}{z} - f'h''\right] + ab\right\}y = 0.$$

1.11.4 Confluent hypergeometric function

The confluent hypergeometric differential equation is

$$x\frac{d^2}{dx^2}u(x) + (c-x)\frac{d}{dx}u(x) - au(x) = 0 \tag{1.11.29}$$

or

$$[\delta(\delta+c-1) - z(\delta+a)]u(x) = 0, \quad \delta = z\frac{d}{dz}. \tag{1.11.30}$$

This can be obtained from Gauss hypergeometric equation (1.11.15) by taking $x = bz$ and then making $b \to \infty$. This equation has singularities at $x = 0$ and $x = \infty$. A merged (confluence) of the singularities of equation (1.11.15) at $z = 1$ and $z = \infty$ has occurred. The singularity at $x = 0$ is regular and at $x = \infty$, irregular. To get the general solution of (1.11.29), substitute

$$u(x) = \sum_{k=0}^{\infty} a_k x^{k+s} \tag{1.11.31}$$

so that

$$s(s-1+c)a_0 = 0. \tag{1.11.32}$$

Thus the nontrivial general solution to (1.11.29) is given by

$$u(x) = A_1 F_1(a;c;x) + Bx^{1-c} {}_1F_1(1+a-c;2-c;x). \tag{1.11.33}$$

Exercises 1.11.

1.11.8. Show that

$$y = z^{-\alpha} e^{-f(z)} {}_1F_1(a;c;h(z))$$

satisfies the second order differential equation

$$hy'' + \left\{\frac{2\alpha h}{z} + 2f'h - \frac{hh''}{h'} - hh' + ch'\right\}y' + \left\{h'\left(\frac{\alpha}{z}+f'\right)(c-h)\right.$$
$$\left. +h\left[\frac{\alpha(\alpha-1)}{z^2} + \frac{2\alpha f'}{z} + f'' + (f')^2 - \frac{h''}{h'}\left(\frac{\alpha}{z}+f'\right)\right] - a(h')^2\right\}y = 0.$$

1.11.5 Hermite polynomials

The differential equation

$$\frac{d^2}{dy^2}f(y) - 2y\frac{d}{dy}f(y) + 2nf(y) = 0 \tag{1.11.34}$$

is known as Hermite's equation. For $z = y^2$, (1.11.34) takes the following from

$$z\frac{d^2}{dz^2}f(z) + \left(\frac{1}{2}-z\right)\frac{d}{dz}f(z) + \frac{n}{2}f(z) = 0. \tag{1.11.35}$$

This is of the form of confluent hypergeometric equation (1.11.29) yielding the two solutions for (1.11.34) as

$$f(y) = \frac{(-1)^{-n/2}n!}{\left(\frac{n}{2}\right)!} {}_1F_1\left(-\frac{n}{2};\frac{1}{2};y^2\right) \text{ for } n \text{ even} \tag{1.11.36}$$
$$\equiv H_n(y)$$

and

$$f(y) = \frac{(-1)^{(1-n)/2}2n!y}{\left(\frac{n-1}{2}\right)!} {}_1F_1\left(-\frac{n-1}{2};\frac{3}{2};y^2\right) \text{ for } n \text{ odd} \tag{1.11.37}$$
$$\equiv H_n(y)$$

Exercises 1.11.

1.11.9.   For what values of s the substitution $u(z) = z^s f(z)$ in

$$z^2 \frac{d^2}{dz^2} u(z) - z^2 \frac{d}{dz} u(z) + \frac{3}{16} u(z) = 0$$

will result in confluent hypergeometric equation. Write the general solution for $u(z)$.

1.11.10.   For what value of a the substitution $u(z) = e^{az} f(z)$ in

$$z \frac{d^2}{dz^2} u(z) + (1 - 5z) \frac{d}{dz} u(z) + 6zu(z) = 0$$

will result in confluent hypergeometric equation. Find the solution.

1.11.11.   Find the solution of the type $u(x) = v(x)w(x)$ for the differential equation

$$\frac{d^2}{dx^2} u(x) - \frac{d}{dx} u(x) - \frac{1}{4x^2} \left[a(a-2) + 2ax \right] u(x) = 0.$$

Choose $v(x)$ so that $w(x)$ satisfies the confluent hypergeometric equation. Find the general solution $u(x)$. Show that a particular solution reduces to $x^{a/2} e^x$.

1.11.6 Bessel functions

The differential equation

$$x^2 \frac{d^2 u}{dx^2} + x \frac{du}{dx} + (x^2 - v^2)u = 0 \tag{1.11.38}$$

(v need not to be an integer) is called *Bessel equation*. There are two singularities at $x = 0$ (regular) and $x = \infty$ (irregular). The general solution of (1.11.38), when v is not an integer, is given by

$$u(x) = e^{-ix} \left[A_v x^v {}_1F_1 \left(v + \frac{1}{2}; 2v+1; 2ix \right) + B_v x^{-v} {}_1F_1 \left(-v + \frac{1}{2}; -2v+1; 2ix \right) \right] \tag{1.11.39}$$

where A_v and B_v are arbitrary constants. Using the relation

$$J_v(x) = \frac{e^{-ix}(x/2)^v}{\Gamma(v+1)} {}_1F_1 \left(v + \frac{1}{2}; 2v+1; 2ix \right), \tag{1.11.40}$$

where $J_v(x)$ is Bessel function of the first kind, the solution of (1.11.38) may be written as

$$u(x) = a_v J_v(x) + b_v J_{-v}(x) \tag{1.11.41}$$

where $a_v = 2^v \Gamma(v+1) A_v$, $b_v = 2^{-v} \Gamma(-v+1) B_v$. It may be noted that $J_v(x)$ is regular at $x = 0$, whereas $J_{-v}(x)$ is irregular at $x = 0$. The function

$$N_v(x) = \frac{J_v(x) \cos(v\pi) - J_{-v}(x)}{\sin(v\pi)}, \tag{1.11.42}$$

which is known as Bessel function of the second kind (or Newman's function), is also a solution of (1.11.38) and is linearly independent of $J_v(x)$. The function $N_v(x)$ is irregular at $x = 0$. Thus the general solution of (1.11.38) may also be written as

$$u(x) = a J_v(x) + b N_v(x) \tag{1.11.43}$$

where a and b are arbitrary constants. Two functions frequently encountered in physical applications are

$$H_v^{(1)}(x) = J_v(x) + i N_v(x) \tag{1.11.44}$$

and

$$H_v^{(2)}(x) = J_v(x) - i N_v(x) \tag{1.11.45}$$

which are known as Bessel functions of the third kind (or Hankel functions). If v is zero or a positive integer, then the two independent solutions of (1.11.38) are $J_v(x)$ and a logarithmic solution which has a complicated expression.

Exercises 1.11.

1.11.12. Let $u(x) = \sum_{k=0}^{\infty} c_k x^{k+s}$ be the solution of the Bessel equation

$$x^2 \frac{d^2 u}{dx^2} + x \frac{du}{dx} + (x^2 - v^2) u = 0.$$

Show that for $k > 1$ the coefficient c_k satisfies the equation

$$\{(k+s)^2 - v^2\} c_k + c_{k-2} = 0.$$

For $v^2 \neq \frac{1}{4}$, verify that, if $c_0 \neq 0$, then $c_1 = 0$ and vice versa. Obtain the solutions of the Bessel equation.

1.11.13. With the change of dependent variable $w(x) = e^{bx} f(x)$ show that the differential equation

$$x \frac{d^2}{dx^2} w(x) + 2\lambda \frac{d}{dx} w(x) + xw(x) = 0$$

can be transformed into the confluent hypergeometric equation. Write down the general solution expressed in terms of Bessel functions.

1.11.14. Transform the differential equation

$$\frac{d^2}{dx^2} u(x) + \lambda x^s u(x) = 0$$

($\lambda = $ constant, $s = $ a real positive number) into Bessel equation by using $u(x) = x^p f(x)$ and $z = bx^q$. Find the values of b, p and q. Write down the differential equation for f as a function of z. Find the general solution to the original equation.

1.11.7 Laguerre polynomial

The Laguerre polynomial

$$L_n^{(\alpha)}(x) = \sum_{i=0}^{n} \frac{(1+\alpha)_n(-x)^i}{i!(n-i)!(1+\alpha)_i} \tag{1.11.46}$$

is a solution of the differential equation

$$x\frac{d^2y}{dx^2} + (1+\alpha-x)\frac{d}{dx}y + ny = 0. \tag{1.11.47}$$

Laguerre polynomial (1.11.46) is connected to confluent hypergeometric function by the relation

$$L_n^{(\alpha)}(x) = \frac{(1+\alpha)_n}{n!} {}_1F_1(-n;1+\alpha;x). \tag{1.11.48}$$

1.11.8 Legendre polynomial

The Legendre's equation is

$$(1-x^2)\frac{d^2f}{dx^2} - 2x\frac{df}{dx} + m(m+1)f = 0, \quad m = 0,1,\ldots \tag{1.11.49}$$

which has its solution as

$$P_m(x) = {}_2F_1\left(-m, m+1; 1; \frac{1}{2}(1-x)\right) \tag{1.11.50}$$

which is a polynomial of order m.

Exercises 1.11.

1.11.15. Show that the Legendre equation

$$(1-x^2)\frac{d^2f}{dx^2} - 2x\frac{df}{dx} + n(n+1)f = 0$$

has a solution

$$f(x) = \sum_{k=0}^{[n/2]} \frac{(-1)^k(2n-2k)!x^{n-2k}}{2^n k!(n-2k)!(n-k)!}$$

where n is either even or odd.

1.11.16. Show that the differential equation

$$(1-x^2)\frac{d^2y}{dx^2} - x\frac{dy}{dx} + n^2 y = 0$$

has three regular singularities.

1.11.17. Obtain the general solution to the differential equation

$$(1-x^2)\frac{d^2y}{dx^2} - 3x\frac{dy}{dx} + n(n+2)y = 0$$

in terms of hypergeometric functions. Show that one of the solutions is a polynomial if n is an integer ($n \neq -1$).

1.11.9 Generalized hypergeometric function

The homogeneous linear differential equation

$$\left[\delta \prod_{j=1}^{q}(\delta + b_j - 1) - z \prod_{i=1}^{p}(\delta + a_i) \right] u(z) = 0 \qquad (1.11.51)$$

where $\delta = z\frac{d}{dz}$,

(a) is of order $\max(p, q+1)$,

(b) has singularities at $z = 0$ (regular) and $z = \infty$ (irregular) for $p < q+1$,

(c) has regular singularities at $z = 0, 1$ and ∞ when $p = q+1$.

The $q+1$ linearly independent solutions of (1.11.51) for $p \leq q+1$ near $z = 0$ when no two b_j's differ by an integer or zero, and no b_j is a negative integer or zero, are given by

$$u_h(z) = A_h z^{1-b_h}\, {}_pF_q(1+a_1-b_h, \ldots, 1+a_p-b_h; 1+b_0-b_h, \ldots, 1+b_q-b_h; z) \qquad (1.11.52)$$

for $h = 0, 1, \ldots, q$, $b_0 = 1$, with the term $1+b_j - b_j$, $j = 0, 1, \ldots, q$ omitted where A_h, $h = 0, 1, \ldots, q$ are arbitrary constants. Thus

$$\begin{cases} u_0 = A_0\ {}_pF_q(a_1, \ldots, a_p; b_1, \ldots, b_q; z) \\ u_1 = A_1\ z^{1-b_1}\, {}_pF_q(1+a_1-b_1, \ldots, 1+a_p-b_1; 2-b_1, 1+b_2-b_1, \ldots, 1+b_q-b_1; z) \\ u_2 = A_2\ z^{1-b_2}\, {}_pF_q(1+a_1-b_2, \ldots, 1+a_p-b_2; 1+b_1-b_2, 2-b_2, \\ \qquad\qquad\qquad 1+b_3-b_2, \ldots, 1+b_q-b_2; z) \\ \text{etc} \end{cases}$$

$$(1.11.53)$$

Solution 1.11.2: Substitution of

$$u(z) = \sum_{n=0}^{\infty} c_n z^{n+s} \tag{1.11.54}$$

in (1.11.51) yields

$$\sum_{n=0}^{\infty} c_n \left\{ (s+n) \prod_{j=1}^{q} (s+n+b_j-1) z^n - \prod_{i=1}^{p} (s+n+a_i) z^{n+1} \right\} = 0. \tag{1.11.55}$$

Equating the coefficient of c_0 to zero gives the indicial equation roots as $s_h = 1 - b_h$, $h = 0, 1, \ldots, q$. For the root s_h, we find from (1.11.55) that

$$c_n = \frac{\prod_{i=1}^{p} (s_h + \alpha_i) c_0}{(s_h + 1)_n \prod_{j=1}^{n} (s_h + b_j)_n}. \tag{1.11.56}$$

Thus (1.11.56) and (1.11.54) yield (1.11.53). If $p \geq q+1$ and no two of the a_i's differ by an integer or zero, then there are p linearly independent solutions of the equation (1.11.51) near $z = \infty$:

$$v_h(t) = B_h z^{-a_h}{}_{q+1}F_{p-1}(1 + a_h - b_1, \ldots, 1 + a_h - b_q;$$
$$1 + a_h - a_1, \ldots, 1 + a_h - a_p; (-1)^{q+1-p}/z),$$

$h = 1, 2, \ldots, p$, where B_h, $h = 1, \ldots, p$, are arbitrary constants.

1.11.10 G-function

The G-function

$$y(z) = G_{p,q}^{m,n} \left[z \left| \begin{array}{c} a_1, \ldots, a_p \\ b_1, \ldots, b_q \end{array} \right. \right] \tag{1.11.57}$$

satisfies the homogeneous linear differential equation

$$\left[(-1)^{m+n-p} z \prod_{i=1}^{p} (\delta - a_i + 1) - \prod_{j=1}^{q} (\delta - b_j) \right] y(z) = 0 \tag{1.11.58}$$

where $\delta = z \frac{d}{dz}$. This equation

(a) is of order $\max(p, q)$,
(b) has singularities at $z = 0$ (regular), $z = \infty$ (irregular) when $p < q$,
(c) has regular singularities at $z = 0, \infty$ and $(-1)^{m+n-p}$ if $p = q$.

In view of

$$G_{p,q}^{m,n}\left[z\left|\begin{matrix}a_1,\ldots,a_p\\b_1,\ldots,b_q\end{matrix}\right.\right] = G_{q,p}^{n,m}\left[\frac{1}{z}\left|\begin{matrix}1-b_1,\ldots,1-b_q\\1-a_1,\ldots,1-a_p\end{matrix}\right.\right], \quad -\arg z = \arg\left(\frac{1}{z}\right)$$

(1.11.59)

it is enough to consider $p < q$. The q functions

$$y_h(z) = A_h e^{i\pi(m+n-p+1)b_h} G_{p,q}^{1,p}\left[ze^{-i\pi(m+n-p+1)}\left|\begin{matrix}a_1,\ldots,a_p\\b_h,b_1,\ldots,b_{h-1},b_{h+1},\ldots,b_q\end{matrix}\right.\right],$$

(1.11.60)

$h = 1,2,\ldots,q$ for A_h arbitrary constant, form linearly independent solutions for (1.11.58) around $z = 0$ provided that no two of b_j, $j = 1,\ldots,m$, differ by an integer or zero. Equation (1.11.60) may be written as

$$y_h(z) = \frac{\prod_{i=1}^{p}\Gamma(1+b_h-a_i)}{\prod_{j=1}^{q}\Gamma(1+b_h-b_j)} A_h z^{b_h}$$

(1.11.61)

$$\times {}_pF_{q-1}(1+b_h-a_1,\ldots,1+b_h-a_p;1+b_h-b_1,\ldots,1+b_h-b_q;(-1)^{m+n-p}z),$$

where $p \le q-1$ or $p = q$ and $|z| < 1$ and $1+b_h-b_h$ term is omitted. When two or more b_j's differ by an integer or zero, the corresponding independent solution may involve log, psi and/or zeta functions. The solution of (1.11.58) in the neighborhood of $z = \infty$ (irregular singularity) are rather lengthy to obtain.

Exercises 1.11.

1.11.18. Show that $u(z) = {}_pF_q(a_1,\ldots,a_p;b_1,\ldots,b_q;z)$ satisfies equation (1.11.51).

1.11.19. Show that (1.11.57) satisfies (1.11.58).

References

Luke, Y. L. (1969). *The Special Functions and Their Approximations*, Academic Press, New York.
Mathai, A.M. (1993). *A Handbook of Generalized Special Functions for Statistical and Physical Sciences*, Oxford University Press, Oxford.
Mathai, A.M. (1997). *Jacobians of Matrix Transformations and Functions of Matrix Argument*, World Scientific Publishing, New York.
Mathai, A.M. and Saxena, R.K. (1973). *Generalized Hypergeometric Functions with Applications in Statistics and Physical Sciences*, Springer-Verlag, Lecture Notes 348, Heidelberg.

Mathai, A.M. and Saxena, R.K. (1978). *The H-function with Applications in Statistics and Other Disciplines*, Wiley Halsted, New York.

Rainville, E. D. (1960). *Special Functions*, MacMillan, New York.

Seaborn, J. B. (1991). *Hypergeometric Functions and Their Applications*, Springer-Verlag, New York.

Springer, M. D. (1979). *The Algebra of Random Variables*, Wiley, New York.

Srivastava, H.M. and Karlsson, P.W. (1985). *Multiple Gaussian Hypergeometric Series*, Ellis Horwood, Chichester, U.K.

Chapter 2
Mittag-Leffler Functions and Fractional Calculus

[This chapter is based on the lectures of Professor R.K. Saxena of Jai Narain Vyas University, Jodhpur, Rajasthan, India.]

2.0 Introduction

This section deals with Mittag-Leffler function and its generalizations. Its importance is realized during the last one and a half decades due to its direct involvement in the problems of physics, biology, engineering and applied sciences. Mittag-Leffler function naturally occurs as the solution of fractional order differential equations and fractional order integral equations. Various properties of Mittag-Leffler functions are described in this section. Among the various results presented by various researchers, the important ones deal with Laplace transform and asymptotic expansions of these functions, which are directly applicable in the solution of differential equations and in the study of the behavior of the solution for small and large values of the argument. Hille and Tamarkin in 1920 have presented a solution of Abel-Volterra type integral equation

$$\phi(x) - \frac{\lambda}{\Gamma(\alpha)} \int_0^x \frac{\phi(t)}{(x-t)^{1-\alpha}} dt = f(x), \ 0 < x < 1$$

in terms of Mittag-Leffler function. Dzherbashyan (1966) has shown that both the functions defined by (2.1.1) and (2.1.2) are entire functions of order $p = \frac{1}{\alpha}$ and type $\sigma = 1$. A detailed account of the basic properties of these functions is given in the third volume of Batemann Manuscript Project written by Erdélyi et al (1955) under the heading "Miscellaneous Functions".

2.1 Mittag-Leffler Function

Notation 2.1.1. $E_\alpha(x)$: Mittag-Leffler function

Notation 2.1.2. $E_{\alpha,\beta}(x)$: Generalized Mittag-Leffler function

79

Note 2.1.1:　　According to Erdélyi, et al (1955) both $E_\alpha(\mathbf{x})$ and $E_{\alpha,\beta}(\mathbf{x})$ are called Mittag-Leffler functions.

Definition 2.1.1.

$$E_\alpha(z) = \sum_{k=0}^{\infty} \frac{z^k}{\Gamma(\alpha k + 1)}, \quad \alpha \in C, \Re(\alpha) > 0. \tag{2.1.1}$$

Definition 2.1.2.

$$E_{\alpha,\beta}(z) = \sum_{k=0}^{\infty} \frac{z^k}{\Gamma(\alpha k + \beta)}, \quad \alpha, \beta \in C, \Re(\alpha) > 0, \Re(\beta) > 0. \tag{2.1.2}$$

The function $E_\alpha(z)$ was defined and studied by Mittag-Leffler in the year 1903. It is a direct generalization of the exponential series. For $\alpha = 1$ we have the exponential series. The function defined by (2.1.2) gives a generalization of (2.1.1). This generalization was studied by Wiman in 1905, Agarwal in 1953, Humbert and Agarwal in 1953, and others.

Example 2.1.1.　Prove that $E_{1,2}(z) = \frac{e^z - 1}{z}$.

Solution 2.1.1:　We have

$$E_{1,2}(z) = \sum_{k=0}^{\infty} \frac{z^k}{\Gamma(k+2)} = \sum_{k=0}^{\infty} \frac{z^k}{(k+1)!} = \frac{1}{z} \sum_{k=0}^{\infty} \frac{z^{k+1}}{(k+1)!} = \frac{1}{z}(e^z - 1).$$

Definition 2.1.3. Hyperbolic function of order n.

$$h_r(z,n) = \sum_{k=0}^{\infty} \frac{z^{nk+r-1}}{(nk+r-1)!} = z^{r-1} E_{n,r}(z^n), \quad r = 1, 2, \dots. \tag{2.1.3}$$

Definition 2.1.4. Trigonometric functions of order n.

$$K_r(z,n) = \sum_{k=0}^{\infty} \frac{(-1)^k z^{kn+r-1}}{(kn+r-1)!} = z^{r-1} E_{n,r}(-z^n). \tag{2.1.4}$$

$$E_{\frac{1}{2},1}(z) = \sum_{k=0}^{\infty} \frac{z^k}{\Gamma\left(\frac{k}{2}+1\right)} = e^{z^2} \text{erfc}(-z), \tag{2.1.5}$$

where erfc is complementary to the error function erf.

Definition 2.1.5. Error function.

$$\text{erfc}(z) = \frac{2}{\pi^{\frac{1}{2}}} \int_z^{\infty} e^{-u^2} du = 1 - \text{erf}(z), \quad z \in C. \tag{2.1.6}$$

To derive (2.1.5), we see that Dzherbashyan (1966, P.297, Eq.7.1.) reads as

$$w(z) = e^{-z^2} \text{erfc}(-iz) \tag{2.1.7}$$

whereas Dzherbashyan (1966, P.297, Eq.7.1.8) is

$$w(z) = \sum_{n=0}^{\infty} \frac{(iz)^n}{\Gamma\left(\frac{n}{2}+1\right)}. \tag{2.1.8}$$

From (2.1.7) and (2.1.8) we easily obtain (2.1.5). In passing, we note that $w(z)$ is also an error function (Dzherbashyan (1966)).

Definition 2.1.6. Mellin-Ross function.

$$E_t(v,a) = t^v \sum_{k=0}^{\infty} \frac{(at)^k}{\Gamma(v+k+1)} = t^v E_{1,v+1}(at). \tag{2.1.9}$$

Definition 2.1.7. Robotov's function.

$$R_\alpha(\beta,t) = t^\alpha \sum_{k=0}^{\infty} \frac{\beta^k t^{k(\alpha+1)}}{\Gamma((1+\alpha)(k+1))} = t^\alpha E_{\alpha+1,\alpha+1}(\beta t^{\alpha+1}). \tag{2.1.10}$$

Example 2.1.2. Prove that $E_{1,3}(z) = \frac{e^z-z-1}{z^2}$.

Solution 2.1.2: We have

$$E_{1,3}(z) = \sum_{k=0}^{\infty} \frac{z^k}{\Gamma(k+3)} = \frac{1}{z^2} \sum_{k=0}^{\infty} \frac{z^{k+2}}{(k+2)!}$$
$$= \frac{1}{z^2}(e^z - z - 1).$$

Example 2.1.3. Prove that

$$E_{1,r}(z) = \frac{1}{z^{r-1}} \left\{ e^z - \sum_{k=0}^{r-2} \frac{z^k}{k!} \right\}, \, r = 1,2,\dots.$$

The proof is similar to that in Example 2.1.2.

Revision Exercises 2.1.

2.1.1. Prove that

$$H_{1,2}^{1,1}\left[x \Big|_{(a,A),(0,1)}^{(a,A)}\right] = A^{-1} \sum_{k=0}^{\infty} \frac{(-1)^k x^{(k+a)/A}}{\Gamma(1+(k+a)A)},$$

and write the right side in terms of a generalized Mittag-Leffler function.

2.1.2. Prove that

$$\frac{d}{dx} H_{1,2}^{1,1}\left[x\Big|_{(a,A),(0,1)}^{(a,A)}\right] = H_{1,2}^{1,1}\left[x\Big|_{(a-A,A),(0,1)}^{(a-A,A)}\right].$$

2.1.3. Prove that

$$H_{2,1}^{1,1}\left[\frac{1}{x}\Big|_{(1-a,A)}^{(1-a,A),(1,1)}\right] = A^{-1}\sum_{k=0}^{\infty}\frac{(-1)^k(\frac{1}{x})^{\frac{k+1-a}{A}}}{\Gamma(1-(k+1-a)/A)}.$$

2.2 Basic Properties of Mittag-Leffler Function

As a consequence of the definitions (2.1.1) and (2.1.2) the following results hold:

Theorem 2.2.1. *There hold the following relations:*

$$\text{(i)}\ E_{\alpha,\beta}(z) = zE_{\alpha,\alpha+\beta}(z) + \frac{1}{\Gamma(\beta)} \tag{2.2.1}$$

$$\text{(ii)}\ E_{\alpha,\beta}(z) = \beta E_{\alpha,\beta+1}(z) + \alpha z\frac{d}{dz}E_{\alpha,\beta+1}(z) \tag{2.2.2}$$

$$\text{(iii)}\ \left(\frac{d}{dz}\right)^m\left[z^{\beta-1}E_{\alpha,\beta}(z^\alpha)\right] = z^{\beta-m-1}E_{\alpha,\beta-m}(z^\alpha), \tag{2.2.3}$$

$$\Re(\beta - m) > 0,\ m = 0, 1, \ldots. \tag{2.2.4}$$

Solutions 2.2.1: (i) We have

$$E_{\alpha,\beta}(z) = \sum_{k=0}^{\infty}\frac{z^k}{\Gamma(\alpha k+\beta)} = \sum_{k=-1}^{\infty}\frac{z^{k+1}}{\Gamma(\alpha+\beta+\alpha k)}$$

$$= zE_{\alpha,\alpha+\beta}(z) + \frac{1}{\Gamma(\beta)},\ \Re(\beta) > 0.$$

(ii) We have

$$R.H.S. = \beta E_{\alpha,\beta+1}(z) + \alpha z\frac{d}{dz}\sum_{k=0}^{\infty}\frac{z^k}{\Gamma(\alpha k+\beta+1)}$$

$$= \sum_{k=0}^{\infty}\frac{(\alpha k+\beta)z^k}{\Gamma(\alpha k+\beta+1)} = \sum_{k=0}^{\infty}\frac{z^k}{\Gamma(\alpha k+\beta)}$$

$$= E_{\alpha,\beta}(z) = L.H.S.$$

(iii)

$$L.H.S. = \left(\frac{d}{dz}\right)^m \sum_{k=0}^{\infty} \frac{z^{\alpha k + \beta - 1}}{\Gamma(\alpha k + \beta)}$$

$$= \sum_{k=0}^{\infty} \frac{z^{\alpha k + \beta - m - 1}}{\Gamma(\alpha k + \beta - m)}, \ \Re(\beta - m) > 0,$$

since

$$\sum_{k=0}^{\infty} \left(\frac{d}{dz}\right)^m (z^{\alpha k + \beta - 1}) = \sum_{k=0}^{\infty} \frac{\Gamma(\alpha k + \beta)}{\Gamma(\alpha k + \beta - m)} z^{\alpha k + \beta - m - 1}$$

$$= z^{\beta - m - 1} E_{\alpha, \beta - m}(z^{\alpha}), \ m = 0, 1, 2, \dots$$

$$= R.H.S.$$

Following special cases of (2.2.3) are worth mentioning. If we set $\alpha = \frac{m}{n}$, $m, n = 1, 2, \dots$ then

$$\left(\frac{d}{dz}\right)^m \left[z^{\beta - 1} E_{\frac{m}{n}, \beta}(z^{\frac{m}{n}})\right] = z^{\beta - m - 1} E_{\frac{m}{n}, \beta - m}(z^{\frac{m}{n}})$$

$$= z^{\beta - m - 1} \sum_{k=0}^{\infty} \frac{z^{\frac{mk}{n}}}{\Gamma\left(\frac{mk}{n} + \beta - m\right)}.$$

for $\Re(\beta - m) > 0$, (replacing k by $k + n$)

$$= z^{\beta - m - 1} \sum_{k=-n}^{\infty} \frac{z^{\frac{m(k+n)}{n}}}{\Gamma\left(\beta + \frac{mk}{n}\right)}$$

$$= z^{\beta - 1} E_{\frac{m}{n}, \beta}(z^{\frac{m}{n}}) + z^{\beta - 1} \sum_{k=1}^{n} \frac{z^{-\frac{mk}{n}}}{\Gamma\left(\beta - \frac{mk}{n}\right)}, \ m, n = 1, 2, 3.$$

$$\tag{2.2.5}$$

$$\left(\frac{d}{dz}\right)^m \left[z^{\beta - 1} E_{m, \beta}(z^m)\right] = z^{\beta - 1} E_{m, \beta}(z^m) + \frac{z^{-m}}{\Gamma(\beta - m)}, \ \Re(\beta - m) > 0. \tag{2.2.6}$$

Putting $z = t^{\frac{n}{m}}$ in (2.2.3) it yields

$$\left(\frac{m}{n} t^{1 - \frac{n}{m}} \frac{d}{dt}\right)^m \left[t^{(\beta - 1)\frac{n}{m}} E_{\frac{m}{n}, \beta}(t)\right] = t^{(\beta - 1)\frac{n}{m}} E_{\frac{m}{n}, \beta}(t)$$

$$+ t^{(\beta - 1)\frac{n}{m}} \sum_{k=1}^{n} \frac{t^{-k}}{\Gamma(\beta - \frac{mk}{n})}, \ \Re(\beta - m) > 0, \ m, n = 1, 2, \dots \tag{2.2.7}$$

When $m = 1$, (2.2.7) reduces to

$$\frac{t^{1-n}}{n} \frac{d}{dt} \left[t^{(\beta - 1)n} E_{\frac{1}{n}, \beta}(t)\right] = t^{(\beta - 1)n} E_{\frac{1}{n}, \beta}(t) + t^{(\beta - 1)n} \sum_{k=1}^{n} \frac{t^{-k}}{\Gamma\left(\beta - \frac{k}{n}\right)},$$

for $\Re(\beta) > 1$, which can be written as

$$\frac{1}{n}\frac{d}{dt}\left[t^{(\beta-1)n}E_{\frac{1}{n},\beta}(t)\right] = t^{\beta n-1}E_{\frac{1}{n},\beta}(t) + t^{\beta n-1}\sum_{k=1}^{n}\frac{t^{-k}}{\Gamma\left(\beta-\frac{k}{n}\right)}, \quad \Re(\beta) > 1. \quad (2.2.8)$$

2.2.1 Mittag-Leffler functions of rational order

Now we consider the Mittag-Leffler functions of rational order $\alpha = \frac{p}{q}$ with $p, q = 1, 2, \dots$ relatively prime. The following relations readily follow from the definitions (2.1.1) and (2.1.2).

$$(i) \quad \left(\frac{d}{dz}\right)^{p} E_{p}(z^{p}) = E_{p}(z^{p}) \quad (2.2.9)$$

$$(ii) \quad \left(\frac{d}{dz}\right)^{p} E_{\frac{p}{q}}(z^{\frac{p}{q}}) = E_{\frac{p}{q}}(z^{\frac{p}{q}}) + \sum_{k=1}^{q-1}\frac{z^{k\frac{p}{q}-p}}{\Gamma(k\frac{p}{q}+1-p)}, \quad (2.2.10)$$

$q = 1, 2, 3, \dots$. We now derive the relation

$$(iii) \quad E_{\frac{1}{q}}(z^{\frac{1}{q}}) = e^{z}\left[1 + \sum_{k=1}^{q-1}\frac{\gamma(1-\frac{k}{q},z)}{\Gamma(1-\frac{k}{q})}\right], \quad (2.2.11)$$

where $q = 2, 3, \dots$ and $\gamma(\alpha, z)$ is the incomplete gamma function, defined by

$$\gamma(\alpha, z) = \int_{0}^{z} e^{-u}u^{\alpha-1}du$$

To prove (2.2.11), set $p = 1$ in (2.2.10) and multiply both sides by e^{-z} and use the definition of $\gamma(\alpha, z)$. Thus we have

$$\frac{d}{dz}\left[e^{-z}E_{\frac{1}{q}}(z^{\frac{1}{q}})\right] = e^{-z}\sum_{k=1}^{q-1}\frac{z^{-\frac{k}{q}}}{\Gamma\left(1-\frac{k}{q}\right)}. \quad (2.2.12)$$

Integrating (2.2.12) with respect to z, we obtain (2.2.11).

2.2.2 Euler transform of Mittag-Leffler function

By virtue of beta function formula it is not difficult to show that

$$\int_{0}^{1} z^{\rho-1}(1-z)^{\sigma-1}E_{\alpha,\beta}(xz^{\gamma})dz = \Gamma(\sigma)\,_2\psi_2\left[x\Big|^{(\rho,\gamma),(1,1)}_{(\beta,\alpha),(\sigma+\rho,\gamma)}\right] \quad (2.2.13)$$

where $\Re(\alpha) > 0, \Re(\beta) > 0, \Re(\sigma) > 0, \gamma > 0$. Here $_2\psi_2$ is the generalized Wright function and $\alpha, \beta, \rho, \sigma \in C$.

Special cases of (2.2.13):

(i) When $\rho = \beta, \gamma = \alpha$, (2.2.13) yields

$$\int_0^1 z^{\beta-1}(1-z)^{\sigma-1}E_{\alpha,\beta}(xz^\alpha)dz = \Gamma(\sigma)E_{\alpha,\sigma+\beta}(x), \qquad (2.2.14)$$

where $\alpha > 0; \beta, \sigma \in C, \Re(\beta) > 0, \Re(\sigma) > 0$ and,

(ii)

$$\int_0^1 z^{\sigma-1}(1-z)^{\beta-1}E_{\alpha,\beta}[x(1-z)^\alpha]dz = \Gamma(\sigma)E_{\alpha,\beta+\sigma}(x), \qquad (2.2.15)$$

where $\alpha > 0; \beta, \sigma \in C, \Re(\beta) > 0, \Re(\sigma) > 0$.

(iii) When $\alpha = \beta = 1$ we have

$$\int_0^1 z^{\rho-1}(1-z)^{\sigma-1}\exp(xz^\gamma)dz = \Gamma(\sigma)_2\psi_2\left[x\Big|^{(\rho,\gamma),(1,1)}_{(1,1),(\sigma+\rho,\gamma)}\right]$$
$$= \Gamma(\sigma)_1\psi_1\left[x\Big|^{(\rho,\gamma)}_{(\sigma+\rho,\gamma)}\right], \qquad (2.2.16)$$

where $\gamma > 0, \rho, \sigma \in C, \Re(\rho) > 0, \Re(\sigma) > 0$.

2.2.3 Laplace transform of Mittag-Leffler function

Notation 2.2.1. $F(s) = L\{f(t); s\} = (Lf)(s)$: Laplace transform of $f(t)$ with parameter s.

Notation 2.2.2. $L^{-1}\{F(s); t\}$: Inverse Laplace transform

Definition 2.2.1. The Laplace transform of a function $f(t)$, denoted by $F(s)$, is defined by the equation

$$F(s) = (Lf)(s) = L\{f(t); s\} = \int_0^\infty e^{-st}f(t)dt, \qquad (2.2.17)$$

where $\Re(s) > 0$, which may be symbolically written as

$$F(s) = L\{f(t); s\} \text{ or } f(t) = L^{-1}\{F(s); t\},$$

provided that the function $f(t)$ is continuous for $t \geq 0$, it being tacitly assumed that the integral in (2.2.17) exists.

Example 2.2.1. Prove that

$$L^{-1}\{s^{-\rho}\} = \frac{t^{\rho-1}}{\Gamma(\rho)}, \quad \Re(s) > 0, \ \Re(\rho) > 0. \tag{2.2.18}$$

It follows from the Laplace integral

$$\int_0^\infty e^{-st} t^{\rho-1} dt = \frac{\Gamma(s)}{s^\rho}, \quad \Re(s) > 0, \ \Re(\rho) > 0. \tag{2.2.19}$$

Example 2.2.2. Find the inverse Laplace transform of $\frac{F(s)}{a+s^\alpha}$; $a, \alpha > 0$; where $\Re(s) > 0, F(s) = L\{f(t); s\}$.

Solution 2.2.1: Let

$$G(s) = \frac{1}{a+s^\alpha} = \sum_{r=0}^\infty (-a)^r s^{-\alpha-\alpha r}, \quad |\frac{a}{s^\alpha}| < 1.$$

Therefore,

$$L^{-1}\{G(s)\} = g(t) = L^{-1}\left\{ \sum_{r=0}^\infty (-a)^r s^{-\alpha-\alpha r} \right\}$$

$$= t^{\alpha-1} E_{\alpha,\alpha}(-at^\alpha). \tag{2.2.20}$$

Application of convolution theorem of Laplace transform yields the result

$$L^{-1}\left\{ \frac{F(s)}{a+s^\alpha}; t \right\} = \int_0^x (x-t)^{\alpha-1} E_{\alpha,\alpha}(-a(x-t)^\alpha) f(t) dt \tag{2.2.21}$$

where $\Re(\alpha) > 0$.

By the application of Laplace integral, it follows that

$$\int_0^\infty z^{\rho-1} e^{-az} E_{\alpha,\beta}(xz^\gamma) dz = \frac{1}{a^\rho} {}_2\psi_1\left[\frac{x}{a^\gamma} \Big|_{(\beta,\alpha)}^{(1,1),(\rho,\gamma)} \right], \tag{2.2.22}$$

where $\rho, a, \alpha, \beta \in C, \Re(\alpha) > 0, \Re(\beta) > 0, \Re(\gamma) > 0, \Re(a) > 0, \Re(\rho) > 0$ and $|\frac{z}{a^\gamma}| < 1$. Special cases of (2.2.22) are worth mentioning.

(i) For $\rho = \beta, \gamma = \alpha, \Re(\alpha) > 0$, (2.2.22) gives

$$\int_0^\infty e^{-az} z^{\beta-1} E_{\alpha,\beta}(xz^\alpha) dz = \frac{a^{\alpha-\beta}}{a^\alpha - x}, \tag{2.2.23}$$

where $a, \alpha, \beta \in C, \Re(\alpha) > 0, \Re(\beta) > 0, |\frac{x}{a^\alpha}| < 1$.

When $a = 1$, (2.2.23) yields a known result.

$$\int_0^\infty e^{-z} z^{\beta-1} E_{\alpha,\beta}(xz^\alpha) dz = \frac{1}{1-x}, |x| < 1, \tag{2.2.24}$$

where $\Re(\alpha) > 0, \Re(\beta) > 0$. If we further take $\beta = 1$, (2.2.24) reduces to

$$\int_0^\infty e^{-z} E_\alpha(xz^\alpha) dz = \frac{1}{1-x}, |x| < 1.$$

(ii) When $\beta = 1$, (2.2.23) gives

$$\int_0^\infty e^{-az} E_\alpha(xz^\alpha) dz = \frac{a^{\alpha-1}}{a^\alpha - x}, \tag{2.2.25}$$

where $\Re(a) > 0, \Re(\alpha) > 0, |\frac{x}{a^\alpha}| < 1$.

2.2.4 Application of Lalace transform

From (2.2.23) we find that

$$L\{x^{\beta-1} E_{\alpha,\beta}(ax^\alpha)\} = \frac{s^{\alpha-\beta}}{s^\alpha - a} \tag{2.2.26}$$

where $\Re(\alpha) > 0, \Re(\beta) > 0$. We also have

$$L\{x^{\gamma-1} E_{\alpha,\gamma}(-ax^\alpha)\} = \frac{s^{\alpha-\gamma}}{s^\alpha + a}. \tag{2.2.27}$$

Now

$$\left[\frac{s^{\alpha-\beta}}{s^\alpha - a}\right]\left[\frac{s^{\alpha-\gamma}}{s^\alpha + a}\right] = \frac{s^{2\alpha-(\beta+\gamma)}}{s^{2\alpha} - a^2} \text{ for } \Re(s^2) > \Re(a). \tag{2.2.28}$$

By virtue of the convolution theorem of the Laplace transform, it readily follows that

$$\int_0^t u^{\beta-1} E_{\alpha,\beta}(au^\alpha)(t-u)^{\gamma-1} E_{\alpha,\gamma}(-a(t-u)^\alpha) du = t^{\beta+\gamma-1} E_{2\alpha,\beta+\gamma}(a^2 t^{2\alpha}), \tag{2.2.29}$$

where $\Re(\beta) > 0, \Re(\gamma) > 0$. Further, if we use the identity

$$\frac{1}{s^2} = \frac{s^{\alpha-\beta}}{s^\alpha - 1}\left[s^{\beta-2} - s^{\beta-\alpha-2}\right] \tag{2.2.30}$$

and the relation

$$L\{t^{\rho-1}; s\} = \Gamma(\rho) s^{-\rho}, \tag{2.2.31}$$

where $\Re(\rho) > 0, \Re(s) > 0$, we obtain

$$\int_0^t u^{\beta-1} E_{\alpha,\beta}(u^\alpha) \left[\frac{(t-u)^{1-\beta}}{\Gamma(2-\beta)} - \frac{(t-u)^{\alpha-\beta+1}}{\Gamma(\alpha-\beta+2)} \right] du = t, \tag{2.2.32}$$

where $0 < \beta < 2, \Re(\alpha) > 0$. Next we note that the following result (2.2.34) can be derived by the application of inverse Laplace transform to the identity

$$\left[\frac{s^{2\alpha-\beta}}{s^{2\alpha}-1} \right] [s^{-\alpha}] = -\frac{s^{2\alpha-\beta}}{s^{2\alpha}-1} + \frac{s^{\alpha-\beta}}{s^\alpha-1}, \quad \Re(s^\alpha) > 1. \tag{2.2.33}$$

We have

$$\frac{1}{\Gamma(\alpha)} \int_0^x (x-t)^{\alpha-1} E_{2\alpha,\beta}(t^{2\alpha}) t^{\beta-1} dt = -x^{\beta-1} E_{2\alpha,\beta}(x^{2\alpha}) + x^{\beta-1} E_{\alpha,\beta}(x^\alpha), \tag{2.2.34}$$

where $\Re(\alpha) > 0, \Re(\beta) > 0$. If we set $\beta = 1$ in (2.2.34), it reduces to

$$\frac{1}{\Gamma(\alpha)} \int_0^x (x-t)^{\alpha-1} E_{2\alpha}(t^{2\alpha}) dt = E_\alpha(x^\alpha) - E_{2\alpha}(x^{2\alpha}) \tag{2.2.35}$$

where $\Re(\alpha) > 0$.

2.2.5 Mittag-Leffler functions and the H-function

Both the Mittag-Leffler functions $E_\alpha(z)$ and $E_{\alpha,\beta}(z)$ belong to H-function family. We derive their relations with the H-function.

Lemma 2.2.1: Let $\alpha \in R_+ = (0, \infty)$. Then $E_\alpha(z)$ is represented by the Mellin-Barnes integral

$$E_\alpha(z) = \frac{1}{2\pi i} \int_L \frac{\Gamma(s)\Gamma(1-s)(-z)^{-s}}{\Gamma(1-\alpha s)} ds, |\arg z| < \pi, \tag{2.2.36}$$

where the contour of integration L, beginning at $c - i\infty$ and ending at $c + i\infty, 0 < c < 1$, separates all poles $s = -k, k = 0, 1, 2, \ldots$ to the left and all poles $s = 1+n, n = 0, 1, \ldots$ to the right.

Proof. We now evaluate the integral (2.2.36) as the sum of the residues at the points $s = 0, -1, -2, \ldots$. We find that

$$\frac{1}{2\pi i}\int_L \frac{\Gamma(s)\Gamma(1-s)(-z)^{-s}}{\Gamma(1-\alpha s)}ds = \sum_{k=0}^{\infty}\lim_{s\to -k}\left[\frac{(s+k)\Gamma(s)\Gamma(1-s)(-z)^{-s}}{\Gamma(1-\alpha s)}\right] \quad (2.2.37)$$

$$= \sum_{k=0}^{\infty}\frac{(-1)^k\Gamma(1+k)}{k!\Gamma(1+\alpha k)}(-z)^k$$

$$= E_\alpha(z),$$

which yields (2.2.36) in accordance with the definition (2.1.1). It readily follows from the definition of the H-function and (2.2.36) that $E_\alpha(z)$ can be represented in the form

$$E_\alpha(z) = H_{1,2}^{1,1}\left[-z\Big|_{(0,1),(0,\alpha)}^{(0,1)}\right], \quad (2.2.38)$$

where $H_{1,2}^{1,1}$ is the H-function, which is studied in Chapter 1.

Lemma 2.2.2: Let $\alpha \in R_+ = (0,\infty), \beta \in C$, then

$$E_{\alpha,\beta}(z) = \frac{1}{2\pi i}\int_L \frac{\Gamma(s)\Gamma(1-s)(-z)^{-s}}{\Gamma(\beta-\alpha s)}ds. \quad (2.2.39)$$

The proof of (2.2.39) is similar to that of (2.2.36). Hence the proof is omitted. From (2.2.39) and the definition of the H-function we obtain the relation

$$E_{\alpha,\beta}(z) = H_{1,2}^{1,1}\left[-z\Big|_{(0,1),(1-\beta,\alpha)}^{(0,1)}\right]. \quad (2.2.40)$$

In particular, $E_\alpha(z)$ can be expressed in terms of generalized Wright function in the form

$$E_\alpha(z) = {}_1\psi_1\left[z\Big|_{(1,\alpha)}^{(1,1)}\right]. \quad (2.2.41)$$

Similarly, we have

$$E_{\alpha,\beta}(z) = {}_1\psi_1\left[z\Big|_{(\beta,\alpha)}^{(1,1)}\right]. \quad (2.2.42)$$

Next, if we calculate the residues at the poles of the gamma function $\Gamma(1-s)$ at the points $s = 1+n, n = 0,1,2,\ldots$ it gives

$$\frac{1}{2\pi i}\int_L \frac{\Gamma(s)\Gamma(1-s)}{\Gamma(1-\alpha s)}(-z)^{-s}ds = \sum_{n=0}^{\infty}\lim_{s\to 1+n}\frac{(s-1-n)\Gamma(s)\Gamma(1-s)(-z)^{-s}}{\Gamma(1-\alpha s)}$$

$$= \sum_{n=0}^{\infty}\frac{(-1)^n\Gamma(1+n)(-z)^{-n-1}}{n!\Gamma(1-\alpha(1+n))}$$

$$= -\sum_{n=1}^{\infty}\frac{z^{-n}}{\Gamma(1-\alpha n)}, \quad (2.2.43)$$

for $\alpha \neq 1,2,\cdots$. Similarly for $\alpha \neq 1,2,\cdots, E_{\alpha,\beta}(z)$, gives

$$\frac{1}{2\pi i}\int_L \frac{\Gamma(s)\Gamma(1-s)}{\Gamma(\beta-\alpha s)}(-z)^{-s}ds = -\sum_{n=1}^{\infty}\frac{z^{-n}}{\Gamma(\beta-\alpha n)}. \quad (2.2.44)$$

Exercises 2.2.

2.2.1. Let

$$U_1(t) = t^{\beta-1} E_{\frac{m}{n},\beta}(t^{\frac{m}{n}})$$
$$U_2(t) = t^{\beta-1} E_{m,\beta}(t^m)$$
$$U_3(t) = t^{(\beta-1)\frac{n}{m}} E_{\frac{m}{n},\beta}(t)$$
$$U_4(t) = t^{(\beta-1)n} E_{\frac{1}{n},\beta}(t).$$

Then show that these functions respectively satisfy the following differential equations of Mittag-Leffler functions when m,n are relatively prime.

(i) $\dfrac{d^m}{dt^m} U_1(t) - U_1(t) = t^{\beta-1} \sum_{k=1}^{n} \dfrac{t^{-\frac{m}{n}k}}{\Gamma(\beta - \frac{mk}{n})}$

$\Re(\beta) > m, (m,n = 1,2,3,...);$

(ii) $\dfrac{d^m}{dt^m} U_2(t) - U_2(t) = \dfrac{t^{-m+\beta-1}}{\Gamma(\beta-m)}, \ \Re(\beta) > m, m = 1,2,...;$

(iii) $\left(\dfrac{m}{n} t^{1-\frac{n}{m}} \dfrac{d}{dt}\right)^m U_3(t) - U_3(t) = t^{(\beta-1)\frac{n}{m}} \sum_{k=1}^{n} \dfrac{t^{-k}}{\Gamma(\beta - \frac{mk}{n})}$

$m,n = 1,2,3,..., \Re(\beta) > m;$

(iv) $\dfrac{1}{n}\left[\dfrac{d}{dt} U_4(t)\right] - t^{n-1} U_4(t) = t^{n\beta-1} \sum_{k=1}^{n} \dfrac{t^{-k}}{\Gamma(\beta - \frac{k}{n})}$

$n = 1,2,3,..., \Re(\beta) > 1.$

2.2.2. Prove that

$$\frac{\lambda}{\Gamma(\alpha)} \int_0^x \frac{E_\alpha(\lambda t^\alpha)}{(x-t)^{1-\alpha}} dt = E_\alpha(\lambda x^\alpha) - 1, \Re(\alpha) > 0.$$

2.2.3. Prove that

$$\frac{d}{dx}[x^{\gamma-1} E_{\alpha,\beta}(ax^\alpha)] = x^{\gamma-2} E_{\alpha,\beta-1}(ax^\alpha) + (\gamma-\beta)x^{\gamma-2} E_{\alpha,\beta}(ax^\alpha), \beta \neq \gamma.$$

2.2.4. Prove that

$$\frac{1}{\Gamma(v)} \int_0^z t^{\beta-1}(z-t)^{v-1} E_{\alpha,\beta}(\lambda t^\alpha) dt = z^{\beta+v-1} E_{\alpha,\beta+v}(\lambda z^\alpha),$$
$$\Re(\beta) > 0, \Re(v) > 0, \Re(\alpha) > 0.$$

2.2.5. Prove that

$$\frac{1}{\Gamma(\alpha)} \int_0^z (z-t)^{\alpha-1} \cosh(\sqrt{\lambda}t) dt = z^\alpha E_{2,\alpha+1}(\lambda z^2), \Re(\alpha) > 0.$$

2.2.6. Prove that

$$\frac{1}{\Gamma(\alpha)} \int_0^z e^{\lambda t}(z-t)^{\alpha-1} dt = z^{\alpha} E_{1,\alpha+1}(\lambda z), \Re(\alpha) > 0.$$

2.2.7. Prove that

$$\frac{1}{\Gamma(\alpha)} \int_0^z (z-t)^{\alpha-1} \frac{\sinh(\sqrt{\lambda}t)}{\sqrt{\lambda}} dt = z^{\alpha+1} E_{2,\alpha+2}(\lambda z^2), \Re(\alpha) > 0.$$

2.2.8. Prove that

$$\int_0^{\infty} e^{-sx} x^{\beta-1} E_{\alpha,\beta}(x^{\alpha}) dx = \frac{s^{\alpha-\beta}}{s^{\alpha}-1}, \Re(s) > 1.$$

2.2.9. Prove that

$$\int_0^{\infty} e^{-st} E_{\alpha}(t^{\alpha}) dt = \frac{1}{s-s^{1-\alpha}}, \Re(s) > 1.$$

2.2.10. Prove that

$$\int_0^x u^{\gamma-1} E_{\alpha,\gamma}(yu^{\alpha})(x-u)^{\beta-1} E_{\alpha,\beta}[z(x-u)^{\alpha}] du$$
$$= \frac{y E_{\alpha,\beta+\gamma}(yx^{\alpha}) - z E_{\alpha,\beta+\gamma}(zx^{\alpha})}{y-z} x^{\beta+\gamma-1},$$

where $y, z \in C$; $y \neq z, \gamma > 0, \beta > 0$.

2.3 Generalized Mittag-Leffler Function

Notation 2.3.1. $E_{\beta,\gamma}^{\delta}(z)$: **Generalized Mittag-Leffler function**

Definition 2.3.1.

$$E_{\beta,\gamma}^{\delta}(z) = \sum_{n=0}^{\infty} \frac{(\delta)_n z^n}{\Gamma(\beta n + \gamma) n!}, \tag{2.3.1}$$

where $\beta, \gamma, \delta \in C$ with $\Re(\beta) > 0$. For $\delta = 1$, it reduces to Mittag-Leffler function (2.1.2). This function was introduced by T.R. Prabhakar in 1971. It is an entire function of order $\rho = [\Re(\beta)]^{-1}$.

2.3.1 Special cases of $E_{\beta,\gamma}^{\delta}(z)$

$$\text{(i)} \quad E_{\beta}(z) = E_{\beta,1}^{1}(z) \tag{2.3.2}$$

$$\text{(ii)} \quad E_{\beta,\gamma}(z) = E_{\beta,\gamma}^{1}(z) \tag{2.3.3}$$

$$\text{(iii)} \quad \phi(\gamma,\delta;z) = {}_1F_1(\gamma;\delta;z) = \Gamma(\delta)E_{1,\delta}^{\gamma}(z), \tag{2.3.4}$$

where $\phi(\gamma,\delta;z)$ is Kummer's confluent hypergeometric function.

2.3.2 Mellin-Barnes integral representation

Lemma 2.3.1: Let $\beta \in R_+ = (0,\infty); \gamma,\delta \in C, \gamma \neq 0, \Re(\delta) > 0$. Then $E_{\beta,\gamma}^{\delta}(z)$ is represented by the Mellin-Barnes integral

$$E_{\beta,\gamma}^{\delta}(z) = \frac{1}{\Gamma(\delta)} \frac{1}{2\pi i} \int_L \frac{\Gamma(s)\Gamma(\delta-s)}{\Gamma(\gamma-\beta s)}(-z)^{-s}\mathrm{d}s, \tag{2.3.5}$$

where $|\arg(z)| < \pi$; the contour of integration beginning at $c - i\infty$ and ending at $c + i\infty, 0 < c < \Re(\delta)$, separates all the poles at $s = -k, k = 0,1,\ldots$ to the left and all the poles at $s = n + \delta, n = 0,1,\ldots$ to the right.

Proof. We will evaluate the integral on the R.H.S. of (2.3.5) as the sum of the residues at the poles $s = 0, -1, -2, \ldots$. We have

$$\frac{1}{2\pi i} \int_L \frac{\Gamma(s)\Gamma(\delta-s)}{\Gamma(\gamma-\beta s)}(-z)^{-s}\mathrm{d}s = \sum_{k=0}^{\infty} \lim_{s \to -k} \left[\frac{(s+k)\Gamma(s)\Gamma(\delta-s)(-z)^{-s}}{\Gamma(\gamma-\beta s)} \right]$$

$$= \sum_{k=0}^{\infty} \frac{(-1)^k}{k!} \frac{\Gamma(\delta+k)}{\Gamma(\gamma+\beta k)}(-z)^k$$

$$= \Gamma(\delta) \sum_{k=0}^{\infty} \frac{(\delta)_k}{\Gamma(\beta k+\gamma)} \frac{z^k}{k!} = \Gamma(\delta)E_{\beta,\gamma}^{\delta}(z)$$

which proves (2.3.5).

2.3.3 Relations with the H-function and Wright function

It follows from (2.3.5) that $E_{\beta,\gamma}^{\delta}(z)$ can be represented in the form

$$E_{\beta,\gamma}^{\delta}(z) = \frac{1}{\Gamma(\delta)} H_{1,2}^{1,1} \left[-z \Big|_{(0,1),(1-\gamma,\beta)}^{(1-\delta,1)} \right] \tag{2.3.6}$$

where $H_{1,2}^{1,1}(z)$ is the H-function, the theory of which can be found in Chapter 1. This function can also be represented by

$$E_{\beta,\gamma}^{\delta}(z) = \frac{1}{\Gamma(\delta)} {}_1\Psi_1 \left[z \Big|_{(\gamma,\beta)}^{(\delta,1)} \right] \tag{2.3.7}$$

where ${}_1\Psi_1$ is the Wright hypergeometric function ${}_p\Psi_q(z)$.

2.3.4 Cases of reducibility

In this subsection we present some interesting cases of reducibility of the function $E_{\beta,\gamma}^{\delta}(z)$. The results are given in the form of five theorems. The results are useful in the investigation of the solutions of certain fractional order differential and integral equations. The proofs of the following theorems can be developed on similar lines to that of equation (2.2.1).

Theorem 2.3.1. If $\beta, \gamma, \delta \in C$ with $\Re(\beta) > 0, \Re(\gamma) > 0, \Re(\gamma - \beta) > 0$, then there holds the relation

$$zE_{\beta,\gamma}^{\delta}(z) = E_{\beta,\gamma-\beta}^{\delta}(z) - E_{\beta,\gamma-\beta}^{\delta-1}(z). \tag{2.3.8}$$

Corollary 2.3.1: If $\beta, \gamma \in C, \Re(\gamma) > \Re(\beta) > 0$, then we have

$$zE_{\beta,\gamma}^{1}(z) = E_{\beta,\gamma-\beta}(z) - \frac{1}{\Gamma(\gamma-\beta)}. \tag{2.3.9}$$

Theorem 2.3.2. If $\beta, \gamma, \delta \in C, \Re(\beta) > 0, \Re(\gamma) > 1$, then there holds the formula

$$\beta E_{\beta,\gamma}^{2}(z) = E_{\beta,\gamma-1}(z) + (1 + \beta - \gamma)E_{\beta,\gamma}(z). \tag{2.3.10}$$

Theorem 2.3.3. If $\Re(\beta) > 0, \Re(\gamma) > 2 + \Re(\beta)$, then there holds the formula

$$zE_{\beta,\gamma}^{3}(z) = \frac{1}{2\beta^2}[E_{\beta,\gamma-\beta-2}(z) - (2\gamma - 3\beta - 3)E_{\beta,\gamma-\beta-1}(z)$$
$$+ (2\beta^2 + \gamma^2 - 3\beta\gamma + 3\beta - 2\gamma + 1)E_{\beta,\gamma-\beta}(z)]. \tag{2.3.11}$$

Theorem 2.3.4. If $\Re(\beta) > 0, \Re(\gamma) > 2$, then there holds the formula

$$E_{\beta,\gamma}^{3}(z) = \frac{1}{2\beta^2}[E_{\beta,\gamma-2}(z) + (3 + 3\beta - 2\gamma)E_{\beta,\gamma-1}(z)$$
$$+ (2\beta^2 + \gamma^2 + 3\beta - 3\beta\gamma - 2\gamma + 1)E_{\beta,\gamma}(z)]. \tag{2.3.12}$$

2.3.5 Differentiation of generalized Mittag-Leffler function

Theorem 2.3.5. *Let* $\beta, \gamma, \delta, \rho, w \in C$. *Then for any* $n = 1, 2, \ldots$ *there holds the formula, for* $\Re(\gamma) > n$,

$$\left(\frac{d}{dz}\right)^n [z^{\gamma-1} E^{\delta}_{\beta,\gamma}(wz^{\beta})] = z^{\gamma-n-1} E^{\delta}_{\beta,\gamma-n}(wz^{\beta}). \qquad (2.3.13)$$

In particular, for $\Re(\gamma) > n$,

$$\left(\frac{d}{dz}\right)^n [z^{\gamma-1} E_{\beta,\gamma}(wz^{\beta})] = z^{\gamma-n-1} E_{\beta,\gamma-n}(wz^{\beta}) \qquad (2.3.14)$$

and for $\Re(\gamma) > n$,

$$\left(\frac{d}{dz}\right)^n [z^{\gamma-1} \phi(\delta; \gamma; wz)] = \frac{\Gamma(\gamma)}{\Gamma(\gamma-n)} z^{\gamma-n-1} \phi(\delta; \gamma-n; wz). \qquad (2.3.15)$$

Proof. Using (2.3.1) and taking term by term differentiation under the summation sign, which is possible in accordance with uniform convergence of the series in (2.3.1) in any compact set of C, we obtain

$$\left(\frac{d}{dz}\right)^n [z^{\gamma-1} E^{\delta}_{\beta,\gamma}(wz^{\beta})] = \sum_{k=0}^{\infty} \frac{(\delta)_k}{\Gamma(\beta k + \gamma)} \left(\frac{d}{dz}\right)^n \left[\frac{w^k z^{\beta x + \gamma - 1}}{k!}\right]$$
$$= z^{\gamma-n-1} E^{\delta}_{\beta,\gamma-n}(wz^{\beta}), \quad \Re(\gamma) > n,$$

which establishes (2.3.13). Note that (2.3.14) follows from (2.3.13) when $\delta = 1$ due to (2.3.3), and (2.3.15) follows from (2.3.13) when $\beta = 1$ on account of (2.3.4).

2.3.6 Integral property of generalized Mittag-Leffler function

Corollary 2.3.2: *Let* $\beta, \gamma, \delta, w \in C, \Re(\gamma) > 0, \Re(\beta) > 0, \Re(\delta) > 0$. *Then*

$$\int_0^z t^{\gamma-1} E^{\delta}_{\beta,\gamma}(wt^{\beta}) dt = z^{\gamma} E^{\delta}_{\beta,\gamma+1}(wz^{\beta}) \qquad (2.3.16)$$

and (2.3.16) follows from (2.3.13). In particular,

$$\int_0^z t^{\gamma-1} E_{\beta,\gamma}(wt^{\beta}) dt = z^{\gamma} E_{\beta,\gamma+1}(wz^{\beta}) \qquad (2.3.17)$$

and

$$\int_0^z t^{\delta-1} \phi(\gamma, \delta; wt) dt = \frac{1}{\delta} z^{\delta} \phi(\gamma, \delta+1; wz) \qquad (2.3.18)$$

Remark 2.3.1: The relations (2.3.15) and (2.3.18) are well known.

2.3.7 Integral transform of $E_{\beta,\gamma}^{\delta}(z)$

By appealing to the Mellin inversion formula, (2.3.5) yields the Mellin transform of the generalized Mittag-Leffler function.

$$\int_0^\infty t^{s-1} E_{\beta,\gamma}^{\delta}(-wt)dt = \frac{\Gamma(s)\Gamma(\delta-s)w^{-s}}{\Gamma(\delta)\Gamma(\gamma-s\beta)}. \tag{2.3.19}$$

If we make use of the integral

$$\int_0^\infty t^{v-1} e^{-\frac{t}{2}} W_{\lambda,\mu}(t)dt = \frac{\Gamma\left(\frac{1}{2}+\mu+v\right)\Gamma\left(\frac{1}{2}-\mu+v\right)}{\Gamma(1-\lambda+v)} \tag{2.3.20}$$

where $\Re(v\pm\mu) > -\frac{1}{2}$, we obtain the Whittaker transform of the Mittag-Leffler function

$$\int_0^\infty t^{\rho-1} e^{-\frac{1}{2}pt} W_{\lambda,\mu}(pt) E_{\beta,\gamma}^{\delta}(wt^\alpha)dt = \frac{p^{-\rho}}{\Gamma(\delta)} {}_3\psi_2\left[\frac{w}{p^\alpha}\Big|{}^{(\delta,1),(\frac{1}{2}\pm\mu+\rho,\alpha)}_{(\gamma,\beta),(1-\lambda+\rho,\alpha)}\right] \tag{2.3.21}$$

where ${}_3\psi_2$ is the generalized Wright function, and $\Re(\rho) > |\Re(\mu)| - \frac{1}{2}, \Re(p) > 0, |\frac{w}{p^\alpha}| < 1$. When $\lambda = 0$ and $\mu = \frac{1}{2}$, then by virtue of the identity

$$W_{\pm\frac{1}{2},0}(t) = \exp\left(-\frac{t}{2}\right), \tag{2.3.22}$$

the Laplace transform of the generalized Mittag-Leffler function is obtained.

$$\int_0^\infty t^{\rho-1} e^{-pt} E_{\beta,\gamma}^{\delta}(wt^\alpha)dt = \frac{p^{-\rho}}{\Gamma(\delta)} {}_2\psi_1\left[\frac{w}{p^\alpha}\Big|{}^{(\delta,1),(\rho,\alpha)}_{(\gamma,\beta)}\right] \tag{2.3.23}$$

where $\Re(\beta) > 0, \Re(\gamma) > 0, \Re(\rho) > 0, \Re(p) > 0, p > |w|^{\frac{1}{\Re(\alpha)}}$. In particular, for $\rho = \gamma$ and $\alpha = \beta$ we obtain a result given by Prabhakar (1971, Eq.2.5).

$$\int_0^\infty t^{\gamma-1} e^{-pt} E_{\beta,\gamma}^{\delta}(wt^\beta)dt = p^{-\gamma}(1-wp^{-\beta})^{-\delta} \tag{2.3.24}$$

where $\Re(\beta) > 0, \Re(\gamma) > 0, \Re(p) > 0$ and $p > |w|^{\frac{1}{\Re(\beta)}}$.

The Euler transform of the generalized Mittag-Leffler function follows from the beta function.

$$\int_0^1 t^{a-1}(1-t)^{b-1} E_{\beta,\gamma}^{\delta}(xt^\alpha)dt = \frac{\Gamma(b)}{\Gamma(\delta)} {}_2\psi_2\left[x\Big|{}^{(\delta,1),(a,\alpha)}_{(\gamma,\beta),(a+b,\alpha)}\right], \tag{2.3.25}$$

where $\Re(a) > 0, \Re(b) > 0, \Re(\delta) > 0, \Re(\beta) > 0, \Re(\gamma) > 0, \Re(\alpha) > 0$.

Theorem 2.3.6. *We have*

$$\int_0^\infty e^{-pt} t^{\alpha k + \beta - 1} E_{\alpha,\beta}^{(k)}(at^\alpha) dt = \frac{k! p^{\alpha-\beta}}{(p^\alpha - a)^{k+1}}, \tag{2.3.26}$$

where $\Re(p) > |a|^{\frac{1}{\Re(\alpha)}}, \Re(\alpha) > 0, \Re(\beta) > 0,$ *and* $E_{\alpha,\beta}^{(k)}(y) = \frac{d^k}{dy^k} E_{\alpha,\beta}(y).$

Proof: We will use the following result:

$$\int_0^\infty e^{-t} t^{\beta-1} E_{\alpha,\beta}(zt^\alpha) dt = \frac{1}{1-z}, |z| < 1. \tag{2.3.27}$$

The given integral

$$= \frac{d^k}{da^k} \int_0^\infty e^{-pt} t^{\beta-1} E_{\alpha,\beta}(\pm at^\alpha) dt$$

$$= \frac{d^k}{da^k} \frac{p^{\alpha-\beta}}{(p^\alpha - a)} = \frac{k! p^{\alpha-\beta}}{(p^\alpha - a)^{k+1}}, \Re(\beta) > 0.$$

Corollary 2.3.3:

$$\int_0^\infty e^{-pt} t^{\frac{k-1}{2}} E_{\frac{1}{2},\frac{1}{2}}^{(k)}(a\sqrt{t}) dt = \frac{k!}{(\sqrt{p}-a)^{k+1}} \tag{2.3.28}$$

where $\Re(p) > a^2.$

Exercises 2.3.

2.3.1. Prove that

$$\frac{1}{\Gamma(\alpha)} \int_0^1 u^{\gamma-1}(1-u)^{\alpha-1} E_{\beta,\gamma}^\delta(zu^\beta) du = E_{\beta,\gamma+\alpha}^\delta(z), \Re(\alpha) > 0, \Re(\beta) > 0, \Re(\gamma) > 0.$$

2.3.2. Prove that

$$\frac{1}{\Gamma(\alpha)} \int_t^x (x-u)^{\alpha-1}(u-t)^{\gamma-1} E_{\beta,\gamma}^\delta[\lambda(u-t)^\beta] du = (x-t)^{\gamma+\alpha-1} E_{\beta,\gamma+\alpha}^\delta[\lambda(x-t)^\beta]$$

where $\Re(\alpha) > 0, \Re(\gamma) > 0, \Re(\beta) > 0.$
2.3.3. Prove that for $n = 1, 2, \ldots$

$$E_{n,\gamma}^\delta(z) = \frac{1}{\Gamma(\gamma)} {}_1F_n(\delta; \Delta(n;\gamma); n^{-n}z),$$

where $\Delta(n;\gamma)$ represents the sequence of n parameters $\frac{\gamma}{n}, \frac{\gamma+1}{n}, \ldots, \frac{\gamma+n-1}{n}.$

2.3.4. Show that for $\Re(\beta) > 0, \Re(\gamma) > 0,$

$$\left(\frac{d}{dz}\right)^m E_{\beta,\gamma}^{\delta}(z) = (\delta)_m E_{\beta,\gamma+m\beta}^{\delta+m}(z).$$

2.3.5. Prove that for $\Re(\beta) > 0, \Re(\gamma) > 0$,

$$\left(z\frac{d}{dz} + \delta\right) E_{\beta,\gamma}^{\delta}(z) = \delta E_{\beta,\gamma}^{\delta+1}(z).$$

2.3.6. Prove that for $\Re(\gamma) > 1$,

$$(\gamma - \beta\delta - 1)E_{\beta,\gamma}^{\delta}(z) = E_{\beta,\gamma-1}^{\delta}(z) - \beta\delta E_{\beta,\gamma}^{\delta+1}(z).$$

2.3.7. Prove that

$$\int_0^x t^{\nu-1}(x-t)^{\mu-1}E_{\rho,\mu}^{\gamma}[w(x-t)^{\rho}]E_{\rho,\nu}^{\sigma}(wt^{\rho})dt = x^{\mu+\nu-1}E_{\rho,\mu+\nu}^{\gamma+\sigma}(wx^{\rho}),$$

where $\rho,\mu,\gamma,\nu,\sigma,w \in C; \Re(\rho),\Re(\mu),\Re(\nu) > 0$.

2.3.8. Find

$$L^{-1}\left[s^{-\lambda}\left(1 - \frac{z}{s^{\rho}}\right)^{-\alpha}\right]$$

and give the conditions of validity.

2.3.9. Prove that

$$L^{-1}\left[s^{-\lambda}\left(1 - \frac{z_1}{s}\right)^{-\alpha_1}\left(1 - \frac{z_2}{s}\right)^{-\alpha_2}\right] = \frac{t^{\lambda-1}}{\Gamma(\lambda)}\Phi_2(\alpha_1,\alpha_2;\lambda;z_1t,z_2t),$$

where $\Re(\lambda) > 0, \Re(s) > \max[0,\Re(z_1),\Re(z_2)]$ and Φ_2 is the confluent hypergeometric function of two variables defined by

$$\Phi_2(b,b';c;u,z) = \sum_{k,j=0}^{\infty} \frac{(b)_k(b')_j u^k z^j}{(c)_{k+j}k!j!}. \tag{2.3.29}$$

2.3.10. From the above result deduce the formula

$$L^{-1}\left[s^{-\lambda}(1 - \frac{z}{s})^{-\alpha}\right] = \frac{t^{\lambda-1}}{\Gamma(\lambda)}\phi(\alpha,\lambda;zt), \tag{2.3.30}$$

where $\Re(\lambda) > 0, \Re(s) > \max[0,|z|]$.

2.4 Fractional Integrals

This section deals with the definition and properties of various operators of fractional integration and fractional differentiation of arbitrary order. Among the various operators studied are the Riemann-Liouville fractional integral operators, Riemann-Liouville fractional differential operators, Weyl operators, Kober operators etc. Besides the basic properties of these operators, their behaviors under Laplace, Fourier and Mellin transforms are also presented. Application of Riemann-Liouville operators in the solution of fractional order differential and fractional order integral equations is demonstrated.

2.4.1 Riemann-Liouville fractional integrals of arbitrary order

Notation 2.4.1. $_aI_x^n, _aD_x^{-n}, n \in \mathbb{N} \cup 0$: **Fractional integral of integer order n**

Definition 2.4.1.

$$_aI_x^n f(x) = _aD_x^{-n} f(x) = \frac{1}{\Gamma(n)} \int_a^x (x-t)^{n-1} f(t) dt \qquad (2.4.1)$$

where $n \in \mathbb{N} \cup 0$.

We begin our study of fractional calculus by introducing a fractional integral of integer order n in the form of Cauchy formula.

$$_aD_x^{-n} f(x) = \frac{1}{\Gamma(n)} \int_a^x (x-t)^{n-1} f(t) dt. \qquad (2.4.2)$$

It will be shown that the above integral can be expressed in terms of n-fold integral, that is,

$$_aD_x^{-n} f(x) = \int_0^x dx_1 \int_a^{x_1} dx_2 \int_a^{x_2} dx_3 \dots \int_a^{x_{n-1}} f(t) dt. \qquad (2.4.3)$$

Proof. When $n = 2$, by using the well-known Dirichlet formula, namely

$$\int_a^b dx \int_a^x f(x,y) dy = \int_a^b dy \int_y^b f(x,y) dx \qquad (2.4.4)$$

(2.4.3) becomes

$$\int_a^x dx_1 \int_a^{x_1} f(t) dt = \int_a^x dt f(t) \int_t^x dx_1$$
$$= \int_a^x (x-t) f(t) dt. \qquad (2.4.5)$$

This shows that the two-fold integral can be reduced to a single integral with the help of Dirichlet formula. For $n = 3$, the integral in (2.4.3) gives

$$_aD_x^{-3} f(x) = \int_a^x dx_1 \int_a^{x_1} dx_2 \int_a^{x_2} f(t) dt$$
$$= \int_a^x dx_1 \left[\int_a^{x_1} dx_2 \int_a^{x_2} f(t) dt \right]. \qquad (2.4.6)$$

By using the result in (2.4.5) the integrals within big brackets simplify to yield

$$_aD_x^{-3} f(x) = \int_a^x dx_1 \left[\int_a^{x_1} (x_1 - t) f(t) dt \right]. \qquad (2.4.7)$$

If we use (2.4.4), then the above expression reduces to

$$_aD_x^{-3}f(x) = \int_a^x dt\, f(t) \int_x^t (x_1-t)dx_1 = \int_a^x \frac{(x-t)^2}{2!} f(t)dt. \qquad (2.4.8)$$

Continuing this process, we finally obtain

$$_aD_x^{-n}f(x) = \frac{1}{(n-1)!} \int_a^x (x-t)^{n-1} f(t)dt. \qquad (2.4.9)$$

It is evident that the integral in (2.4.9) is meaningful for any number n provided its real part is greater than zero.

2.4.2 Riemann-Liouville fractional integrals of order α

Notation 2.4.2. $_xI_b^\alpha, _xD_b^{-\alpha}, I_{b-}^\alpha$: **Riemann-Liouville right-sided fractional integral of order α.**

Definition 2.4.2. Let $f(x) \in L(a,b), \alpha \in C, \Re(\alpha) > 0$, then

$$_aI_x^\alpha f(x) = _aD_x^{-\alpha} f(x) = I_{a+}^\alpha f(x) = \frac{1}{\Gamma(\alpha)} \int_a^x \frac{f(t)}{(x-t)^{1-\alpha}} dt, x > a \qquad (2.4.10)$$

is called Riemann-Liouville left-sided fractional integral of order α.

Definition 2.4.3. Let $f(x) \in L(a,b), \alpha \in C, \Re(\alpha) > 0$, then

$$_xI_b^\alpha f(x) = _xD_b^{-\alpha} f(x) = I_{b-}^\alpha f(x) = \frac{1}{\Gamma(\alpha)} \int_x^b \frac{f(t)}{(t-x)^{1-\alpha}} dt, x < b \qquad (2.4.11)$$

is called Riemann-Liouville right-sided fractional integral of order α.

Example 2.4.1. If $f(x) = (x-a)^{\beta-1}$, then find the value of $_aI_x^\alpha f(x)$.

Solution 2.4.1: We have

$$_aI_x^\alpha f(x) = \frac{1}{\Gamma(\alpha)} \int_a^x (x-t)^{\alpha-1}(t-a)^{\beta-1} dt.$$

If we substitute $t = a + y(x-a)$ in the above integral, it reduces to

$$\frac{\Gamma(\beta)}{\Gamma(\alpha+\beta)}(x-a)^{\alpha+\beta-1}$$

where $\Re(\beta) > 0$. Thus

$$_aI_x^\alpha f(x) = \frac{\Gamma(\beta)}{\Gamma(\alpha+\beta)}(x-a)^{\alpha+\beta-1}. \qquad (2.4.12)$$

Example 2.4.2.   It can be similarly shown that

$$x I_b^\alpha g(x) = \frac{\Gamma(\beta)}{\Gamma(\alpha+\beta)} (b-x)^{\alpha+\beta-1}, x < b \tag{2.4.13}$$

where $\Re(\beta) > 0$ and $g(x) = (b-x)^{\beta-1}$.

Note 2.4.1:   It may be noted that (2.4.12) and (2.4.13) give the Riemann-Liouville integrals of the power functions $f(x) = (x-a)^{\beta-1}$ and $g(x) = (b-x)^{\beta-1}, \Re(\beta) > 0$.

2.4.3 Basic properties of fractional integrals

Property: Fractional integrals obey the following properties:

$$\begin{aligned}
{}_a I_x^\alpha \, {}_a I_x^\beta \phi &= {}_a I_x^{\alpha+\beta} \phi = {}_a I_x^\beta \, {}_a I_x^\alpha \phi, \\
x I_b^\alpha \, x I_b^\beta \phi &= x I_b^{\alpha+\beta} \phi = x I_b^\beta \, x I_b^\alpha \phi.
\end{aligned} \tag{2.4.14}$$

Proof: By virtue of the definition(2.4.10), it follows that

$$\begin{aligned}
{}_a I_x^\alpha \, {}_a I_x^\beta \phi &= \frac{1}{\Gamma(\alpha)} \int_a^x \frac{dt}{(x-t)^{1-\alpha}} \frac{1}{\Gamma(\beta)} \int_a^t \frac{\phi(u)du}{(t-u)^{1-\beta}} \\
&= \frac{1}{\Gamma(\alpha)\Gamma(\beta)} \int_a^x du\phi(u) \int_u^x \frac{dt}{(x-t)^{1-\alpha}(t-u)^{1-\beta}}.
\end{aligned} \tag{2.4.15}$$

If we use the substitution $y = \frac{t-u}{x-u}$, the value of the second integral is

$$\frac{1}{\Gamma(\alpha)\Gamma(\beta)(x-u)^{1-\alpha-\beta}} \int_0^1 y^{\beta-1}(1-y)^{\alpha-1} dy = \frac{(x-u)^{\alpha+\beta-1}}{\Gamma(\alpha+\beta)},$$

which, when substituted in (2.4.15) yields the first part of (2.4.14). The second part can be similarly established. In particular,

$$_a I_x^{n+\alpha} f = {}_a I_x^n \, {}_a I_x^\alpha f, n \in \mathbb{N}, \Re(\alpha) > 0 \tag{2.4.16}$$

which shows that the n-fold differentiation

$$\frac{d^n}{dx^n} {}_a I_x^{n+\alpha} f(x) = {}_a I_x^\alpha f(x), n \in \mathbb{N}, \Re(\alpha) > 0 \tag{2.4.17}$$

for all x. When $\alpha = 0$, we obtain

$$_a I_x^0 f(x) = f(x); \; _a I_x^{-n} f(x) = \frac{d^n}{dx^n} f(x) = f^{(n)}(x). \tag{2.4.18}$$

Note 2.4.2:   The property given in (2.4.14) is called semigroup property of fractional integration.

Notation 2.4.3. $L(a,b)$: **space of Lebesgue measurable real or complex valued functions.**

Definition 2.4.4. $L(a,b)$, **consists of Lebesgue measurable real or complex valued functions** $f(x)$ **on** $[a,b]$:

$$L(a,b) = \{f : ||f||_1 \sim \int_a^b |f(t)|dt < \infty\}. \tag{2.4.19}$$

Note 2.4.3: The operators $_aI_x^\alpha$ and $_xI_b^\alpha$ are defined on the space $L(a,b)$.

Property: The following results hold:

$$\int_a^b f(x)(_aI_x^\alpha g)dx = \int_a^b g(x)(_xI_b^\alpha f)dx. \tag{2.4.20}$$

(2.4.20) can be established by interchanging the order of integration in the integral on the left-hand side of (2.4.20) and then using the Dirichlet formula (2.4.4).

The above property is called the property of "integration by parts" for fractional integrals.

2.4.4 A useful integral

We now evaluate the following integral given by Saxena and Nishimoto [Journal of Fractional Calculus, Vol. 6, 1994, 65-75].

$$\int_a^b (t-a)^{\alpha-1}(b-t)^{\beta-1}(ct+d)^\gamma dt = (ac+d)^\gamma(b-a)^{\alpha+\beta-1}$$
$$\times B(\alpha,\beta)_2F_1\left[\alpha,-\gamma;\alpha+\beta;\frac{(a-b)c}{(ac+d)}\right], \tag{2.4.21}$$

where $\Re(\alpha) > 0, \Re(\beta) > 0, |\arg\frac{(d+bc)}{(d+ac)}| < \pi, a, c$ and d are constants.

Solution Let

$$I = \int_a^b (t-a)^{\alpha-1}(b-t)^{\beta-1}(ct+d)^\gamma dt$$
$$= (ac+d)^\gamma \sum_{k=0}^\infty \frac{(-1)^k(-\gamma)_k c^k}{(ac+d)^k}\int_a^b (t-a)^{\alpha+k-1}(b-t)^{\beta-1}dt$$
$$= (ac+d)^\gamma(b-a)^{\alpha+\beta-1}B(\alpha,\beta)_2F_1\left(-\gamma,\alpha;\alpha+\beta;\frac{(a-b)c}{(ac+d)}\right).$$

In evaluating the inner integral the modified form of the beta function, namely

$$\int_a^b (t-a)^{\alpha-1}(b-t)^{\beta-1}\mathrm{d}t = (b-a)^{\alpha+\beta-1}B(\alpha,\beta), \qquad (2.4.22)$$

where $\Re(\alpha) > 0, \Re(\beta) > 0$, is used.

Example 2.4.3. As a consequence of (2.4.21) it follows that

$$_aI_x^{\alpha}[(x-a)^{\beta-1}(cx+d)^{\gamma}] = (ac+d)^{\gamma}(x-a)^{\alpha+\beta-1}\frac{\Gamma(\beta)}{\Gamma(\alpha+\beta)}$$

$$\times {}_2F_1\left(\beta,-\gamma;\alpha+\beta;\frac{(a-x)c}{(ac+d)}\right), \qquad (2.4.23)$$

where $\Re(\alpha) > 0, \Re(\beta) > 0, |\arg\frac{(a-x)c}{(ac+d)}| < \pi, a, c$ and d being constants. In a similar manner we obtain the following result:

Example 2.4.4. We also have

$$_xI_b^{\alpha}[(b-x)^{\beta-1}(cx+d)^{\gamma} = (cx+d)^{\gamma}](b-x)^{\alpha+\beta-1}\frac{\gamma(\beta)}{\Gamma(\alpha+\beta)}$$

$$\times {}_2F_1\left(\alpha,-\gamma;\alpha+\beta;\frac{(x-b)c}{(cx+d)}\right), \qquad (2.4.24)$$

where $\Re(\alpha) > 0, \Re(\beta) > 0, |\arg\frac{(x-b)c}{(cx+d)}| < \pi$.

Example 2.4.5. On the other hand if we set $\gamma = -\alpha - \beta$ in (2.4.21) it is found that

$$_aD_x^{-\alpha}[(x-a)^{\beta-1}(cx+d)^{-\alpha-\beta}] = \frac{\Gamma(\beta)}{\Gamma(\alpha+\beta)}(ac+d)^{-\alpha}(x-a)^{\alpha+\beta-1}(d+cx)^{-\beta},$$

$$(2.4.25)$$

where $\Re(\alpha) > 0, \Re(\beta) > 0$.

Example 2.4.6. Similarly, we have

$$_xI_b^{\alpha}[(b-x)^{\beta-1}(cx+d)^{-\alpha-\beta}] = \frac{\Gamma(\beta)}{\Gamma(\alpha+\beta)}(cx+d)^{-\beta}(bc+d)^{-\alpha}(b-x)^{\alpha+\beta-1}$$

$$(2.4.26)$$

where $\Re(\alpha) > 0, \Re(\beta) > 0$.

2.4.5 The Weyl integral

Notation 2.4.4. $_xW_\infty^\alpha$, $_xI_\infty^\alpha$: **Weyl integral of order** α.

Definition 2.4.5. **The Weyl fractional integral of** $f(x)$ **of order** α, **denoted by** $_xW_\infty^\alpha$, **is defined by**

$$_xW_\infty^\alpha f(x) = \frac{1}{\Gamma(\alpha)} \int_x^\infty (t-x)^{\alpha-1} f(t)dt, \quad -\infty < x < \infty \qquad (2.4.27)$$

where $\alpha \in C, \Re(\alpha) > 0$. (2.4.27) is also denoted by $_xI_\infty^\alpha f(x)$.

Example 2.4.7. Prove that

$$_xW_\infty^\alpha e^{-\lambda x} = \frac{e^{-\lambda x}}{\lambda^\alpha} \text{ where } \Re(\alpha) > 0. \qquad (2.4.28)$$

Solution: We have

$$\begin{aligned}
_xW_\infty^\alpha e^{-\lambda x} &= \frac{1}{\Gamma(\alpha)} \int_x^\infty (t-x)^{\alpha-1} e^{-\lambda t} dt, \quad \lambda > 0 \\
&= \frac{e^{-\lambda x}}{\Gamma(\alpha)\lambda^\alpha} \int_0^\infty u^{\alpha-1} e^{-u} du \\
&= \frac{e^{-\lambda x}}{\lambda^\alpha}, \quad \Re(\alpha) > 0.
\end{aligned}$$

Notation 2.4.5. $_xD_\infty^\alpha$, D_-^α **Weyl fractional derivative.**

Definition 2.4.6. **The Weyl fractional derivative of order** α, **denoted by** $_xD_\infty^\alpha$, **is defined by**

$$_xD_\infty^\alpha f(x) = D_-^\infty f(x) = (-1)^m \left(\frac{d}{dx}\right)^m \left(_xW_\infty^{m-\alpha} f(x)\right)$$

$$= (-1)^m \left(\frac{d}{dx}\right)^m \frac{1}{\Gamma(m-\alpha)} \int_x^\infty \frac{f(t)}{(t-x)^{1+\alpha-m}} dt, \quad -\infty < x < \infty \qquad (2.4.29)$$

where $m-1 \le \alpha < m$, $\alpha \in C, m = 0, 1, 2, \ldots$.

Example 2.4.8. Find $_xD_\infty^\alpha e^{-\lambda x}$, $\lambda > 0$.

Solution: We have

$$\begin{aligned}
_xD_\infty^\alpha e^{-\lambda x} &= (-1)^m \left(\frac{d}{dx}\right)^m {}_xW_\infty^{m-\alpha} e^{-\lambda x} \\
&= (-1)^m \left(\frac{d}{dx}\right)^m \lambda^{-(m-\alpha)} e^{-\lambda x} \qquad (2.4.30) \\
&= \lambda^\alpha e^{-\lambda x}.
\end{aligned}$$

2.4.6 Basic properties of Weyl integral

Property: The following relation holds:

$$\int_0^\infty \phi(x) \left({}_0I_x^\alpha \psi(x) \right) dx = \int_0^\infty \left({}_xW_\infty^\alpha \phi(x) \right) \psi(x) dx. \tag{2.4.31}$$

(2.4.31) is called the formula for fractional integration by parts. It is also called Parseval equality. (2.4.31) can be established by interchanging the order of integration.

Property: Weyl fractional integral obeys the semigroup property. That is,

$$\left({}_xW_\infty^\alpha {}_xW_\infty^\beta f \right) = \left({}_xW_\infty^{\alpha+\beta} f \right) = \left({}_xW_\infty^\beta {}_xW_\infty^\alpha f \right). \tag{2.4.32}$$

Proof: We have

$$\begin{aligned}
{}_xW_\infty^\alpha {}_xW_\infty^\beta f(x) &= \frac{1}{\Gamma(\alpha)} \int_x^\infty dt(t-x)^{\alpha-1} \\
&\quad \times \frac{1}{\Gamma(\beta)} \int_x^\infty (u-t)^{\beta-1} f(u) du.
\end{aligned}$$

By using the modified form of the Dirichlet formula (2.4.4), namely

$$\int_x^a dt(t-x)^{\alpha-1} \int_t^a (u-t)^{\beta-1} f(u)du = B(\alpha,\beta) \int_t^a (u-t)^{\alpha+\beta-1} f(u)du, \tag{2.4.33}$$

and letting $a \to \infty$, (2.4.33) yields the desired result:

$$\left({}_xW_\infty^\alpha {}_xW_\infty^\beta f \right) = \left({}_xW_\infty^{\alpha+\beta} f \right). \tag{2.4.34}$$

Notation 2.4.6.　$_{-\infty}W_x^\alpha$, I_+^α : **Weyl integral with lower limit** $-\infty$.

Definition 2.4.7. Another companion to the operator (2.4.27) is the following:

$$_{-\infty}W_x^\alpha f(x) = I_+^\alpha f(x) = \frac{1}{\Gamma(\alpha)} \int_{-\infty}^x (x-t)^{\alpha-1} f(t)dt, \quad -\infty < x < \infty \tag{2.4.35}$$

where $\Re(\alpha) > 0$.

Note 2.4.4:　The operator defined by (2.4.35) is useful in fractional diffusion problems in astrophysics and related areas.

Example 2.4.9.　Prove that

$$_{-\infty}W_x^\alpha e^{ax} = \frac{e^{ax}}{a^\alpha}. \tag{2.4.36}$$

Solution: We have the result by setting $x - t = u$.

Note 2.4.5: An alternative form of (2.4.35) in terms of convolution is given by

$$-_\infty W_x^\alpha f(x) = \frac{1}{\Gamma(\alpha)} \int_{-\infty}^\infty t_+^{\alpha-1} f(x-t) dt \qquad (2.4.37)$$

where

$$t_+^{\alpha-1} = \begin{cases} t^{\alpha-1}, t > 0 \\ 0, t < 0 \end{cases}$$

Example 2.4.10. Prove that

$$_x W_\infty^\nu(\cos ax) = a^{-\nu} \cos\left(ax + \frac{1}{2}\pi\nu\right) \qquad (2.4.38)$$

where $a > 0, 0 < \Re(\nu) < 1$.

Solution: The result follows from the known integral

$$\int_u^\infty (x-u)^{\nu-1} \cos ax \, dx = \frac{\Gamma(\nu)}{a^\nu} \cos\left(au + \frac{\nu\pi}{2}\right). \qquad (2.4.39)$$

Example 2.4.11. Prove that

$$_x W_\infty^\nu(\sin ax) = a^{-\nu} \sin\left(ax + \frac{1}{2}\pi\nu\right). \qquad (2.4.40)$$

Hint: Use the integral

$$\int_u^\infty (x-u)^{\nu-1} \sin ax \, dx = \frac{\Gamma(\nu)}{a^\nu} \sin\left(au + \frac{1}{2}\pi\nu\right) \qquad (2.4.41)$$

where $a > 0, 0 < \Re(\nu) < 1$.

Exercises 2.4.

2.4.1. Prove that

$$\left(_a I_x^\alpha (x-a)^{\beta-1}\right) = \frac{\Gamma(\beta)}{\Gamma(\alpha+\beta)} (x-a)^{\alpha+\beta-1}, \; \Re(\beta) > 0.$$

2.4.2. Prove that

$$\left(_a I_x^\alpha (x \pm c)^{\gamma-1}\right) = \frac{(a \pm c)^{\gamma-1}}{\Gamma(\alpha+1)} (x-a)^\alpha {}_2F_1\left(1, 1-\gamma; \alpha+1; \frac{a-x}{a \pm c}\right)$$

where $\Re(\beta) > 0, \gamma \in C, a \neq c, \left|\frac{a-x}{a \pm c}\right| < 1$.

2.4.3. Prove that

$$\left({}_aI_x^\alpha[(x-a)^{\beta-1}(b-x)^{\gamma-1}] \right) = \frac{\Gamma(\beta)}{\Gamma(\alpha+\beta)} \frac{(x-a)^{\alpha+\beta-1}}{(b-a)^{1-\gamma}}$$
$$\times {}_2F_1\left(\beta, 1-\gamma; \alpha+\beta; \frac{x-a}{b-a} \right)$$

where $\Re(\beta) > 0, \gamma \in C, \ a < x < b$.

2.4.4. Prove that

$$\left({}_aI_x^\alpha\left[\frac{(x-a)^{\beta-1}}{(b-x)^{\alpha+\beta}} \right] \right) = \frac{\Gamma(\beta)}{\Gamma(\alpha+\beta)} \frac{(x-a)^{\alpha+\beta-1}}{(b-a)^\alpha(b-x)^\beta}$$

where $\Re(\beta) > 0, \ a < x < b$.

2.4.5. Prove that

$$\left({}_aI_x^\alpha\left[(x-a)^{\beta-1}(x\pm c)^{\gamma-1} \right] \right) = \frac{\Gamma(\beta)}{\Gamma(\alpha+\beta)} \frac{(x-a)^{\alpha+\beta-1}}{(a\pm c)^{1-\gamma}}$$
$$\times {}_2F_1\left(\beta, 1-\gamma; \alpha+\beta; \frac{(a-x)}{a\pm c} \right),$$

where $\Re(\beta) > 0, \ \gamma \in C, a \neq c, |\frac{a-x}{a\pm c}| < 1$.

2.4.6. Prove that for $\Re(\beta) > 0$,

$$\left({}_aI_x^\alpha\left[\frac{(x-a)^{\beta-1}}{(x\pm c)^{\alpha+\beta}} \right] \right) = \frac{\Gamma(\beta)}{\Gamma(\alpha+\beta)} \frac{(x-a)^{\alpha+\beta-1}}{(a\pm c)^\alpha(x\pm c)^\beta}, \left| \frac{a-x}{a\pm c} \right| < 1.$$

2.4.7. Prove that

$$\left({}_aI_x^\alpha[e^{\lambda x}] \right) = e^{\lambda a}(x-a)^\alpha E_{1,\alpha+1}(\lambda x - \lambda a).$$

2.4.8. Prove that

$$\left({}_aI_x^\alpha[e^{\lambda x}(x-a)^{\beta-1}] \right) = \frac{\Gamma(\beta)}{\Gamma(\alpha+\beta)} e^{\lambda a}(x-a)^{\alpha+\beta-1} {}_1F_1(\beta; \alpha+\beta; \lambda x - \lambda a),$$

where $\Re(\beta) > 0, \Re(\alpha) > 0$.

2.4.9. Prove that

$$\left({}_aI_x^\alpha\left[(x-a)^{\frac{\nu}{2}} J_\nu(\lambda\sqrt{x-a}) \right] \right) = \left(\frac{2}{\lambda} \right)^\alpha (x-a)^{\frac{\alpha+\nu}{2}} J_{\alpha+\nu}(\lambda\sqrt{x-a}),$$

where $\Re(\nu) > -1$.

2.4.10. Prove that

$$\left({}_aI_x^{\nu}\left[(x-a)^{\beta-1}{}_2F_1(\mu,\nu;\beta;\lambda(x-a)) \right] \right)$$
$$= \frac{\Gamma(\beta)}{\Gamma(\nu+\beta)}(x-a)^{\nu+\beta-1}{}_2F_1(\mu,\nu;\nu+\beta;\lambda x - \lambda a),$$

where $\Re(\beta) > 0$.

2.4.7 Laplace transform of the fractional integral

We have

$$_0I_x^{\nu}f(x) = \frac{1}{\Gamma(\nu)}\int_0^x (x-t)^{\nu-1}f(t)dt, \tag{2.4.42}$$

where $\Re(\nu) > 0$. Application of convolution theorem of the Laplace transform gives

$$L\{_0I_x^{\nu}f(x)\};s = L\left\{ \frac{t^{\nu-1}}{\Gamma(\nu)} \right\}L\{f(t);s\}$$
$$= s^{-\nu}F(s), \tag{2.4.43}$$

where $\Re(s) > 0$, $\Re(\nu) > 0$.

2.4.8 Laplace transform of the fractional derivative

If $n \in \mathbb{N}$, then by the theory of the Laplace transform, we know that

$$L\left\{ \frac{d^n}{dx^n}f;s \right\} = s^n F(s) - \sum_{k=0}^{n-1} s^{n-k-1}f^{(k)}(0+) \tag{2.4.44}$$

$$= s^n F(s) - \sum_{k=0}^{n-1} s^k f^{(n-k-1)}(0+), \ (n-1 \le \alpha < n) \tag{2.4.45}$$

where $\Re(s) > 0$ and $F(s)$ is the Laplace transform of $f(t)$. By virtue of the definition of the derivative, we find that

$$L\{_0D_x^{\alpha}f;s\} = L\left\{ \frac{d^n}{dx^n}{}_0I_x^{n-\alpha}f;s \right\}$$

$$= s^n L\{_0I_x^{n-\alpha}f;s\} - \sum_{k=0}^{n-1} s^k \frac{d^{n-k-1}}{dx^{n-k-1}} {}_0I_x^{n-\alpha}f(0+)$$

$$= s^{\alpha} F(s) - \sum_{k=0}^{n-1} s^k D^{\alpha-k-1} f(0+), \left(D = \frac{d}{dx} \right) \tag{2.4.46}$$

$$= s^{\alpha} F(s) - \sum_{k=1}^{n} s^{k-1} D^{\alpha-k} f(0+) \tag{2.4.47}$$

where $\Re(s) > 0$.

2.4.9 Laplace transform of Caputo derivative

Notation 2.4.7. $\underset{0}{C} {}_a D_x^{\alpha}$

Definition 2.4.8. The Caputo derivative of a casual function $f(t)$ (that is $f(t) = 0$ for $t < 0$) with $\alpha > 0$ was defined by Caputo (1969) in the form

$$\underset{0}{C} {}_a D_x^{\alpha} f(x) = {}_a I_x^{n-\alpha} \frac{d^n}{dx^n} f(x) = {}_a D_t^{-(n-\alpha)} f^{(n)}(t) \tag{2.4.48}$$

$$= \frac{1}{\Gamma(n-\alpha)} \int_a^x (x-t)^{n-\alpha-1} f^{(n)}(t) dt, (n-1 < \alpha < n) \tag{2.4.49}$$

where $n \in \mathbb{N}$.

From (2.4.43) and (2.4.49), it follows that

$$L\{\underset{0}{C} {}_0 D_t^{\alpha} f(t); s\} = s^{-(n-\alpha)} L\{f^{(n)}(t)\}. \tag{2.4.50}$$

On using (2.4.44), we see that

$$L\{\underset{0}{C} {}_0 D_t^{\alpha} f(t); s\} = s^{-(n-\alpha)} \left[s^n F(s) - \sum_{k=0}^{n-1} s^{n-k-1} f^{(k)}(0+) \right]$$

$$= s^{\alpha} F(s) - \sum_{k=0}^{n-1} s^{\alpha-k-1} f^{(k)}(0+), \ (n-1 < \alpha \le n), \tag{2.4.51}$$

where $\Re(s) > 0$ and $\Re(\alpha) > 0$.

Note 2.4.6: From (2.4.48), it can be seen that

$$\underset{0}{C} {}_0 D_t^{\alpha} A = 0, \text{ where } A \text{ is a constant,}$$

whereas the Riemann-Liouville derivative

$$_0 D_t^{\alpha} A = \frac{A t^{-\alpha}}{\Gamma(1-\alpha)}, \ (\alpha \neq 1, 2, \cdots), \tag{2.4.52}$$

which is a surprising result.

Exercises 2.4.

2.4.11. Prove that

$$(_0I_x^v f(x)) = L^{-1}s^{-v}L\{f(x);s\},\tag{2.4.53}$$

where $\Re(v) > 0$.

2.4.12. Prove that the solution of Abel integral equation of the second kind

$$\phi(x) - \frac{\lambda}{\Gamma(\alpha)}\int_0^x \frac{\phi(t)dt}{(x-t)^{1-\alpha}} = f(x), \ 0 < x < 1$$

$\alpha > 0$, is given by

$$\phi(x) = \frac{d}{dx}\int_0^x E_\alpha[\lambda(x-t)^\alpha]\, f(t)\, dt,\tag{2.4.54}$$

where $E_\alpha(x)$ is the Mittag-Leffler function defined by equation (2.1.1).

2.4.13. Show that

$$\frac{\lambda}{\Gamma(\alpha)}\int_0^x \frac{E_\alpha(\lambda t^\alpha)}{(x-t)^{1-\alpha}}dt = E_\alpha(\lambda x^\alpha) - 1, \ \alpha > 0.\tag{2.4.55}$$

2.5 Mellin Transform of the Fractional Integrals and the Fractional Derivatives

2.5.1 Mellin transform

Notation 2.5.1. $m\{f(x);\, s\}, f^*(s)$: The Mellin transform

Notation 2.5.2. $m^{-1}\{f^*(s);\, x\}$: Inverse Mellin transform

Definition 2.5.1. The Mellin transform of a function $f(x)$, denoted by $f^*(s)$, is defined by

$$f^*(s) = m\{f(x);\, s\} = \int_0^\infty x^{s-1} f(x)dx, \ x > 0.\tag{2.5.1}$$

The inverse Mellin transform is given by the contour integral

$$f(x) = m^{-1}\{f^*(s);\, x\} = \frac{1}{2\pi i}\int_{\gamma-i\infty}^{\gamma+i\infty} f^*(s)x^{-s}ds, \ i = \sqrt{-1}\tag{2.5.2}$$

where γ is real.

2.5.2 Mellin transform of the fractional integral

Theorem 2.5.1. *The following result holds true.*

$$m(_0I_x^\alpha f)(s) = \frac{\Gamma(1-\alpha-s)}{\Gamma(1-\alpha)} f^*(s+\alpha), \tag{2.5.3}$$

where $\Re(\alpha) > 0$ *and* $\Re(\alpha+s) < 1$.

Proof 2.5.1: We have

$$m(_0I_x^\alpha f)(s) = \int_0^\infty z^{s-1} \frac{1}{\Gamma(\alpha)} \int_0^z (z-t)^{\alpha-1} f(t)\, dt dz$$

$$= \frac{1}{\Gamma(\alpha)} \int_0^\infty f(t)\, dt \int_t^\infty z^{s-1}(z-t)^{\alpha-1}\, dz. \tag{2.5.4}$$

On setting $z = \frac{t}{u}$, the z-integral becomes

$$t^{\alpha+s-1} \int_0^1 u^{-\alpha-s}(1-u)^{\alpha-1} du = t^{\alpha+s-1} B(\alpha, 1-\alpha-s), \tag{2.5.5}$$

where $\Re(\alpha) > 0$, $\Re(\alpha+s) < 1$. Putting the above value of z-integral, the result follows.

Similarly we can establish

Theorem 2.5.2. *The following result holds true.*

$$m(_xI_\infty^\alpha f)(s) = \frac{\Gamma(s)}{\Gamma(s+\alpha)} m\left\{ t^\alpha f(t); s \right\}$$

$$= \frac{\Gamma(s)}{\Gamma(s+\alpha)} f^*(s+\alpha), \tag{2.5.6}$$

where $\Re(\alpha) > 0$, $\Re(s) > 0$.

Note 2.5.1: If we set $f(x) = x^{-\alpha}\phi(x)$, then using the property of the Mellin transform

$$x^\alpha \phi(x) \leftrightarrow \phi^*(s+\alpha), \tag{2.5.7}$$

the results (2.5.3) and (2.5.6) become

$$(_0I_x^\alpha x^{-\alpha} f(x))(s) = \frac{\Gamma(1-\alpha-s)}{\Gamma(1-s)} f^*(s), \tag{2.5.8}$$

where $\Re(\alpha) > 0$, $\Re(\alpha+s) < 1$ and

$$(_xI_\infty^\alpha x^{-\alpha} f(x))(s) = \frac{\Gamma(s)}{\Gamma(s+\alpha)} f^*(s), \tag{2.5.9}$$

where $\Re(\alpha) > 0$, and $\Re(s) > 0$, respectively.

2.5.3 Mellin transform of the fractional derivative

Theorem 2.5.3. *If $n \in \mathbb{N}$ and $\lim_{t \to \infty} t^{s-1} f^{(v)}(t) = 0, \ v = 0, 1, \cdots, n$, then*

$$m\{f^{(n)}(t); (s)\} = (-1)^n \frac{\Gamma(s)}{\Gamma(s-n)} m\{f(t); s-n\}, \qquad (2.5.10)$$

where $\Re(s) > 0, \Re(s-n) > 0$.

Proof 2.5.2: Integrate by parts and using the definition of the Mellin transform, the result follows.

Example 2.5.1. Find the Mellin transform of the fractional derivative.

Solution 2.5.1: We have

$$_0D_x^\alpha f = {_0D_x^n} \, _0D_x^{\alpha-n} f = {_0D_x^n} \, _0I_x^{n-\alpha} f.$$

Therefore,

$$m(_0D_x^\alpha f)(s) = \frac{(-1)^n \Gamma(s)}{\Gamma(s-n)} m\{_0I_x^{n-\alpha} f\}(s-n), (n-1 \le \Re(\alpha) < n) \qquad (2.5.11)$$

$$= \frac{(-1)^n \Gamma(s)\Gamma(1-(s-\alpha))}{\Gamma(s-n)\Gamma(1-s+n)} m\{f(t); s-\alpha\}, \qquad (2.5.12)$$

where $\Re(s) > 0, \Re(s) < 1 + \Re(\alpha)$.

Remark 2.5.1: An alternative form of (2.5.12) is given in Exercise 2.5.2.

Exercises 2.5.

2.5.1. Prove Theorem 2.5.2.

2.5.2. Prove that the Mellin transform of fractional derivative is given by

$$m(_0D_x^\alpha f)(s) = \frac{(-1)^n \Gamma(s) \sin[\pi(s-n)]}{\Gamma(s-\alpha) \sin[\pi(s-\alpha)]} m\{f(t); s-\alpha\}, \qquad (2.5.13)$$

where $\Re(s) > 0, \Re(\alpha - s) > -1$.

2.5.3. Find the Mellin transform of $(1+x^a)^{-b}; a, b > 0$.

2.6 Kober Operators

Kober operators are the generalization of Riemann-Liouville and Weyl operators. These operators have been used by many authors in deriving the solution of single, dual and triple integral equations possessing special functions of mathematical physics, as their kernels.

Notation 2.6.1. Kober operator of the first kind

$$\mathbb{I}[f(x)],\ \mathbb{I}[\alpha,\eta:f(x)],\ \mathbb{I}(\alpha,\eta)f(x), E_{0,x}^{\alpha,\eta}\, f,\ \mathbb{I}_x^{\eta,\alpha}\, f.$$

Notation 2.6.2. Kober operator of the second kind

$$\mathbb{R}[f(x)],\ \mathbb{R}[\alpha,\zeta:f(x)],\ \mathbb{R}(\alpha,\zeta)f(x), K_{x,\infty}^{\alpha,\zeta}\, f, K_x^{\zeta,\alpha}\, f.$$

Definition 2.6.1.

$$\mathbb{I}[f(x)] = \mathbb{I}[\alpha,\eta:f(x)] = \mathbb{I}(\alpha,\eta)f(x) = E_{0,x}^{\alpha,\eta}f$$
$$= \mathbb{I}_x^{\eta,\alpha}\, f = \frac{x^{-\eta-\alpha}}{\Gamma(\alpha)} \int_0^x (x-t)^{\alpha-1} t^\eta f(t)\mathrm{d}t, \tag{2.6.1}$$

where $\Re(\alpha) > 0$.

Definition 2.6.2.

$$\mathbb{R}[f(x)] = \mathbb{R}[\alpha,\zeta:f(x)] = \mathbb{R}(\alpha,\zeta)f(x) = K_{x,\infty}^{\alpha,\zeta}f$$
$$= K_x^{\zeta,\alpha}\, f = \frac{x^\zeta}{\Gamma(\alpha)} \int_x^\infty (t-x)^{\alpha-1} t^{-\zeta-\alpha} f(t)\mathrm{d}t, \tag{2.6.2}$$

where $\Re(\alpha) > 0$.

(2.6.1) and (2.6.2) hold true under the following conditions:

$$f \in L_p(0,\infty), \Re(\alpha) > 0, \Re(\eta) > -\frac{1}{q}, \Re(\zeta) > -\frac{1}{p}, \frac{1}{p} + \frac{1}{q} = 1, p \geq 1.$$

When $\eta = 0$, (2.6.1) reduces to Riemann-Liouville operator. That is,

$$I_x^{0,\alpha}f = x^{-\alpha}\, {}_0I_x^\alpha f. \tag{2.6.3}$$

For $\zeta = 0$, (2.6.2) yields the Weyl operator of $t^{-\alpha}f(t)$. That is,

$$K_x^{0,\alpha}f = {}_xW_\infty^\alpha\, t^{-\alpha}f(t). \tag{2.6.4}$$

Theorem 2.6.1. *[Kober (1940)].*

If $\Re(\alpha) > 0, \Re(\eta - s) > -1, f \in L_p(o,\infty), 1 \leq p \leq 2$ (or $f \in M_p(o,\infty)$, a subspace of $L_p(o,\infty)$ and $p > 2$), $\Re(\eta) > -\frac{1}{q}, \frac{1}{p} + \frac{1}{q} = 1$, then there holds the formula

$$m\{\mathbb{I}(\alpha,\eta)f\}(s) = \frac{\Gamma(1+\eta-s)}{\Gamma(\alpha+\eta+1-s)} m\{f(x);s\}. \tag{2.6.5}$$

Proof 2.6.1: It is similar to the proof of Theorem 2.6.1.

In a similar manner, we can establish

Theorem 2.6.2. *[Kober (1940)].*

If $\Re(\alpha) > 0, \Re(s + \zeta) > 0, f \in L_p(o,\infty), 1 \leq p \leq 2$ (or $f \in M_p(o,\infty)$, a subspace of $L_p(o,\infty)$ and $p > 2$)

$$\Re(\zeta) > -\frac{1}{p}, \frac{1}{p} + \frac{1}{q} = 1,$$

then,

$$m\{\Re(\alpha,\zeta)f\}(s) = \frac{\Gamma(\zeta+s)}{\Gamma(\alpha+\zeta+s)}m\{f(x);s\}. \tag{2.6.6}$$

Semigroup property of the Kober operators has been given in the form of

Theorem 2.6.3. If $f \in L_p(o,\infty), g \in L_q(o,\infty), \frac{1}{p} + \frac{1}{q} = 1, \Re(\eta) > -\frac{1}{q}, \Re(\zeta) > -\frac{1}{p}, 1 \leq p \leq 2$, (or $f \in M_p(o,\infty)$, a subspace of $L_p(o,\infty)$ and $p > 2$), then

$$\int_0^\infty g(x)(\mathbb{I}(\alpha,\eta:f))(x)dx = \int_0^\infty f(x)(\mathbb{R}(\alpha,\eta:g))(x)dx. \tag{2.6.7}$$

Proof 2.6.2: Interchange the order of integration.

Remark 2.6.1: Operators defined by (2.6.1.) and (2.6.2) are also called Erdélyi-Kober operators.

Exercises 2.6.

2.6.1. Prove Theorem 2.6.1.

2.6.2. For the modified Erdélyi-Kober operators, defined by the following equations for $m > 0$:

$$\mathbb{I}(\alpha,\eta:m)f(x) = \mathbb{I}(f(x):\alpha,\eta,m)$$
$$= \frac{m}{\Gamma(\alpha)}x^{-\eta-m\alpha+m-1}\int_o^x t^\eta(x^m - t^m)^{\alpha-1}f(t)dt, \tag{2.6.8}$$

and

$$\mathbb{R}(\alpha,\zeta:m)f(x) = \mathbb{R}(f(x):\alpha,\zeta,m)$$
$$= \frac{mx^\zeta}{\Gamma(\alpha)}\int_x^\infty t^{-\zeta-m\alpha+m-1}(t^m - x^m)^{\alpha-1}f(t)dt, \tag{2.6.9}$$

where $f \in L_p(0, \infty), \Re(\alpha) > 0, \Re(\eta) > -\frac{1}{q}, \Re(\zeta) > -\frac{1}{p}, \frac{1}{p} + \frac{1}{q} = 1$, find the Mellin transforms of (i) $\mathbb{I}(\alpha, \eta : m) f(x)$ and (ii) $\mathbb{R}(\alpha, \zeta : m) f(x)$, giving the conditions of validity.

2.6.3. For the operators defined by (2.6.8) and (2.6.9.), show that

$$\int_0^\infty \mathbb{R}(f(x) : \alpha, \eta, m) g(x) \mathrm{d}x = \int_0^\infty f(x) \mathbb{I}(g(x) : \alpha, \eta, m) \mathrm{d}x, \tag{2.6.10}$$

where the parameters α, η, m are the same in both the operators $\mathbb{I}$ and $\mathbb{R}$. Give conditions of validity of (2.6.10).

2.6.4. For the Erdélyi-Kober operator, defined by

$$I_{\eta,\alpha} f(x) = \frac{2x^{-2\alpha-2\eta}}{\Gamma(\alpha)} \int_0^x (x^2 - t^2)^{\alpha-1} t^{2\eta+1} f(t) \mathrm{d}t, \tag{2.6.11}$$

where $\Re(\alpha) > 0$, establish the following results (Sneddon (1975)):

$$\begin{align}
\text{(i) } & I_{\eta,\alpha} x^{2\beta} f(x) = x^{2\beta} I_{\eta+\beta,\alpha} f(x) \tag{2.6.12} \\
\text{(ii) } & I_{\eta,\alpha} I_{\eta+\alpha,\beta} = I_{\eta,\alpha+\beta} = I_{\eta+\alpha,\beta} I_{\eta,\alpha} \tag{2.6.13} \\
\text{(iii) } & I_{\eta,\alpha}^{-1} = I_{\eta+\alpha,-\alpha}. \tag{2.6.14}
\end{align}$$

Remark 2.6.2: The results of Exercise 2.6.4 also hold for the operator, defined by

$$\mathbb{K}_{\eta,\alpha} f(x) = \frac{2x^{2\eta}}{\Gamma(\alpha)} \int_x^\infty (t^2 - x^2)^{\alpha-1} t^{-2\alpha-2\eta+1} f(t) \mathrm{d}t, \tag{2.6.15}$$

where $\Re(\alpha) > 0$.

Remark 2.6.3: Operators more general than the operators defined by (2.6.11) and (2.6.15) are recently defined by Galué et al [Integral Transform & Spec. Funct. Vol. 9 (2000), No. 3, pp. 185-196] in the form

$$_a I_x^{\eta,\alpha} f(x) = \frac{x^{-\eta-\alpha}}{\Gamma(\alpha)} \int_a^x (x-t)^{\alpha-1} t^\eta f(t) \mathrm{d}t, \tag{2.6.16}$$

where $\Re(\alpha) > 0$.

2.7 Generalized Kober Operators

Notation 2.7.1. $\mathbb{I}[\alpha, \beta, \gamma : m, \mu, \eta, a : f(x)], \mathbb{I}[f(x)]$

Notation 2.7.2. $\mathbb{I}[\alpha, \beta, \gamma : m, \mu, \delta, a : f(x)], \mathbb{I}[f(x)]$

Notation 2.7.3. $\mathbb{R}[f(x)], \mathbb{R}\left[\begin{smallmatrix}\alpha,\beta,\gamma;\\ \sigma,\rho,a;\end{smallmatrix} : f(x)\right]$

Notation 2.7.4. $\mathbb{K}[f(x)], \mathbb{K}\left[\begin{smallmatrix}\alpha,\beta,\gamma;\\ \delta,\rho,a;\end{smallmatrix} : f(x)\right]$

Notation 2.7.5. $I_{0,x}^{\alpha,\beta,\eta;} f(x)$ (Saigo, 1978)

Notation 2.7.6. $J_{x,\alpha}^{\alpha,\beta,\eta;} f(x)$ (Saigo, 1978)

Definition 2.7.1.

$$\mathbb{I}[f(x)] = \mathbb{I}[\alpha, \beta, \gamma : m, \mu, \eta, a : f(x)]$$
$$= \frac{\mu x^{-\eta-1}}{\Gamma(1-\alpha)} \int_0^x {}_2F_1\left(\alpha, \beta+m, \gamma; \frac{at^\mu}{x^\mu}\right) t^\eta f(t) \mathrm{d}t, \qquad (2.7.1)$$

where ${}_2F_1(\cdot)$ is the Gauss hypergeometric function.

Definition 2.7.2.

$$\mathbb{I}[f(x)] = \mathbb{I}[\alpha, \beta, \gamma : m, \mu, \delta, a : f(x)]$$
$$= \frac{\mu x^\delta}{\Gamma(1-\alpha)} \int_x^\infty {}_2F_1\left(\alpha, \beta+m; \gamma; \frac{ax^\mu}{t^\mu}\right) t^{-\delta-1} f(t) \mathrm{d}t. \qquad (2.7.2)$$

Operators defined by (2.7.1) and (2.7.2) exist under the following conditions:

(i) $1 \leq p, \ q < \infty, \ p^{-1}+q^{-1} = 1, \ |\arg(1-a)| < \pi$
(ii) $\Re(1-\alpha) > m, \Re(\eta) > -\frac{1}{q}, \Re(\delta) > -\frac{1}{p}, \Re(\gamma - \alpha - \beta - m) > -1, m \in \mathbb{N}_0;$
$\gamma \neq 0, -1, -2, \cdots$
(iii) $f \in L_p(0, \infty)$

Equations (2.7.1) and (2.7.2) are introduced by Kalla and Saxena (1969).

For $\gamma = \beta$, (2.7.1) and (2.7.2) reduce to generalized Kober operators, given by Saxena (1967).

Definition 2.7.3.

$$\mathbb{R}[f(x)] = \mathbb{R}\left[\begin{smallmatrix}\alpha,\beta,\gamma;\\ \sigma,\rho,a;\end{smallmatrix} f(x)\right]$$
$$= \frac{x^{-\sigma-\rho}}{\Gamma(\rho)} \int_0^x t^\sigma (x-t)^{\rho-1} {}_2F_1\left[\alpha, \beta; \gamma; a\left(1-\frac{t}{x}\right)\right] f(t) \mathrm{d}t. \qquad (2.7.3)$$

Definition 2.7.4.

$$\mathbb{K}[f(x)] = \mathbb{K}\left[\begin{smallmatrix}\alpha,\beta,\gamma;\\ \delta,\rho,a;\end{smallmatrix} f(x)\right]$$
$$= \frac{x^\delta}{\Gamma(\rho)} \int_x^\infty t^{-\delta-\rho}(t-x)^{\rho-1} {}_2F_1\left[\alpha, \beta; \gamma; a\left(1-\frac{x}{t}\right)\right] f(t) \mathrm{d}t. \qquad (2.7.4)$$

The conditions of validity of the operators (2.7.3) and (2.7.4) are given below:

(i) $p \geq 1$, $q < \infty$, $p^{-1} + q^{-1} = 1$, $|\arg(1-a)| < \pi$.
(ii) $\Re(\sigma) > -\frac{1}{q}$, $\Re(\delta) > -\frac{1}{p}$, $\Re(\rho) > 0$.
(iii) $\gamma \neq 0, -1, -2, \cdots$; $\Re(\gamma - \alpha - \beta) > 0$.
(iv) $f \in L_p(0, \infty)$.

The operators defined by (2.7.3) and (2.7.4) are given by Saxena and Kumbhat (1973). When a is replaced by $\frac{a}{\alpha}$ and α tends to infinity, the operators defined by (2.7.3) and (2.7.4) reduce to the following operators associated with confluent hypergeometric functions.

Definition 2.7.5.

$$\mathbb{R}\left[\begin{smallmatrix}\beta,\gamma;\\\sigma,\rho,a;\end{smallmatrix} f(x)\right] = \lim_{\alpha \to \infty} \mathbb{R}\left[\begin{smallmatrix}\alpha,\beta,\gamma;\\\sigma,\rho,\frac{a}{\alpha};\end{smallmatrix} f(x)\right]$$

$$= \frac{x^{-\sigma-\rho}}{\Gamma(\rho)} \int_0^x \Phi\left[\beta,\gamma;a\left(1-\frac{t}{x}\right)\right] t^\sigma (x-t)^{\rho-1} f(t)\mathrm{d}t. \qquad (2.7.5)$$

Definition 2.7.6.

$$\mathbb{K}\left[\begin{smallmatrix}\beta,\gamma;\\\sigma,\rho,a;\end{smallmatrix} f(x)\right] = \lim_{\alpha \to \infty} \mathbb{K}\left[\begin{smallmatrix}\alpha,\beta,\gamma;\\\delta,\rho,\frac{a}{\alpha};\end{smallmatrix} f(x)\right]$$

$$= \frac{x^\delta}{\Gamma(\rho)} \int_x^\infty \Phi\left[\beta,\gamma;a\left(1-\frac{x}{t}\right)\right] t^{-\delta-\rho} (t-x)^{\rho-1} f(t)\mathrm{d}t, \qquad (2.7.6)$$

where $\Re(\rho) > 0$, $\Re(\delta) > 0$.

Remark 2.7.1: Many interesting and useful properties of the operators defined by (2.7.3) and (2.7.4) are investigated by Saxena and Kumbhat (1975), which deal with relations of these operators with well-known integral transforms, such as Laplace, Mellin and Hankel transforms. Equation (2.7.3) was first considered by Love (1967).

Remark 2.7.2: In the special case, when α is replaced by $\alpha + \beta$, γ by α, σ by zero, ρ by α and β by $-\eta$, then (2.7.3) reduces to the operator (2.7.7) considered by Saigo (1978). Similarly, (2.7.4) reduces to another operator (2.7.9) introduced by Saigo (1978).

Definition 2.7.7. Let $\alpha, \beta, \eta \in \mathbb{C}$, and let $x \in \mathbb{R}_+$ the fractional integral ($\Re(\alpha) > 0$) and the fractional derivative ($\Re(\alpha) < 0$) of the first kind of a function $f(x)$ on $\mathbb{R}_+$ are defined by Saigo (1978) in the form

$$I_{0,x}^{\alpha,\beta,\eta} f(x) = \frac{x^{-\alpha-\beta}}{\Gamma(\alpha)} \int_0^x (x-t)^{\alpha-1}$$

$$\times {}_2F_1\left(\alpha+\beta,-\eta;\alpha;1-\frac{t}{x}\right) f(t)\mathrm{d}t, \quad \Re(\alpha) > 0 \qquad (2.7.7)$$

$$= \frac{\mathrm{d}^n}{\mathrm{d}x^n} I_{0,x}^{\alpha+n,\beta-n,\eta-n} f(x), \quad 0 < \Re(\alpha)+n \leq 1, \quad (n \in \mathbb{N}_0). \qquad (2.7.8)$$

Definition 2.7.8. The fractional integral $(\Re(\alpha) > 0)$ and fractional derivative $(\Re(\alpha) < 0)$ of the second kind of a function $f(x)$ on $\mathbb{R}_+$ are given by Saigo (1978) in the form

$$J_{x,\infty}^{\alpha,\beta,\eta} f(x) = \frac{1}{\Gamma(\alpha)} \int_x^\infty (t-x)^{\alpha-1} t^{-\alpha-\beta}$$

$$\times {}_2F_1\left(\alpha+\beta, -\eta; \alpha; 1-\frac{x}{t}\right) f(t)dt, \ \Re(\alpha) > 0 \qquad (2.7.9)$$

$$= (-1)^n \frac{d^n}{dx^n} J_{x,\infty}^{\alpha+n,\beta-n,\eta} f(x), \ 0 < \Re(\alpha)+n \le 1, \ (n \in \mathbb{N}_0). \quad (2.7.10)$$

Example 2.7.1. Find the value of

$$I_{0,x}^{\alpha,\beta,\eta} \left\{ x^{\sigma-1} \, {}_2F_1(a,b;c;-a'x) \right\}.$$

Solution 2.7.1: We have

$$K = I_{0,x}^{\alpha,\beta,\eta} \left\{ x^{\sigma-1} \, {}_2F_1(a,b;c;-a'x) \right\}$$

$$= \sum_{r=0}^\infty \frac{(a)_r (b)_r (-1)^r (a')^r}{(c)_r r!} I_{0,x}^{\alpha,\beta,\eta} x^{r+\sigma-1}.$$

Applying the result of Exercise 2.7.1, we obtain

$$K = x^{\sigma-\beta-1} \sum_{r=0}^\infty (-1)^r \frac{(a)_r (b)_r}{(c)_r r!} \frac{\Gamma(\sigma+r)\Gamma(\sigma-\beta+\eta+r)(a')^r}{\Gamma(\sigma-\beta+r)\Gamma(\alpha+\eta+\sigma+r)} x^r$$

$$= x^{\sigma-\beta-1} \frac{\Gamma(\sigma)\Gamma(\sigma+\eta-\beta)}{\Gamma(\sigma-\beta)\Gamma(\sigma+\alpha+\eta)}$$

$$\times {}_4F_3(a,b,\sigma,\sigma+\eta-\beta;c,\sigma-\beta,\sigma+\alpha+\eta;-a'x),$$

where $\Re(\alpha) > 0, \Re(\sigma) > 0, \Re(\sigma+\eta-\beta) > 0, c \ne 0, -1, -2, \cdots; |a'x| < 1.$

Example 2.7.2. Find the value of

$$J_{x,\infty}^{\alpha,\beta,\eta} \left(x^\lambda \, {}_2F_1\left(a,b;c; \frac{a'}{x}\right) \right).$$

Solution 2.7.2: Following a similar procedure and using the result of Exercise 2.7.3, it gives

$$J_{x,\infty}^{\alpha,\beta,\eta} \left(x^\lambda \, {}_2F_1\left(a,b;c; \frac{a'}{x}\right) \right) = \frac{\Gamma(\beta-\lambda)\Gamma(\eta-\lambda)}{\Gamma(-\lambda)\Gamma(\alpha+\beta+\eta-\lambda)} x^{\lambda-\beta}$$

$$\times {}_4F_3\left(a,b,\beta-\lambda,\eta-\lambda;c,-\lambda,\alpha+\beta+\eta-\lambda; \frac{a'}{x}\right),$$

where $\Re(\alpha) > 0, \Re(\beta-\lambda) > 0, \Re(\eta-\lambda) > 0, x > 0, c \ne 0, -1, -2, \cdots; |x| > |a'|.$

Remark 2.7.3: Special cases of the operators $I_{0,x}^{\alpha,\beta,\eta}$ and $J_{x,\infty}^{\alpha,\beta,\eta}$ are the operators of Riemann -Liouville:

$$I_{0,x}^{\alpha,-\alpha,\eta} f(x) = {}_0D_x^{-\alpha}f(x) = \frac{1}{\Gamma(\alpha)} \int_0^x (x-t)^{\alpha-1}f(t)dt, \ (\Re(\alpha) > 0) \qquad (2.7.11)$$

the Weyl:

$$J_{x,\infty}^{\alpha,-\alpha,\eta} f(x) = {}_xW_\infty^\alpha f(x) = \frac{1}{\Gamma(\alpha)} \int_x^\infty (t-x)^{\alpha-1}f(t)dt, \ (\Re(\alpha) > 0) \qquad (2.7.12)$$

and the Erdélyi-Kober operators:

$$I_{0,x}^{\alpha,0,\eta} f(x) = E_{0,x}^{\alpha,\eta} f(x) = \frac{x^{-\alpha-\eta}}{\Gamma(\alpha)} \int_0^x (x-t)^{\alpha-1}t^\eta f(t)dt, \ (\Re(\alpha) > 0) \quad (2.7.13)$$

and

$$J_{x,\infty}^{\alpha,0,\eta} f(x) = K_{x,\infty}^{\alpha,\eta} f(x) = \frac{x^\eta}{\Gamma(\alpha)} \int_x^\infty (t-x)^{\alpha-1}t^{-\alpha-\eta} f(t)dt, \ (\Re(\alpha) > 0)$$
$$(2.7.14)$$

Example 2.7.3. Prove the following theorem.

If $\Re(\alpha) > 0$ and $\Re(s) < 1 + \min[0, \Re(\eta - \beta)]$, then the following formula holds for $f(x) \in L_p(0,\infty)$ with $1 \le p \le 2$ or $f(x) \in M_p(0,\infty)$ with $p > 2$:

$$m\left\{ x^\beta I_{0,x}^{\alpha,\beta,\eta} f \right\} = \frac{\Gamma(1-s)\Gamma(\eta-\beta+1-s)}{\Gamma(1-s-\beta)\Gamma(\alpha+\eta+1-s)} m\{f(x)\}. \qquad (2.7.15)$$

Solution 2.7.3: Use the integral

$$\int_x^\infty u^{-\sigma-\gamma}(u-x)^{\gamma-1} {}_2F_1\left(\alpha,\beta;\gamma;1-\frac{x}{u}\right)du = \frac{\Gamma(\gamma)\Gamma(\sigma)\Gamma(\gamma+\sigma-\alpha-\beta)}{\Gamma(\gamma+\sigma-\alpha)\Gamma(\gamma+\sigma-\beta)},$$
$$(2.7.16)$$

where $\Re(\gamma) > 0$, $\Re(\sigma) > 0$, $\Re(\gamma+\sigma-\alpha-\beta) > 0$.

Exercises 2.7.

2.7.1. Prove that

$$I_{0,x}^{\alpha,\beta,\eta} x^\lambda = \frac{\Gamma(1+\lambda)\Gamma(1+\lambda+\eta-\beta)}{\Gamma(1+\lambda-\beta)\Gamma(1+\lambda+\alpha+\eta)} x^{\lambda-\beta}, \qquad (2.7.17)$$

and give the conditions of validity.

2.7.2. Find the Mellin transform of $x^\beta\, J_{x,\infty}^{\alpha,\beta,\eta}\, f(x)$, giving conditions of its validity.

2.7.3. Prove that

$$J_{x,\infty}^{\alpha,\beta,\eta}\, x^\lambda = \frac{\Gamma(\beta-\lambda)\Gamma(\eta-\lambda)}{\Gamma(-\lambda)\Gamma(\alpha+\beta+\eta-\lambda)}x^{\lambda-\beta} \qquad (2.7.18)$$

and give the conditions of validity.

2.7.4. Prove that

$$I_{0,x}^{\alpha,\beta,\eta}\, (x^k e^{-\lambda x}) = \frac{\Gamma(k+1)\Gamma(\eta+k-\beta+1)}{\Gamma(k-\beta+1)\Gamma(\alpha+\eta+k+1)}x^{k-\beta}$$
$$\times\ {}_2F_2(k+1,\eta+k-\beta+1;\ k-\beta+1,\alpha+\eta+k+1;-\lambda x),$$
$$(2.7.19)$$

and give the conditions of validity.

2.7.5. Prove that

$$J_{x,\infty}^{\alpha,\beta,\eta}\, e^{-sx} = s^\eta x^{\eta-\beta}\frac{\Gamma(\beta-\eta)}{\Gamma(\alpha+\beta)}\Phi(1-\alpha-\beta,1+\eta-\beta;-sx)$$
$$+ s^\beta\frac{\Gamma(\eta-\beta)}{\Gamma(\alpha+\eta)}\Phi(1-\alpha-\eta,1+\beta-\eta;-sx), \qquad (2.7.20)$$

and give the conditions of its validity. Deduce the results for $L[\,{}_x W_\infty^\alpha\, f](s)$ and $L[\,K_{x,\infty}^{\alpha,\eta}\, f](s)$.

2.7.6. Prove that [Saxena and Nishimoto (2002)]

$$I_{0,x}^{\alpha,\beta,\eta}\, [x^{\sigma-1}(a+bx)^c] = a^c\frac{\Gamma(\sigma)\Gamma(\sigma+\eta-\beta)}{\Gamma(\sigma-\beta)\Gamma(\sigma+\alpha+\eta)}x^{\sigma-\beta-1}$$
$$\times\ {}_3F_2\left(\sigma,\sigma+\eta-\beta,-c;\ \sigma-\beta,\sigma+\alpha+\eta;-\frac{bx}{a}\right),$$
$$(2.7.21)$$

where $\Re(\sigma) > \max[0,\Re(\beta-\eta)], |\frac{bx}{a}| < 1$.

2.7.7. Evaluate

$$I_{0,x}^{\alpha,\beta,\eta}\,\left\{x^{\sigma-1}H_{p,q}^{m,n}\left[ax^\lambda\,\big|_{(b_q,B_q)}^{(a_p,A_p)}\right]\right\},\ \lambda > 0, \qquad (2.7.22)$$

and give the conditions of its validity.

2.7.8. Evaluate

$$J_{x,\infty}^{\alpha,\beta,\eta}\,\left\{x^{\sigma-1}H_{p,q}^{m,n}\left[ax^{-\lambda}\,\big|_{(b_q,B_q)}^{(a_p,A_p)}\right]\right\},\ \lambda > 0, \qquad (2.7.23)$$

and give the conditions of its validity.

2.7.9. Establish the following property of Saigo operators called "Integration by parts".

$$\int_0^\infty f(x) \left(I_{0,x}^{\alpha,\beta,\eta} g \right)(x)\mathrm{d}x = \int_0^\infty g(x) \left(J_{x,\infty}^{\alpha,\beta,\eta} f \right)(x)\mathrm{d}x.$$

2.7.10. From Exercise 2.7.6, deduce the formula for

$$I_{0,x}^{\alpha,-\alpha,\eta}(a+bx)^c, \tag{2.7.24}$$

given by B. Ross (1993).

2.7.11. Prove that

$$_0I_x^\alpha x^k = \frac{\Gamma(k+1)}{\Gamma(\alpha+k+1)} x^{k+\alpha}, \tag{2.7.25}$$

where $\Re(\alpha) > 0$, $\Re(k) > -1$,

2.7.12. Prove that

$$W_{x,\infty}^\alpha x^k = \frac{\Gamma(-\alpha-k)}{\Gamma(-k)} x^{k+\alpha}, \tag{2.7.26}$$

where $\Re(\alpha) > 0$, $\Re(k) < -\Re(\alpha)$.

2.7.13. Show that

$$J_{x,\infty}^{\alpha,\beta,\eta}(x^\lambda e^{-px}) = x^{\lambda-\beta} G_{2,3}^{3,0}\left[px\Big|_{0,\beta-\lambda,\eta-\lambda}^{-\lambda,\alpha+\beta+\eta-\lambda}\right], \tag{2.7.27}$$

where $G_{2,3}^{3,0}(\cdot)$ is the Meijer's G-function, $\Re(px) > 0$, $\Re(\alpha) > 0$.

Hint: Use the integral

$$e^{-px} = \frac{1}{2\pi i} \int_L \Gamma(-s)(px)^s \mathrm{d}s. \tag{2.7.28}$$

2.7.14. Evaluate

$$I_{0,x}^{\alpha,\beta,\eta} x^{\sigma-1} H_{p,q}^{m,n}\left[ax^{-\lambda}\Big|_{(b_q,B_q)}^{(a_p,A_p)}\right], \quad \lambda > 0, \tag{2.7.29}$$

giving the conditions of its validity.

2.7.15. Evaluate

$$J_{x,\infty}^{\alpha,\beta,\eta} x^{\sigma-1} H_{p,q}^{m,n}\left[ax^{\lambda}\Big|_{(b_q,B_q)}^{(a_p,A_p)}\right], \quad \lambda > 0 \tag{2.7.30}$$

and give the conditions of validity of the result.

2.7.16. With the help of the following chain rules for Saigo operators (Saigo, 1985)

$$I_{0,x}^{\alpha,\beta,\eta} I_{0,x}^{\gamma,\delta,\alpha+\eta} f = I_{0,x}^{\alpha+\gamma,\beta+\delta,\eta} f, \tag{2.7.31}$$

and

$$J_{x,\infty}^{\alpha,\beta,\eta} J_{x,\infty}^{\gamma,\delta,\alpha+\eta} f = J_{x,\infty}^{\alpha+\gamma,\beta+\delta,\eta} f, \tag{2.7.32}$$

derive the inverses

$$(I_{0,x}^{\alpha,\beta,\eta})^{-1} = I_{0,x}^{-\alpha,-\beta,\alpha+\eta}. \tag{2.7.33}$$

and

$$(J_{x,\infty}^{\alpha,\beta,\eta})^{-1} = J_{x,\infty}^{-\alpha,-\beta,\alpha+\eta}. \tag{2.7.34}$$

2.8 Compositions of Riemann-Liouville Fractional Calculus Operators and Generalized Mittag-Leffler Functions

In this section, composition relations between Riemann-Liouville fractional calculus operators and generalized Mittag-Leffler functions are derived. These relations may be useful in the solution of fractional differintegral equations. For details, one can refer to the work of Saxena and Saigo (2005). For ready reference some of the definitions are repeated here.

2.8.1 Composition Relations Between R-L Operators and $E_{\beta,\gamma}^{\delta}(z)$

Notation 2.8.1. $E_{\alpha}(x)$: Mittag-Leffler function.

Notation 2.8.2. $E_{\alpha,\beta}(x)$: Generalized Mittag-Leffler function.

Notation 2.8.3. $I_{0+}^{\alpha} f$: Riemann-Liouville left-sided integral.

Notation 2.8.4. $I_{-}^{\alpha} f$: Riemann-Liouville right-sided integral.

Notation 2.8.5. $D_{0+}^{\alpha} f$: Riemann-Liouville left-sided derivative.

Notation 2.8.6. $D_{-}^{\alpha} f$: Riemann-Liouville right-sided derivative.

Notation 2.8.7. $E_{\beta,\gamma}^{\delta}(z)$: Generalized Mittag-Leffler function (Prabhakar, 1971).

Definition 2.8.1.

$$E_\alpha(z) = \sum_{k=0}^{\infty} \frac{z^k}{\Gamma(\alpha k + 1)}, \quad (\alpha \in C, \Re(\alpha) > 0). \tag{2.8.1}$$

Definition 2.8.2.

$$E_{\alpha,\beta}(z) = \sum_{k=0}^{\infty} \frac{z^k}{\Gamma(\alpha k + \beta)}, \quad (\alpha,\beta \in C, \Re(\alpha) > 0, \Re(\beta) > 0). \tag{2.8.2}$$

Definition 2.8.3.

$$(I_{0+}^\alpha f)(x) = \frac{1}{\Gamma(\alpha)} \int_0^x \frac{f(t)}{(x-t)^{1-\alpha}} dt, \quad \Re(\alpha) > 0. \tag{2.8.3}$$

Definition 2.8.4.

$$(I_-^\alpha f)(x) = \frac{1}{\Gamma(\alpha)} \int_x^\infty \frac{f(t)}{(t-x)^{1-\alpha}} dt, \quad \Re(\alpha) > 0. \tag{2.8.4}$$

Definition 2.8.5.

$$(D_{0+}^\alpha f)(x) = \left(\frac{d}{dx}\right)^{[\alpha]+1} \left(I_{0+}^{1-\{\alpha\}}\right)(x); \quad \Re(\alpha) > 0 \tag{2.8.5}$$

$$= \frac{1}{\Gamma(1-\{\alpha\})} \left(\frac{d}{dx}\right)^{[\alpha]+1} \int_0^x \frac{f(t)}{(x-t)^{\{\alpha\}}} dt, \quad \Re(\alpha) > 0. \tag{2.8.6}$$

Definition 2.8.6.

$$(D_-^\alpha f)(x) = \left(\frac{d}{dx}\right)^{[\alpha]+1} (I_-^{1-\{\alpha\}} f)(x), \quad \Re(\alpha) > 0 \tag{2.8.7}$$

$$= \frac{1}{\Gamma(1-\{\alpha\})} \left(-\frac{d}{dx}\right)^{[\alpha]+1} \int_x^\infty \frac{f(t)}{(t-x)^{\{\alpha\}}} dt, \quad \Re(\alpha) > 0. \tag{2.8.8}$$

Remark 2.8.1: Here $[\alpha]$ means the maximal integer not exceeding α and $\{\alpha\}$ is the fractional part of α. Note that $\Gamma(1-\{\alpha\}) = \Gamma(m-\alpha), [\alpha]+1 = m, \{\alpha\} = 1+\alpha-m$.

Definition 2.8.7.

$$E_{\beta,\gamma}^\delta(z) = \sum_{k=0}^{\infty} \frac{(\delta)_k z^k}{\Gamma(\beta k + \gamma)k!}, \quad (\beta,\gamma,\delta \in C; \Re(\gamma) > 0, \Re(\beta) > 0). \tag{2.8.9}$$

For $\delta = 1$, (2.8.9) reduces to (2.8.2).

Theorem 2.8.1. *Let $\alpha > 0$, $\beta > 0$, $\gamma > 0$ and $\alpha \in \mathbb{R}$. Let I_{0+}^{α} be the left-sided operator of Riemann-Liouville fractional integral (2.8.3). Then there holds the formula*

$$(I_{0+}^{\alpha}[t^{\gamma-1}E_{\beta,\gamma}^{\delta}(at^{\beta})])(x) = x^{\alpha+\gamma-1}E_{\beta,\alpha+\gamma}^{\delta}(ax^{\beta}). \tag{2.8.10}$$

Proof 2.8.1: By virtue of (2.8.3) and (2.8.9), we have

$$K \equiv (I_{0+}^{\alpha}[t^{\gamma-1}E_{\beta,\gamma}^{\delta}(at^{\beta})])(x) = \frac{1}{\Gamma(\alpha)}\int_0^x (x-t)^{\alpha-1}\sum_{n=0}^{\infty}\frac{(\delta)_n a^n t^{n\beta+\gamma-1}}{\Gamma(\beta n+\gamma)n!}\,dt.$$

Interchanging the order of integration and summation and evaluating the inner integral by means of beta-function formula, it gives

$$K \equiv x^{\alpha+\gamma-1}\sum_{n=0}^{\infty}\frac{(\delta)_n(ax^{\beta})^n}{\Gamma(\alpha+\beta n+\gamma)(n)!} = x^{\alpha+\gamma-1}E_{\beta,\alpha+\gamma}^{\delta}(ax^{\beta}).$$

This completes the proof of Theorem 2.8.1.

Corollary 2.8.1: *For $\alpha > 0, \beta > 0, \gamma > 0$ and $\alpha \in \mathbb{R}$, there holds the formula*

$$(I_{0+}^{\alpha}[t^{\gamma-1}E_{\beta,\gamma}(at^{\beta})])(x) = x^{\alpha+\gamma-1}E_{\beta,\alpha+\gamma}(ax^{\beta}). \tag{2.8.11}$$

Remark 2.8.2: For $\beta = \alpha$, (2.8.11) reduces to

$$(I_{0+}^{\alpha}[t^{\gamma-1}E_{\alpha,\gamma}(at^{\alpha})])(x) = \frac{x^{\gamma-1}}{a}\left[E_{\alpha,\gamma}(ax^{\alpha}) - \frac{1}{\Gamma(\gamma)}\right], (a \neq 0) \tag{2.8.12}$$

by virtue of the identity

$$E_{\alpha,\gamma}(x) = \frac{1}{\Gamma(\gamma)} + xE_{\alpha,\alpha+\gamma}(x), (a \neq 0). \tag{2.8.13}$$

Theorem 2.8.2. *Let $\alpha > 0, \beta > 0, \gamma > 0$ and $\alpha \in \mathbb{R}$, $(a \neq 0)$ and let I_{0+}^{α} be the left-sided operator of Riemann-Liouville fractional integral (2.8.3). Then there holds the formula*

$$(I_{0+}^{\alpha}[t^{\gamma-1}E_{\beta,\gamma}^{\delta}(at^{\beta})])(x) = \frac{1}{a}x^{\alpha+\gamma-\beta-1}[E_{\beta,\alpha+\gamma-\beta}^{\delta}(ax^{\beta}) - E_{\beta,\alpha+\gamma-\beta}^{\delta-1}(ax^{\beta})]. \tag{2.8.14}$$

Proof. Use Theorem 2.8.1.

The following two theorems can be established in the same way.

Theorem 2.8.3. *Let $\alpha > 0, \beta > 0, \gamma > 0$ and $\alpha \in \mathbb{R}$ and let I_-^α be the right-sided operator of Riemann-Liouville fractional integral (2.8.4). Then we arrive at the following result:*

$$(I_-^\alpha [t^{-\alpha-\gamma} E_{\beta,\gamma}^\delta (at^{-\beta})])(x) = x^{-\gamma} [E_{\beta,\alpha+\gamma}^\delta (ax^{-\beta})] \qquad (2.8.15)$$

Corollary 2.8.2: *For $\alpha > 0, \beta > 0, \gamma > 0$ and $\alpha \in \mathbb{R}$, there holds the formulas:*

$$(I_-^\alpha [t^{-\alpha-\gamma} E_{\beta,\gamma}(at^{-\beta})])(x) = x^{-\gamma} [E_{\beta,\alpha+\gamma}(ax^{-\beta})] \qquad (2.8.16)$$

and

$$(I_-^\alpha t^{-\alpha-1} E_\beta (at^{-\beta}))(x) = x^{-1} [E_{\beta,\alpha+1}(ax^{-\beta})]. \qquad (2.8.17)$$

Theorem 2.8.4. *Let $\alpha > 0, \beta > 0, \gamma > 0, \alpha \in \mathbb{R}$, $(a \neq 0), \alpha + \gamma > \beta$ and let I_-^α be the right-sided operator of Riemann-Liouville fractional integral (2.8.4). Then there holds the formula*

$$(I_-^\alpha [t^{-\alpha-\gamma} E_{\beta,\gamma}^\delta (at^{-\beta})])(x) = \frac{1}{a} x^{\beta-\gamma} [E_{\beta,\alpha+\gamma-\beta}^\delta (ax^{-\beta}) - E_{\beta,\alpha+\gamma-\beta}^{\delta-1}(ax^{-\beta})]. \qquad (2.8.18)$$

Corollary 2.8.3: *For $\alpha > 0, \beta > 0, \gamma > 0$ with $\alpha + \gamma > \beta$ and for $\alpha \in \mathbb{R}$, $(a \neq 0)$, there holds the formula*

$$(I_-^\alpha [t^{-\alpha-\gamma} E_{\beta,\gamma}(at^{-\beta})])(x) = \frac{1}{a} x^{\beta-\gamma} \left[E_{\beta,\alpha+\gamma-\beta}(ax^{-\beta}) - \frac{1}{\Gamma(\alpha+\gamma-\beta)} \right]. \qquad (2.8.19)$$

Remark 2.8.3: (Kilbas and Saigo, (1998))

$$(I_-^\alpha [t^{-\alpha-\gamma} E_{\alpha,\gamma}(at^{-\alpha})])(x) = \frac{x^{\alpha-\gamma}}{a} \left[E_{\alpha,\gamma}(ax^{-\alpha}) - \frac{1}{\Gamma(\gamma)} \right], \; (a \neq 0) \qquad (2.8.20)$$

$$(I_-^\alpha [t^{-\alpha-1} E_\alpha (at^{-\alpha})])(x) = \frac{x^{\alpha-1}}{a} \left[E_\alpha (ax^{-\alpha}) - 1 \right], \; (a \neq 0). \qquad (2.8.21)$$

Theorem 2.8.5. *Let $\alpha > 0, \beta > 0, \gamma > 0, \gamma > \alpha, \alpha \in \mathbb{R}$ and let D_{0+}^α be the left-sided operator of Riemann -Liouville fractional derivative (2.8.6). Then there holds the formula.*

$$(D_{0+}^\alpha [t^{\gamma-1} E_{\beta,\gamma}^\delta (at^\beta)])(x) = x^{\gamma-\alpha-1} E_{\beta,\gamma-\alpha}^\delta (ax^\beta). \qquad (2.8.22)$$

Proof 2.8.2: By virtue of (2.8.9) and (2.8.6), we have

$$K \equiv (D_{0+}^{\alpha}[t^{\gamma-1}E_{\beta,\gamma}^{\delta}(at^{\beta})])(x) = \left(\frac{d}{dx}\right)^{[\alpha]+1}\left(I_{0+}^{1-\{\alpha\}}\left[t^{\gamma-1}E_{\beta,\gamma}^{\delta}(at^{\beta})\right]\right)(x)$$

$$= \sum_{n=0}^{\infty}\frac{a^n(\delta)_n}{\Gamma(\gamma+n\beta)\Gamma(1-\{a\})n!}\left(\frac{d}{dx}\right)^{[\alpha]+1}\int_0^x t^{n\beta+\gamma-1}(x-t)^{-\{\alpha\}}dt$$

$$= \sum_{n=0}^{\infty}\frac{a^n(\delta)_n}{\Gamma(\gamma+n\beta+1-\{\alpha\})n!}\left(\frac{d}{dx}\right)^{[\alpha]+1}x^{n\beta+\gamma-\{\alpha\}}$$

$$= \sum_{n=0}^{\infty}\frac{a^n(\delta)_n x^{\gamma+n\beta-\alpha-1}}{\Gamma(n\beta+\gamma-\alpha)n!} = x^{\gamma-\alpha-1}E_{\beta,\gamma-\alpha}^{\delta}(ax^{\beta}),$$

which proves the theorem.

By using a similar procedure, we arrive at the following theorem.

Theorem 2.8.6. *Let $\alpha > 0, \gamma > \beta > 0, \alpha \in \mathbb{R}$, $(a \neq 0)$, $\gamma > \alpha + \beta$ and let D_{0+}^{α} be the left-sided operator of Riemann-Liouville fractional derivative (2.8.6). Then there holds the formula*

$$\left(D_{0+}^{\alpha}[t^{\gamma-1}E_{\beta,\gamma}^{\delta}(at^{\beta})]\right)(x) = \frac{1}{a}x^{\gamma-\alpha-\beta-1}\left[E_{\beta,\gamma-\alpha-\beta}^{\delta}(ax^{\beta}) - E_{\beta,\gamma-\alpha-\beta}^{\delta-1}(ax^{\beta})\right].$$
(2.8.23)

Corollary 2.8.4: *Let $\alpha > 0, \gamma > \beta > 0, \alpha \in \mathbb{R}$, $(a \neq 0)$, $\gamma > \alpha + \beta$, then there holds the formula.*

$$\left(D_{0+}^{\alpha}[t^{\gamma-1}E_{\beta,\gamma}(at^{\beta})]\right)(x) = \frac{1}{a}x^{\gamma-\alpha-\beta-1}\left[E_{\beta,\gamma-\alpha-\beta}(ax^{\beta}) - \frac{1}{\Gamma(\gamma-\alpha-\beta)}\right].$$
(2.8.24)

Theorem 2.8.7. *Let $\alpha > 0, \gamma > 0, \gamma - \alpha > 0$ with $\gamma - \alpha + \{\alpha\} > 1, \alpha \in \mathbb{R}$, and let D_-^{α} be the right-sided operator of Riemann-Liouville fractional derivative (2.8.8). Then there holds the formula.*

$$\left(D_-^{\alpha}[t^{\alpha-\gamma}E_{\beta,\gamma}^{\delta}(at^{-\beta})]\right)(x) = x^{-\gamma}E_{\beta,\gamma-\alpha}^{\delta}(ax^{-\beta}).$$
(2.8.25)

Theorem 2.8.8. *Let $\alpha > 0, \beta > 0$ with $\gamma - \{\alpha\} > 1$, $\alpha \in \mathbb{R}$, $\gamma > \alpha + \beta$, $(a \neq 0)$ and let D_-^{α} be the right-sided operator of Riemann-Liouville fractional derivative (2.8.8). Then there holds the formula*

$$\left(D_-^{\alpha}[t^{\alpha-\gamma}E_{\beta,\gamma}^{\delta}(at^{-\beta})]\right)(x) = \frac{x^{\beta-\gamma}}{a}\left[E_{\beta,\gamma-\alpha-\beta}^{\delta}(ax^{-\beta}) - E_{\beta,\gamma-\alpha-\beta}^{\delta-1}(ax^{-\beta})\right].$$
(2.8.26)

Exercises   2.8.

2.8.1. Show that

$$ax^\beta E^\delta_{\beta,\gamma}(ax^\beta) = E^\delta_{\beta,\gamma-\beta}(ax^\beta) - E^{\delta-1}_{\beta,\gamma-\beta}(ax^\beta), (a \neq 0) \qquad (2.8.27)$$

2.8.2. Show that

$$\left(I^\alpha_{0+}[t^{\gamma-1}E_{\alpha,\gamma}(at^\alpha)]\right)(x) = \frac{x^{\gamma-1}}{a}\left[E_{\alpha,\gamma}(ax^\alpha) - \frac{1}{\Gamma(\gamma)}\right], (a \neq 0). \qquad (2.8.28)$$

2.8.3. Prove Theorem 2.8.3.

2.8.4. Prove Theorem 2.8.4.

2.8.5. Prove Theorem 2.8.6.

2.8.6. Prove Theorem 2.8.7.

2.8.7. Prove Theorem 2.8.8.

2.8.8. Prove that

$$\left(I^\alpha_{0+}t^\omega H^{m,n}_{p,q}\left[t^\sigma\Big|^{(a_p,A_p)}_{(b_q,B_q)}\right]\right)(x) = x^{\omega+\alpha}H^{m,n+1}_{p+1,q+1}\left[x^\sigma\Big|^{(-\omega,\sigma),(a_p,A_p)}_{(b_q,B_q),(-\omega-\alpha,\sigma)}\right], \qquad (2.8.29)$$

giving conditions of validity.

2.8.9. Evaluate

$$\left(I^\alpha_- t^\omega H^{m,n}_{p,q}\left[t^\sigma\Big|^{(a_p,A_p)}_{(b_q,B_q)}\right]\right)(x), \qquad (2.8.30)$$

and give the conditions of validity.

2.9 Fractional Differential Equations

Differential equations contain integer order derivatives, whereas fractional differential equations involve fractional derivatives, like $\frac{d^\alpha}{dx^\alpha}$, which are defined for $\alpha > 0$. Here α is not necessarily an integer and can be rational, irrational or even complex-valued. Today, fractional calculus models find applications in physical, biological, engineering, biomedical and earth sciences. Most of the problems discussed involve relaxation and diffusion models in the so called complex or disordered systems. Thus, it gives rise to the generalization of initial value problems involving ordinary differential equations to generalized fractional-order differential equations and Cauchy problems involving partial differential equations to fractional reaction, fractional diffusion and fractional reaction-diffusion equations. Fractional calculus plays a dominant role in the solution of all these physical problems.

2.9.1 Fractional relaxation

In order to formulate a relaxation process, we require a physical law, say the relaxation equation

$$\frac{d}{dt}f(t) + \frac{1}{c}f(t) = 0, t > 0, c > 0, \tag{2.9.1}$$

to be solved for the initial value $f(t = 0) = f_0$. The unique solution of (2.9.1) is given by

$$f(t) = f_0 \, e^{-\frac{t}{c}}, t \geq 0, c > 0. \tag{2.9.2}$$

Now the problem is as to how we can generalize the initial-value problem (2.9.1) into a fractional value problem with physical motivation. If we incorporate the initial value f_0 into the integrated relaxation equation (2.9.1), we find that

$$f(t) - f_0 = -\frac{1}{c} \, _0D_t^{-1} f(t), \tag{2.9.3}$$

where $_0D_t^{-1}$ is the standard Riemann integral of $f(t)$. On replacing $\frac{1}{c} \, _0D_t^{-1} f(t)$ by $\frac{1}{c^\alpha} \, _0D_t^{-\alpha} f(t)$, it yields the fractional integral equation

$$f(t) - f_0 = - \left(\frac{1}{c^\alpha} \right) \, _0D_t^{-\alpha} f(t), \alpha > 0 \tag{2.9.4}$$

with initial value

$$f_0 = f(t = 0).$$

Applying the Riemann-Liouville differential operator $_0D_t^\alpha$ from the left and making use of the formula (2.4.16), we arrive at

$$_0D_t^\alpha [f(x) - f_0] = -c^{-\alpha} f(t), \ \alpha > 0, c > 0, \tag{2.9.5}$$

with initial condition $f_0 = f(t = 0)$.

Theorem 2.9.1. *The solution of the fractional differential equation (2.9.4) is given by*

$$f(t) = f_0 H_{1,2}^{1,1} \left[\left(\frac{t}{c} \right)^\alpha \Big|_{(0,1),(0,\alpha)}^{(0,1)} \right], \tag{2.9.6}$$

where $\alpha > 0, c > 0$.

Proof 2.9.1: If we apply the Laplace transform to equation (2.9.4), it gives

$$F(s) - f_0 s^{-1} = -\frac{1}{c^\alpha} s^{-\alpha} F(s), \tag{2.9.7}$$

where we have used the result (2.4.7) and $F(s)$ is the Laplace transform of $f(t)$. Solving for $F(s)$, we have

$$F(s) = L\{f(t)\} = f_0 \left[\frac{s^{-1}}{1 + (cs)^{-\alpha}} \right]. \tag{2.9.8}$$

Taking inverse Laplace transform, (2.9.8) gives

$$\begin{aligned}
f(t) = L^{-1}\{F(s)\} &= f_0 L^{-1} \left[\frac{s^{-1}}{1 + (cs)^{-\alpha}} \right] \\
&= f_0 L^{-1} \left[\sum_{k=0}^{\infty} (-1)^k c^{-\alpha k} s^{-\alpha k - 1} \right] \\
&= f_0 \sum_{k=0}^{\infty} \frac{(-1)^k (\frac{t}{c})^{\alpha k}}{\Gamma(\alpha k + 1)} \\
&= f_0 E_\alpha \left[-\left(\frac{t}{c}\right)^\alpha \right], \tag{2.9.9}
\end{aligned}$$

where $E_\alpha(\cdot)$ is the Mittag-Leffler function. (2.9.9) can be written in terms of the H-function as

$$f(t) = f_0 H_{1,2}^{1,1} \left[\left(\frac{t}{c}\right)^\alpha \Big|_{(0,1),(0,\alpha)}^{(0,1)} \right], \tag{2.9.10}$$

where $c > 0, \alpha > 0$. This completes the proof of the Theorem 2.9.1.

Alternative form of the solution. By virtue of the identity

$$H_{p,q}^{m,n} \left[x^\mu \Big|_{(b_q,B_q)}^{(a_p,A_p)} \right] = \frac{1}{\mu} H_{p,q}^{m,n} \left[x \Big|_{(b_q,\frac{B_q}{\mu})}^{(a_p,\frac{A_p}{\mu})} \right], \ (\mu > 0) \tag{2.9.11}$$

the solution (2.9.10) can be written as

$$f(t) = \frac{f_0}{\alpha} H_{1,2}^{1,1} \left[\frac{t}{c} \Big|_{(0,\frac{1}{\alpha}),(0,1)}^{(0,\frac{1}{\alpha})} \right], \tag{2.9.12}$$

where $\alpha > 0, c > 0$.

Remark 2.9.1: In the limit as $\alpha \to 1$, one recovers the result (2.9.2)

$$f(t) = f_0 \exp\left(-\frac{t}{c}\right) = f_0 E_1 \left(\frac{t}{c}\right). \tag{2.9.13}$$

Remark 2.9.2: In terms of Wright's function, the solution (2.9.10) can be expressed in the form

$$f(t) = f_0 \, {}_1\psi_1 \left[{}_{(1,\alpha)}^{(1,1)} \Big| ; (\frac{t}{c})^\alpha \right], \tag{2.9.14}$$

where $\alpha > 0, c > 0$.

In a similar manner, we can establish Theorems 2.9.2 and 2.9.3 given below.

Theorem 2.9.2. *The solution of the fractional integral equation*

$$N(t) - N_0 t^{\mu-1} = -c^\nu {}_0 D_t^{-\nu} N(t), \qquad (2.9.15)$$

is given by

$$N(t) = N_0 \Gamma(\mu) t^{\mu-1} E_{\nu,\mu}(-c^\nu t^\mu), \qquad (2.9.16)$$

where $E_{\nu,\mu}(\cdot)$ is the generalized Mittag-Leffler function (2.1.2), $\nu > 0, \mu > 0$.

Remark 2.9.3: When $\mu = 1$, we obtain the result given by Haubold and Mathai (2000).

Theorem 2.9.3. *If $c > 0, \nu > 0, \mu > 0$, then for the solution of the integral equation*

$$N(t) - N_0 t^{\mu-1} E_{\nu,\mu}^\gamma[-(ct)^\nu] = -c^\nu {}_0 D_t^{-\nu} N(t), \qquad (2.9.17)$$

there holds the formula

$$N(t) = N_0 t^{\mu-1} E_{\nu,\mu}^{\gamma+1}[-(ct)^\nu]. \qquad (2.9.18)$$

Hint: Use the formula

$$L^{-1}\left\{s^{-\beta}(1 - as^{-\alpha})^{-\gamma}\right\} = t^{\beta-1} E_{\alpha,\beta}^\gamma(at^\alpha), \qquad (2.9.19)$$

where $\Re(\alpha) > 0, \Re(\beta) > 0, \Re(s) > |a|^{\frac{1}{\Re(\alpha)}}, \Re(s) > 0$.

Corollary 2.9.1: *If $c > 0, \mu > 0, \nu > 0$, then for the solution of*

$$N(t) - N_0 t^{\mu-1} E_{\nu,\mu}[-c^\nu t^\nu] = -c^\nu {}_0 D_t^{-\nu} N(t), \qquad (2.9.20)$$

there holds the relation

$$N(t) = \frac{N_0}{\nu} t^{\mu-1} \left[E_{\nu,\mu-1}(-c^\nu t^\nu) + (1 + \nu - \mu) E_{\nu,\mu}(-c^\nu t^\nu) \right]. \qquad (2.9.21)$$

Theorem 2.9.4. *The Cauchy problem for the integro-differential equation*

$$_0 D_x^\mu f(x) + \lambda \, _0 D_x^{-\nu} f(x) = h(x), \quad (\lambda, \mu, \nu \in \mathbb{C}) \qquad (2.9.22)$$

with the initial condition

$$D_x^{\mu-k-1} f(0) = a_k, k = 0, 1, \cdots, [\mu], \qquad (2.9.23)$$

where $\Re(\nu) > 0, \Re(\mu) > 0$ and $h(x)$ is any integrable function on the finite interval $[0,b]$ has the unique solution, given by

$$f(x) = \int_0^x (x-t)^{\mu-1} E_{\mu+\nu,\mu}[-\lambda(x-t)^{\mu+\nu}] h(t) dt$$

$$+ \sum_{k=0}^{n-1} a_k x^{\mu-k-1} E_{\mu+\nu,\mu-k}(-\lambda x^{\mu+\nu}) \qquad (2.9.24)$$

Proof 2.9.2: Exercise.

Theorem 2.9.5. *The solution of the equation*

$$_0D_t^{\frac{1}{2}} f(t) + bf(t) = 0; \quad \left[_0D_t^{-\frac{1}{2}} f(t) \right]_{t=0} = C, \tag{2.9.25}$$

where C is a constant is given by

$$f(t) = C t^{-\frac{1}{2}} E_{\frac{1}{2},\frac{1}{2}} \left(-bt^{\frac{1}{2}} \right), \tag{2.9.26}$$

where $E_{\frac{1}{2},\frac{1}{2}}(\cdot)$ is the Mittag-Leffler function.

Proof 2.9.3: Exercise see (2.4.47).

Remark 2.9.4: Theorem 2.9.5 gives the generalized form of the equation solved by Oldham and Spanier (1974).

Exercises 2.9.

2.9.1. Prove that if $c > 0, v > 0, \mu > 0$, then the solution of

$$N(t) - N_0 t^{\mu-1} E_{v,\mu}^2(c^v t^v) = -c^v {}_0D_t^{-v} N(t), \tag{2.9.27}$$

is given by

$$N(t) = N_0 t^{\mu-1} E_{v,\mu}^3(-c^v t^v) = \frac{N_0 t^{\mu-1}}{2v^2} \Big[E_{v,\mu-2}(-c^v t^v)$$
$$+ \{3(v+1) - 2\mu\} E_{v,\mu-1}(-c^v t^v)$$
$$+ \{2v^2 + \mu^2 + 3v - 2\mu - 3v\mu + 1\} E_{v,\mu}(-c^v t^v) \Big], \tag{2.9.28}$$

where $\Re(v) > 0, \ \Re(\mu) > 2$.

2.9.2. Prove that if $v > 0, c > 0, d > 0, \mu > 0, c \neq d$, then for the solution of the equation

$$N(t) - N_0 t^{\mu-1} E_{v,\mu}(-d^v t^v) = -c^v {}_0D_t^{-v} N(t), \tag{2.9.29}$$

there holds the formula.

$$N(t) = N_0 \frac{t^{\mu-v-1}}{c^v - d^v} \left[E_{v,\mu-v}(-d^v t^v) - E_{v,\mu-v}(-c^v t^v) \right]. \tag{2.9.30}$$

2.9.3. Prove that if $c > 0, v > 0, \mu > 0$, then for the solution of the equation

$$N(t) - N_0 t^{\mu-1} E_{v,\mu}(-c^v t^v) = -c^v {}_0 D_t^{-v} N(t),\tag{2.9.31}$$

the following result holds:

$$N(t) = \frac{N_0}{v} t^{\mu-1} \left[E_{v,\mu-1}(-c^v t^v) + (1+v-\mu) E_{v,\mu}(-c^v t^v) \right].\tag{2.9.32}$$

2.9.4. Solve the equation

$${}_0 D_t^Q f(t) + {}_0 D_t^q f(t) = g(t),$$

where $q - Q$ is not an integer or a half integer and the initial condition is

$$\left[{}_0 D_t^{q-1} f(t) + {}_0 D_t^{Q-1} f(t) \right]_{t=0} = C\tag{2.9.33}$$

where C is a constant.

2.9.5. Solve the equation

$${}_0 D_t^\alpha x(t) - \lambda x(t) = h(t), \quad (t > 0),\tag{2.9.34}$$

subject to the initial conditions

$$\left[{}_0 D_t^{\alpha-k} h(t) \right]_{t=0} = b_k, \quad (k = 1, \cdots, n)\tag{2.9.35}$$

where $n - 1 < \alpha < n$.

2.9.6. Prove Theorem 2.9.4.

2.9.7. Prove Theorem 2.9.5.

2.9.2 Fractional diffusion

Theorem 2.9.6. *The solution of the following initial value problem for the fractional diffusion equation in one dimension*

$${}_0 D_t^\alpha U(x,t) = \lambda^2 \frac{\partial^2 U(x,t)}{\partial x^2}, \quad (t > 0, -\infty < x < \infty)\tag{2.9.36}$$

with initial conditions :

$$\lim_{x \to \pm\infty} U(x,t) = 0; \left[{}_0 D_t^{\alpha-1} U(x,t) \right]_{t=0} = \phi(x)\tag{2.9.37}$$

is given by

$$U(x,t) = \int_{-\infty}^{\infty} G(x-\zeta,t)\phi(\zeta)d\zeta, \tag{2.9.38}$$

where

$$G(x,t) = \frac{1}{\pi} \int_0^{\infty} t^{\alpha-1} E_{\alpha,\alpha}(-k^2\lambda^2 t^{\alpha}) \cos kx \, dk. \tag{2.9.39}$$

Solution 2.9.1: Let $0 < \alpha < 1$. Using the boundary conditions (2.9.37), the Fourier transform of (2.9.36) with respect to variable x gives

$$_0D_x^{\alpha} \bar{U}(k,t) + \lambda^2 k^2 \bar{U}(k,t) = 0 \tag{2.9.40}$$

$$\left[_0D_t^{\alpha-1} \bar{U}(k,t)\right]_{t=0} = \bar{\phi}(k), \tag{2.9.41}$$

where k is a Fourier transform parameter and ' $-$ ' indicates Fourier transform. Applying the Laplace transform to (2.9.40) and using (2.9.41), it gives

$$\tilde{\bar{U}}(k,s) = \frac{\bar{\phi}(k)}{s^{\alpha} + k^2\lambda^2}, \tag{2.9.42}$$

where ' $\sim$ ' indicates Laplace transform. The inverse Laplace transform of (2.9.42) yields

$$\bar{U}(k,t) = t^{\alpha-1}\bar{\phi}(k)E_{\alpha,\alpha}(-\lambda^2 k^2 t^2), \tag{2.9.43}$$

and then the solution is obtained by taking inverse Fourier transform. By taking inverse Fourier transform of (2.9.43) and using the formula

$$\frac{1}{2\pi} \int_{-\infty}^{\infty} e^{-ikx} f(k)dk = \frac{1}{\pi} \int_0^{\infty} f(k)\cos(kx)dk \tag{2.9.44}$$

we have

$$U(x,t) = \int_{-\infty}^{\infty} G(x-\zeta,t)\phi(\zeta)d\zeta, \tag{2.9.45}$$

where

$$G(x,t) = \frac{1}{\pi} \int_0^{\infty} t^{\alpha-1} E_{\alpha,\alpha}(-k^2\lambda^2 t^{\alpha}) \cos(kx)dk \tag{2.9.46}$$

with $\Re(\alpha) > 0, k > 0$.

Exercises 2.9.

2.9.8. Evaluate the integral in (2.9.46).

2.9.9. Find the solution of the Fick's diffusion equation

$$\frac{\partial}{\partial t}P(x,t) = \lambda \frac{\partial^2}{\partial x^2}P(x,t),$$

with the initial condition $P(x,t=0) = \delta(x)$, where $\delta(x)$ is the Dirac delta function.

References

Agarwal, R. P. (1953). A Propos d'une note de M. Pierre Humbert, *C. R. Acad. Sci. Paris*, **296**, 2031-2032.

Agarwal, R. P. (1963). *Generalized Hypergeometric Series*, Asia Publishing House, Bombay, London and New York.

Caputo, M. (1969). *Elasticitá e Dissipazione*, Zanichelli, Bologna.

Dzherbashyan, M.M. (1966). *Integral Transforms and Representation of Functions in Complex Domain* (in Russian), Nauka, Moscow.

Erdélyi, A. (1950-51). On some functional transformations, *Univ. Politec. Torino, Rend. Sem. Mat.* **10**, 217-234.

Erdélyi, A., Magnus, W., Oberhettinger, F. and Tricomi, F. G. (1953). *Higher Transcendental Functions*, Vol. 1, McGraw - Hill, New York, Toronto and London.

Erdélyi, A., Magnus, W., Oberhettinger, F. and Tricomi, F. G. (1954). *Tables of Integral Transforms*, Vol. 1, McGraw - Hill, New York, Toronto and London.

Erdélyi, A., Magnus, W., Oberhettinger, F. and Tricomi, F. G. (1954a). *Tables of Integral Transforms*, Vol. 2, McGraw - Hill, New York, Toronto and London.

Erdélyi, A., Magnus, W., Oberhettinger, F. and Tricomi, F. G. (1955). *Higher Transcendental Functions*, Vol. 3, McGraw - Hill, New York, Toronto and London.

Fox, C. (1963). Integral transforms based upon fractional integration, *Proc. Cambridge Philos. Soc.*, **59**, 63-71.

Haubold, H. J. and Mathai, A. M. (2000). The fractional kinetic equation and thermonuclear functions, *Astrophysics and Space Science*, **273**, 53-63.

Hilfer, R. (Ed.). (2000). *Applications of Fractional Calculus in Physics*, World Scientific, Singapore.

Kalla, S. L. and Saxena, R. K. (1969). Integral operators involving hypergeometric functions, *Math. Zeitschr.*, **108**, 231-234.

Kilbas. A. A. and Saigo, M. (1998). Fractional calculus of the H-function, *Fukuoka Univ. Science Reports*, **28**, 41-51.

Kilbas, A. A. and Saigo, M. (1996). On Mittag- Leffler type function, fractional calculus operators and solutions of integral equations, *Integral Transforms and Special Functions*, **4**, 355-370.

Kilbas, A. A, Saigo, M. and Saxena, R. K. (2002). Solution of Volterra integrodifferential equations with generalized Mittag-Leffler function in the kernels, *J. Integral Equations and Applications*, **14**, 377-396.

Kilbas, A. A., Saigo, M. and Saxena, R. K. (2004). Generalized Mittag-Leffler function and generalized fractional calculus operators, *Integral Transforms and Special Functions*, **15**, 31-49.

Kober, H. (1940). On fractional integrals and derivatives, *Quart. J. Math. Oxford*, Ser. ll, 193-211.

Love, E. R (1967). Some integral equations involving hypergeometric functions, *Proc. Edin. Math. Soc.*, **15(2)**, 169-198.

Mathai, A. M. and Saxena, R. K. (1973). *Generalized Hypergeometric Functions with Applications in Statistics and Physical Sciences*, Lecture Notes in Mathematics, **348**, Springer- Verlag, Berlin, Heidelberg.

Mathai, A. M. and Saxena, R. K. (1978). *The H-function with Applications in Statistics and Other Disciplines*, John Wiley and Sons, New York - London - Sydney.

Miller, K. S. and Ross, B. (1993). *An Introduction to the Fractional Calculus and Fractional Differential Equations*, Wiley, New York.

Mittag-Leffler, G. M. (1903). Sur la nouvelle fonction $E_\alpha(x)$, *C. R. Acad. Sci. Paris*, (Ser. II) **137**, 554-558.

Oldham, K. B. and Spanier, J. (1974). *The Fractional Calculus: Theory and Applications of Differentiation and Integration to Arbitrary Order*, Academic Press, New York.

Podlubny, 1. (1999). *Fractional Differential Equations*, Academic Press, San Diego.

Podlubny, 1. (2002). Geometric and physical interpretations of fractional integration and fractional differentiation, *Frac. Calc. Appl. Anal.*, **5(4)**, 367-386.

Prabhakar, T. R. (1971). A singular integral equation with a generalized Mittag- Leffler function in the kernel, *Yokohama Math. J.*, **19**, 7-15.

Ross, B. (1994). A formula for the fractional integration and differentiation of $(a+bx)^c$, *J. Fract. Calc.*, **5**, 87-89.

Saigo, M. (1978). A remark on integral operators involving the Gauss hypergeometric function, *Math. Reports of College of Gen. Edu., Kyushu University*, **11**, 135-143.

Saigo, M. and Raina, R. K. (1988). Fractional calculus operators associated with a general class of polynomials, *Fukuoka Univ. Science Reports,* **18**, 15-22.

Samko, S. G., Kilbas, A. A. and Marichev, 0. 1. (1993). *Fractional Integrals and Derivatives, Theory and Applications*, Gordon and Breach, Reading.

Saxena, R. K. (1967). On fractional integration operators, *Math. Zeitsch.*, **96**, 288-291.

Saxena, R. K. (2002). Certain properties of generalized Mittag-Leffler function, *Proceedings of the Third Annual Conference of the Society for Special Functions and Their Applications*, Varanasi, March 4-6, 75-81.

Saxena, R. K. (2003). Alternative derivation of the solution of certain integro-differential equations of Volterra-type, *Ganita Sandesh,* **17(1)**, 51-56.

Saxena,R. K. (2004). On a unified fractional generalization of free electron laser equation, *Vijnana Parishad Anusandhan Patrika,* **47(l)**, 17-27.

Saxena, R. K. and Kumbhat, R. K. (1973). A generalization of Kober operators, *Vijnana Parishad Anusandhan Patrika*, **16**, 31-36.

Saxena, R. K. and Kumbhat, R. K. (1974). Integral operators involving H-function, *Indian J. Pure appl. Math.*, **5**, 1-6.

Saxena, R. K. and Kumbhat, R. K. (1975). Some properties of generalized Kober operators, *Vijnana Parishad Anusandhan Patrika,* **18**, 139-150.

Saxena, R. K, Mathai, A. M and Haubold, H. J. (2002). On fractional kinetic equations, *Astrophysics and Space Science,* **282**, 281-287.

Saxena, R. K, Mathai, A. M and Haubold, H. J. (2004). On generalized fractional kinetic equations, *Physica A* , **344**, 657-664.

Saxena, R. K, Mathai, A. M. and Haubold, H. J. (2004). Unified fractional kinetic equations and a fractional diffusion equation, *Astrophysics and Space Science,* **290**, 241-245.

Saxena, R. K. and Nishimoto, K. (2002). On a fractional integral formula of Saigo operator, *J. Fract. Calc.*, **22**, 57-58.

Saxena, R. K. and Saigo, M. (2005). Certain properties of fractional calculus operators associated with generalized Mittag-Leffler function. *Frac. Calc. Appl. Anal.*, **8(2)**, 141-154.

Sneddon, I. N. (1975). *The Use in Mathematical Physics of Erdélyi- Kober Operators and Some of Their Applications*, Lecture Notes in Mathematics (Edited by B. Ross), **457**, 37-79.

Srivastava, H. M. and Saxena, R. K. (2001). Operators of fractional integration and their applications, *Appl. Math. Comput.*, **118**, 1-52.

Srivastava, H. M. and Karlsson, P. W. (1985). *Multiple Gaussian Hypergeometric Series*, Ellis Horwood, Chichester, U.K.

Stein, E. M. (1970). *Singular Integrals and Differential Properties of Functions*, Princeton University Press, New Jersey.

Wiman, A. (1905). Uber den Fundamental satz in der Theorie de Funktionen $E_\alpha(x)$. *Acta Math.*, **29**, 191-201.

Weyl, H. (1917). Bemerkungen zum Begriff des Differentialquotienten gebrochener Ordnung, *Vierteljahresschr. Naturforsch. Gen. Zurich,* **62**, 296-302.

Chapter 3
An Introduction to q-Series

[This chapter is based on the lectures of Dr. R. Jagannathan of the Institute of Mathematical Sciences, Chennai, India, at the 2nd SERC School]

3.0 Introduction

The development of quantum groups and their applications in mathematics and physics, starting from 1980's, has lead to renewed interest in the subject of q-series with a history starting in the 19th century. Here we shall start learning how to deal with the q-series.

3.1 Hypergeometric Series

Hypergeometric series is a systematic generalization of the geometric series $1 + z + z^2 + \cdots$. The shifted factorial, or the Pochhammer symbol, is defined by

$$(\alpha)_n = \begin{cases} 1, \ n = 0, \ \alpha \neq 0 \\ \alpha(\alpha+1)(\alpha+2)\dots(\alpha+n-1), \ n = 1,2,\dots. \end{cases} \tag{3.1.1}$$

Gauss' hypergeometric series is given by

$$\begin{aligned} {}_2F_1(\alpha,\beta;\gamma;z) &= 1 + \frac{\alpha\beta}{(1)(\gamma)}z + \frac{\alpha(\alpha+1)\beta(\beta+1)}{(1)(2)\gamma(\gamma+1)}z^2 + \cdots \\ &= \sum_{n=0}^{\infty} \frac{(\alpha)_n(\beta)_n}{(\gamma)_n}\frac{z^n}{n!}, \end{aligned} \tag{3.1.2}$$

where it is assumed that $\gamma \neq 0, -1, -2, \dots$ so that no zero factors appear in the denominator terms of the series. This series (3.1.2) converges absolutely for $|z| < 1$, and for $|z| = 1$ when $\Re(\gamma - \alpha - \beta) > 0$. The geometric series is a special case when $\alpha = 1, \gamma = \beta$:

$$_2F_1(1,\beta;\beta;z) = 1 + z + z^2 + \cdots \tag{3.1.3}$$

From this it is clear that we can write

$$_2F_1(1,\beta;\beta;z) = \frac{1}{1-z}, \text{ for } |z| < 1. \tag{3.1.4}$$

The generalized hypergeometric series with r numerator parameters and s denominator parameters is defined by

$$_rF_s(\alpha_1,\alpha_2,\ldots,\alpha_r;\beta_1,\beta_2,\ldots,\beta_s;z) = \sum_{n=0}^{\infty} \frac{(\alpha_1)_n(\alpha_2)_n\ldots(\alpha_r)_n}{(\beta_1)_n(\beta_2)_n\ldots(\beta_s)_n} \frac{z^n}{n!}. \tag{3.1.5}$$

This series converges absolutely for all z if $r \leq s$, and for $|z| < 1$ if $r = s+1$. It converges absolutely for $|z| = 1$ if $r = s+1$ and $\Re\left[(\beta_1 + \beta_2 + \cdots + \beta_s) - (\alpha_1 + \alpha_2 + \cdots + \alpha_r)\right] > 0$. If $r > s+1$ and $z \neq 0$ or $r = s+1$ and $|z| > 1$, then it diverges, unless it terminates. The simplest case of an $_rF_s$ series is

$$_0F_0(\ ;\ ;z) = e^z \tag{3.1.6}$$

where a space indicates the absence of numerator/denominator parameters.

Exercises 3.1.

3.1.1. Show that an $_rF_s$ series terminates if one of its numerator parameters is a negative integer, or zero (trivial case).

3.1.2. Verify that

$$_1F_0(-\alpha;\ ;-z) = (1+z)^\alpha.$$

3.1.3. Show that

$$\lim_{\alpha\to\infty} {_1F_0}\left(\alpha;\ ;\frac{z}{\alpha}\right) = e^z.$$

3.1.4. Verify that

$$_2F_1(1,1;2;-z) = \frac{1}{z}\ln(1+z).$$

3.1.5. Verify that

$$_0F_1\left(\ ;\frac{3}{2};-\frac{z^2}{4}\right) = \frac{1}{z}\sin z.$$

3.1.6. Verify that

$$_0F_1\left(;\frac{1}{2};-\frac{z^2}{4}\right) = \cos z.$$

3.1.7. Verify that

$$_2F_1\left(\frac{1}{2},\frac{1}{2};\frac{3}{2};z^2\right) = \frac{1}{z}\sin^{-1}z.$$

3.1.8. Verify that

$$_2F_1\left(\frac{1}{2},1;\frac{3}{2};-z^2\right) = \frac{1}{z}\tan^{-1}z.$$

3.1.9. Show that for $n = 0,1,2,\ldots$

$$H_n(z) = (2z)^n {_2F_0}\left(-\frac{n}{2},\frac{(1-n)}{2};\ ;-z^{-2}\right)$$

represents the Hermite polynomial given by

$$H_n(z) = \sum_{k=0}^{[n/2]} \frac{(-1)^k n!}{k!(n-2k)!}(2z)^{n-2k}$$

where $[n/2]$ is the integer part of $n/2$.

3.1.10. Show that for $n = 0,1,2,\ldots$

$$L_n(z) = {_1F_1}(-n;1;z)$$

represents the Laguerre polynomial given by

$$L_n(z) = \sum_{k=0}^{n} \binom{n}{k}\frac{(-z)^k}{k!},$$

where $\binom{n}{k}$ is the binomial coefficient $n(n-1)\ldots(n-k+1)/k!$.

3.1.11. Show that for $n = 0,1,2,\ldots$

$$P_n(z) = \frac{(2n)!}{2^n(n!)^2}z^n\ {_2F_1}\left(-\frac{n}{2},\frac{(1-n)}{2};\frac{1}{2}-n;z^{-2}\right)$$

represents the Legendre polynomial given by

$$P_n(z) = \frac{1}{2^n}\sum_{k=0}^{[n/2]}\frac{(-1)^k(2n-2k)!}{k!(n-k)!(n-2k)!}z^{n-2k}.$$

3.1.12. Verify for $n = 0, 1, 2, 3, 4$ that

$$T_n(z) = {}_2F_1\left(-n, n; \frac{1}{2}; \frac{(1-z)}{2}\right)$$

represents the Chebyshev polynomial of the first kind given by

$$T_n(z) = \cos(n\cos^{-1}z) = \frac{n}{2}\sum_{k=0}^{\lfloor n/2 \rfloor} \frac{(-1)^k(n-k-1)!}{k!(n-2k)!}(2z)^{n-2k}, n \geq 1.$$

3.1.13. Verify for $n = 0, 1, 2, 3, 4$ that

$$U_n(z) = (n+1){}_2F_1\left(-n, n+2; \frac{3}{2}; \frac{(1-z)}{2}\right)$$

represents the Chebyshev polynomial of the second kind given by

$$U_n(z) = \frac{\sin(n\cos^{-1}z)}{\sqrt{1-z^2}} = \sum_{k=0}^{\lfloor (n-1)/2 \rfloor} \frac{(-1)^k(n-k-1)!}{k!(n-2k-1)!}(2z)^{n-2k-1}, n \geq 1.$$

3.1.14. Verify for $n = 0, 1, 2, 3, 4$ that

$$C_n^m(z) = \frac{(2m)_n}{n!}{}_2F_1\left(-n, n+2m; m+\frac{1}{2}; \frac{(1-z)}{2}\right)$$

represents the Gegenbauer, or the ultraspherical, polynomial given by

$$C_n^m(z) = \frac{1}{(m-1)!}\sum_{k=0}^{\lfloor n/2 \rfloor} \frac{(-1)^k(m+n-k-1)!}{k!(n-2k)!}(2z)^{n-2k}$$

where m is a positive integer.

3.1.15. Show that for any power series $\sum_{n=0}^{\infty} u_n z^n$ with $u_0 = 1$ and $\frac{u_{n+1}}{u_n}$ as a rational function of n can be written as a hypergeometric series.

3.2 Basic Hypergeometric Series (q-Series)

A process of q- generalization of the hypergeometric series started in the 19th century itself. Thus, the subject of q-hypergeometric series, or q-series, has a rich history. For a modern introduction to the subject one should refer to the book of Gasper and Rahman (1990) which would serve as an excellent textbook for any study of basic hypergeometric series. Let

Notation 3.2.1.

$$[\alpha]_q = \frac{1-q^\alpha}{1-q}. \tag{3.2.1}$$

It is easy to see that

$$\lim_{q\to 1}[\alpha]_q = \alpha. \tag{3.2.2}$$

Definition 3.2.1. The q-shifted factorial. It is defined as

$$(a;q)_n = \begin{cases} 1, \ n=0 \\ (1-a)(1-aq)(1-aq^2)\ldots\left(1-aq^{n-1}\right), \\ \qquad n=1,2,3,\ldots. \end{cases}$$

Then note that

$$\frac{(q^\alpha;q)_n}{(1-q)^n} = [\alpha]_q[\alpha+1]_q[\alpha+2]_q\ldots[\alpha+n-1]_q,$$

$$n=1,2,3,\ldots, \tag{3.2.3}$$

and

$$\lim_{q\to 1}\frac{(q^\alpha;q)_n}{(1-q)^n} = (\alpha)_n, \ n=1,2,3,\ldots \tag{3.2.4}$$

Unless otherwise stated, throughout we shall have n,m,k, to take nonnegative integer values. Equation (3.2.4) suggests an obvious q-generalization of the hypergeometric series by the replacements

$$(\alpha)_n \to \frac{(q^\alpha;q)_n}{(1-q)^n}, \ n! = (1)_n \to \frac{(q;q)_n}{(1-q)^n}. \tag{3.2.5}$$

This process of q-generalization has finally led to a standard definition of q-hypergeometric series as follows:

Notation 3.2.2. $_r\phi_s(a_1,a_2,\ldots,a_r;b_1,b_2,\ldots,b_s;q,z)$: The q-hypergeometric function.

Definition 3.2.2.

$$_r\phi_s(a_1,a_2,\ldots,a_r;b_1,b_2,\ldots,b_s;q,z) = \sum_{n=0}^{\infty}\frac{(a_1;q)_n(a_2;q)_n\ldots(a_r;q)_n}{(q;q)_n(b_1;q)_n(b_2;q)_n\ldots(b_s;q)_n}$$

$$\times\left[(-1)^n\,q^{\binom{n}{2}}\right]^{1+s-r}z^n, \tag{3.2.6}$$

where $q \neq 0$ when $r > s+1$. This series (3.2.6) is called also the basic hypergeo-
metric series (in view of the base q) or simply the q-series. If $0 < |q| < 1$, the $_r\phi_s$
series converges absolutely for all z if $r \leq s$, and for $|z| < 1$ if $r = s+1$. This series
also converges absolutely if $|q| > 1$ and $|z| < |b_1 b_2 \ldots b_s|/|a_1 a_2 \ldots a_r|$. It diverges for
$z \neq 0$ if $0 < |q| < 1$ and $r > s+1$, and if $|q| > 1$ and $|z| > |b_1 b_2 \ldots b_s|/|a_1 a_2 \ldots a_r|$,
unless it terminates. It is customary to use the notation $_r\phi_s$ (as in the case of $_r F_s$) also
for the sum of the series inside the circle of convergence and for its analytic contin-
uation (called the basic hypergeometric function) outside the circle of convergence.
Since for $m = 0, 1, 2, \ldots$

$$\left(q^{-m}; q\right)_n = 0, \quad n = m+1, m+2, \ldots, \tag{3.2.7}$$

in (3.2.7) it is assumed that none of the denominator parameters is of the form
q^{-m} with $m = 0, 1, 2, \ldots$. Equation (3.2.7) also shows that an $_r\phi_s$ series terminates
if one of its numerator parameters is of the form q^{-m} with $m = 0, 1, 2, \ldots$. Unless
stated otherwise, when dealing with non-terminating basic hypergeometric series it
is usually assumed that $|q| < 1$ and that the parameters and variables are such that
the series converges absolutely. Further, with the definitions:

Notation 3.2.3.

$$(a; q)_\infty = \prod_{k=0}^{\infty} \left(1 - aq^k\right), \tag{3.2.8}$$

we can write

$$(a; q)_n = \frac{(a; q)_\infty}{(aq^n; q)_\infty}. \tag{3.2.9}$$

Since products of q-shifted factorials occur so often, to simplify the writing, the
following more compact notations are used frequently:

$$(a_1, a_2, \ldots, a_m; q)_n = (a_1; q)_n (a_2; q)_n \ldots (a_m; q)_n, \tag{3.2.10}$$

$$(a_1, a_2, \ldots, a_m; q)_\infty = (a_1; q)_\infty (a_2; q)_\infty \ldots (a_m; q)_\infty. \tag{3.2.11}$$

Exercises 3.2.

3.2.1.   Show that any power series $\sum_{n=0}^{\infty} \upsilon_n z^n$ with $\upsilon_0 = 1$ and $\frac{\upsilon_{n+1}}{\upsilon_n}$ as a rational
function of q^n can be written as a basic hypergeometric series.

3.2.2.   Show that

$$\lim_{a_r \to \infty} {}_r\phi_s(a_1, a_2, \ldots, a_r; b_1, b_2, \ldots, b_s; q, z/a_r)$$

$$= {}_{r-1}\phi_s(a_1, a_2, \ldots, a_{r-1}; b_1, b_2, \ldots, b_s; q, z).$$

3.2.3. Show that

$$(a;q)_n = \left(a^{-1};q^{-1}\right)_n (-a)^n \, q^{\binom{n}{2}}.$$

3.2.4. Show that

$$\left(a^{-1}q^{1-n};q\right)_n = (a;q)_n(-a^{-1})^n \, q^{-\binom{n}{2}}.$$

3.2.5. Show that

$$(a;q)_{n-k} = \frac{(a;q)_n}{(a^{-1}q^{1-n};q)_k} \left(-qa^{-1}\right)^k q^{\binom{k}{2}-nk}.$$

3.2.6. Show that

$$(a;q)_{n+k} = (a;q)_n(aq^n;q)_k.$$

3.2.7. Show that

$$(aq^n;q)_k = \frac{(a;q)_k(aq^k;q)_n}{(a;q)_n}.$$

3.2.8. Show that

$$\left(aq^k;q\right)_{n-k} = \frac{(a;q)_n}{(a;q)_k}.$$

3.2.9. Show that

$$\left(aq^{2k};q\right)_{n-k} = \frac{(a;q)_n\,(aq^n;q)_k}{(a;q)_{2k}}.$$

3.2.10. Show that

$$(q^{-n};q)_k = \frac{(q;q)_n}{(q;q)_{n-k}}(-1)^k q^{\binom{k}{2}-nk}.$$

3.2.11. Show that

$$\left(aq^{-n};q\right)_k = \frac{(a;q)_k \left(qa^{-1};q\right)_n}{(a^{-1}q^{1-k};q)_n}q^{-nk}.$$

3.2.12. Show that

$$(a;q)_{2n} = (a;q^2)_n\,(aq;q^2)_n.$$

3.2.13. Show that

$$\left(a^2;q^2\right)_n = (a;q)_n\,(-a;q)_n.$$

3.2.14. Show that

$$\left(aq^{-n};q\right)_n = \left(a^{-1}q;q\right)_n \left(-aq^{-1}\right)^n q^{-\binom{n}{2}}.$$

3.2.15. Show that

$$\left(aq^{-k-n};q\right)_n = \frac{(a^{-1}q;q)_{n+k}}{(a^{-1}q;q)_k}\left(-aq^{-1}\right)^n q^{-nk-\binom{n}{2}}.$$

3.2.16. Show that

$$\frac{(q\sqrt{a}, -q\sqrt{a}; q)_n}{(\sqrt{a}, -\sqrt{a}; q)_n} = \frac{1 - aq^{2n}}{1 - a}.$$

3.2.17. Show that

$$(a; q)_\infty = (\sqrt{a}, -\sqrt{a}, \sqrt{aq}, -\sqrt{aq}; q)_\infty.$$

3.2.1 The q-binomial theorem

The binomial expansion, for $|z| < 1$,

$$(1 - z)^{-\alpha} = 1 + \frac{\alpha}{1!}z + \frac{\alpha(\alpha + 1)}{2!}z^2 + \frac{\alpha(\alpha + 1)(\alpha + 2)}{3!}z^3 + \cdots$$

$$= \sum_{n=0}^{\infty} \frac{(\alpha)_n}{n!} z^n = {}_1F_0(\alpha; ; z), \tag{3.2.12}$$

is one of the most important summation formulas for hypergeometric series. This formula has the following q-analogue known as the q-binomial theorem:

Definition 3.2.3. q-binomial series:

$$_1\phi_0(a; ; q, z) = \sum_{n=0}^{\infty} \frac{(a; q)_n}{(q; q)_n} z^n = \frac{(az; q)_\infty}{(z; q)_\infty}, \quad |z| < 1, \ |q| < 1. \tag{3.2.13}$$

To derive the q-binomial theorem let us first prove the binomial theorem (3.2.12) and then carry out the analogous steps for the q case. Let

$$f_\alpha(z) = \sum_{n=0}^{\infty} \frac{(\alpha)_n}{n!} z^n, \ |z| < 1. \tag{3.2.14}$$

Since this series is uniformly convergent in $|z| < \varepsilon$ where $0 < \varepsilon < 1$, we may differentiate it term-wise to get

$$f'_\alpha(z) = \sum_{n=1}^{\infty} \frac{n(\alpha)_n}{n!} z^{n-1} = \sum_{n=0}^{\infty} \frac{(\alpha)_{n+1}}{n!} z^n$$

$$= \alpha \sum_{n=0}^{\infty} \frac{(\alpha + 1)_n}{n!} z^n = \alpha f_{\alpha+1}(z). \tag{3.2.15}$$

Also

$$f_\alpha(z) - f_{\alpha+1}(z) = \sum_{n=1}^{\infty} \frac{(\alpha)_n - (\alpha+1)_n}{n!} z^n$$

$$= \sum_{n=1}^{\infty} \frac{(\alpha+1)_{n-1}[\alpha - (\alpha+n)]}{n!} z^n$$

$$= -\sum_{n=1}^{\infty} \frac{(\alpha+1)_{n-1}}{(n-1)!} z^n$$

$$= -\sum_{n=0}^{\infty} \frac{(\alpha+1)_n}{n!} z^{n+1} = -z f_{\alpha+1}(z). \qquad (3.2.16)$$

Thus we have from (3.2.15) and (3.2.16)

$$f'_\alpha(z) = \alpha f_{\alpha+1}(z), \quad f_{\alpha+1}(z) = \frac{f_\alpha(z)}{(1-z)}. \qquad (3.2.17)$$

Now, eliminating $f_{\alpha+1}(z)$ from (3.2.17) we get a first order differential equation for $f_\alpha(z)$

$$f'_\alpha(z) = \frac{\alpha}{1-z} f_\alpha(z). \qquad (3.2.18)$$

The initial condition for $f_\alpha(z)$ is

$$f_\alpha(0) = 1, \qquad (3.2.19)$$

as seen from (3.2.14). Solving (3.2.18), subject to (3.2.19), one has

$$f_\alpha(z) = (1-z)^{-\alpha}, \; |z| < 1, \qquad (3.2.20)$$

proving the binomial theorem

$$_1F_0(\alpha; ;z) = \sum_{n=0}^{\infty} \frac{(\alpha)_n}{n!} z^n = (1-z)^{-\alpha}. \qquad (3.2.21)$$

To derive the q-binomial theorem let us carry out analogous steps. Let

$$h_a(z) = \sum_{n=0}^{\infty} \frac{(a;q)_n}{(q;q)_n} z^n, \; |z| < 1, \; |q| < 1. \qquad (3.2.22)$$

Now, observe that for any function $g(z)$

$$\lim_{q \to 1} \frac{g(z) - g(qz)}{(1-q)z} = \frac{d}{dz} g(z), \qquad (3.2.23)$$

if $g(z)$ is differentiable at z. This suggests that to get the q-analogue of (3.2.15) we should compute the difference $h_a(z) - h_a(qz)$. This leads to

$$
\begin{aligned}
h_a(z) - h_a(qz) &= \sum_{n=1}^{\infty} \frac{(a;q)_n}{(q;q)_n} (z^n - q^n z^n) \\
&= \sum_{n=1}^{\infty} \frac{(a;q)_n}{(q;q)_{n-1}} z^n \\
&= \sum_{n=0}^{\infty} \frac{(a;q)_{n+1}}{(q;q)_n} z^{n+1} \\
&= \sum_{n=0}^{\infty} \frac{(1-a)(aq;q)_n}{(q;q)_n} z^{n+1} \\
&= (1-a)z\, h_{aq}(z). \tag{3.2.24}
\end{aligned}
$$

Next, to get an analogue of (3.2.16) let us compute the difference $h_a(z) - h_{aq}(z)$. This leads to

$$
\begin{aligned}
h_a(z) - h_{aq}(z) &= \sum_{n=1}^{\infty} \frac{(a;q)_n - (aq;q)_n}{(q;q)_n} z^n \\
&= \sum_{n=1}^{\infty} \frac{(aq;q)_{n-1}[(1-a)-(1-aq^n)]}{(q;q)_n} z^n \\
&= -a \sum_{n=1}^{\infty} \frac{(aq;q)_{n-1}(1-q^n)}{(q;q)_n} z^n \\
&= -a \sum_{n=1}^{\infty} \frac{(aq;q)_{n-1}}{(q;q)_{n-1}} z^n \\
&= -a \sum_{n=0}^{\infty} \frac{(aq;q)_n}{(q;q)_n} z^{n+1} = -az\, h_{aq}(z). \tag{3.2.25}
\end{aligned}
$$

From (3.2.24) and (3.2.25) we have

$$
h_a(z) - (1-a)z\, h_{aq}(z) = h_a(qz), \quad h_{aq}(z) = \frac{h_a(z)}{(1-az)}. \tag{3.2.26}
$$

Now, eliminating $h_{aq}(z)$ from (3.2.26) we get

$$
h_a(z) = \frac{1-az}{1-z} h_a(qz). \tag{3.2.27}
$$

Iterating this relation $(n-1)$ times

$$h_a(z) = \frac{(1-az)(1-azq)}{(1-z)(1-zq)} h_a(zq^2)$$

$$= \frac{(1-az)(1-azq)(1-azq^2)}{(1-z)(1-zq)(1-zq^2)} h_a(zq^3)$$

$$\cdots$$

$$= \frac{(1-az)(1-azq)(1-azq^2)\ldots(1-azq^{n-1})}{(1-z)(1-zq)(1-zq^2)\ldots(1-zq^{n-1})} h_a(zq^n)$$

$$= \frac{(az;q)_n}{(z;q)_n} h_a(zq^n). \tag{3.2.28}$$

Taking the limit $n \to \infty$ we can write

$$h_a(z) = \lim_{n\to\infty}\left\{\frac{(az;q)_n}{(z;q)_n} h_a(zq^n)\right\} = \frac{(az;q)_\infty}{(z;q)_\infty} h_a(0) = \frac{(az;q)_\infty}{(z;q)_\infty}, \tag{3.2.29}$$

since as $n \to \infty$, $h_a(zq^n) \to h_a(0)$ for $|q| < 1$, and $h_a(0) = 1$ as seen from (3.2.22). Thus, we have the q-binomial theorem

$$_1\phi_0(a;\;;q,z) = \sum_{n=0}^{\infty} \frac{(a;q)_n}{(q;q)_n} z^n = \frac{(az;q)_\infty}{(z;q)_\infty}, \tag{3.2.30}$$

$$|z| < 1, |q| < 1.$$

Equation (3.2.30) is one of the most important summation formulas for basic hypergeometric series. The observations

$$_0F_0(\;;\;;z) = e^z, \quad \lim_{\alpha\to\infty} {}_1F_0\left(\alpha;\;;\frac{z}{\alpha}\right) = e^z \tag{3.2.31}$$

suggest q-generalization of the exponential function using $_0\phi_0$ or $_1\phi_0$. It should be noted that for a given function there can be several q-analogues if the only condition to be met is that the q-analogue tends to the original function in the limit $q \to 1$. So, the choice of a q-analogue for a given function depends on the particular application. For the exponential function there are two standard q-analogues defined in the literature. These are the following:

Notation 3.2.4. $e_q(z), E_q(z)$: q-exponential functions.

Definition 3.2.4.

$$e_q(z) = {}_1\phi_0(0;\;;q,z) = \sum_{n=0}^{\infty} \frac{z^n}{(q;q)_n} = \frac{1}{(z;q)_\infty}, \quad |z| < 1. \tag{3.2.32}$$

$$E_q(z) = \sum_{n=0}^{\infty} \frac{q^{n(n-1)/2}}{(q;q)_n} z^n$$

$$= {}_0\phi_0(\ ;\ ;q,-z) = \lim_{a\to\infty} {}_1\phi_0\left(a;\ ;q,-\frac{z}{a}\right)$$

$$= \lim_{a\to\infty} \frac{(a(-z/a);q)_\infty}{(-z/a;q)_\infty} = (-z;q)_\infty. \tag{3.2.33}$$

Note that the last steps in (3.2.32) and (3.2.33) follow from the q-binomial theorem.

Exercises 3.2.

3.2.18. Show that

$$\lim_{q\to 1} \frac{(q^\alpha z : q)_\infty}{(z;q)_\infty} = \lim_{q\to 1} {}_1\phi_0(q^\alpha;\ ;q,z)$$

$$\to {}_1F_0(\alpha;\ ;z) = (1-z)^{-\alpha}.$$

3.2.19. Show that

$${}_1\phi_0(a;\ ;q,z)\,{}_1\phi_0(b;\ ;q,az) = {}_1\phi_0(ab;\ ;q,z)$$

which is a q analogue of the relation

$$(1-z)^{-\alpha}(1-z)^{-\beta} = (1-z)^{-(\alpha+\beta)}.$$

3.2.20. Show that

$$e_q(z)E_q(-z) = 1$$

which is the q-analogue of the relation $e^z e^{-z} = 1$.

3.2.21. Show that

$$\lim_{q\to 1} e_q((1-q)z) = e^z.$$

3.2.22. Show that

$$\lim_{q\to 1} E_q((1-q)z) = e^z.$$

3.2.2 The q-binomial coefficients

The binomial theorem

$${}_1F_0(\alpha;\ ;z) = (1-z)^{-\alpha} \tag{3.2.34}$$

becomes, when $\alpha = -n$, a negative integer,

$$(1-z)^n = {}_1F_0(-n;\,;z)$$

$$= \sum_{k=0}^{\infty} \frac{(-n)_k}{k!} z^k = \sum_{k=0}^{n} \frac{(-n)_k}{k!} z^k$$

$$= \sum_{k=0}^{n} \frac{(-1)^k n(n-1)(n-2)\dots(n-k+1)}{k!} z^k$$

$$= \sum_{k=0}^{n} \binom{n}{k} (-z)^k. \tag{3.2.35}$$

In the q-binomial theorem

$$_1\phi_0(a;\,;q,z) = \frac{(az;q)_\infty}{(z;q)_\infty} \tag{3.2.36}$$

let $a = q^{-n}$ and replace z by zq^n where n is a non-negative integer. The result is

$$_1\phi_0(q^{-n};\,;q,zq^n) = \frac{(z;q)_\infty}{(zq^n;q)_\infty} = (z;q)_n. \tag{3.2.37}$$

Thus

$$(z;q)_n = {}_1\phi_0(q^{-n};\,;q,zq^n)$$

$$= \sum_{k=0}^{\infty} \frac{(q^{-n};q)_k}{(q;q)_k} (zq^n)^k$$

$$= \sum_{k=0}^{n} \frac{(q^{-n};q)_k}{(q;q)_k} q^{nk} z^k. \tag{3.2.38}$$

From Exercise 3.2.10 we know that

$$(q^{-n};q)_k = \frac{(q;q)_n (-1)^k q^{\binom{k}{2}-nk}}{(q;q)_{n-k}}. \tag{3.2.39}$$

Substituting this relation in (3.2.38) we have

$$(z;q)_n = \sum_{k=0}^{n} \frac{(q;q)_n}{(q;q)_k (q;q)_{n-k}} q^{\binom{k}{2}} (-z)^k. \tag{3.2.40}$$

Now, note that

$$\lim_{q\to 1} (z;q)_n = (1-z)^n. \tag{3.2.41}$$

Thus, we realize that (3.2.40) is the q-analogue of the binomial expansion

$$(1-z)^n = \sum_{k=0}^{n} \binom{n}{k} (-z)^k. \tag{3.2.42}$$

Comparing (4.2.42) with (4.2.40) the q-binomial coefficient is defined as follows:

Notation 3.2.5. q-**binomial coefficient**

$$\begin{bmatrix} n \\ k \end{bmatrix}_q = \frac{(q;q)_n}{(q;q)_k(q;q)_{n-k}}. \tag{3.2.43}$$

Definition 3.2.5.   The q-binomial expansion (3.2.40) is written as

$$(z;q)_n = \sum_{k=0}^{n} \begin{bmatrix} n \\ k \end{bmatrix}_q q^{\binom{k}{2}} (-z)^k. \tag{3.2.44}$$

Exercises   3.2.

3.2.23.   Show that

$$\lim_{q \to 1} \begin{bmatrix} n \\ k \end{bmatrix}_q = \binom{n}{k}.$$

3.2.24.   Show that

$$\begin{bmatrix} n \\ n-k \end{bmatrix}_q = \begin{bmatrix} n \\ k \end{bmatrix}_q.$$

3.2.25.   Show that

$$\begin{bmatrix} n+k \\ k \end{bmatrix}_q = \frac{(q^{n+1};q)_k}{(q;q)_k}.$$

3.2.26.   Show that

$$\begin{bmatrix} n+1 \\ k \end{bmatrix}_q = \begin{bmatrix} n \\ k \end{bmatrix}_q q^k + \begin{bmatrix} n \\ k-1 \end{bmatrix}_q = \begin{bmatrix} n \\ k \end{bmatrix}_q + \begin{bmatrix} n \\ k-1 \end{bmatrix}_q q^{n+1-k}.$$

3.2.27.   Show that if X and Y are such that $XY = qYX$ then

$$(X+Y)^n = \sum_{k=0}^{n} \begin{bmatrix} n \\ k \end{bmatrix}_q Y^k X^{n-k} = \sum_{k=0}^{n} \begin{bmatrix} n \\ k \end{bmatrix}_{q^{-1}} X^k Y^{n-k}.$$

where $\begin{bmatrix} n \\ k \end{bmatrix}_{q^{-1}}$ is obtained from $\begin{bmatrix} n \\ k \end{bmatrix}_q$ by replacing q by q^{-1}.

3.2.28. Show that if X and Y are such that $XY = qYX$ where $q = e^{2\pi i/n}$ then

$$(X+Y)^n = X^n + Y^n.$$

3.3 q-Calculus

We have seen that for any function $g(z)$

$$\lim_{q\to 1} \frac{g(z) - g(qz)}{(1-q)z} = \frac{d}{dz} g(z)$$

if $g(z)$ is differentiable at z.

Notation 3.3.1. $D_q = q$-derivative.

Definition 3.3.1. For a fixed q a q-derivative operator D_q is defined by

$$D_q\, g(z) = \frac{g(z) - g(qz)}{(1-q)z}. \tag{3.3.1}$$

Note that

$$D_q\, z^n = \frac{z^n - (qz)^n}{(1-q)z} = \frac{1-q^n}{1-q} z^{n-1} = [n]_q\, z^{n-1}, \tag{3.3.2}$$

analogous to the relation

$$\frac{d}{dz} z^n = n z^{n-1}. \tag{3.3.3}$$

The differential equations of special functions get q-deformed into q-differential equations, involving the q-derivative D_q, satisfied by the corresponding q-special functions. For example, $_2\phi_1(a,b;c;q,z)$ satisfies (for $|z| < 1$ and in the formal power series sense) the following second order q-differential equation:

$$z(c - abqz)D_q^2\phi + \left[\frac{1-c}{1-q} + \frac{(1-a)(1-b) - (1-abq)}{1-q} z\right] D_q\phi$$
$$- \frac{(1-a)(1-b)}{(1-q)^2} \phi = 0. \tag{3.3.4}$$

By replacing a, b, c respectively, by $q^\alpha, q^\beta, q^\gamma$ and then letting $q \to 1$ it is seen that (3.3.4) tends to the second order differential equation satisfied by $_2F_1(\alpha, \beta; \gamma; z)$, namely,

$$z(1-z)\frac{d^2F}{dz^2} + [\gamma - (\alpha + \beta + 1)z]\frac{dF}{dz} - \alpha\beta F = 0 \tag{3.3.5}$$

where $|z| < 1$. Once we have a q-differentiation the question of a q-integration arises naturally.

Definition 3.3.2.   The q-integral is defined by

$$\int_a^b f(z)\mathrm{d}_q z = \int_0^b f(z)\mathrm{d}_q z - \int_0^a f(z)\mathrm{d}_q z, \tag{3.3.6}$$

where

$$\int_0^a f(z)\mathrm{d}_q z = a(1-q)\sum_{n=0}^\infty f(aq^n)q^n. \tag{3.3.7}$$

The q-integral (3.3.7) can be viewed as an infinite Riemann sum with non-equidistant mesh widths. In the limit $q \to 1$ the right side of (3.3.7) tends to the usual integral $\int_0^a f(z)\mathrm{d}z$. As an example consider, with q and α real and $|q| < 1$, $\alpha > -1$,

$$\int_0^1 z^\alpha \mathrm{d}_q z = (1-q)\sum_{n=0}^\infty (q^n)^\alpha\, q^n.$$

$$= (1-q)\sum_{n=0}^\infty q^{n(\alpha+1)} = \frac{1-q}{1-q^{\alpha+1}}. \tag{3.3.8}$$

In the limit $q \to 1$ we have $\frac{1-q}{1-q^{\alpha+1}} \to \frac{1}{\alpha+1}$, as should be since

$$\int_0^1 z^\alpha \mathrm{d}z = \frac{1}{\alpha+1}. \tag{3.3.9}$$

Standard operations of classical analysis like differentiation and integration do not fit well with q-series and these are to be replaced by q-differentiation and q-integration. Already we saw that $_2\phi_1(a,b;c;q,z)$ satisfies a second order of q-differential equation analogous to the second order differential equation satisfied by $_2F_1(\alpha,\beta;\gamma;z)$. Similarly, integral representation of hypergeometric functions get q-deformed into q-integral representations for q-hypergeometric functions. For example, the integral representation

$$_2F_1(\alpha,\beta;\gamma;z) = \frac{\Gamma(\gamma)}{\Gamma(\beta)\Gamma(\gamma-\beta)} \int_0^1 t^{\beta-1}(1-t)^{\gamma-\beta-1}(1-tz)^{-\alpha}\mathrm{d}t, \tag{3.3.10}$$

with $|\arg(1-z)| < \pi$ and $\Re(\gamma) > \Re(\beta) > 0$, has the q-analogue

$$_2\phi_1(q^\alpha,q^\beta;q^\gamma;q,z) = \frac{\bar{\Gamma}_q(\gamma)}{\bar{\Gamma}_q(\beta)\bar{\Gamma}_q(\gamma-\beta)} \int_0^1 t^{\beta-1}\frac{(tzq^\alpha,tq;q)_\infty}{(tz,tq^{\gamma-\beta};q)_\infty}\mathrm{d}_q t, \tag{3.3.11}$$

where $\bar{\Gamma}_q(z)$ is the q-gamma function to be defined below. The q-integral notation is often quite useful in simplifying and manipulating various formulas involving sums of series.

Exercises 3.3.

3.3.1. Show that

$$D_q\, e_q((1-q)\alpha z) = \alpha\, e_q((1-q)\alpha z).$$

3.3.2. Show that

$$D_q E_q((1-q)\alpha z) = \alpha\, Eq((1-q)\alpha qz).$$

3.3.3. Show that for any positive integer n

$$D_{q^2}^n {}_2\phi_1(a,b;c;q,z) = \frac{(a,b;q)_n}{(c;q)_n(1-q)^n}\, {}_2\phi_1\left(aq^n, bq^n; cq^n; q, z\right).$$

3.4 The q-Gamma and q-Beta Functions

Notation 3.4.1. $\bar{\Gamma}_q(x)$: The q-gamma function.

Definition 3.4.1. The q-gamma function is defined by

$$\bar{\Gamma}_q(x) = \frac{(q;q)_\infty}{(q^x;q)_\infty}(1-q)^{1-x}, \quad 0 < q < 1. \tag{3.4.1}$$

When $x = n+1$ with n a non-negative integer this definition reduces to

$$\bar{\Gamma}_q(n+1) = [1]_q[2]_q[3]_q\ldots[n]_q = [n]_q! \tag{3.4.2}$$

which clearly tends to $n!$ as $q \to 1$. Hence

$$\lim_{q\to 1}\bar{\Gamma}_q(n+1) = \Gamma(n+1) = n!. \tag{3.4.3}$$

One can also show that

$$\bar{\Gamma}_q(x+1) = [x]_q\, \bar{\Gamma}_q(x), \quad \bar{\Gamma}_q(1) = 1, \tag{3.4.4}$$

analoguous to the relation

$$\Gamma(x+1) = x\Gamma(x), \quad \Gamma(1) = 1. \tag{3.4.5}$$

Analogous to the case of $\Gamma(x)$, $\bar{\Gamma}_q(x)$ has poles at $x = 0, -1, -2, \ldots$. Since the beta function is defined by

$$B(x,y) = \frac{\Gamma(x)\Gamma(y)}{\Gamma(x+y)} \tag{3.4.6}$$

it is natural to define the q-beta function by the following:

Notation 3.4.2. $\bar{B}_q(x,y)$: q-beta function.

Definition 3.4.2.

$$\bar{B}_q(x,y) = \frac{\bar{\Gamma}_q(x)\bar{\Gamma}_q(y)}{\bar{\Gamma}_q(x+y)} \tag{3.4.7}$$

which tends to $B(x,y)$ as $q \to 1$. One can show that

$$\bar{B}_q(x,y) = (1-q)\sum_{n=0}^{\infty} \frac{(q^{n+1};q)_\infty}{(q^{n+y};q)_\infty} q^{nx}, \; \Re(x) > 0, \; \Re(y) > 0. \tag{3.4.8}$$

3.5 Transformation and Summation Formulas for q-Series

Heine (1847) showed that

$$_2\phi_1(a,b;c;q,z) = \frac{(b,az;q)}{(c,z;q)_\infty} {}_2\phi_1(c/b,z;az;q,b) \tag{3.5.1}$$

where $|z| < 1$ and $|b| < 1$. This is an example of a transformation formula. To prove this, first observe from the q-binomial theorem that

$$\frac{(cq^n;q)_\infty}{(bq^n;q)_\infty} = \sum_{m=0}^{\infty} \frac{(c/b;q)_m}{(q;q)_m}(bq^n)^m. \tag{3.5.2}$$

Hence, for $|z| < 1$ and $|b| < 1$,

$$\begin{aligned}
_2\phi_1(a,b;c;q,z) &= \frac{(b;q)_\infty}{(c;q)_\infty}\sum_{n=0}^{\infty}\frac{(a,q)_n(cq^n;q)_\infty}{(q;q)_n(bq^n;q)_\infty}z^n \\
&= \frac{(b;q)_\infty}{(c;q)_\infty}\sum_{n=0}^{\infty}\frac{(a;q)_n}{(q;q)_n}z^n\sum_{m=0}^{\infty}\frac{(c/b;q)_m}{(q;q)_m}(zq^m)^n \\
&= \frac{(b;q)_\infty}{(c;q)_\infty}\sum_{m=0}^{\infty}\frac{(c/b;q)_m}{(q;q)_m}b^m\sum_{n=0}^{\infty}\frac{(a;q)_n}{(q;q)_n}(zq^m)^n \\
&= \frac{(b;q)_\infty}{(c;q)_\infty}\sum_{m=0}^{\infty}\frac{(c/b;q)_m}{(q;q)_m}b^m\frac{(azq^m;q)_\infty}{(zq^m;q)_\infty} \\
&= \frac{(b,az;q)_\infty}{(c,z;q)_\infty}{}_2\phi_1(c/b,z;az;q,b),
\end{aligned} \tag{3.5.3}$$

which proves (3.5.1). Heine also showed that Euler's transformation formula

$$_2F_1(\alpha,\beta;\gamma;z) = (1-z)^{\gamma-\alpha-\beta}{}_2F_1(\gamma-\alpha,\gamma-\beta;\gamma;z) \tag{3.5.4}$$

has a q-analogue of the form

$$2\phi_1(a,b;c;q,z) = \frac{(abz/c;q)_\infty}{(z;q)_\infty} {}_2\phi_1(c/a,c/b;c;q,abz/c). \tag{3.5.5}$$

A short way to prove this formula is just to iterate (3.5.1) by interchanging the two numerator parameters as follows:

$$2\phi_1(a,b;c;q,z) = \frac{(b,az;q)_\infty}{(c,z;q)_\infty} {}_2\phi_1(c/b,z;az;q,b)$$

$$= \frac{(c/b,bz;q)_\infty}{(c,z;q)_\infty} {}_2\phi_1(abz/c,b;bz;q,c/b)$$

$$= \frac{(abz/c;q)_\infty}{(z;q)_\infty} {}_2\phi_1(c/a,c/b;c;q,abz/c). \tag{3.5.6}$$

Take $a = q^\alpha$, $b = q^\beta$, $c = q^\gamma$ and let $q \to 1$ to see that (3.5.5) becomes Euler transformation formula. If we set in (3.5.1), $z = \frac{c}{ab}$, assume that $|b| < 1, |\frac{c}{ab}| < 1$, and observe that the series on the right side of

$$2\phi_1\left(a,b;c;q,\frac{c}{ab}\right) = \frac{(b,c/b;q)_\infty}{(c,c/ab;q)_\infty} {}_1\phi_0\left(\frac{c}{ab}; ;q,\right), \tag{3.5.7}$$

can be summed by the q-binomial theorem, we get

$$2\phi_1\left(a,b;c;q,\frac{c}{ab}\right) = \frac{\left(\frac{c}{a},\frac{c}{b};q\right)_\infty}{\left(c,\frac{c}{ab};q\right)_\infty}. \tag{3.5.8}$$

This is the q-analogue of the Gauss summation formula

$$2F_1(\alpha,\beta;\gamma;1) = \frac{\Gamma(\gamma)\Gamma(\gamma-\alpha-\beta)}{\Gamma(\gamma-\alpha)\Gamma(\gamma-\beta)}, \quad \Re(\gamma-\alpha-\beta) > 0. \tag{3.5.9}$$

For the terminating case when $a = q^{-n}$, (3.5.8) reduces to

$$2\phi_1\left(q^{-n},b;c;q,\frac{cq^n}{b}\right) = \frac{(c/b;q)_n}{(c;q)_n}, \quad n = 0,1,2,\ldots, \tag{3.5.10}$$

which is a q-analogue of the Chu-Vandermonde formula

$$2F_1(-n,\beta;\gamma;1) = \frac{(\gamma-\beta)_n}{(\gamma)_n}, \quad n = 0,1,2,\ldots. \tag{3.5.11}$$

Exercises 3.5.

3.5.1. Set $c = bzq^{\frac{1}{2}}$ in (3.5.8) and then let $b \to 0$ and $a \to \infty$ to obtain

$$\sum_{n=0}^{\infty} \frac{(-1)^n q^{n^2/2}}{(q;q)_n} z^n = (zq^{\frac{1}{2}};q)_\infty.$$

3.5.2. Set $c = bzq^{\frac{1}{2}}$ in (3.5.8) and then let $a \to 0$ and $a \to \infty$ to obtain

$$\sum_{n=0}^{\infty} \frac{q^{n^2}}{(q,zq;q)_n} z^n = \frac{1}{(zq;q)_\infty}.$$

3.5.3. Let Δ_θ denote the q-difference operator defined for a fixed q by

$$\Delta_\theta \, g(z) = \theta \, g(qz) - g(z)$$

and let $\Delta = \Delta_1$. Then show that

$$\vartheta_n(z) = \frac{(a_1, a_2, \ldots, a_r; q)_n}{(q, b_1, b_2, \ldots, b_s; q)_n} \left[(-1)^n q^{\binom{n}{2}} \right]^{1+s-r} z^n$$

satisfies the relation

$$\left(\Delta \Delta_{b_1/q} \Delta_{b_2/q} \cdots \Delta_{b_s/q} \right) \vartheta_n(z) = z(\Delta_{a_1} \Delta_{a_2} \cdots \Delta_{a_r}) \vartheta_{n-1} \left(zq^{1+s-r} \right), n = 1, 2, \ldots.$$

Hence show that $_r\phi_s(a_1, a_2, \ldots, a_r; b_1, b_2, \ldots, b_s; q, z)$ satisfies (in the sense of formal power series) the q-defference equation

$$\left(\Delta \Delta_{b_1/q} \Delta_{b_2/q} \cdots \Delta_{b_s/q} \right) {}_r\phi_s(z) = z(\Delta_{a_1} \Delta_{a_2} \cdots \Delta_{a_r}) \, {}_r\phi_s(zq^{1+s-r}).$$

3.5.4. Let $|x| < 1$ and define

$$\sin_q(x) = \frac{1}{2i} \left[e_q(ix) - e_q(-ix) \right]$$

$$\cos_q(x) = \frac{1}{2} \left[e_q(ix) + e_q(-ix) \right]$$

$$\mathrm{Sin}_q(x) = \frac{1}{2i} \left[E_q(ix) - E_q(-ix) \right]$$

$$\mathrm{Cos}_q(x) = \frac{1}{2} \left[E_q(ix) + E_q(-ix) \right].$$

3.5.5. Show that

$$\sin_q(x)\mathrm{Sin}_q(x) + \cos_q(x)\mathrm{Cos}_q(x) = 1$$
$$\sin_q(x)\mathrm{Cos}_q(x) - \mathrm{Sin}_q(x) \cos_q(x) = 0.$$

3.6 Jacobi's Triple Product and Rogers-Ramanujan Identities

Using the results of Exercises 3.5.1 and 3.5.2 it can be shown that

$$(zq^{\frac{1}{2}}, q^{\frac{1}{2}}/z, q; q)_\infty = \sum_{n=-\infty}^{\infty} (-1)^n q^{\frac{n^2}{2}} z^n \tag{3.6.1}$$

which is well known as the Jacobi triple product identity. This is a limiting case of Ramanujan's $_1\psi_1$ summation formula

$$_1\psi_1(a;b;q,z) = \frac{(q,b/a,az,q/az;q)_\infty}{(b,q/a,z,b/az;q)_\infty}, |b/a| < |z| < 1,$$

which extends the q-binomial theorem to bilateral q-hypergeometric series defined by

$$\begin{aligned}
_r\psi_s(z) &= \sum_{n=0}^{\infty} \frac{(a_1,a_2,\ldots,a_r;q)_n}{(b_1,b_2,\ldots,b_s;q)_n} \\
&\quad \times (-1)^{(s-r)n} q^{(s-r)\binom{n}{2}} z^n \\
&\quad + \sum_{n=1}^{\infty} \frac{(q/b_1,q/b_2,\ldots,q/b_s;q)_n}{(q/a_1,q/a_2,\ldots,q/a_r;q)_n} \\
&\quad \times \left(\frac{(b_1 b_2 \ldots b_s)}{(a_1 a_2 \ldots a_r z)}\right)^n,
\end{aligned} \tag{3.6.2}$$

with suitable assumptions on the parameters and z to ensure convergence.

The theory of q-series is full of transformation formulas (or identities) of which we have seen so far a few elementary examples. Any introduction to q-series will not be complete without mentioning the famous Rogers-Ramanujan identities: for $|q| < 1$,

$$\sum_{n=0}^{\infty} \frac{q^{n^2}}{(q;q)_n} = \frac{(q^2,q^3,q^5;q^5)_\infty}{(q;q)_\infty}, \tag{3.6.3}$$

$$\sum_{n=0}^{\infty} \frac{q^{n(n+1)}}{(q;q)_n} = \frac{(q,q^4,q^5;q^5)_\infty}{(q;q)_\infty}. \tag{3.6.4}$$

An $_r\phi_s$ series is also denoted by the notation

$$_r\phi_s(a_1,a_2,\ldots,a_r,b_1,b_2,\ldots,b_s,q,z) = {}_r\phi_s\begin{bmatrix} a_1, a_2, \ldots, a_r \\ b_1, b_2, \ldots, b_s \end{bmatrix};q,z\end{bmatrix}$$

which is more compact when the number of parameters is large. Using this notation a transformation formula due to Watson looks like

$$_8\phi_7\begin{bmatrix} a, qa^{\frac{1}{2}}, -qa^{\frac{1}{2}}, b, c, d, e, q^{-n} \\ a^{\frac{1}{2}}, -a^{\frac{1}{2}}, aq/b, aq/c, aq/d, aq/e, aq^{n+1} \end{bmatrix};q, \frac{a^2 q^{2+n}}{bcde}\end{bmatrix}$$

$$= \frac{(aq,aq/de;q)_n}{(aq/d,aq/e;q)_n} {}_4\phi_3\begin{bmatrix} q^{-n}, d, e, aq/bc \\ aq/b, aq/c, deq^{-n}/a \end{bmatrix};q,q\end{bmatrix} \tag{3.6.5}$$

with certain conditions on the parameters. Such a transformation formula can be used to give a simple proof of the Rogers-Ramanujan identities (3.6.4)-(3.6.5).

So far we have seen the richness of the theory of q-series. The recent interest in the subject with a history starting in 19th century is due to the fact that q-series has popped up in such diverse areas as statistical mechanics, quantum groups, transcendental number theory, etc. The famous example in statistical mechanics is Baxter's beautiful solution of the hard hexagon model wherein the Rogers-Ramanujan identities first arose in physics. Now we have several examples for the applications of q-series outside pure mathematics. Lie algebras and their representations are well known to provide a unifying framework for special functions. It is found that quantum groups play a similar role for q-special functions. For example the representation theory of the quantum group $U_q(sl(2))$ involves $_2\phi_1$ functions and some q-Jacobi polynomials. So it is to be expected that whenever quantum groups are relevant to the description of physical models the q-series will arise.

TEST

on Basic Hypergeometric Series

Time: 1 hour

3.1. Show that

$$i \quad (a;q)_{n-k} = \frac{(a;q)_n}{(a^{-1}q^{1-n};q)_k}(-qa^{-1})^k q^{\binom{k}{2}-nk}$$

$$ii \quad (a;q)_\infty = (\sqrt{a}, -\sqrt{a}, \sqrt{aq}, -\sqrt{aq}; q)_\infty$$

3.2. Show that

$$i \quad \lim_{q\to 1} e_q((1-q)z) = e^z.$$

$$ii \quad \lim_{q\to 1} E_q((1-q)z) = e^z.$$

$$iii \quad e_q(z)E_q(-z) = 1.$$

3.3. Show that

$$\begin{bmatrix} n+1 \\ k \end{bmatrix}_q = \begin{bmatrix} n \\ k \end{bmatrix}_q q^k + \begin{bmatrix} n \\ k-1 \end{bmatrix}_q = \begin{bmatrix} n \\ k \end{bmatrix}_q + \begin{bmatrix} n \\ k-1 \end{bmatrix}_q q^{n+1-k}.$$

3.4. Using q-binomial theorem show that

$$_1\phi_0\left(q^{-n}; ;q,z\right) = \left(zq^{-n};q\right)_n, \quad n = 0,1,2,\dots.$$

3.5. Find

$$(i) \quad D_q E_q((1-q)\alpha z), \quad (ii) \quad D_q(z;q)_n$$

where D_q is the q-differential operator. What are the relations in these two cases in the limit $q \to 1$?

(R. Jagannathan)

References

Agarwal, R.P. (1963): *Generalised Hypergeometric series*, Asia Publishing House, Bombay.

Gasper, G. and Rahman, M. (1990): *Basic Hypergeometric Series,* Cambridge University Press, Cambridge.

Andrews, G.E. (1986): *q-series: Their Development and Applications in Analysis, Number Theory, Combinatorics, Physics and Computer Algebra*, Regional Conference Series in Mathematics No.66, American Mathematical Society, Providence, Rhode Island, U.S.A.

Floreanini, R. and Vinet, L. (1993): Quantum Algebras and q-Special Functions, *Annals of Physics*, **221**, 53-70.

Chapter 4
Ramanujan's Theories of Theta and Elliptic Functions

[This chapter is based on the lectures of Dr. S. Bhargava of the Department of Post-graduate Studies and Research in Mathematics of the University of Mysore, India]

4.0 Introduction

Ramanujan develops the classical theories of theta and elliptic functions in his own unique way. Moreover, he has his own theories of theta and elliptic functions beyond the classical theories. Ramanujan's approach to establishing his theorems, even the deep theorems, is elementary.

The purpose of the present lectures is to acquaint the audience with some selected parts (mile stones) in Ramanujan's development to enable them for further reading and research.

4.1 Ramanujan's Theory of Classical Theta Functions

4.1.1 Series definition and additive results

Definition 4.1.1. Theta function: Ramanujan defines the theta function (algebraically) by

$$f(a,b) = \sum_{n=-\infty}^{\infty} a^{n(n+1)/2} b^{n(n-1)/2}, \ |ab| < 1. \tag{4.1.1}$$

Remark 4.1.1: This is indeed Ramanujan's version of Jacobi's theta function

$$\theta_3(q,z) = f(qe^{iz}, qe^{-iz}), \ |q| < 1$$

(which is trigonometric in nature).

159

It turns out that the (algebraic) definition by (4.1.1) of the theta function renders the statements as well as proofs of most of the theorems elegant. The following special cases, with $|q| < 1$, are frequently employed:

$$\phi(q) = f(q,q) = \sum_{n=-\infty}^{\infty} q^{n^2}$$

$$\psi(q) = \frac{1}{2}f(1,q) = \sum_{n=0}^{\infty} q^{n(n+1)/2}$$

and

$$f(-q) = f(-q,-q^2) = \sum_{n=-\infty}^{\infty} (-1)^n q^{n(3n-1)/2}. \tag{4.1.2}$$

The quantities $n^2, n(n+1)/2$ and $n(3n-1)/2$ being square, triangular, and pentagonal numbers, we observe that these numbers can be regarded as generated by the theta functions $\phi(q)$, $\psi(q)$ and $f(-q)$.

Exercise 4.1.1:   Discuss the convergence of the series in (4.1.1) and (4.1.2).

Exercise 4.1.2:   Justify the terminologies triangular numbers and pentagonal numbers for $n(n+1)/2$ and $n(3n-1)/2$ respectively.

Exercise 4.1.3:   Prove the following properties of $f(a,b)$:

$$f(-1,a) = 0$$
$$f(1,a) = 2\,f(a,a^3)$$
$$f(a,b) = f(b,a). \tag{4.1.3}$$

Exercise 4.1.4:   Obtain the series expansions for

$$f(qe^{iz}, qe^{-iz}) = \theta_3(q,z)$$

and for

$$f(q,q) = \phi(q), \quad \frac{1}{2}f(1,q) = \psi(q)$$

and

$$f(-q,-q^2) = f(-q) \text{ of } (4.1.2).$$

The following theorem is Ramanujan's (algebraic) version of the quasiperiodicity theorem of the classical theory.

Theorem 4.1.1. *For* $|ab| < 1$,

$$f(a,b) = a^{n(n+1)/2}\, b^{n(n-1)/2} f(a(ab)^n,\ b(ab)^{-n}). \tag{4.1.4}$$

Proof 4.1.1: We leave the proof as an exercise which is no more difficult than the proof given of the following theorem.

The following theorem is one of Ramanujan's tools for proving many of his so-called modular equations. It provides a formula for decomposing the theta function $f(a,b)$, $(ab = modulus)$ into theta functions of different moduli. Note again the algebraic nature of the formula. The proof consists of simple algebraic manipulations.

Theorem 4.1.2. Let $|ab| < 1$ and $U_n = a^{n(n+1)/2} b^{n(n-1)/2}$ and $V_n = U_{-n}$, where n is an integer. Then

$$f(U_1, V_1) = \sum_{r=0}^{n-1} U_r f\left(\frac{U_{n+r}}{U_r}, \frac{V_{n-r}}{U_r}\right). \tag{4.1.5}$$

Proof 4.1.2: On slight calculations, we can write the right side of (4.1.5) as

$$\sum_{r=0}^{n-1} \sum_{m=-\infty}^{\infty} U_r^{1-m^2} U_{n+r}^{m(m+1)/2} V_{n-r}^{m(m-1)/2}$$

or, which is the same as,

$$\sum_{m=-\infty}^{\infty} \sum_{r=0}^{n-1} a^{\sigma(r,m,n)} b^{\sigma'(r,m,n)} = \sum_{-\infty}^{\infty} a^{k(k+1)/2} b^{k(k-1)/2}$$

$$= f(U_1, V_1),$$

where

$$\sigma(r,m,n) = (mn + r)(mn + r + 1)/2$$

and

$$\sigma'(r,m,n) = (nm + r)(nm + r - 1)/2.$$

Thus the theorem is proved.

Exercise 4.1.5:   Complete the details of the proof of (4.1.5).

Exercise 4.1.6:   Prove (4.1.4).

Many additive properties of $f(a,b)$ can be proved quite simply and elegantly from the definitions in (4.1.1) and (4.1.2).

Theorem 4.1.3. If $|q| < 1$, then

$$\phi(q) + \phi(-q) = 2\,\phi(q^4)$$

and

$$\phi(q) - \phi(-q) = 4q\,\psi(q^8). \tag{4.1.6}$$

Proof 4.1.3: Exercise.

Exercise 4.1.7: Prove the addition results in (4.1.6) as well as the following addition results.

$$f(a,b) + f(-a,-b) = 2 f(a^3b, ab^3)$$

$$f(a,b) - f(-a,-b) = 2a f\left(\frac{b}{a}, \frac{a}{b}a^4b^4\right).$$

Exercise 4.1.8: The following results for products and sums of products of theta functions also follow directly from the definition. If $ab = cd$ and $|ab| < 1$, then

$$f(a,b)f(c,d) + f(-a,-b)f(-c,-d) = 2 f(ac,bd)f(ad,bc)$$

$$f(a,b)f(-c,-d) - f(-a,-b)f(-c,-d) = 2a f\left(\frac{b}{c}, ac^2d\right) f\left(\frac{b}{d}, acd^2\right)$$

$$f(a,b)f(c,d) = \sum_{-\infty}^{\infty}\sum_{-\infty}^{\infty} p^{\frac{(m^2+n^2)}{2} - \frac{(m+n)}{2}} a^m c^n$$

$$f(a,b)f(-a,-b) = f(-a^2,-b^2)\phi(-ab)$$

$$f^2(a,b) + f^2(-a,-b) = 2 f(a^2,b^2)\phi(ab)$$

$$f^2(a,b) - f^2(-a,-b) = 4a f\left(\frac{b}{a}, a^3b\right) \psi(a^2b^2).$$

Exercise 4.1.9: If $|q| < 1$, prove (special cases of Exercise 4.1.8)

$$\phi^2(q) + \phi^2(-q) = 2 \phi^2(q^2)$$

$$\phi^2(q) - \phi^2(-q) = 8q \psi^2(q^4)$$

$$\phi^4(q) - \phi^4(-q) = 16q \phi^2(q^2)\psi^2(q^4) = 16q \psi^4(q^2).$$

Remark 4.1.2: The last formula of Exercise 4.1.9 is the Ramanujan's version of *Jacobi's quartic identity*:

$$\theta_3^4(q) = \theta_2^4(q) + \theta_4^4(q)$$

where

$$\theta_3(q) = \phi(q), \; \theta_2(q) = \phi(-q)$$

and

$$\theta_4(q) = 2q^{\frac{1}{4}} \, \psi(q^2).$$

This identity of Jacobi plays a crucial role in proving the so called *inversion theorems* in the classical theory of elliptic functions.

The following theorem is an important tool for Ramanujan in proving his modular equations. Its variant is known as Schröter's formula. Generalizations of Schröter's formula have also been found.

Theorem 4.1.4. *If* $p = \frac{ab}{cd}$, $|ab| < 1$, $|cd| < 1$, *then*

$$S = \frac{1}{2}\{f(a,b)f(c,d) + f(-a,-b)f(-c,-d)\}$$

$$= \sum_{n=-\infty}^{\infty} (ad)^{n(n+1)/2}(bc)^{n(n-1)/2} f(ac/p^n, bd/p^n)$$

$$D = \frac{1}{2}\{f(a,b)f(c,d) - f(-a,-b)f(-c,-d)\}$$

$$= \sum_{n=-\infty}^{\infty} a^{2n+1}(ad)^{n(n-1)/2}(bc)^{n(n+1)/2} f\left(\frac{c}{ap^n}, \frac{ap^n}{c}abcd\right).$$

Proof 4.1.4: The proof is elementary, and is left as an exercise to the readers as it is quite mechanical.

Remark 4.1.3: Note that S and D above are in terms of infinite series of theta functions.

Exercise 4.1.10: Complete the proof of the above theorem which is no different from that of the special cases $(p = 1)$ given in Exercise 4.1.8.

4.2 Ramanujan's $_1\psi_1$ Summation Formula and Multiplicative Results for Theta Functions

The following $_1\psi_1$-sum formula of Ramanujan (termed *remarkable* by G.H. Hardy) contains as special cases the so called *Jacobi's Triple Product Identity* and *Euler-Cauchy's q-binomial Theorem* which are very important in the classical development of theories of theta and elliptic functions and number theory. Besides, the $_1\psi_1$-sum is still another very important tool for Ramanujan throughout the development of his theories: For $|q| < 1$, and $|\beta q| < |z| < |\alpha q|^{-1}$, we have

$$1 + \sum_{n=1}^{\infty} \frac{(\alpha^{-1};q^2)_n(-\alpha qz)^n}{(\beta q^2;q^2)_n} + \sum_{n=1}^{\infty} \frac{(\beta^{-1};q^2)_n(-\beta q/z)^n}{(\alpha q^2;q^2)_n}$$

$$= \frac{(-qz;q^2)_\infty(-q/z;q^2)_\infty(q^2;q^2)_\infty(\alpha\beta q^2;q^2)_\infty}{(-\alpha qz;q^2)_\infty(-\beta q/z;q^2)_\infty(\alpha q^2;q^2)_\infty(\beta q^2;q^2)_\infty}. \tag{4.2.1}$$

Here, the notations used are the following:

$$(a)_0 = (a;q)_0 = 1;$$
$$(a)_n = (a;q)_n = (1-a)(1-aq)...(1-aq^{n-1})$$

for n (integer) ≥ 1 and

$$(a)_\infty = (a;q)_\infty = \lim_{n\to\infty}(a;q)_n.$$

Before we sketch a proof of this, we will demonstrate many *multiplicative* results for $f(a,b)$, $\phi(q)$, $\psi(q)$ and $f(-q)$ as consequences of (4.2.1). With $\alpha = 0 = \beta$ and $qz = a$, $q/z = b$, we have

$$f(a,b) = (-a;ab)_\infty(-b;ab)_\infty(ab;ab)_\infty$$

$$\phi(q) = \frac{(-q;-q)_\infty}{(q;-q)_\infty}$$

$$\psi(q) = \frac{(q^2;q^2)_\infty}{(q;q^2)_\infty}$$

$$f(-q) = (q;q)_\infty. \tag{4.2.2}$$

The first of (4.2.2) is the famous Jacobi's triple product identity.

Exercise 4.2.1:   Recast, on defining $(a)_\lambda = \frac{(a)_\infty}{(aq^\lambda)_\infty}$ for any real λ, the left side of (4.2.1) as

$$\sum_{n=-\infty}^{\infty} \frac{(\alpha^{-1};q^2)_n(-\alpha qz)^n}{(\beta q^2;q^2)_n} \text{ or } \sum_{n=-\infty}^{\infty} \frac{(\beta^{-1};q^2)_n(-\beta q/z)^n}{(\alpha q^2;q^2)_n}.$$

Exercise 4.2.2:   Prove the trivial identity (due to Euler)

$$(q;q^2)_\infty = (-q;q)_\infty^{-1}.$$

Exercise 4.2.3:   Prove that

$$(a)_\infty = \prod_{k=0}^{n-1}(aq^k;q^n)_\infty$$

and

$$(a)_\infty = \frac{(a;\sqrt{q})_\infty}{(a\sqrt{q};q)_\infty}, \quad |q| < 1.$$

Exercise 4.2.4: Complete the derivations of (4.2.2).

Exercise 4.2.5: If $|q| < 1$, prove that

$$\phi(q)\psi(q^2) = \psi^2(q)$$

and thus complete the proof of the third identity of Exercise 4.2.4 (Jacobi's quartic identity):

$$\phi^4(q) - \phi^4(-q) = 16q\,\psi^4(q^2). \tag{4.2.3}$$

Exercise 4.2.6: Recast (4.2.1) into the form

$$\sum_{n=-\infty}^{\infty} \frac{(a)_n z^n}{(b)_n} = \frac{(az)_\infty (q/(az))_\infty (q)_\infty (b/a)}{(z)_\infty (b/(az))_\infty (b)_\infty (q/a)_\infty}.$$

Exercise 4.2.7: If $|q| < 1$, prove the following multiplicative results:

$$\frac{f(q)}{f(-q)} = \frac{\psi(q)}{\psi(-q)} = \frac{\chi(q)}{\chi(-q)} = \sqrt{\frac{\phi(q)}{\phi(-q)}} \tag{i}$$

$$f^3(-q) = \phi^2(-q)\psi(q) = \sum_{n=0}^{\infty} (-1)^n (2n+1) q^{n(n+1)/2} \tag{ii}$$

$$\chi(q) = \frac{f(q)}{f(-q^2)} = \left[\frac{\phi(q)}{\psi(-q)}\right]^{\frac{1}{3}} = \frac{\phi(q)}{f(q)} = \frac{f(-q^2)}{\psi(-q)} \tag{iii}$$

$$f^3(-q^2) = \phi(-q)\psi^2(q), \quad \chi(q)\chi(-q) = \chi(-q^2),$$

$$\phi^2(-q^2) = \phi(q)\phi(-q), \quad \psi(q)\psi(-q) = \psi(q^2)\phi(-q^2). \tag{iv}$$

Putting $\beta = 1$ in (4.2.1) we get, after some trivial transformations, the famous *Euler-Cauchy q-binomial theorem*:

$$\sum_{n=0}^{\infty} \frac{(a)_n}{(q)_n} t^n = \frac{(at)_\infty}{(t)_\infty}. \tag{4.2.4}$$

Exercise 4.2.8: Justify the terminology q-binomial theorem for (4.2.4).

Exercise 4.2.9: Show that

$$\frac{(q)_\infty (1-q)^{-x}}{(q^{x+1})_\infty} \to \Gamma(x+1) \text{ as } q \to 1.$$

We repeat here for convenience the result of Exercise 4.1.8. For $p = \frac{ab}{cd}$,

$$\frac{1}{2}[f(a,b)f(c,d) \pm f(-a,-b)f(-c,d)]$$
$$= \begin{cases} \sum_{-\infty}^{\infty}(ad)^{n(n+1)/2}(bc)^{n(n-1)/2}f(acp^n, bd/p^n) \\ a\sum_{-\infty}^{\infty}(ad)^{n(n-1)/2}(bc)^{n(n+1)/2}f\left(\frac{c}{ap^n}, \frac{ap^n}{c}abcd\right). \end{cases}$$

In the following section we sketch a proof of (4.2.1) as well as another very useful *quintuple product identity*. For $|q| < 1$ and $z \neq 0$, we have

$$(qz;q^2)_\infty(q/z;q^2)_\infty(q^2;q^2)_\infty(z^2;q^4)_\infty(q^4/z^2;q^4)_\infty$$
$$= \sum_{n=-\infty}^{\infty} q^{3n^2+n}\left[\left(\frac{z}{q}\right)^{3n} - \left(\frac{q}{z}\right)^{3n+1}\right]. \quad (4.2.5)$$

Proofs of $_1\psi_1$-sum and the quintuple product identity: [Proof of $_1\psi_1$-sum by K. Venkatachaliengar]: Put

$$g(z) = \frac{(-qz;q^2)_\infty(-q/z;q^2)_\infty}{(-\alpha qz;q^2)_\infty(-\beta q/z;q^2)_\infty}.$$

We immediately have the functional relation

$$(1+\alpha qz)g(z) = (\beta + qz)g(q^2 z).$$

The problem is one of finding constants c_n such that

$$g(z) = \sum_{-\infty}^{\infty} c_n z^n \text{ in } |\beta q| < |z| < |\alpha q|^{-1}.$$

Assuming tentatively $|\alpha\beta| < 1$ and employing this in the functional relation we get recurrence relations for c_n and c_{-n}, $n = 1,2,\dots$ which, on iteration give

$$c_n = \frac{(\alpha^{-1};q^2)_n(-\alpha q)^n c_0}{(\beta q^2;q^2)_n}, \quad n = 1,2,\dots$$

and

$$c_{-n} = \frac{(\beta^{-1};q^2)_n(-\beta q)^n c_0}{(\alpha q^2;q^2)_n}, \quad n = 1,2,\dots$$

We have thus proved so far:

$$g(z) = \frac{(-qz;q^2)_\infty(-q/z;q^2)_\infty}{(-\alpha qz;q^2)_\infty(-\beta q/z;q^2)_\infty}$$
$$= c_0\left[1 + \sum_{n=1}^{\infty}\frac{(\alpha^{-1};q^2)_n(-\alpha qz)^n}{(\beta q^2;q^2)_n} + \sum_{n=1}^{\infty}\frac{(\beta^{-1};q^2)_n(-\beta q/z)^n}{(\alpha q^2;q^2)_n}\right].$$

Comparing this with (4.2.1) it remains only to prove

$$c_0 = \frac{(\alpha q^2; q^2)_\infty (\beta q^2; q^2)_\infty}{(q^2; q^2)_\infty (\alpha \beta q^2; q^2)_\infty}.$$

Now, on the one hand, from our definition above of $g(z)$ by means of infinite products we have

$$(1 + \alpha qz)\, g(z) = \frac{(-qz; q^2)_\infty (-q/z; q^2)_\infty}{(-\alpha q^3 z; q^2)_\infty (-\beta q/z; q^2)_\infty}.$$

Thus

$$\lim_{z \to -(\alpha q)^{-1}} (1 + \alpha qz)\, g(z) = \frac{(\alpha^{-1}; q^2)_\infty (\alpha q^2; q^2)_\infty}{(q^2; q^2)_\infty (\alpha \beta q^2; q^2)_\infty}.$$

On the otherhand, we have from the infinite series expansion for $g(z)$,

$$\lim_{z \to -(\alpha q)^{-1}} \{(1 + \alpha qz)\, g(z)\} = \lim_{z \to -(\alpha q)^{-1}} \left\{ (1 + \alpha qz) \sum_{n=0}^{\infty} c_n z^n \right\}$$

$$+ \lim_{z \to -(\alpha q)^{-1}} \left\{ (1 + \alpha qz) \sum_{n=1}^{\infty} c_{-n} z^{-n} \right\}$$

$$= \lim_{z \to -(\alpha q)^{-1}} \left\{ (1 + \alpha qz) \sum_{0}^{\infty} c_n z^n \right\}$$

(since the second series is analytic at $z = -(\alpha q)^{-1}$)

$$= \lim_{n \to \infty} \left\{ c_n \left(-\frac{1}{\alpha q} \right)^n \right\}$$

(by using Abel's theorem)

$$= \frac{(\alpha^{-1}; q^2)_\infty}{(\beta q^2; q^2)_\infty} c_0$$

on using the expression for c_n, $n > 0$ obtained above. Equating the two limits we just obtained of $(1 + \alpha qz)\, g(z)$ as $z \to -(\alpha q)^{-1}$, we have the required expression for c_0. The condition $|\alpha \beta| < 1$ can now be removed by analytic continuation.

Exercise 4.2.10: Show in detail:

$$(1 + \alpha qz)\, g(z) = (\beta + qz)\, g(q^2 z).$$

Exercise 4.2.11: Show in detail

$$c_n = \frac{(\alpha^{-1}; q^2)_n (-\alpha q)^n c_0}{(\beta q^2; q^2)_n}, \quad n = 1, 2, \ldots$$

and

$$c_{-n} = \frac{(\beta^{-1};q^2)_n(-\beta q)^n c_0}{(\alpha q^2;q^2)_n}, \quad n = 1,2,\dots$$

Exercise 4.2.12: Explain the need of the tentative assumption $|\alpha\beta| < 1$ for the proof.

Exercise 4.2.13: Putting $\alpha = \beta = -1$ and $z = 1$ in (4.2.1) show that:

$$1 + \sum_1^\infty r_2(n)q^n = 1 + 4 \sum_{m=0}^\infty \sum_{k=1}^\infty (-1)^m q^{k(2m+1)}$$

where $r_2(n)$ denotes the number of representations of n as a sum of two squares. Note that

$$\phi^2(q) = 1 + \sum_{n=1}^\infty r_2(n)q^n = \sum_{-\infty}^\infty \sum_{-\infty}^\infty q^{m^2+n^2}.$$

Exercise 4.2.14: If $d_i(n)$ stands for the number of positive divisors of n that are congruent to $i(\bmod 4)$ then deduce from Exercise 4.2.13 the Jacobi's two square theorem,

$$r_2(n) = 4[d_1(n) - d_3(n)].$$

(Separate $m = 2n$ and $m = 2n + 1$ terms and then compare coefficients of q^n).

Exercise 4.2.15: By putting $\alpha = \beta = -1$ in the $_1\psi_1$-sum (4.2.1) and manipulating the series show that:

$$\frac{(-qz;q^2)_\infty(-q/z;q^2)_\infty(q^2;q^2)_\infty^2}{(qz;q^2)_\infty(q/z;q^2)_\infty(-q^2;q^2)_\infty^2}$$
$$= \frac{1+qz}{1-qz} - 2\sum_1^\infty \frac{q^{3n}z^n}{1+q^{2n}} + 2\sum_1^\infty \frac{q^n z^{-n}}{1+q^{2n}}.$$

Exercise 4.2.16: Dividing the result of Exercise 4.2.15 by

$$\frac{1+qz}{1-qz}$$

and then letting z to $-q^{-1}$, obtain

$$\phi^4(q) = 1 + 8\sum_1^\infty \frac{q^n}{(1+(-q)^n)^2}.$$

Exercise 4.2.17: Show that the result of Exercise 4.2.16 is equivalent to Jacobi's four square theorem: $r_4(n) = 8\sum d$ where $r_4(n)$ stands for the number of representations of the positive integer n as sum of 4 squares and the summation is taken over all positive divisors d of n such that 4 is not divided by d.

Exercise 4.2.18: (A generalization of Jacobi's theorem for power series of $(q)_\infty^3$ as a limiting case of $_1\psi_1$-sum). If $|q|, |\alpha|, |\beta|$ are each < 1, show that

$$\frac{(q)_\infty^3 (\alpha\beta)_\infty}{(\alpha)_\infty^2 (\beta)_\infty^2} = \sum_0^\infty \left\{ \frac{(k+1)(q/\alpha)_k \, \alpha^k}{(\beta)_{k+1}} + \frac{k(q/\beta)_k \, \beta^k}{(\alpha)_{k+1}} \right\}.$$

Proof of the quintuple product identity (4.2.5): Left side of (4.2.5) is nothing but $f(-qz, -q/z) \, f(-z^2, -q^4/z^2)$ but for an additional factor $(q^4; q^4)_\infty$. Now,

$$f(-qz, -q/z) \, f(-z^2, -q^4/z^2)$$
$$= \sum_{-\infty}^\infty (-1)^k q^{k^2} z^k \sum_{-\infty}^\infty (-1)^j q^{2j(j-1)} z^{2j}$$
$$= \sum_{-\infty}^\infty \sum_{-\infty}^\infty (-1)^{k+j} q^{k^2 + 2j(j-1)} z^{k+2j}.$$

On the other hand, denoting by $g(z)$ the left side of (4.2.5), or, which is the same as, putting

$$(q^4; q^4)_\infty \, g(z) = f(-qz, -q/z) \, f(-z^2, -q^4/z^2)$$

we have

$$g(z) = -z^2 \, g\left(\frac{1}{z}\right) \text{ and } g(z) = qz^3 \, g(q^2 z).$$

These imply, on seeking

$$g(z) = \sum_{-\infty}^\infty c_n z^n$$

$$c_2 = -c_0, \ c_1 = 0, \ c_n = q^{2n-5} c_{n-3}$$

and hence

$$c_{3n} = q^{3n^2 - 2n} c_0, \ c_{3n+1} = 0, \ c_{3n+2} = -q^{3n^2 + 2n} c_0.$$

We thus have

$$g(z) = c_0 \left[\sum_{-\infty}^\infty q^{3n^2 - 2n} z^{3n} - \sum_{-\infty}^\infty q^{3n^2 + 2n} z^{3n+2} \right].$$

Equating this with the series already obtained in the beginning of the proof and comparing the constant terms we get

$$c_0 (q^4; q^4)_\infty = \sum_{j=-\infty}^\infty (-1)^j (q^4)^{3j(j-1)/2}.$$

The series on the right side being equal to $(q^4; q^4)_\infty$, we have $c_0 = 1$. This completes the proof of (4.2.5).

4.3 Modular Equations

We need the following Lambert series developments for later use.

$$\phi(q)\phi(q^3) = 1 + 2 \sum_{m=0}^{\infty} \left[\frac{q^{3m+1}}{1-(-1)^m q^{3m+1}} - \frac{q^{3m+2}}{1+(-1)^m q^{3m+2}} \right] \tag{4.3.1}$$

and

$$q\,\psi(q^2)\psi(q^6) = \sum_{n=1}^{\infty} \left[\frac{q^{2n-1} - q^{5(2n-1)}}{1 - q^{6(2n-1)}} \right]. \tag{4.3.2}$$

These are obtainable from the $_1\psi_1$-sum. Indeed putting $\alpha = \beta = -1$, $q = \sqrt{ab}$, $z = \sqrt{a/b}$ and $\alpha = ab = q$, $\beta = (ab)^{-1}$, $z = -a^{-2}$ in (4.2.1) we respectively get

$$\frac{f(a,b)}{f(-a,-b)}\phi^2(-ab) = 1 + 2\sum_{m=0}^{\infty}(-1)^m$$
$$\times \left[\frac{a^{m+1}b^m}{1-a^{m+1}b^m} + \frac{a^m b^{m+1}}{1-a^m b^{m+1}} \right] \tag{4.3.3}$$

and

$$a\,\frac{f(-b/a,-a^3 b)}{f(-a^2,-b^2)}\psi^2(ab) = \sum_{n=1}^{\infty} \frac{a^{2n-1} - b^{2n-1}}{1-(ab)^{2n-1}}. \tag{4.3.4}$$

Putting $a = q$, $b = -q^2$ in the former identity and $a = q$, $b = q^5$ in the latter we get the desired identities. From the above we have

$$\sum_{-\infty}^{\infty}\sum_{-\infty}^{\infty} q^{m^2+mn+n^2} = \phi(q)\,\phi(q^3) + 4q\,\psi(q^2)\psi(q^6)$$
$$= 1 + 6\sum_{n=0}^{\infty}\left[\frac{q^{3n+1}}{1-q^{3n+1}} - \frac{q^{3n+2}}{1-q^{3n+2}} \right] \tag{4.3.5}$$

in which the first equality is yet to be established. But this is easily done. For,

$$\sum_{-\infty}^{\infty}\sum_{-\infty}^{\infty} q^{m^2+mn+n^2} = \left[\sum_{m:\,even}\sum + \sum_{m:\,odd}\sum \right] q^{m^2+mn+n^2}$$
$$= \sum_{-\infty}^{\infty}\sum q^{4k^2+2kn+n^2} + \sum_{-\infty}^{\infty}\sum q^{4k^2+2kn+n^2+4k+n+1}$$
$$= \left(\sum_{l=-\infty}^{\infty} q^{l^2} \right)\left(\sum_{k=-\infty}^{\infty} q^{3k^2} \right)$$
$$+ q\sum_{l=-\infty}^{\infty}(q^2)^{l(l+1)/2}\sum_{k=-\infty}^{\infty}(q^6)^{k(k+1)/2}, \quad l = k+n$$
$$= \phi(q)\,\phi(q^3) + q\,\psi(q^2)\,\psi(q^6)$$

as required.

Exercise 4.3.1: Fill in all the details in the above derivations.

Exercise 4.3.2: Prove the following modular equation of degree 3.

$$\phi(q)\,\phi(q^3) - \phi(-q)\,\phi(-q^3) = 4q\,\psi(q^2)\,\psi(q^6). \tag{4.3.6}$$

Dividing both sides of the quintuple product identity by $1 - (q/z)$ and then letting z to q and then changing q to $q^{\frac{1}{2}}$ we get

$$\sum_{-\infty}^{\infty}(6n+1)q^{(3n^2+n)/2} = \phi^2(-q)\,f(-q). \tag{4.3.7}$$

This is indeed equivalent to the Lambert series development

$$\frac{\phi^3(q)}{\phi(q^3)} = 1 + 6\sum_{0}^{\infty}\left[\frac{(-1)^n q^{3n+1}}{1+(-q)^{3n+1}} + \frac{(-1)^n q^{3n+2}}{1+(-q)^{3n+2}}\right]. \tag{4.3.8}$$

We need only to realize that

$$\sum_{-\infty}^{\infty}(6n+1)q^{(3n^2+n)/2} = \frac{d}{dz}\left(\sum_{-\infty}^{\infty}q^{(3n^2+n)/2}z^{6n+1}\right) \text{ at } z=1$$

$$= \frac{d}{dz}\left\{z\,f(q^2z^6, q/z^6)\right\} \text{ at } z=1$$

$$= f(q^2, q)\frac{d}{dz}\ln\left\{z\,f(q^2z^6, q/z^6)\right\} \text{ at } z=1$$

$$= f(q, q^2)\left[1 + 6\sum_{0}^{\infty}\left[\frac{(-1)^n q^{3n+1}}{1+(-q)^{3n+1}} + \frac{(-1)^n q^{3n+2}}{1+(-q)^{3n+2}}\right]\right].$$

We also need the fact that

$$\phi^2(-q)\,f(-q) = \frac{\phi^3(q)}{\phi(q^3)}f(q, q^2),$$

on employing the triple product identity.

Exercise 4.3.3: Fill in the missing details in the derivation of the Lambert series above for $\phi^3(q)/\phi(q^3)$. Dividing both sides of the quintuple product identity (4.2.5) by $(1 - z^2)$ and then letting $z \to 1$ we get similarly,

$$\frac{\psi^3(q)}{\psi(q^3)} = 1 + 3\sum_{0}^{\infty}\frac{q^{6n+1}}{1-q^{6n+1}} - 3\sum_{0}^{\infty}\frac{q^{6n+5}}{1-q^{6n+5}}. \tag{4.3.9}$$

Exercise 4.3.4: Prove the following modular identities of degree 3.

$$2\frac{\psi^3(q)}{\psi(q^3)} = \frac{\phi(q)}{\phi(q^3)} + \frac{\phi^3(-q^2)}{\phi(-q^6)} \tag{4.3.10}$$

and

$$\frac{\phi^3(q)}{\phi(q^3)} = 3\,\phi(q)\,\phi(q^3) - 2\frac{\phi^3(-q^2)}{\phi(-q^6)}. \tag{4.3.11}$$

4.4 Inversion Formulas and Evaluations

Definition 4.4.1. Gauss' hypergeometric function: If $0 \leq x < 1$, define

$$2F_1\left(\frac{1}{2},\frac{1}{2};1;x\right) = \sum_0^\infty \frac{\left(\frac{1}{2}\right)_n^2 x^n}{(n!)^2}$$

where

$$(a)_0 = 1, \ (a)_n = a(a+1)...(a+n-1), \ a \neq 0 \ n = 1,2,3,...$$

Theorem 4.4.1. *Hypergeometric transformation:* The function $2F_1$ has the property

$$(1+x) \, 2F_1\left(\frac{1}{2},\frac{1}{2};1;x^2\right)$$
$$= 2F_1\left(\frac{1}{2},\frac{1}{2};1;\frac{4x}{(1+x)^2}\right) \tag{4.4.1}$$

or, which is the same as,

$$(1+x) \, 2F_1\left(\frac{1}{2},\frac{1}{2};1;x^2\right)$$
$$= 2F_1\left(\frac{1}{2},\frac{1}{2};1;1-\frac{(1-x)^2}{(1+x)^2}\right). \tag{4.4.2}$$

Proof 4.4.1: For a proof, one can show that each side of (4.4.1) and (4.4.2) satisfies the same second order differential equation and the same set of initial conditions. Details are left as an exercise.

Exercise 4.4.1: Complete the proofs of (4.4.1) and (4.4.2).

Exercise 4.4.2: Prove

$$2F_1\left(\frac{1}{2},\frac{1}{2};1;\frac{\phi^4(-q)}{\phi^4(q)}\right)$$
$$= \frac{\phi^2(q)}{2\,\phi^2(q^2)} 2F_1\left(\frac{1}{2},\frac{1}{2};1;\frac{\phi^4(-q^2)}{\phi^4(q^2)}\right). \tag{4.4.3}$$

Exercise 4.4.3: Prove the transformation

$$2F_1\left(\frac{1}{2},\frac{1}{2};1;1-\frac{\phi^4(-q)}{\phi^4(q)}\right)$$
$$= \frac{\phi^2(q)}{\phi^2(q^2)} 2F_1\left(\frac{1}{2},\frac{1}{2};1;1-\frac{\phi^4(-q^2)}{\phi^4(q^2)}\right). \tag{4.4.4}$$

Theorem 4.4.2. *Evaluation of $_2F_1(\cdot)$ and of $F(\cdot)$ defined below by (4.4.5), inversion of mappings: Let*

$$t = \frac{\phi^2(-q)}{\phi^2(q)} \; and \; x = 1 - \frac{\phi^4(-q)}{\phi^4(q)}.$$

If

$$F(\cdot) = \exp\left\{ -\pi \frac{_2F_1\left(\frac{1}{2},\frac{1}{2};1;1-\cdot\right)}{_2F_1\left(\frac{1}{2},\frac{1}{2};1;\cdot\right)} \right\}, \tag{4.4.5}$$

then

$$F\left(1 - \frac{\phi^4(-q)}{\phi^4(q)}\right) = q. \tag{4.4.6}$$

Further,

$$_2F_1\left(\frac{1}{2},\frac{1}{2};1;1-\frac{\phi^4(-q)}{\phi^4(q)}\right) = \phi^2(q). \tag{4.4.7}$$

Proof 4.4.2: From (4.4.3) and (4.4.4) we get

$$F\left[1 - \frac{\phi^4(-q)}{\phi^4(q)}\right] = \left\{ F\left[1 - \frac{\phi^4(-q^n)}{\phi^4(q^n)}\right] \right\}^{\frac{1}{n}} \tag{4.4.8}$$

for $n = 2^m$, $m = 1, 2, \ldots$ Further, it is not hard to show that

$$F(x) \sim \frac{x}{10} \; as \; x \to 0+.$$

Using this and the Jacobi's quartic identity (of Exercise 4.1.9) in (4.4.8) and then letting $n \to \infty$ we get (4.4.6). Further, iterating (4.4.4) and letting $m \to \infty$ we get (4.4.7).

Exercise 4.4.4: Fill in the details in the proof of (4.4.6) and (4.4.7).

Theorem 4.4.3. *(Evaluation of $\phi^2(\cdot)$, inversion of $q = F(x)$): We have*

$$x = 1 - \frac{\phi^4(-F(x))}{\phi^4(F(x))} \tag{4.4.9}$$

and

$$\phi^2(F(x)) = {_2F_1}\left(\frac{1}{2},\frac{1}{2};1;x\right). \tag{4.4.10}$$

Proof 4.4.3: Given $0 \leq x < 1$, define q by $q = F(x)$ where $F(x)$ is as in (4.4.5). From this and (4.4.6) we have

$$\frac{{}_2F_1\left(\frac{1}{2},\frac{1}{2};1;\frac{\phi^4(-q)}{\phi^4(q)}\right)}{{}_2F_1\left(\frac{1}{2},\frac{1}{2};1;1-\frac{\phi^4(-q)}{\phi^4(q)}\right)} = \frac{{}_2F_1\left(\frac{1}{2},\frac{1}{2};1;1-x\right)}{{}_2F_1\left(\frac{1}{2},\frac{1}{2};1;x\right)}. \tag{4.4.11}$$

Suppose now that

$$_2F_1\left(\frac{1}{2},\frac{1}{2};1;1-\frac{\phi^4(-q)}{\phi^4(q)}\right) \leq {}_2F_1\left(\frac{1}{2},\frac{1}{2};1;1-x\right). \tag{4.4.12}$$

Then this and (4.4.11) imply that

$$_2F_1\left(\frac{1}{2},\frac{1}{2};1;\frac{\phi^4(-q)}{\phi^4(q)}\right) \leq {}_2F_1\left(\frac{1}{2},\frac{1}{2};1;1-x\right).$$

This in turn, by monotonicity of ${}_2F_1\left(\frac{1}{2},\frac{1}{2};1;x\right)$ implies that

$$\frac{\phi^4(-q)}{\phi^4(q)} \leq 1-x \text{ or } x \leq 1 - \frac{\phi^4(-q)}{\phi^4(q)}$$

which implies that

$$_2F_1\left(\frac{1}{2},\frac{1}{2};1;x\right) \leq {}_2F_1\left(\frac{1}{2},\frac{1}{2};1;1-\frac{\phi^4(-q)}{\phi^4(q)}\right). \tag{4.4.13}$$

Now, (4.4.12) and (4.4.13) imply equality of both and hence

$$x = 1 - \frac{\phi^4(-q)}{\phi^4(q)}. \tag{4.4.14}$$

This proves (4.4.9) since we started with $q = F(x)$. Using (4.4.14) in (4.4.7) we have (4.4.10).

Remark 4.4.1: Given q, we have just evaluated $\phi(q)$ as $\phi(q) = \sqrt{z}$ where

$$z = {}_2F_1\left(\frac{1}{2},\frac{1}{2};1;x\right)$$

with

$$x = 1 - \frac{\phi^4(-q)}{\phi^4(q)}.$$

Also, given x we have evaluated $\phi(q)$ as $\phi(q) = \sqrt{z}$ with $q = F(x)$, where

$$z = {}_2F_1\left(\frac{1}{2},\frac{1}{2},;1;x\right).$$

Similarly, one could evaluate $\phi(\cdot)$ and $\psi(\cdot)$ and $f(\cdot)$ at other arguments such as $\pm q, \pm q^2, \ldots$ in terms of x and z. Ramanujan gives scores of such evaluations employing additive and multiplicative theorems, hypergeometric transforms for ${}_2F_1\left(\frac{1}{2},\frac{1}{2};1;x\right)$ and the ${}_1\psi_1$-sum and possibly by other means.

4.5 Modular Identities (Classical Theory)

Definition 4.5.1. A *modular identity (equation) of degree n* is an explicit (algebraic) relation between the *moduli* $\sqrt{\alpha}$ and $\sqrt{\beta}$ and the *multiplier m* implied by a given relation

$$m = \frac{{}_2F_1\left(\frac{1}{2},\frac{1}{2};1;\alpha\right)}{{}_2F_1\left(\frac{1}{2},\frac{1}{2};1;\beta\right)}$$
$$= n\frac{{}_2F_1\left(\frac{1}{2},\frac{1}{2};1;1-\alpha\right)}{{}_2F_1\left(\frac{1}{2},\frac{1}{2};1;1-\beta\right)}. \tag{4.5.1}$$

Or, equivalently, by virtue of the results of the previous section, a modular equation of degree n is an identity involving theta functions at arguments q and q^n. β in (4.5.1) is said to be of degree n (over α).

Ramanujan employed the basic multiplicative and additive results of the theta functions $f(a,b)$, $\phi(q)$, $\psi(q)$, $f(-q)$, the triple product identity and more generally his ${}_1\psi_1$-sum, the quintuple product identity, Schröter's type formulas, Lambert series representations and other properties and techniques to generate identities relating to $\phi(t)$, $\psi(t)$ and $f(t)$ at $t = q$, q^n. Then he used his evaluations of these functions at various arguments as algebraic functions of q, $\alpha(=x)$ and

$$z = {}_2F_1\left(\frac{1}{2},\frac{1}{2};1;x\right).$$

For example we give the following theorem:

Theorem 4.5.1. *The following modular equations of degree* 3 *hold: If* β *is of degree* 3 *over* α, *that is if*

$$m = \frac{{}_2F_1\left(\frac{1}{2},\frac{1}{2};1;\alpha\right)}{{}_2F_1\left(\frac{1}{2},\frac{1}{2};1;\beta\right)} = 3\frac{{}_2F_1\left(\frac{1}{2},\frac{1}{2};1;1-\alpha\right)}{{}_2F_1\left(\frac{1}{2},\frac{1}{2};1;1-\beta\right)}, \tag{4.5.2}$$

then

$$(i)\quad (\alpha\beta)^{\frac{1}{4}} + [(1-\alpha)(1-\beta)]^{\frac{1}{4}} = 1$$

$$(ii)\quad (\alpha^3/\beta)^{\frac{1}{8}} - [(1-\alpha)^3/(1-\beta)]^{\frac{1}{8}} = 1$$
$$= [(1-\beta)^3/(1-\alpha)]^{\frac{1}{8}} - [\beta^3/\alpha]^{\frac{1}{8}} \tag{4.5.3}$$

$$(iii)\quad m = 1 + 2(\beta^3/\alpha)^{\frac{1}{8}}, \ (3/m) = 1 + 2[(1-\alpha)^3/(1-\beta)]^{\frac{1}{8}} \tag{4.5.4}$$

and

$$(iv)\quad m^2\left[(\alpha^3/\beta)^{\frac{1}{8}} - \alpha\right] = \left[(\alpha^3/\beta)^{\frac{1}{8}} - \beta\right]. \tag{4.5.5}$$

Proof 4.5.1: We have

$$\phi^2(q) = {}_2F_1\left(\frac{1}{2}, \frac{1}{2}; 1; \alpha\right)$$

with

$$q = \exp\left(-\pi \frac{{}_2F_1\left(\frac{1}{2}, \frac{1}{2}; 1; 1-\alpha\right)}{{}_2F_1\left(\frac{1}{2}, \frac{1}{2}; 1; \alpha\right)}\right).$$

So,

$$q^3 = \exp\left(-3\pi \frac{{}_2F_1\left(\frac{1}{2}, \frac{1}{2}; 1; 1-\alpha\right)}{{}_2F_1\left(\frac{1}{2}, \frac{1}{2}; 1; \alpha\right)}\right)$$

$$= \exp\left(-\pi \frac{{}_2F_1\left(\frac{1}{2}, \frac{1}{2}; 1; 1-\beta\right)}{{}_2F_1\left(\frac{1}{2}, \frac{1}{2}; 1; \beta\right)}\right)$$

and hence

$$\phi^2(q^3) = {}_2F_1\left(\frac{1}{2}, \frac{1}{2}; 1; \beta\right).$$

Using standard evaluations namely

$$\phi(q) = \left[{}_2F_1\left(\frac{1}{2}, \frac{1}{2}; 1; \alpha\right)\right]^{\frac{1}{2}} = \sqrt{z_1}, \text{ say,}$$

$$\phi(q^3) = \left[{}_2F_1\left(\frac{1}{2}, \frac{1}{2}; 1; \beta\right)\right]^{\frac{1}{2}} = \sqrt{z_3}, \text{ say}$$

$$\phi(-q) = (1-\alpha)^{\frac{1}{4}}\sqrt{z_1}, \ \phi(-q^3) = (1-\beta)^{\frac{1}{4}}\sqrt{z_3}$$

$$\psi(q^2) = \frac{1}{2}\sqrt{z_1}(\alpha/q)^{\frac{1}{4}}, \ \psi(q^6) = \frac{1}{2}\sqrt{z_3}(\beta/q^3)^{\frac{1}{4}} \qquad (4.5.6)$$

in the identity

$$4q\,\psi(q^2)\,\psi(q^6) = \phi(q)\,\phi(q^3) - \phi(-q)\,\phi(-q^3)$$

obtained in an earlier section we have

$$\sqrt{z_1 z_3}(\alpha\beta)^{\frac{1}{4}} = \sqrt{z_1 z_3} - \sqrt{z_1 z_3}(1-\alpha)^{\frac{1}{4}}(1-\beta)^{\frac{1}{4}}$$

which is nothing but (4.5.2). Similarly the identities

$$2\frac{\psi^3(q)}{\psi(q^3)} = \frac{\phi(q)}{\phi(q^3)} + \frac{\phi^3(-q^2)}{\phi(-q^6)}$$

and

$$\frac{\phi^3(q)}{\phi(q^3)} = 3\,\phi(q)\,\phi(q^3) - 2\frac{\phi^3(-q^2)}{\phi(-q^6)}$$

respectively give the first of (4.5.3) and the first of (4.5.4). The other parts of (4.5.3) and (4.5.4) follow similarly. Now, (4.5.3) and (4.5.4) give

$$\alpha = \frac{(m-1)(3+m)^3}{16m^3} \text{ and } \beta = \frac{(m-1)^3(3+m)}{16m}. \tag{4.5.7}$$

Employing (4.5.7) we get (4.5.5).

Exercise 4.5.1: Work out the details in the proof of the last theorem.

4.6 Ramanujan's Theory of Cubic Theta Functions

4.6.1 The cubic theta functions

Definition 4.6.1. We define the cubic theta functions by

$$a(q) = \sum_{-\infty}^{\infty}\sum_{-\infty}^{\infty} q^{m^2+mn+n^2} \tag{4.6.1}$$

$$b(q) = \sum_{-\infty}^{\infty}\sum_{-\infty}^{\infty} w^{m-n} q^{m^2+mn+n^2}, \; w = e^{2\pi i/3} \tag{4.6.2}$$

and

$$c(q) = \sum_{-\infty}^{\infty}\sum_{-\infty}^{\infty} q^{\left(m+\frac{1}{3}\right)^2+\left(m+\frac{1}{3}\right)\left(n+\frac{1}{3}\right)+\left(n+\frac{1}{3}\right)^2}, |q| < 1. \tag{4.6.3}$$

The following lemma tells that $a(q)$ can be expressed in terms of the classical theta functions and that $b(q)$ and $c(q)$ are expressible in terms of $a(q)$ itself.

Lemma 4.6.1: *We have, with* $|q| < 1$,

$$a(q) = \phi(q)\,\phi(q^3) + 4q\,\psi(q^2)\,\psi(q^6) \tag{4.6.4}$$

$$b(q) = \frac{1}{2}[3\,a(q^3) - a(q)] \tag{4.6.5}$$

and

$$c(q) = \frac{1}{2}\{a(q^{\frac{1}{3}}) - a(q)\}. \tag{4.6.6}$$

Proof 4.6.1: Identity (4.6.4), which is the same as

$$\sum_{-\infty}^{\infty}\sum_{-\infty}^{\infty} q^{m^2+mn+n^2} = \phi(q)\,\phi(q^3) + 4q\,\psi(q^2)\,\psi(q^6)$$

has already been proved in Section 4.3. The remaining identities follow by similar series manipulations. We may regard (4.6.7) below as the cubic analogue of the Jacobi's quartic identity of Section 4.1.

Theorem 4.6.1. *(Cubic modular identity of J.M. Borwein and P.B. Borwein): We have*

$$a^3(q) = b^3(q) + c^3(q). \tag{4.6.7}$$

Proof 4.6.2: The facts of the above lemma that $a(q)$ is expressible in terms of the ordinary theta functions $\phi(\cdot)$ and $\psi(\cdot)$ and that $b(q)$ and $c(q)$ are expressible in terms of $a(\cdot)$ along with the evaluations in Section 4.5 and parametrizations (4.6.7), we have

$$a(q) = \sqrt{z_1 z_3}\left(1 + (\alpha\beta)^{\frac{1}{4}}\right)$$
$$= \sqrt{z_1 z_3}\left(1 + \frac{(m-1)(3+m)}{4m}\right)$$
$$= \sqrt{z_1 z_3}\left(\frac{m^2+6m-3}{4m}\right),$$

and similarly

$$b(q) = \frac{1}{2}\sqrt{z_1 z_3}\left(\frac{(3-m)(9-m^2)^{\frac{1}{3}}}{2m^{\frac{2}{3}}}\right)$$

and

$$c(q) = \frac{1}{2}\sqrt{z_1 z_3}\left(\frac{3(m^2-1)^{\frac{1}{3}}(m+1)}{2m}\right).$$

These immediately yield (4.6.7).

Exercise 4.6.1: Work out the details of proof of the lemma above.

Exercise 4.6.2: Work out the details of proof of the theorem above.

Remark 4.6.1: Borwein's proof employs theory of modular forms on the group generated by the transformations $t \to \frac{1}{t}$ and $t \to t + i\sqrt{3}$. They later gave a direct proof, different from the proof given above.

Theorem 4.6.2. *(Ramanujan: Cubic hypergeometric transformation): The hypergeometric function*

$$2F_1\left(\frac{1}{3},\frac{2}{3};1;x\right) = \sum_0^\infty \frac{\left(\frac{1}{3}\right)_n\left(\frac{2}{3}\right)_n x^n}{(n!)^2}$$

satisfies the relation

$$2F_1\left(\frac{1}{3},\frac{2}{3};1;1-\left(\frac{1-x}{1+2x}\right)^3\right)$$

$$= (1+2x)\,2F_1\left(\frac{1}{3},\frac{2}{3};1;x^3\right). \tag{4.6.8}$$

Proof 4.6.3: Each side of (4.6.8) satisfies the differential equation

$$x(1-x)(1+x+x^2)(1+2x)^2y''$$
$$- (1+2x)[(4x^3-1)(x+1)+3x]y' - 2(1-x)^2y = 0.$$

This differential equation has a regular singular point at $x = 0$ and the roots of the associated indicial equation consists of a double zero at $x = 0$. Thus to verify that (4.6.8) holds, we must show that the values at $x = 0$ of the functions and their derivatives on each side are equal. But this is easily seen to be true.

Exercise 4.6.3: Work out the details of the proof of the above theorem.

4.6.2 Inversion formulas and evaluations (cubic theory)

The following theorem gives the cubic analogue of the inversion theorems in the classical case.

Theorem 4.6.3. *Given $0 < q < 1$, define*

$$x = 1 - \frac{b^3(q)}{a^3(q)} = \frac{c^3(q)}{a^3(q)}. \tag{4.6.9}$$

Then we have the inversion

$$q = \exp\left\{-\frac{2\pi}{\sqrt{3}}\frac{2F_1\left(\frac{1}{3},\frac{2}{3};1;1-x\right)}{2F_1\left(\frac{1}{3},\frac{2}{3};1;x\right)}\right\}. \tag{4.6.10}$$

Similarly, given $0 < x < 1$, define

$$q = \exp\left\{-\frac{2\pi}{\sqrt{3}}\frac{2F_1\left(\frac{1}{3},\frac{2}{3};1;1-x\right)}{2F_1\left(\frac{1}{3},\frac{2}{3};1;x\right)}\right\}. \tag{4.6.11}$$

Then

$$x = 1 - \frac{b^3(q)}{a^3(q)} = \frac{c^3(q)}{a^3(q)}. \tag{4.6.12}$$

In both cases we have the evaluation

$$a(q) = {}_2F_1\left(\frac{1}{3}, \frac{2}{3}; 1; x\right), (= z = z(x)), \tag{4.6.13}$$

where, given x we have q as in (4.6.11) and given q we have x as in (4.6.9).

Proof 4.6.4: Putting

$$x = \frac{a(q) - b(q)}{a(q) + 2 b(q)}$$

in (4.6.8) and using the lemma of Section 4.6 we have (with a free use of the cubic modular identity $a^3 = b^3 + c^3$ (4.6.7) throughout).

$${}_2F_1\left(\frac{1}{3}, \frac{2}{3}; 1; \frac{c^3(q)}{a^3(q)}\right)$$
$$= \frac{a(q)}{a(q^3)} \, {}_2F_1\left(\frac{1}{3}, \frac{2}{3}; 1; \frac{c^3(q^3)}{a^3(q^3)}\right).$$

Iterating this we have

$${}_2F_1\left(\frac{1}{3}, \frac{2}{3}; 1; \frac{c^3(q)}{a^3(q)}\right)$$
$$= \frac{a(q)}{a(q^n)} \, {}_2F_1\left(\frac{1}{3}, \frac{2}{3}; 1; \frac{c^3(q^n)}{a^3(q^n)}\right) \tag{4.6.14}$$

$n = 3^m$, $m = 1, 2, \dots$ Letting $m \to \infty$ in this we get

$$a(q) = {}_2F_1\left(\frac{1}{3}, \frac{2}{3}; 1; \frac{b^3(q)}{c^3(q)}\right),$$

incidently proving (4.6.13) with x as in (4.6.9), given q. Similarly, putting $x = \frac{b(q)}{a(q)}$ in (4.6.8) and using the lemma of Section 4.6 we get

$${}_2F_1\left(\frac{1}{3}, \frac{2}{3}; 1; \frac{b^3(q)}{a^3(q)}\right)$$
$$= \frac{a(q)}{n \, a(q^n)} \, {}_2F_1\left(\frac{1}{3}, \frac{1}{3}; 1; \frac{b^3(q^n)}{a^3(q^n)}\right) \tag{4.6.15}$$

with $n = 3^m$, $m = 1, 2, \dots$ Dividing (4.6.15) by (4.6.14) we have

$$F\left(1 - \frac{b^3(q)}{a^3(q)}\right) = \left[F\left(\frac{c^3(q^n)}{a^3(q^n)}\right)\right]^{\frac{1}{n}},\tag{4.6.16}$$

where $F(x)$ is defined by

$$F(x) = \exp\left\{-\frac{2\pi}{\sqrt{3}}\frac{{}_2F_1\left(\frac{1}{3},\frac{2}{3},1;1-x\right)}{{}_2F_1\left(\frac{1}{3},\frac{2}{3};1;x\right)}\right\}.\tag{4.6.17}$$

Letting $n \to \infty$ in (4.6.16) and using the fact that $F(x) \sim \frac{x}{27}$ as $x \to 0+$, we get

$$F\left(\frac{c^3(q)}{a^3(q)}\right) = F\left(1 - \frac{b^3(q)}{c^3(q)}\right) = q\tag{4.6.18}$$

proving (4.6.10). For the other part of the theorem, given x put $q = F(x)$ where $F(x)$ is as in (4.6.17). From this and (4.6.18) we have

$$\frac{{}_2F_1\left(\frac{1}{3},\frac{2}{3};1;\frac{b^3(q)}{a^3(q)}\right)}{{}_2F_1\left(\frac{1}{3},\frac{2}{3};1;1-\frac{b^3(q)}{a^3(q)}\right)} = \frac{{}_2F_1\left(\frac{1}{3},\frac{2}{3};1;1-x\right)}{{}_2F_1\left(\frac{1}{3},\frac{2}{3};1;x\right)}.\tag{4.6.19}$$

Suppose now,

$$_2F_1\left(\frac{1}{3},\frac{2}{3};1;1-\frac{b^3(q)}{a^3(q)}\right) \leq {}_2F_1\left(\frac{1}{3},\frac{1}{3};1;x\right).\tag{4.6.20}$$

This in (4.6.19) gives

$$_2F_1\left(\frac{1}{3},\frac{2}{3};1;\frac{b^3(q)}{a^3(q)}\right) \leq {}_2F_1\left(\frac{1}{3},\frac{2}{3};1;1-x\right)\tag{4.6.21}$$

which implies, by monotonicity of $_2F_1$, that

$$\frac{b^3(q)}{a^3(q)} \leq 1 - x \text{ or } x \leq 1 - \frac{b^3(q)}{a^3(q)}.$$

This in turn gives

$$_2F_1\left(\frac{1}{3},\frac{1}{3};1;x\right) \leq {}_2F_1\left(\frac{1}{3},\frac{2}{3};1;1-\frac{b^3(q)}{a^3(q)}\right).\tag{4.6.22}$$

From (4.6.20) and (4.6.21) we have equality in each of (4.6.20) and (4.6.21) and hence (4.6.4) holds. Now (4.6.13) follows from (4.6.12) on applying the first part of the theorem.

We may regard (4.6.13) as evaluations of $a(\cdot)$ and $z(\cdot)$ as follows:

$$a(F(x)) = z(x) \text{ given } x$$

and

$$z\left(1 - \frac{b^3(q)}{a^3(q)}\right) = a(q), \text{ given } q.\tag{4.6.23}$$

4.6.3 Triplication and trimediation formulas

The following theorem tells us how a given relation among x, q and z gets translated if we replace q by q^3 or $q^{\frac{1}{3}}$. The theorem will be useful in obtaining further evaluations.

Theorem 4.6.4. *Let q, x and z be related as in Theorem 4.6.3. Suppose we are given a relation among x, q and z:*

$$g(x,q,z) = 0. \tag{4.6.24}$$

Then the following relations also hold:

$$g\left[\left\{\frac{1-(1-x)^{\frac{1}{3}}}{1+2(1-x)^{\frac{1}{3}}}\right\}^3, q^3, \frac{1}{3}\left\{1+2(1-x)^{\frac{1}{3}}\right\}z\right] = 0 \tag{4.6.25}$$

and

$$g\left[1-\left\{\frac{1-x^{\frac{1}{3}}}{1+2x^{\frac{1}{3}}}\right\}^3, q^{\frac{1}{3}}, \left(1+2x^{\frac{1}{3}}\right)z\right] = 0. \tag{4.6.26}$$

Proof 4.6.5: Putting

$$x' = 1 - \left(\frac{1-x^{\frac{1}{3}}}{1+2x^{\frac{1}{3}}}\right)^3, \tag{4.6.27}$$

we have

$$x^{\frac{1}{3}} = \frac{1-(1-x')^{\frac{1}{3}}}{1+2(1-x')^{\frac{1}{3}}}. \tag{4.6.28}$$

From the cubic hypergeometric transformation theorem of Section 4.6 we have, on substituting the above and simplifying,

$$z' = z(x') = (1+2x^{\frac{1}{3}})z(x) = (1+2x^{\frac{1}{3}})z \tag{4.6.29}$$

and

$$q' = q(x') = q^{\frac{1}{3}}(x) = q^{\frac{1}{3}}. \tag{4.6.30}$$

Now, suppose (4.6.24) holds. Then, on using (4.6.28), (4.6.29) and (4.6.30) in (4.6.24) we get (4.6.25) with x', q' and z' in place of x, q and z. We merely drop primes to get (4.6.29). Similarly, starting with (4.6.24) with x', q' and z' in place of x, q and z we get (4.6.26) on using (4.6.27), (4.6.29) and (4.6.30).

Corollary 4.6.1: With x, q and z as in the Theorem 4.6.3 we have

$$b(q) = (1-x)^{\frac{1}{3}}z, \quad c(q) = x^{\frac{1}{3}}z \qquad (4.6.31)$$

(in addition to $a(q) = z$ of the Theorem 4.6.3).

Proof 4.6.6: By Lemma 4.6.1 and the Theorem 4.6.3 and Theorem 4.6.4 we have

$$b(q) = \frac{1}{2}\left\{3\frac{1}{3}(1+2(1-x)^{\frac{1}{3}})z - z\right\} = (1-x)^{\frac{1}{3}}z$$

and similarly

$$c(q) = \frac{1}{2}\left\{(1+2x^{\frac{1}{3}})z - z\right\} = x^{\frac{1}{3}}z.$$

Exercise 4.6.4: Complete the details of proofs of the theorem and the corollary.

4.6.4 Further evaluations

Theorem 4.6.5. *For any* $0 \leq q < 1$, *we have*

$$q\, f^{24}(-q) = \frac{1}{27}b^9(q)\, c^3(q). \qquad (4.6.32)$$

Proof 4.6.7: We will start with a known evaluation in classical theory namely,

$$q\, f^{24}(-q) = \frac{1}{16}z_1^{12}\alpha(1-\alpha)^4 \qquad (4.6.33)$$

where

$$z_1 = {}_2F_1\left(\frac{1}{2}, \frac{1}{2}; 1; \alpha\right)$$

with

$$\alpha = 1 - \frac{\phi^4(-q)}{\phi^4(q)}.$$

We also know that

$$a(q) = \sqrt{z_1 z_3}\frac{(m^2 + 6m - 3)}{4m},$$

$$b(q) = \sqrt{z_1 z_3}\frac{(3-m)(9-m^2)^{\frac{1}{3}}}{4m^{\frac{2}{3}}},$$

and

$$c(q) = \sqrt{z_1 z_3} \frac{3(m+1)(m^2-1)^{\frac{1}{3}}}{4m}$$

with

$$z_3 = {}_2F_1\left(\frac{1}{2}, \frac{1}{2}; 1; \beta\right),$$

β being the modulus of order 3 over α in the classical theory, and with $m = z_1/z_3$. Further, we know that

$$\alpha = \frac{(m-1)(3+m)^3}{16m^3}, \quad \beta = \frac{(m-1)^3(3+m)}{16m}.$$

With these, we get,

$$b(q) = z_1 m^{\frac{1}{2}} \alpha^{\frac{1}{8}} (1-\alpha)^{\frac{1}{2}} / \left[2^{\frac{1}{3}} \beta^{\frac{1}{24}} (1-\beta)^{\frac{1}{6}}\right]$$

and

$$c(q) = 3z_1 \beta^{\frac{1}{8}} (1-\beta)^{\frac{1}{2}} / \left[2^{\frac{1}{3}} m^{\frac{1}{2}} \alpha^{\frac{1}{24}} (1-\alpha)^{\frac{1}{6}}\right]$$

whence the right side of (4.6.33) becomes $\frac{1}{16} z_1^{12} \alpha (1-\alpha)^4$ which is the right side of (4.6.33) also. Hence the proof of (4.6.33) is complete.

Corollary 4.6.2: Given $0 \leq q < 1$, we have

$$q^{\frac{1}{24}} f(-q) = \sqrt{z} 3^{-\frac{1}{8}} x^{\frac{1}{24}} (1-x)^{\frac{1}{8}} \tag{4.6.34}$$

where x is as in the corollary of Section 4.6.3

Proof 4.6.8: From Theorem 4.6.5 we have

$$q^{\frac{1}{24}} f(-q) = 3^{-\frac{1}{8}} b^{\frac{3}{8}}(q) \, c^{\frac{1}{8}}(q).$$

On using the corollary of Section 4.6.3 this becomes

$$q^{\frac{1}{24}} f(-q) = 3^{-\frac{1}{8}} (1-x)^{\frac{1}{8}} z^{\frac{3}{8}} x^{\frac{1}{24}} z^{\frac{1}{8}}$$

which reduces to the required result (4.6.34).

Corollary 4.6.3:

$$q^{\frac{1}{8}} f(-q^3) = \sqrt{z} 3^{-\frac{3}{8}} x^{\frac{1}{8}} (1-x)^{\frac{1}{24}} \tag{4.6.35}$$

where x, q and z are as in the corollary of Section 4.6.3.

Proof 4.6.9: Changing q to q^3 in Corollary 4.6.2 and using the triplication formula we have the required result.

4.6.5 Evaluations of Ramanujan-eisenstein series (L, M, N or P, Q, R)

Definition 4.6.2. Following series are of importance in number theory and theory of elliptic functions.

$$P(q) = L(q) = 1 - 24 \sum_{1}^{\infty} \frac{nq^n}{1-q^n}$$

$$Q(q) = M(q) = 1 + 240 \sum_{1}^{\infty} \frac{n^3 q^n}{1-q^n}$$

$$R(q) = N(q) = 1 - 504 \sum_{1}^{\infty} \frac{n^5 q^n}{1-q^n}.$$

Theorem 4.6.6. *If q, x and z are related by*

$$q = \exp\left(-\frac{2\pi}{\sqrt{3}} \frac{{}_2F_1\left(\frac{1}{3}, \frac{2}{3}; 1; 1-x\right)}{{}_2F_1\left(\frac{1}{3}, \frac{2}{3}; 1; x\right)}\right)$$

or, equivalently

$$x = 1 - \frac{b^3(q)}{a^3(q)} = \frac{c^3(q)}{a^3(q)}$$

and

$$z = {}_2F_1\left(\frac{1}{3}, \frac{2}{3}; 1; x\right),$$

then

$$L(q) = (1-4x)z^2 + 12x(1-x)z\frac{dz}{dx}$$

$$M(q) = z^4(1+8x)$$

$$N(q) = z^6(1-20x-8x^2)$$

$$M(q^3) = z^4\left(1-\frac{8}{9}x\right)$$

$$N(q^3) = z^6\left(1-\frac{4}{3}x+\frac{8}{27}x^2\right).$$

Proof 4.6.10: Since $f(-q) = \prod_{n=1}^{\infty}(1-q^n)$ we have

$$L(q) = q\frac{d}{dq}\ln\left(q\, f^{24}(-q)\right)$$

$$= q\frac{d}{dx}\ln\left(\frac{1}{27}z^{12}x(1-x)^3\right)\frac{dx}{dq}$$

(on using a result proved earlier). This, along with a known result from the theory of hypergeometric series, namely,

$$_2F_1\left(\frac{1}{3},\frac{2}{3};1;x^2\right)\frac{d}{dx}\left[\frac{2\pi}{\sqrt{3}}\frac{_2F_1\left(\frac{1}{3},\frac{2}{3};1;1-x\right)}{_2F_1\left(\frac{1}{3},\frac{2}{3};1;x\right)}\right]$$
$$= -\frac{1}{x(1-x)}$$

gives the required result for L. (First obtain from the hypergeometric result $\frac{dq}{dx} = -\frac{1}{x(1-x)z^2}$.). The hypergeometric differential equation for

$$z = {}_2F_1\left(\frac{1}{3},\frac{2}{3};1;x\right)$$

namely

$$\frac{2}{9}z = \frac{d}{dx}\left\{x(1-x)\frac{dz}{dx}\right\},$$

the Ramanujan's differential equation

$$q\frac{dL}{dq} = \frac{1}{12}\{L^2 - M\}$$

and the results just obtained, give the required evaluation for $M(q)$. If we use the Ramanujan differential equation

$$q\frac{dM}{dq} = \frac{1}{12}\{LM - N\}$$

and proceed similarly we get the result for $N(q)$. The evaluations for $M(q^3)$ and $N(q^3)$ are now obtained on applying the triplication formulas.

Exercise 4.6.5:   Fill in the details in the proof above.

4.6.6 The cubic analogue of the Jacobian elliptic functions

The familiar Jacobian elliptic function $\phi = \phi(\theta)$ can be defined by means of the integral

$$\theta = \int_0^\phi \frac{_2F_1\left(\frac{1}{2},\frac{1}{2};\frac{1}{2};x\sin^2 t\right)}{_2F_1\left(\frac{1}{2},\frac{1}{2};\frac{1}{2};x\right)}dt$$

$0 \le \phi \le \frac{\pi}{2}$, $0 \le x < 1$, or equivalently

$$\theta = \frac{1}{{}_2F_1\left(\frac{1}{2},\frac{1}{2};1;x\right)} \int_0^\phi \frac{dt}{\sqrt{1-x\sin^2 t}}.$$

In analogy with this we have the cubic analogue of the Jacobian elliptic function given by Ramanujan by means of the following theorem:

Theorem 4.6.7. *(Ramanujan): If* $0 \le \phi \le \frac{\pi}{2}$, $0 < x \le 1$ *and*

$$\theta = \int_0^\phi \frac{{}_2F_1\left(\frac{1}{3},\frac{1}{3};\frac{1}{2};x\sin^2 t\right)}{{}_2F_1\left(\frac{1}{3},\frac{1}{3};1;x\right)} dt$$

or, which is the same as

$$\theta = \frac{1}{{}_2F_1\left(\frac{1}{3},\frac{2}{3};1;x\right)} \int_0^\phi \frac{\cos\left(\frac{\sin^{-1}(\sqrt{x}\sin t)}{3}\right)}{\sqrt{1-x\sin^2 t}} dt$$

then

$$\phi = \theta + 3\sum_{n=1}^\infty \frac{q^n \sin 2n\theta}{n(1+q^n+q^{2n})},$$

$0 \le \theta \le \frac{\pi}{2}$, *where*

$$q = \exp\left[-\frac{2\pi}{\sqrt{3}} \frac{{}_2F_1\left(\frac{1}{3},\frac{2}{3};1;1-x\right)}{{}_2F_1\left(\frac{1}{3},\frac{2}{3};1;x\right)}\right].$$

Proof 4.6.11: Refer to the paper: B.C. Berndt, S. Bhargava, F.G. Garvan, Ramanujan's Theories of Elliptic Functions to Alternative Bases, *Transactions of the American Mathematical Society*, **347** (1995), 4163–4244.

TEST

on Ramanujan's work

(Time : 1 hour)

4.1. If $f(a,b) = \sum_{-\infty}^{\infty} a^{k(k+1)/2} b^{k(k-1)/2}$, $|ab| < 1$, then show:

$$(i) \quad f(1,a) = 2\sum_0^\infty a^{k(k+1)/2}, \ |a| < 1,$$

$$(ii) \quad f(1,a) = 2 f(a,a^3), \ |a| < 1.$$

4.2. If $\phi(q) = \sum_{-\infty}^{\infty} q^{n^2}$ and $\psi(q) = \sum_{0}^{\infty} q^{k(k+1)/2}$ with $|q| < 1$, then show that

$$\phi^2(q) - \phi^2(-q) = 8q \, \psi(q^8) \, \phi(q^4).$$

4.3. Given that $f(a,b) = (-a, ab)_\infty (-b, ab)_\infty (ab, ab)_\infty$, show that

$$\phi(-q) = (q;q)_\infty/(-q;q)_\infty, \quad |q| < 1$$

and

$$\psi(q) = (q^2; q^2)_\infty/(q; q^2)_\infty.$$

Hint: Prove and use

$$(q; q^2)_\infty = (-q; q)_\infty^{-1}$$

where

$$\phi(q) = f(q,q) \text{ and } \psi(q) = \frac{1}{2} f(1,q).$$

4.4. Given that

$$\sum_{-\infty}^{\infty} \frac{(a)_n z^n}{(b)_n} = \frac{(az)_\infty \left(\frac{q}{az}\right)_\infty (q)_\infty \left(\frac{b}{a}\right)_\infty}{(z)_\infty \left(\frac{b}{az}\right)_\infty (b)_\infty \left(\frac{q}{a}\right)_\infty},$$

show that

$$(i) \quad \sum_{0}^{\infty} \frac{(a)_n z^n}{(q)_n} = \frac{(az)_\infty}{(z)_\infty}, \quad |q| < 1$$

$$(ii) \quad \lim_{q \to 1} \sum_{0}^{\infty} \frac{(q^\alpha)_n z^n}{(q)_n} = (1-z)^{-\alpha}, \quad |q| < 1.$$

(S. Bhargava)

4.7 The One-variable Cubic Theta Functions

For the sake of completeness some definitions and some basic properties will be repeated here.

4.7.1 Cubic theta functions and some properties

Definition 4.7.1. One variable cubic theta functions.
We define for $|q| < 1$,

$$a(q) = \sum_{m,n=-\infty}^{\infty} q^{m^2+mn+n^2},$$

$$b(q) = \sum_{m,n=-\infty}^{\infty} \omega^{m-n} q^{m^2+mn+n^2}, \quad (\omega = e^{\frac{2\pi i}{3}}),$$

$$c(q) = \sum_{m,n=-\infty}^{\infty} q^{(m+\frac{1}{3})^2+(m+\frac{1}{3})(n+\frac{1}{3})+(n+\frac{1}{3})^2}.$$

Remark 4.7.1: We note the analogy between the above functions and the one variable classical theta functions (in Ramanujan's and Jacobi's notations, respectively).

$$\phi(q) = \Theta_3(q) = \sum_{n=-\infty}^{\infty} q^{n^2},$$

$$\phi(-q) = \Theta_4(q) = \sum_{n=-\infty}^{\infty} (-1)^n q^{n^2},$$

$$2q^{\frac{1}{4}}\psi(q^2) = \Theta_2(q) = \sum_{n=-\infty}^{\infty} q^{(n+\frac{1}{2})^2}.$$

Exercises 4.7.

4.7.1. Discuss the convergence of the series in Definition 4.7.1 and Remark 4.7.1.

4.7.2. Prove

(i) $a(q) = \phi(q)\phi(q^3) + 4q\psi(q^2)\psi(q^6)$

(ii) $a(q^4) = \frac{1}{2}[\phi(q)\phi(q^3) + \phi(-q)\phi(-q^3)]$

(iii) $b(q) = \frac{3}{2}a(q^3) - \frac{1}{2}a(q)$

(iv) $c(q) = \frac{1}{2}a(q^{\frac{1}{3}}) - \frac{1}{2}a(q)$

(v) $c(q^3) = \frac{1}{3}[a(q) - b(q)]$

(vi) $b(q) = a(q^3) - c(q^3)$.

4.7.2 Product representations for $b(q)$ and $c(q)$

Theorem 4.7.1. *We have*

$$\text{(i)} \quad b(q) = \frac{f^3(-q)}{f(-q^3)} = \frac{(q;q)_\infty^3}{(q^3;q^3)_\infty}$$

$$\text{(ii)} \quad c(q) = 3q^{\frac{1}{3}} \frac{f^3(-q^3)}{f(-q)} = 3q^{\frac{1}{3}} \frac{(q^3;q^3)_\infty^3}{(q;q)_\infty},$$

where, following Ramanujan,

$$f(-q) = \prod_{n=1}^{\infty}(1-q^n) = (q;q)_\infty.$$

Proof: (Borwein, et al (1994)):

The Euler-Cauchy q-binomial theorem says that, for $|q| < 1$,

$$\frac{(a;q)_\infty}{(b;q)_\infty} = \sum_{n=0}^{\infty} \frac{(a/b)_n}{(q)_n} b^n$$

where, as usual,

$$(a)_\infty = (a;q)_\infty = \prod_{n=0}^{\infty}(1-aq^n),$$

$$(a)_n = (a;q)_n = \prod_{k=1}^{n}(1-aq^{k-1}).$$

Letting b to 0 in the q-binomial theorem we have (Euler)

$$(a;q)_\infty = \sum_{n=0}^{\infty} \frac{(-1)^n a^n q^{n(n-1)/2}}{(q)_n}.$$

This gives, since

$$(a^3;q^3)_\infty = (a;q)_\infty (a\omega;q)_\infty (a\omega^2;q)_\infty,$$

$$\sum_{n=0}^{\infty} \frac{a^{3n} q^{3n(n-1)/2}}{(q^3;q^3)_\infty} = \sum_{n_1,n_2,n_3=0}^{\infty} \omega^{n_1+2n_2} a^{n_0+n_1+n_3}$$

$$\times \frac{q^{[n_0(n_0-1)+n_1(n_1-1)+n_2(n_2-1)]/2}}{(q)_{n_0}(q)_{n_1}(q)_{n_2}}.$$

Equating the coefficients of like powers of a, we have,

$$\frac{1}{(q^3;q^3)_\infty} = \sum_{n_0+n_1+n_2=3n} \omega^{n_1-n_2} \frac{q^{[n_0(n_0-1)+n_1(n_1-1)+n_2(n_2-1)]/2}}{(q)_{n_0}(q)_{n_1}(q)_{n_2}}.$$

Or, changing n_i to $m_i + n$, $i = 0, 1, 2$,

$$\frac{1}{(q^3; q^3)_\infty} = \sum_{m_0 + m_1 + m_2 = 0} \omega^{m_1 - m_2} \frac{q^{\frac{1}{2}(m_0^2 + m_1^2 + m_2^2)}}{(q)_{m_0 + n}(q)_{m_1 + n}(q)_{m_2 + n}}.$$

Letting n to ∞, this gives

$$\sum_{m_1, m_2 = -\infty}^{\infty} \omega^{m_1 - m_2} q^{m_1^2 + m_1 m_2 + m_2^2} = \frac{(q)_\infty^3}{(q^3; q^3)_\infty}.$$

This indeed is the first of the required results.

We leave the proof of part (ii) as an exercise.

Exercise 4.7.3: Prove the q-binomial theorem.

Exercise 4.7.4: Complete the proof to the second part of Theorem 4.7.1. (see Exercise 4.7.3 for a proof.)

4.7.3 The cubic analogue of Jacobi's quartic modular equations

The following theorem gives two versions of the cubic counterpart of the Jacobi's modular equation

$$\phi^4(q) = \phi^4(-q) + 16q\psi^4(q^2).$$

The first version is Entry (iv) of Chapter 20 in Ramanujan's Second Notebook, and the second is due to Borwein, et al (1994).

Theorem 4.7.2.

$$3 + \frac{f^3(-q^{\frac{1}{3}})}{q^{\frac{1}{3}} f^3(-q^3)} = \left(27 + \frac{f^{12}(-q)}{q f^{12}(-q^3)}\right)^{\frac{1}{3}},$$

or, what is the same,

$$a^3(q) = b^3(q) + c^3(q).$$

Proof of first version (Berndt (1985).)

We need the following results concerning the classical theta function and its restrictions as found in Chapter 16 of Ramanujan's Second Notebook, see Ramanujan (1957) and Adiga, et al (1985).

$$f(a, b) = \sum_{-\infty}^{\infty} a^{n(n+1)/2} b^{n(n-1)/2}$$

$$= (-a; ab)_\infty (-b; ab)_\infty (ab; ab)_\infty, \ |ab| < 1,$$

$$\phi(q) = f(q, q) = (-q; -q)_\infty (-q; q^2)_\infty^2 = \frac{(-q; -q)_\infty}{(q; -q)_\infty},$$

$$\chi(q) = (-q; q^2)_\infty = \left[\frac{\phi(q)}{\psi(-q)}\right]^{\frac{1}{3}} = \frac{\phi(q)}{f(q)},$$

$$\psi(q) = f(q, q^3) = f(q^3, q^6) + q\psi(q^9).$$

These follow by elementary series and product manipulations (Adiga, et al (1985)). Now, setting

$$v = q^{\frac{1}{3}}\chi(-q)/\chi^3(-q^3),$$

We have

$$v^{-1} = \frac{(-q;q)_\infty \phi(-q^3)}{q^{\frac{1}{3}}\psi(q^3)}$$

$$= \frac{(-q^3;q^3)_\infty f(q,q^2)\phi(-q^3)}{q^{\frac{1}{3}}\psi(q^3)(q^3;q^3)_\infty}$$

$$= \frac{f(q,q^2)}{q^{\frac{1}{3}}\psi(q^3)} = \frac{\psi(q^{\frac{1}{3}})}{q^{\frac{1}{3}}\psi(q^3)} - 1.$$

Thus,

$$1 + v^{-1} = \frac{\psi(q^{\frac{1}{3}})}{q^{\frac{1}{3}}\psi(q^3)}.$$

Similarly,

$$1 - 2v = \frac{\phi(-q^{\frac{1}{3}})}{\phi(-q^3)}.$$

From the last two identities, we have

$$\frac{f^3(-q^{\frac{1}{3}})}{q^{\frac{1}{3}}f^3(-q^3)} = \frac{\phi^2(-q^{\frac{1}{3}})\psi(q^{\frac{1}{3}})}{q^{\frac{1}{3}}\phi^2(-q^3)\psi(q^3)}$$

$$= (1-2v)^2\left(1+\frac{1}{v}\right) = 4v^2 + \frac{1}{v} - 3.$$

Changing $q^{\frac{1}{3}}$ to $\omega q^{\frac{1}{3}}$ and $q^{\frac{1}{3}}$ to $\omega^2 q^{\frac{1}{3}}$ in this equation and multiplying it with the resulting two equations we have, on some manipulations,

$$\frac{f^{12}(-q)}{qf^{12}(-q^3)} = \left(4v^2 + \frac{1}{v}\right)^3 - 27$$

$$= \left(\frac{f^3(-q^{\frac{1}{3}})}{q^{\frac{1}{3}}f^3(-q^3)} + 3\right)^3 - 27,$$

which is the desired identity.

Proof of the second version (Borwein, et al (1994).)

Firstly, we have, under the transformations $q \to \omega q$,

$$(i) \quad a(q^3) = \sum_{m,n=-\infty}^{\infty} q^{3(m^2+mn+n^2)} \to a(q^3)$$

and similarly,

$$\text{(ii)} \quad b(q^3) \to b(q^3) \quad \text{and} \quad \text{(iii)} \quad c(q^3) \to \omega c(q^3).$$

Similarly, we have $a(q^3)$, $b(q^3)$ and $c(q^3)$ going respectively to $a(q^3)$, $b(q^3)$ and $\omega^2 c(q^3)$ under $q \to \omega^2 q$. Thus, for these and result (v) of Exercise 4.7.2, we have

$$b(\omega q) = a(q^3) - \omega c(q^3),$$
$$b(\omega^2 q) = a(q^3) - \omega^2 c(q^3)$$

and, therefore,

$$b(q)b(\omega q)b(\omega^2 q) = (a(q^3) - c(q^3))(a(q^3) - \omega c(q^3))$$
$$\times (a(q^3) - \omega^2 c(q^3))$$
$$= a^3(q^3) - c^3(q^3).$$

However, on using Part (i) of Theorem 4.7.1, the left side of the last equality equals $b^3(q^3)$ on some manipulations. This completes the proof of the desired identity.

Remark 4.7.2: For still another proof of the theorem within Ramanujan's repertoire, one may see Berndt, et al (1995).

Remark 4.7.3: (Chan (1995)). That the two versions of the theorem are equivalent follows on employing Theorem 4.7.1 and result (v) of Exercise 4.7.2.

Exercise 4.7.5: Prove the various identities quoted in the proof to Theorem 4.7.2.

Exercise 4.7.6: Complete the manipulations indicated throughout the proof of Theorem 4.7.2.

Exercise 4.7.7: Work out the details under Remark 4.7.3.

4.8 The Two-variable Cubic Theta Functions

4.8.1 Series definitions and some properties

Definition 4.8.1. Two-variable cubic theta functions.
For $|q| < 1, z \neq 0$, we define

$$a(q,z) = \sum_{m,n=-\infty}^{\infty} q^{m^2+mn+n^2} z^{m-n}$$

$$b(q,z) = \sum_{m,n=-\infty}^{\infty} \omega^{m-n} q^{m^2+mn+n^2} z^n, \quad (\omega = e^{2\pi i/3}),$$

$$c(q,z) = \sum_{m,n=-\infty}^{\infty} q^{(m+\frac{1}{3})^2+(m+\frac{1}{3})(n+\frac{1}{3})+(n+\frac{1}{3})^2} z^{m-n}$$

$$a'(q,z) = \sum_{m,n=-\infty}^{\infty} q^{m^2+mn+n^2} z^n.$$

Remark 4.8.1: We note that the one-variable functions of Section 4.7 are restrictions of the above functions at $z = 1$. In fact

$$a(q) = a(q,1) = a'(q,1);$$
$$b(q) = b(q,1); \ c(q) = c(q,1).$$

Remark 4.8.2: We note the analogy between the above functions and the two-variable classical theta functions (in Ramanujan's and Jacobi's notations, respectively);

$$f(qz,qz^{-1}) = \Theta_3(q,z) = \sum_{n=-\infty}^{\infty} q^{n^2} z^n,$$

$$f(-qz,-qz^{-1}) = \Theta_4(q,z) = \sum_{n=-\infty}^{\infty} (-1)^n q^{n^2} z^n,$$

$$q^{\frac{1}{4}} f(q^2 z, z^{-1}) = \Theta_2(q,z) = \sum_{n=-\infty}^{\infty} q^{(n+\frac{1}{2})^2} z^n,$$

where, as before, following Ramanujan,

$$f(a,b) = \sum_{-\infty}^{\infty} a^{n(n+1)/2} b^{n(n-1)/2}.$$

Exercises   4.8.

4.8.1 Discuss the convergence of the series in Definition 4.8.1.

4.8.2. Prove,

(i) $\ a'(q,z) = z^2 q^3 a'(q,zq^3),$

(ii) $\ a(q,z) = z^2 qa(q,zq),$

(iii) $\ b(q,z) = z^2 q^3 b(q,zq),$

(iv) $\ (q,z) = z^2 qc(q,zq),$

(v) $\ a'(q,z) = a(q^3,z) + 2qc(q^3,z),$

(vi) $\ b(q,z) = a(q^3,z) - qc(q^3,z),$

(vii) $\ a'(q,z) = b(q,z) + 3qc(q^3,z).$

4.8.2 Product representations for $b(q,z)$ and $c(q,z)$

Theorem 4.8.1. *We have,*

$$\text{(i)} \quad b(q,z) = (q)_\infty (q^3;q^3)_\infty \frac{(zq)_\infty (z^{-1}q)_\infty}{(zq^3;q^3)_\infty (z^{-1}q^3;q^3)_\infty}$$

$$\text{(ii)} \quad c(q,z) = q^{\frac{1}{3}}(1+z+z^{-1})(q)_\infty (q^3;q^3)_\infty$$
$$\times \frac{(z^3q^3;q^3)_\infty (z^{-1}q^3;q^3)_\infty}{(zq)_\infty (z^{-1}q)_\infty}.$$

Here, $(a)_\infty = (a;q)_\infty$, for brevity.

Proof (Part (i): Hirschhorn, et al.) We have, by definition of $b(q,z)$,

$$b(q,z) = \sum_{n:even,-\infty}^{\infty} \omega^{m-n} q^{m^2+mn+n^2} z^n + \sum_{n:odd,-\infty}^{\infty} \omega^{m-n} q^{m^2+mn+n^2} z^n.$$

Setting $n = 2k$ in the first sum and $n = 2k+1$ in the second, we get, after slight manipulations,

$$b(q,z) = \sum_{m=-\infty}^{\infty} \omega^m q^{m^2} \sum_{k=-\infty}^{\infty} q^{3k^2} z^{2k}$$
$$+ \omega^{-1} qz \sum_{m=-\infty}^{\infty} \omega^m q^{m^2+m} \sum_{k=-\infty}^{\infty} q^{3k^2+3k} z^{2k}.$$

Applying Jacobi's triple product identity (the product form of $f(a,b)$ met with in the proof of Theorem 4.7.2) to each of the sums we have, after some recombining of the various products involved,

$$b(q,z) = \frac{(-q^3;q^6)_\infty (q^2;q^2)_\infty (q^6;q^6)_\infty}{(-q;q^2)_\infty} (-z^2 q^3;q^6)_\infty$$
$$\times (-z^{-2}q^3;q^6)_\infty$$
$$-q(z+z^{-1})\frac{(-q^6;q^6)_\infty (q^2;q^2)_\infty (q^6;q^6)_\infty}{(-q^2;q^2)_\infty}$$
$$\times (-z^2 q^6;q^6)_\infty (-z^{-2}q^6;q^6)_\infty.$$

On the other hand, one can prove that the right side, say $G(z)$, of the required identity for $b(q,z)$ is precisely the right side of the above identity. One has to first observe that $G(q^3 z) = z^{-2}q^{-3}G(z)$ and then use it to show that we can have

$$G(z) = C_0 \sum_{n=-\infty}^{\infty} q^{3n^2} z^{2n} + C_1 z \sum_{n=-\infty}^{\infty} q^{3n^2+3n} z^{2n}, \text{ or,}$$
$$G(z) = C_0(-z^2 q^3;q^6)_\infty (-z^2 q^3;q^6)_\infty (q^6;q^6)_\infty$$
$$+ C_1(z+z^{-1})(-z^2 q^6;q^6)_\infty (-z^{-2}q^6;q^6)_\infty (q^6;q^6)_\infty.$$

It is now enough to evaluate C_0 and C_1. For this, one can put successively $z = i$ and $z = iq^{-\frac{3}{2}}$ in the last equation. Now, for Part (ii) of the theorem, we have, by definition of $c(q,z)$,

$$c(q,z) = q^{\frac{1}{3}} \sum_{m,n=-\infty}^{\infty} q^{m^2+mn+n^2+m+n} z^{m-n}$$

$$= q^{\frac{1}{3}} \sum_{n,k=-\infty}^{\infty} q^{(n+2k)^2+(n+2k)n+n^2+2n+2k} z^{2k}$$

$$+ q^{\frac{1}{3}} z \sum_{n,k=-\infty}^{\infty} q^{(n+2k+1)^2+(n+2k+1)n+n^2+2n+2k+1} z^{2k+1}$$

(on separating even and odd powers of z),

$$= q^{\frac{1}{3}} \sum_{n,k=-\infty}^{\infty} q^{3(n+k)^2+2(n+k)+k^2} z^{2k}$$

$$+ q^{\frac{1}{3}} z \sum_{n,k=-\infty}^{\infty} q^{3(n+k)^2+5(n+k)+2+k^2+k} z^{2k}$$

$$= q^{\frac{1}{3}} \sum_{t,k=-\infty}^{\infty} q^{3t^2+2t} q^k z^{2k} + q^{\frac{1}{3}} z$$

$$\times \sum_{t',k=-\infty}^{\infty} q^{(3t'-1)(t'+1)} q^{k^2} (z^2 q)^k$$

(setting $n+k = t$ and $n+k+1 = t'$),

$$= q^{\frac{1}{3}} \sum_{t=-\infty}^{\infty} q^{3t^2+2t} \sum_{k=-\infty}^{\infty} q^{k^2} z^{2k} + q^{\frac{1}{3}} z$$

$$\times \sum_{t=-\infty}^{\infty} q^{3t^2-t} \sum_{k=-\infty}^{\infty} q^{k^2} (z^2 q)^k$$

$$= q^{\frac{1}{3}} f(q^5,q) f(qz^2, qz^{-2}) + q^{\frac{1}{3}} z f(q^2,q^4) f(q^2 z^2, z^{-2}).$$

Comparing this with the required identity for $c(q,z)$, it is now enough to prove

$$G(z) = (1+z+z^{-1}) \frac{(z^3 q^3; q^3)_\infty (z^{-3} q^3; q^3)_\infty}{(zq; q)_\infty (z^{-1} q; q)_\infty}$$

$$= \frac{f(q^5,q) f(qz^2, qz^{-2})}{(q;q)_\infty (q^3; q^3)_\infty} + z \frac{f(q^4,q^2) f(q^2 z^2, z^{-2})}{(q;q)_\infty (q^3; q^3)_\infty}. \qquad (4.8.1)$$

Now, we can write

$$G(z) = \frac{1}{z} \frac{(z^3; q^3)_\infty}{(z;q)_\infty} \frac{(z^{-3} q^3; q^3)_\infty}{(z^{-1} q; q)_\infty}.$$

We have,

$$G(qz) = \frac{1}{qz} \frac{(z^3 q^3; q^3)_\infty}{(zq; q)_\infty} \frac{(z^{-3}; q^3)_\infty}{(z^{-1}; q)_\infty}$$

$$= \frac{1}{qz^2} G(z)$$

so that

$$z^2 q G(qz) = G(z).$$

This gives, on seeking

$$G(z) = \sum_{n=-\infty}^{\infty} C_n z^n,$$

$$\sum_{n=-\infty}^{\infty} q^{n+1} C_n z^{n+2} = \sum_{n=-\infty}^{\infty} C_n z^n,$$

and hence

$$C_n = q^{n-1} C_{n-2}, \ n = 0, \pm 1, \pm 2, \ldots$$

This gives, on iteration,

$$C_{2n} = q^{n^2} C_0, \ C_{2n+1} = q^{n(n+1)} C_1, \ n = 0, \pm 1, \pm 2, \ldots$$

Thus, the power series sought for $G(z)$ becomes

$$G(z) = C_0 \sum_{n=-\infty}^{\infty} q^{n^2} z^{2n} + C_1 z \sum_{n=-\infty}^{\infty} q^{n^2} (z^2 q)^n, \ \text{or,}$$

$$G(z) = C_0 f(qz^2, qz^{-2}) + C_1 z f(q^2 z^2, z^{-2}). \tag{4.8.2}$$

Comparing this with (4.8.1), it is enough to prove,

$$C_0 = \frac{f(q^5, q)}{(q;q)_\infty (q^3; q^3)_\infty}, \ \text{and}$$

$$C_1 = \frac{f(q^4, q^2)}{(q;q)_\infty (q^3; q^3)_\infty}.$$

Putting $z = i$ in (4.8.2) and using the definition of $G(z)$ in (4.8.1), we have

$$G(i) = \frac{(-iq^3; q^3)_\infty (iq^3; q^3)_\infty}{(iq;q)_\infty (-iq;q)_\infty}$$

$$= C_0 f(-q, q).$$

or

$$C_0 = \frac{(-q^6; q^6)_\infty}{(-q^2; q^2)_\infty (q; q^2)_\infty (q^2; q^2)_\infty}$$

$$= \frac{(q^2; q^4)_\infty (-q^6; q^6)_\infty}{(q;q)_\infty (q; q^2)_\infty}$$

$$= \frac{(-q; q^2)_\infty (-q^6; q^6)_\infty}{(q;q)_\infty}$$

$$= \frac{(-q; q^6)_\infty (-q^3; q^6)_\infty (-q^5; q^6)_\infty}{(q;q)_\infty}$$

$$= \frac{(-q;q^6)_\infty(-q^5;q^6)_\infty(-q^3;q^3)_\infty(-q^6;q^6)_\infty}{(q;q)_\infty}$$

$$= \frac{(-q;q^6)_\infty(-q^5;q^6)_\infty(q^6;q^6)_\infty}{(q;q)_\infty(q^3;q^3)_\infty} = \frac{f(q,q^5)}{(q;q)_\infty(q^3;q^3)_\infty}$$

as required, as regards C_0. Now, for C_1, putting $z = iq^{-\frac{1}{2}}$ in (4.8.2) and using the defintion of $G(z)$ in (4.8.1), we have.

$$G(iq^{-\frac{1}{2}}) = -iq^{\frac{1}{2}}\frac{(-iq^{-\frac{3}{2}};q^3)_\infty(iq^{\frac{9}{2}};q^3)_\infty}{(iq^{-\frac{1}{2}};q)_\infty(-iq^{\frac{3}{2}};q)_\infty}$$

$$= C_1 iq^{-\frac{1}{2}}f(-q,-q).$$

or,

$$C_1 = \frac{-q(-iq^{-\frac{3}{2}};q^3)_\infty(iq^{\frac{9}{2}};q^3)_\infty}{(iq^{-\frac{1}{2}};q)_\infty(-iq^{\frac{3}{2}};q)_\infty(q;q^2)^2_\infty(q^2;q^2)_\infty}$$

$$= \frac{-q(1+iq^{\frac{1}{2}})(1+iq^{-\frac{3}{2}})(-iq^{\frac{3}{2}};q^3)_\infty(iq^{\frac{3}{2}};q^3)_\infty}{(1-iq^{-\frac{1}{2}})(1-iq^{\frac{3}{2}})(iq^{\frac{1}{2}};q)_\infty(-iq^{\frac{1}{2}};q)_\infty(q;q)_\infty(q;q^2)_\infty}$$

$$= \frac{(-q^3;q^6)_\infty}{(-q;q^2)_\infty(q;q^2)_\infty(q;q)_\infty} = \frac{(-q^3;q^6)_\infty}{(q^2;q^4)_\infty(q;q)_\infty}$$

$$= \frac{(-q^3;q^6)_\infty(-q^2;q^2)_\infty}{(q;q)_\infty}$$

$$= \frac{(-q^3;q^6)_\infty(-q^6;q^6)_\infty(-q^4;q^6)_\infty(-q^2;q^6)_\infty}{(q;q)_\infty}$$

$$= \frac{(-q^3;q^3)_\infty(-q^4;q^6)_\infty(-q^2;q^4)_\infty}{(q;q)_\infty}$$

$$= \frac{(q^6;q^6)_\infty(-q^4;q^2)_\infty(-q^2;q^4)_\infty}{(q;q)_\infty(q^3;q^3)_\infty} = \frac{f(q^2,q^4)}{(q;q)_\infty(q^3;q^3)_\infty},$$

as required. This completes the proof of Part (ii) of the theorem and hence that of the theorem.

Exercise 4.8.3: Deduce from Theorem 4.8.1, product representations for $b(q)$ and $c(q)$.

Exercise 4.8.4: If

$$G(a) = f(azq^{\frac{1}{2}},a^{-1}z^{-1}q^{\frac{1}{2}})f\left(az^{-1}q^{\frac{1}{2}},a^{-1}zq^{\frac{1}{2}}\right)f\left(aq^{\frac{1}{2}},a^{-1}q^{\frac{1}{2}}\right)$$

then show successively that

(i) $G(aq) = a^3 q^{-\frac{3}{2}} f(a)$,

(ii) $G(a) = C_0 f\left(a^3 q^{\frac{3}{2}}, a^{-3} q^{\frac{3}{2}}\right) + C_1 \left[f\left(a^3 q^{\frac{5}{2}}, a^{-3} q^{\frac{1}{2}}\right) + a^{-1} f\left(a^3 q^{\frac{1}{2}}, a^{-3} q^{\frac{5}{2}}\right) \right]$.

[Hint: Seek $G(a) = \sum_{n=-\infty}^{\infty} C_n a^n$ and apply Part (i) for a recurrence relation for C_n].

(iii) $3C_0 = \dfrac{z^{-1}(q)_\infty^3}{(q^3;q^3)_\infty^3}$

$$\times \left[\frac{f(-z^{-3}q^3, -z^3)}{f(-z^{-1}q, -z)} \left\{ 1 + 6 \sum_{n=1}^{\infty} \left(\frac{q^{3n-2}}{1-q^{3n-2}} - \frac{q^{3n-1}}{1-q^{3n-1}} \right) \right\} \right.$$
$$\left. - f^2(-z^{-1}q, -z) \right].$$

[Hint: Change a to $a^2 q^{\frac{1}{2}}$ in (ii), multiply the resulting equation by a^3, apply operator $\Theta_a = a\frac{d}{da}$ after rewriting the equation suitably and let a tend to i in the resulting equation.]

$$(iv)\quad C_1 = z^{-1} q^{\frac{1}{2}} (q)_\infty^2 \frac{f(-z^3, -z^{-3}q^3)}{f(-z, -z^{-1}q)}.$$

[Hint: Change a to $a^2 q^{\frac{1}{2}}$ in Part (ii), multiply by a^3 and then let a to $e^{\frac{\pi i}{6}}$.]

(v) $a(q,z)$ equals constant term in the expansion of $G(a)$ as power series in a. Hence show that

$$a(q,z) = \frac{1}{3}(1+z+z^{-1})$$
$$\times \left\{ 1 + 6 \sum_{n=1}^{\infty} \left(\frac{q^{3n-2}}{1-q^{3n-2}} - \frac{q^{3n-1}}{1-q^{3n-1}} \right) \right\}$$
$$\times \frac{(q)_\infty^2}{(q^3;q^3)_\infty^2} \frac{(z^3 q^3;q^3)_\infty (z^{-3} q^3;q^3)_\infty}{(zq;q)_\infty (z^{-1}q;q)_\infty}$$
$$+ \frac{1}{3}(2-z-z^{-1}) \frac{(q)_\infty^5}{(q^3;q^3)_\infty^3} (zq;q)_\infty^2 (z^{-1}q;q)_\infty^2.$$

[Hint: $a(q,z) = \sum_{m+n+p=0, m,n,p=-\infty}^{\infty} q^{(m^2+n^2+p^2)/2} z^{m-n}$.]

(vi) $c(q,z) = q^{\frac{1}{3}}$ constant term in the expansion of $aG(aq^{\frac{1}{2}})$ as power series in a. Hence prove the product representation for $c(q,z)$ given in Theorem 4.8.1.

Exercise 4.8.5: Letting $z = 1$ in Part (v) of Exercise 4.8.4, obtain the "Lambert series" for $a(q)$:

$$a(q) = 1 + 6 \sum_{n=1}^{\infty} \left(\frac{q^{3n-2}}{1-q^{3n-2}} - \frac{q^{3n-1}}{1-q^{3n-1}} \right).$$

4.8.3 A two-variable cubic counterpart of Jacobi's quartic modular equation

Theorem 4.8.2. *We have*

$$a^3(q,z) = b^2(q)b(q,z^3) + qc^3(q,z).$$

Proof: By Part (vi) of Exercise (4.8.2),

$$b(q,z)b(q\omega,z)b(q\omega^2,z) = a^3(q^3,z) - q^3c^3(q^3,z).$$

But, by Part (i) of Theorem 4.8.1, we have the left side of this to be, on slight manipulation, equal to $b^2(q^3)b(q^3,z^3)$. This proves the theorem with q^3 instead of q.

Exercise 4.8.6: Show that $\frac{b(q,z)c(q,z)}{b(q^3,z^3)c(q^3,z)}$ is independent of z.

4.9 The Three-variable Cubic Theta Functions

4.9.1 Unification of one and two-variable cubic theta functions

Definition 4.9.1. Bhargava (1995).
If $|q| < 1, \tau, z \neq 0$, we define

$$a(q,\tau,z) = \sum_{m,n=-\infty}^{\infty} q^{m^2+mn+n^2}\tau^{m+n}z^{m-n}$$

$$b(q,\tau,z) = \sum_{m,n=-\infty}^{\infty} \omega^{m-n}q^{m^2+mn+n^2}\tau^m z^n$$

$$c(q,\tau,z) = \sum_{m,n=-\infty}^{\infty} q^{(m+\frac{1}{3})^2+(m+\frac{1}{3})(n+\frac{1}{3})+(n+\frac{1}{3})^2}\tau^{n+m}z^{n-m}$$

$$a'(q,\tau,z) = \sum_{m,n=-\infty}^{\infty} q^{m^2+mn+n^2}\tau^m z^n.$$

Exercises   4.9.

4.9.1. Show that

$$a(q,1,z) = a(q,z),\ a'(q,1,z) = a'(q,z)$$
$$b(q,1,z) = b(q,z),\ c(q,1,z) = c(q,z).$$

4.9.2. Show that

$$a(a) = a(q, 1, 1), = a'(q, 1, 1); \ b(q) = b(q, 1, 1)$$
$$c(q) = c(q, 1, 1).$$

Theorem 4.9.1. *(Bhargava (1995)).*
The four two-variable theta functions are equivalent. In fact,

$$a'(q, \tau, z) = a(q, \sqrt{\tau z}, \sqrt{z/\tau}),$$
$$b(q, \tau, z) = a(q, \sqrt{\tau z}, \omega^2 \sqrt{z/\tau}),$$
$$c(q, \tau, z) = q^{1/3} a(q, \tau q, z).$$

Proof: Exercise

Theorem 4.9.2. *(Bhargava (1995).)*
For any integers λ and μ, we have

$$a(q, \tau, z) = q^{3\lambda^2 + 3\lambda\mu + \mu^2} \tau^{2\lambda + \mu} \ z^{\mu} \ a(q, \tau q^{3(2\lambda + \mu)/2}, z q^{\mu/2})$$

Proof: Exercise.

Theorem 4.9.3. *(Bhargava (1995).)*

$$a(q, \tau, z) = a(q^3, \sqrt{\tau^3/z^3}, \sqrt{\tau z^3}) + q\tau z^{-1} a(q^3, q^3 \sqrt{\tau^3/z^3}, \sqrt{\tau z^3}).$$

Proof: Exercise.

Exercise 4.9.3: Complete the proof of Theorem 4.9.1. For example,

$$a(q, \sqrt{\tau z}, \omega^2 \sqrt{z/\tau}) = \sum_{m,n=-\infty}^{\infty} q^{m^2 + mn + n^2} (\tau z)^{\frac{m+n}{2}}$$
$$\times \omega^{2(m-n)} (z/\tau)^{\frac{m-n}{2}}$$
$$= \sum_{m,n=-\infty}^{\infty} \omega^{n-m} q^{m^2 + mn + n^2} \tau^n z^m$$
$$= b(q, \tau, z).$$

Exercise 4.9.4: Complete the proof of Theorem 4.9.2.
[For a start, expand the right side to get, after some manipulation,

$$\sum_{m,n=-\infty}^{\infty} q^{3(m+\lambda)^2 + 3(m+\lambda)(n+\mu) + (n+\mu)^2} \tau^{2(m+\lambda) + (n+\mu)} z^{n+\mu}.$$

Then change $m + \lambda$ to m and $n + \mu$ to n.]

Exercise 4.9.5:   Complete the proof of Theorem 4.9.3.
[In fact, we can write

$$a(q,\tau,z) = S_0 + S_1 + S_{-1}$$

where

$$S_r = \sum q^{i^2+ij+j^2} \tau^{i+j} z^{j-i}$$

with $i-j = r(mod\ 3)$. Now put $i = j+3m+r$ and manipulate.]

Exercise 4.9.6:   Obtain the following identities as special cases of Theorem 4.9.2:

$$a'(q,z) = z^2 q^3 a'(q,zq^3),$$
$$a(q,z) = z^2 qa(q,zq),$$
$$b(q,z) = z^2 q^3 b(q,zq^3),$$
$$c(q,z) = z^2 qc(q,zq).$$

Exercise 4.9.7:   Obtain the counterparts of Theorem 4.9.2 for $a'(q,\tau,z)$, $b(q,\tau,z)$
and $c(q,\tau,z)$.

Exercise 4.9.8:   Obtain the following identities as special cases of Theorem 4.9.3.

$$a'(q,z) = a(q^3,z) + 2qc(q^3,z),$$
$$b(q,z) = a(q^3,z) - qc(q^3,z).$$

4.9.2 Generalization of Hirschhorn-Garvan-Borwein identity

Theorem 4.9.4. *(Bhargava (1995).)*

$$
\begin{aligned}
a(q,\tau,z) = & \frac{q^{\frac{1}{2}} f(-q\tau^2, -q^2\tau^{-2}) C_1'(\tau,z)}{6\prod_{n=1}^{\infty}(1-q^{3n}\tau^2)(1-q^{3n}\tau^{-2})(1-q^{3n})S_0(\tau)} \\
& \times \left[1 + 6\sum_{n=1}^{\infty}\left(\frac{q^{3n-2}\tau^2}{1-q^{3n-2}\tau^2} - \frac{q^{3n-1}\tau^{-2}}{1-q^{3n-1}\tau^{-2}} \right) \right] \\
& + \frac{q^{-\frac{1}{2}}\tau^2 f(-q\tau^{-2}, -q^2\tau^2) C_1'(\tau^{-1},z)}{6\prod_{n=1}^{\infty}(1-q^{3n}\tau^2)(1-q^{3n}\tau^{-2})(1-q^{3n})S_0(\tau)} \\
& \times \left[1 + 6\sum_{n=1}^{\infty}\left(\frac{q^{3n-2}\tau^{-2}}{1-q^{3n-2}\tau^{-2}} - \frac{q^{3n-1}\tau^2}{1-q^{3n-1}\tau^2} \right) \right] \\
& + \frac{1}{3}\left[\frac{\tau + \tau^{-1} - z - z^{-1}}{S_0(\tau)} \right] \prod_{n=1}^{\infty} \frac{(1-q^n)^5}{(1-q^{3n})} \\
& \times \prod_{n=1}^{\infty} \frac{(1-q^n\tau z)(1-q^n\tau^{-1}z)(1-q^n\tau z^{-1})}{(1-q^{3n}\tau^2)} \\
& \times \frac{(1-q^n\tau^{-1}z^{-1})}{(1-q^{3n}\tau^{-2})},
\end{aligned}
$$

where,

$$C_1'(\tau,z) = \frac{[\omega D(\tau,z) + D(\tau^{-1},z)]\tau^{\frac{2}{3}}q^{\frac{1}{2}}}{\omega+1}$$

with

$$D(\tau,z) = \frac{(-\omega^{-1}\tau^{\frac{1}{3}} + \omega\tau^{-\frac{1}{3}})}{(-\omega^{-1}+\omega)}(z+z^{-1}-\tau^{\frac{1}{3}}\omega^2 - \tau^{-\frac{1}{3}}\omega)$$

$$\times \prod_{n=1}^{\infty} \frac{(1-q^{3n})^3}{(1-q^n)} \prod_{n=1}^{\infty} \frac{(1-q^n\omega\tau^{\frac{2}{3}})(1-q^n\omega^2\tau^{-\frac{2}{3}})}{(1+q^n+q^{2n})}$$

$$\times \prod_{n=1}^{\infty} \frac{(1-q^n\omega\tau^{-\frac{1}{3}}z)}{(1+q^n+q^{2n})^2}$$

$$\times (1-q^n\omega^2\tau^{\frac{1}{3}}z)(1-q^n\omega\tau^{-\frac{1}{3}}z^{-1})(1-q^n\omega^2\tau^{\frac{1}{3}}z^{-1})$$

and

$$\tau^{-1}S_0(\tau) = \frac{\tau+\tau^{-1}}{2} - (\tau - \tau^{-1})\sum_{n=1}^{\infty}\left(\frac{q^{3n}\tau^2}{1-q^{3n}\tau^2} - \frac{q^{3n}\tau^{-2}}{1-q^{3n}\tau^{-2}}\right).$$

Proof: Proof is similar to that of Part (v) of Exercise 4.8.2 but more elaborate due to the presence of extra variable τ. We therefore only sketch the proof by indicating the steps involved.

Step (1). Show that

$$g(aq,\tau,z) = a^{-3}\tau^{-2}q^{-\frac{3}{2}}g(a,\tau,z),$$

where

$$g(a,\tau,z) = f(a\tau zq^{\frac{1}{2}}, a^{-1}\tau^{-1}z^{-1}q^{\frac{1}{2}})f(a\tau z^{-1}q^{\frac{1}{2}}, a^{-1}\tau^{-1}zq^{\frac{1}{2}})$$
$$\times f(aq^{\frac{1}{2}}, a^{-1}q^{\frac{1}{2}})$$

Step (2). Show that

$$g(a,\tau,z) = C_0(\tau,z)\sum_{n=-\infty}^{\infty} a^{3n}\tau^{2n}q^{3n^2/2}$$

$$+C_1(\tau,z)\sum_{n=-\infty}^{\infty} a^{3n}\tau^{2n}q^{(3n^2+2n)/2}$$

$$+C_1(\tau^{-1},z)a^{-1}\sum_{n=-\infty}^{\infty} a^{3n}\tau^{2n}q^{(3n^2-2n)/2}.$$

Step (3). Put each summation in Step (2) in product form.
Step (4). Now, it is easy to see,

$$a(q,\tau,z) = \sum_{m+n+p=0} q^{(m^2+n^2+p^2)/2}\tau^{m+n}z^{m-n}$$

equals the coefficient of a^0 in

$$\left(\sum_{m=-\infty}^{\infty} a^m \tau^m z^m q^{\frac{m^2}{2}} \right) \left(\sum_{n=-\infty}^{\infty} a^n \tau^n z^{-n} q^{\frac{n^2}{2}} \right) \left(\sum_{p=-\infty}^{\infty} a^p q^{\frac{p^2}{2}} \right)$$

equals the coefficient of a^0 in $g(q,\tau,z)$ which is equal to $C_0(\tau,z)$. Now, Similarly,

$$a(q,\tau q,z) = \text{Coefficient of } a^0 \text{ in } ag(aq^{\frac{1}{2}},\tau,z)$$
$$= q^{-\frac{1}{2}}C_1(\tau^{-1},z).$$

Step (5). We have thus proved (after replacing a by $a^2 \tau^{\frac{2}{3}} q^{\frac{1}{2}}$ and then multiplying by $a^3 \tau^2$),

$$a^3 \tau^2 f(a^2 \tau^{\frac{5}{3}} zq, a^{-2} \tau^{-\frac{5}{3}} z^{-1}) f(a^2 \tau^{\frac{5}{3}} z^{-1} q, a^{-2} \tau^{-\frac{5}{3}} z) f(a^2 \tau^{\frac{2}{3}} q, a^{-2} \tau^{-\frac{2}{3}})$$
$$= a(q,\tau,z) a^3 \tau^2 f(a^6 \tau^4 q^3, a^{-6} \tau^{-4})$$
$$+ a(q,q\tau,z) \tau^{\frac{4}{3}} af(a^6 \tau^4 q^2, a^{-6} \tau^{-4} q)$$
$$+ a(q,q\tau^{-1},z) \tau^{-\frac{4}{3}} a^{-1} f(a^6 \tau^4 q, a^{-6} \tau^{-4} q^2).$$

Step (6). Set $a_1 = -i\omega \tau^{-\frac{2}{3}}$ and $a_2 = i\omega^2 \tau^{-\frac{2}{3}}$.

Substitute $a = a_j, j = 1,2$ in the identity of Step (5) to obtain two linear simultaneous equations in $a(q,\tau q,z)$ and $a(q,\tau q^{-1},z)$. Eliminating $a(q,\tau q^{-1},z)$, we get $a(q,\tau q,z) = C_1'(\tau,z)$ where $C_1'(\tau,z)$ is as in the statement of the theorem.

Step (7). Now set $a = a_0 = i\tau^{-\frac{1}{3}}$. We have from the identity in Step (5),

$$a_0^3 \tau^2 f(-\tau z^{-1} q, -\tau^{-1} z) f(-\tau zq, -\tau^{-1} z^{-1}) \lim_{a \to a_0} \frac{f(a^2 \tau^{\frac{2}{3}} q, a^{-2} \tau^{-\frac{2}{3}})}{a - a_0}$$
$$= 3a_0^2 \tau^2 a(q,\tau,z) f(-\tau^{-2}, \tau^{-2} q^3)$$
$$\times \left[1 + \frac{1}{3} a_0 \frac{d}{da_0} \log f(a^{-6} \tau^{-4}, a^6 \tau^4 q^3) \right]$$
$$+ a(q,\tau q,z) \tau^{4/3} f(-\tau^2 q^2, -\tau^{-2} q)$$
$$\times \left[1 + a_0 \frac{d}{da_0} \log f(a^6 \tau^4 q^2, a^{-6} \tau^{-4} q) \right]$$
$$- a(q,\tau^{-1} q,z) a_0^{-2} \tau^{-4/3} f(-\tau^{-2} q^2, -\tau^2 q)$$
$$\times \left[1 - a_0 \frac{d}{da_0} \log f(a^6 \tau^4 q, a^{-6} \tau^{-4} q^2) \right].$$

This yields the expression for $a(q,\tau,z)$ stated in the theorem.

Exercise 4.9.9: Letting $\tau \to 1$ in Theorem 4.9.4, deduce Hirschhorn-Garvan-Borwein representations for $a(q,\tau)$ and $c(q,z)$.

Exercise 4.9.10: Complete the proof of each step in Theorem 4.9.4 following the directions therein.

4.9.3 Laurent's expansions for two-parameter cubic theta functions

Theorem 4.9.5. *(Bhargava and Fathima (2004).)*

(i) $\quad a(q,\tau,z) = f(q^3\tau^2, q^3\tau^{-2})f(qz^2, qz^{-2})$

$$+ q\tau z f(q^6\tau^2, \tau^{-2})f(q^2z^2, z^{-2}),$$

(ii) $\quad b(q,\tau,z) = f(q\tau\omega, q\tau^{-1}\omega^2)f(q^3\tau^{-1}z^2, q^3\tau z^{-2})$

$$+ \omega^2 qz f(q^3\tau\omega, \tau^{-1}\omega^2)f(q^6\tau^{-1}z^2, \tau z^{-2}),$$

(iii) $\quad c(q,\tau,z) = q^{\frac{1}{3}}f(q^5\tau^2, q\tau^{-2})f(qz^2, qz^{-2})$

$$+ q^{\frac{7}{3}}\tau z f(q^8\tau^2, q^{-2}\tau^{-2})f(q^2z^2, z^{-2})$$

(iv) $\quad a'(q,\tau,z) = f(q\tau, q\tau^{-1})f(q^3\tau^{-1}z^2, q^3\tau z^{-2})$

$$+ qz f(q^2\tau, \tau^{-1})f(q^6\tau^{-1}z^2, \tau z^{-2}).$$

Proof: We have,

$$(i)\ a(q,\tau,z) = \sum_{m,n=-\infty}^{\infty} q^{n^2+nm+m^2}\tau^{n+m}z^{n-m}$$

equals the sum of terms with even powers z^{2k} + sum of terms with odd powers z^{2k+1}, which is

$$= \sum_{m,k=\infty}^{\infty} q^{3(m+k)^2+k^2}\tau^{2(m+k)}z^{2k} + q\tau z \sum_{m,k=-\infty}^{\infty} q^{3(m+k)^2+3(m+k)+k^2+k}$$

$$= \sum_{n=-\infty}^{\infty} q^{3n^2}\tau^{2n} \sum_{k=-\infty}^{\infty} q^{k^2}z^{2k} + q\tau z \sum_{n=-\infty}^{\infty} q^{3n^2+3n}\tau^{2n} \sum_{k=-\infty}^{\infty} q^{k^2+k}z^{2k}$$

$$= f(q^3\tau^2, q^3\tau^{-2})f(qz^2, qz^{-2}) + q\tau z f(q^6\tau^2, \tau^{-2})f(q^2z^2, z^{-2}).$$

$$(ii)\ b(q,\tau,z) = \sum_{m,n=-\infty}^{\infty} \omega^{m-n}q^{m^2+mn+n^2}\tau^m z^n$$

equals the part with even powers of z + part with odd powers of z, which is

$$= \sum_{k,m=-\infty}^{\infty} \omega^{k+m} \tau^{3k^2+(k+m)^2} \tau^{k+m} (z^2\tau^{-1})^k$$

$$+qz\omega^2 \sum_{k,m=-\infty}^{\infty} \omega^{m+k} q^{3k^2+3k+(m+k)^2} (q\tau)^{m+k} (z^2\tau^{-1})^k$$

$$= \sum_{n,k=-\infty}^{\infty} \omega^n q^{3k^2+n^2} \tau^n (z^2\tau^{-1})^k$$

$$+qz\omega^2 \sum_{n,k=-\infty}^{\infty} q^{n^2} (\tau\omega q)^n q^{3k^2} (q^3 z^2 \tau^{-1})^k$$

$$= f(q\omega\tau, q\omega^2\tau^{-1}) f(q^3\tau^{-1} z^2, q^3\tau z^{-2})$$

$$+qz\omega^2 f(q^2\tau\omega, \tau^{-1}\omega^2) f(q^6 z^2\tau^{-1}, z^{-2}\tau), \text{ as desired.}$$

(iii) We have $c(q,\tau,z) = q^{\frac{1}{3}} a(q, q\tau, z)$. Using this in Part (i) we have the required result.

(iv) We have,

$$a'(q,\tau,z) = b(q, \tau\omega^2, z\omega).$$

Using this in Part (ii), we have the required result.

Exercise 4.9.11: Putting $\tau = 1$ in Theorem 4.9.5, obtain the corresponding results [Cooper (2003)) for $a(q,z), b(q,z), c(q,z)$ and $a'(q,z)$.

Exercise 4.9.12: Combining Part (i) of Theorem 4.9.5 with Theorem 4.9.1, get alternative representation for $a'(q,\tau,z), b(q,\tau,z)$ and $c(q,\tau,z)$.

Exercise 4.9.13: (Bhargava (1995).) Show that

$$a(q,\tau,z) = f(q\tau z^{-1}, q\tau^{-1}z) f(q^3\tau z^3, q^3\tau^{-1}z^{-3})$$
$$+q\tau z f(q^2\tau z^{-1}, \tau^{-1}z) f(q^6\tau z^3, \tau^{-1}z^{-3}).$$

$$\left[\text{Hint : Write } a(q,\tau,z) = S_0 + S_1, \ j = 2m+r, \text{ where} \right.$$

$$\left. S_r = \sum q^{i^2+\frac{1}{4}j^2+\frac{3}{4}j^2} (\tau/z)^{i+\frac{1}{2}j} (\tau z^3)^{\frac{j}{2}} \right].$$

Exercise 4.9.14: Write the counterparts of Exercise 4.9.13 for $b(q,\tau,z), c(q,\tau,z)$ and $a'(q,\tau,z)$.

Theorem 4.9.6. *(Bhargava and Fathima (2003).)*

$$a(e^{-2\pi t}, e^{i\phi}, e^{i\theta}) = \frac{1}{t\sqrt{3}} \exp\left[-\left(\frac{\phi^2 + 3\theta^2}{6\pi t} \right) \right] a(e^{-\frac{2\pi}{3t}}, e^{\frac{\theta}{t}}, e^{\frac{\phi}{3t}}).$$

Proof: If, as before,

$$f(a,b) = \sum_{n=-\infty}^{\infty} a^{n(n+1)/2} b^{n(n-1)/2}.$$

We need the following transform [Entry 20, Chapter 16, Adiga et al (1985), Ramanujan (1957)],

$$\sqrt{\alpha} f(e^{-\alpha^2 + n\alpha}, e^{-\alpha^2 - n\alpha}) = e^{n^2/4} \sqrt{\beta} f(e^{-\beta^2 + in\beta}, e^{-\beta^2 - in\beta})$$

provided $\alpha\beta = \pi$ and $\Re(\alpha^2) > 0$. In particular, we need

$$f(e^{-\pi t + i\theta}, e^{-\pi t - i\theta}) = \frac{1}{\sqrt{t}} \exp\left(-\frac{\theta^2}{4\pi t} \right) f(e^{-\frac{\pi + \theta}{t}}, e^{-\frac{\pi - \theta}{t}}),$$

and

$$f(e^{-\pi t + i\theta}, e^{-i\theta}) = \sqrt{\frac{2}{t}} \exp\left(\frac{\pi t}{8} - \frac{i\theta}{2} - \frac{\theta^2}{2\pi t} \right) f\left(-e^{-\frac{2\pi + 2\theta}{t}}, -e^{-\frac{2\pi - 2\theta}{t}} \right).$$

We also need the addition results [Entries 30(ii) and 30(iii), Chapter 16, Adiga et al (1985)]

$$f(a,b) + f(-a,-b) = 2f(a^3 b, ab^3)$$

and

$$f(a,b) - f(-a,-b) = 2af\left(\frac{b}{a}, \frac{a}{b} a^4 b^4 \right).$$

We have from Exercise 4.9.13 and repeated use of the above transforms for $f(a,b)$,

$$\begin{aligned}
a(e^{-2\pi t}, e^{i\phi}, e^{i\theta}) &= f(e^{-2\pi t + i(\phi - \theta)}, e^{-2\pi t - i(\phi - \theta)}) \\
&\quad \times f(e^{-6\pi t + i(\phi + 3\theta)}, e^{-6\pi t - i(\phi + 3\theta)}) \\
&\quad + e^{-2\pi t + i(\phi + \theta)} f(e^{-4\pi t + i(\phi - \theta)}, e^{-i(\phi - \theta)}) \\
&\quad \times f(e^{-12\pi t + i(\phi + 3\theta)}, e^{-i(\phi + 3\theta)}) \\
&= \frac{1}{2\sqrt{3}t} \exp\left[-\left(\frac{\phi^2 + 3\theta^2}{6\pi t} \right) \right] (\alpha\beta + \alpha'\beta')
\end{aligned}$$

where

$$\alpha = f\left(e^{-\frac{\pi + \phi - \theta}{2t}}, e^{-\frac{\pi - \phi + \theta}{2t}} \right),$$
$$\beta = f\left(e^{-\frac{\pi + \phi + 3\theta}{6t}}, e^{-\frac{\pi - \phi - 3\theta}{6t}} \right),$$

$$\alpha' = f(-e^{-\frac{\pi+\phi-\theta}{2t}}, -e^{-\frac{\pi-\phi+\theta}{2t}}),$$

$$\beta' = f(-e^{-\frac{\pi+\phi+3\theta}{6t}}, -e^{-\frac{\pi-\phi-3\theta}{6t}}).$$

This becomes, on using the addition theorems for f(a,b) quoted above and the trivial identity

$$2(\alpha\beta + \alpha'\beta') = (\alpha+\alpha')(\beta+\beta') + (\alpha-\alpha')(\beta-\beta'),$$

$$a(e^{-2\pi t}, e^{i\phi}, e^{i\theta}) = \frac{1}{t\sqrt{3}}\exp\left[-\left(\frac{\phi^2+3\theta^2}{6\pi t}\right)\right]$$

$$\times\left[f\left(e^{-\frac{2\pi-\theta+\phi}{t}}, e^{-\frac{2\pi+\theta-\phi}{t}}\right)f\left(e^{-\frac{2\pi+3\theta+\phi}{3t}}, e^{-\frac{2\pi-3\theta-\phi}{3t}}\right)\right.$$

$$\left.+e^{-\frac{2\pi+2\theta}{3t}}f\left(e^{\frac{\phi-\theta}{t}}, e^{-\frac{4\pi-\theta+\phi}{t}}\right)f\left(e^{\frac{\phi+3\theta}{3t}}, e^{-\frac{4\pi+\phi+3\theta}{3t}}\right)\right]$$

$$= \frac{1}{t\sqrt{3}}\exp\left[-\left(\frac{\phi^2+3\theta^2}{6\pi t}\right)\right]$$

$$\times\left[f(q^3\tau z^3, q^3\tau^{-1}z^{-3})f(q\tau z^{-1}, q\tau^{-1}z)\right.$$

$$\left.+q\tau z f(q^6\tau z^3, \tau^{-1}z^{-3})f(q^2\tau z^{-1}, \tau^{-1}z)\right]$$

with $q = e^{-\frac{2\pi}{3t}}, \tau = e^{-\frac{\theta+\phi}{2t}}, z = e^{\frac{3\theta-\phi}{6t}}$. This reduces to the required identity on using Exercise 4.9.13 once again, the trivial identity $a(q,\tau,z) = a(q,\tau^{-1},z^{-1})$ and the easily verified identity

$$a(q, x^3 y, xy^{-1}) = a(q, y^2, x^2).$$

Exercise 4.9.15: Work out all the details in the proof of Theorem 4.9.6.

Exercise 4.9.16: Prove the mixed transformations

(i)
$$a(e^{-2\pi t}, e^{i\phi}, e^{i\theta}) = \frac{1}{t\sqrt{3}}\exp\left[-\left[\frac{\phi^2+3\theta^2}{6\pi t}\right]\right]$$
$$\times a'\left(e^{-\frac{2\pi}{3t}}, e^{\frac{2\phi}{3t}}, e^{\frac{3\theta+\phi}{3t}}\right),$$

(ii)
$$a'(e^{-2\pi t}, e^{i\phi}, e^{i\theta}) = \frac{1}{t\sqrt{3}}\exp\left[-\left(\frac{\phi^2-\phi\theta+3\theta^2}{6\pi t}\right)\right]$$
$$\times a\left(e^{-\frac{2\pi}{3t}}, e^{\frac{\phi}{2t}}, e^{\frac{2\theta-\phi}{6t}}\right),$$

(iii)
$$b(e^{-2\pi t}, e^{i\phi}, e^{i\theta}) = \frac{1}{t\sqrt{3}}\exp\left[-\left(\frac{\phi^2-\phi\theta+\theta^2}{6\pi t}\right)+\frac{\phi}{3t}\right]$$
$$\times c\left(e^{-\frac{2\pi}{3t}}, e^{\frac{\phi}{2t}}, e^{\frac{2\theta-\phi}{6t}}\right),$$

(iv) $$\exp(-\frac{2i\phi}{3})c(e^{-2\pi t}, e^{i\phi}, e^{i\theta}) = \frac{1}{t\sqrt{3}} \exp\left[-(\frac{\phi^2 + 3\theta^2}{6\pi t})\right]$$
$$\times b\left(e^{-\frac{2\pi}{3t}}, e^{\frac{2\phi}{3t}}, e^{\frac{3\theta + \phi}{3t}}\right).$$

[Hint: Use Theorem 4.9.1 on Theorem 4.9.6].

Exercise 4.9.17: Obtain Cooper (2003) modular transformations by putting $\phi = 0$ in Theorem 4.9.6 and Exercise 4.9.16.

References

Adiga, C., Berndt, B.C., Bhargava, S. and Watson, G.N. (1985). Chapter 16 of Ramanujan's second notebook: Theta functions and q-series, *Memoirs Amer. Math.Soc.*, No. 315, 53, American Mathematical Society, Providence.

Berndt, B.C. (1985). *Ramanujan's Notebooks, Part III*, Springer-Verlag, New York.

Berndt, B.C., Bhargava, S and Garvan, F.G. (1995). Ramanujan's theories of elliptic functions to alternative bases, *Trans. Amer. Math. Soc.*, **347**, 4163-4244.

Bhargava, S. (1995). Unification of the cubic analogues of the Jacobian Theta-function, *J. Math. Anal. Appl.*, **193**, 543-558.

Bhargava, S. and Fathima, S.N. (2003). Laurent coefficients for cubic theta-functions, *South East Asian J. Math. and Math. Sc.*, **1**, 27-31.

Bhargava, and Fathima, S.N. (2004). Unification of modular transformations for cubic theta functions, *New Zealand J. Math.*, **33**, 121-127.

Borwein, J.M., Borwein, P.B. and Garvan, F.G. (1994). Some cubic identities of Ramanujan, *Trans. Amer. Math. Soc.*, **343**, 35-37.

Chan, H.H. (1995). On Ramanujan's cubic continued fraction, *Acta Arithmetica*, **LXXIII A**, 343-355.

Cooper, S. (2003). Cubic theta-functions, *J. Comp. Appl. Math.*, **160**, 77-94.

Hirschhorn, M., Garvan F. and Borwein, *J. Cubic analogues of the Jacobian theta-function* $\theta(z,q)$.

Ramanujan, S. (1957). *Ramanujan's Second notebook, TIFR*, Bombay.

Chapter 5
Lie Group and Special Functions

[*This chapter is based on the lectures of Dr. H.L. Manocha of Indian Institute of Technology (IIT), Delhi, India, on Lie Group and Dr. K.S.S. Nambooripad and Dr. E. Krishnan on general group theory.*]

5.1 General Introduction to Group Theory

Definition 5.1.1. Let S be a nonempty set. By a law of composition of S, we mean a rule for combining pairs a, b of elements of S to get another element say P of S. That is, law of composition is a map from

$$S \times S \to S$$
$$(a, b) \to p. \tag{5.1.1}$$

Example 5.1.1.

1. Addition is a law of composition on $\mathbb{Z}^+$.
2. Matrix multiplication is a law of composition on the set S of $n \times n$ matrices.

Definition 5.1.2. A group is a nonempty set G together with a law of composition satisfying the following axioms:

i. $(ab)c = a(bc)$, for all $a, b, c \in G$ (associativity)

ii. There exists an element $e \in G$ such that $ae = a = ea$, for all $a \in G$ (existence of identity)

iii. For each $a \in G$ there exists an element $b \in G$ such that $ab = e = ba$ (existence of inverse).

Definition 5.1.3. An Abelian group is a group whose law of composition is commutative.

211

Example 5.1.2.

1. Z, the set of integers is an Abelian group with respect to addition.
2. Let X be any nonempty set and let $G(X)$ denote the set of all one-one maps from X onto X. $G(X)$ is a group with respect to composition of functions.

Definition 5.1.4. A permutation on n-symbols is a one-one mapping of the set $I_n = 1, 2, \ldots, n$ onto itself.

Definition 5.1.5. The group of permutations of the set $1, 2, \ldots, n$ of integers from 1 to n is called the symmetric group and is denoted by S_n.

S_n contains $n!$ elements. The symmetric group S_3 contains six elements and it is the smallest non-Abelian group.

$$S_3 = \{\rho_0, \rho_1, \rho_2, \mu_1, \mu_2, \mu_3\}$$

where

$$\rho_0 = \begin{pmatrix} 1\ 2\ 3 \\ 1\ 2\ 3 \end{pmatrix}, \rho_1 = \begin{pmatrix} 1\ 2\ 3 \\ 2\ 3\ 1 \end{pmatrix}, \rho_2 = \begin{pmatrix} 1\ 2\ 3 \\ 3\ 1\ 2 \end{pmatrix},$$

$$\mu_1 = \begin{pmatrix} 1\ 2\ 3 \\ 1\ 3\ 2 \end{pmatrix}, \mu_2 = \begin{pmatrix} 1\ 2\ 3 \\ 3\ 2\ 1 \end{pmatrix}, \mu_3 = \begin{pmatrix} 1\ 2\ 3 \\ 2\ 1\ 3 \end{pmatrix}.$$

Definition 5.1.6. A non-empty subset H of a group G is said to be a subgroup of G if it has the following properties:

i.   Closure : If $a \in H$, and $b \in H$, then $ab \in H$

ii.   Identity : $e \in H$

iii.   Inverse: If $a \in H$ then $a^{-1} \in H$.

Example 5.1.3.

1. The set T of 2×2 nonsingular matrices over R is a subgroup of the general linear group $GL_2(R)$.
2. $S_3 = \{\rho_0, \rho_1, \rho_2, \mu_1, \mu_2, \mu_3\}$. Then $H = \{\rho_0, \rho_1, \rho_2\}$ is a subgroup of S_3.

Definition 5.1.7. Let G be a group and $a \in G$. Then the subgroup $H = \{a^n | n \in Z\}$ is called the cyclic subgroup of G generated by a. A group G is called a cyclic group if for some $a \in G, G = \{a^n | n \in Z\}$. This a is called the generator of G and the group $G = < a >$ is cyclic.

Example 5.1.4.

1. $(Z, +)$ is a cyclic group, 1 and -1 are generators.
2. $G = \{1, -1, i, -i\}$ is a cyclic group.

5.1.1 Isomorphisms

Let G and G' be two groups. The groups are said to be isomorphic if all properties of the group structure of G hold for G' as well, and conversely.

Definition 5.1.8. Let G and G' be groups. A homomorphism $f : G \rightarrow G'$ is any map satisfying the rule

$$f(ab) = f(a)f(b)$$

for all $a, b \in G$. In addition to this, if f is bijective then f is an isomorphism from G onto G' and the two groups are said to be isomorphic.

Example 5.1.5.

1. R under addition is isomorphic to R^+ under multiplication $f : R \rightarrow R^+$ defined by $f(x) = e^x$.
2. The infinite cyclic group $C = \{\ldots, a^{-2}, a^{-1}, 1, a, a^2, \ldots\}$ is isomorphic to the group of integers $f : C \rightarrow \mathbb{Z}$ defined by $f(a^n) = n$ is an isomorphism.

Definition 5.1.9. Let G and G' be groups and $f : G \rightarrow G'$ be a homomorphism. The kernel of f denoted by $Kerf$ is defined as

$$Kerf = \{a : a \in G \text{ and } f(a) = e'\}.$$

Let $GL(n, F) = \{A : A = (a_{ij}), 1 \leq i, j \leq n, a_{ij} \in F \text{ and } |A| \neq 0\}$ and $R^* = \{r \in R : r \neq 0\}$.

Example 5.1.6. The determinant function $\det : GL(n, R) \rightarrow R^*$ is a homomorphism.

$$
\begin{aligned}
Ker \det &= \{A : A = (a_{ij}), a_{ij} \in R, 1 \leq i, j \leq n \text{ and } \det A = 1\} \\
&= SL(n, R).
\end{aligned}
$$

$SL(n, R)$ is called the special linear group.

Note 5.1.1: Ker f is a subgroup of G.

Note 5.1.2: $SL(n, R)$ is a subgroup of $GL(n, R)$.

5.1.2 Symmetry groups

Definition 5.1.10. A map $m : P \rightarrow P$ from the plane P to itself is called a rigid motion or an isometry if it is distance preserving. That is for any $p, q \in P$, $d(p, q) = d(m(p), m(q))$.

Every isometry is a one-one onto mapping of a plane onto itself. The set of all isometries M form a group under composition of functions. Symmetry is some type of transformation in which the original figure gets reflected through itself. A group of symmetries of a plane figure is possible only if an isometry is possible.

5.1.3 Isometries of the Euclidean plane

Let us identify the plane with the space R^2 of column vectors, by choosing a co-ordinate system.

(a) Translation: A translation of the plane is a transformation that moves each point a fixed distance in a fixed direction. The translation t_a by a vector a moves a point x to $x + a$. That is,

$$t_a(x) = \begin{pmatrix} x_1 + a \\ x_2 + a \end{pmatrix}.$$

Clearly, $t_a t_b = t_{a+b}$.

(b) Rotation: A rotation $\rho_{(P,\theta)}$ is a transformation that rotates the plane about the point P counterclockwise through the angle θ, $0 \le \theta \le 2\pi$.

Rotating the plane by an angle θ about the origin is denoted by ρ_θ

$$\rho_\theta(x) = \begin{bmatrix} \cos\theta & -\sin\theta \\ \sin\theta & \cos\theta \end{bmatrix} \begin{bmatrix} x_1 \\ x_2 \end{bmatrix}.$$

In R^3, the matrix representing a rotation through the angle θ about the vector e_i is

$$A = \begin{bmatrix} 1 & 0 & 0 \\ 0 & \cos\theta & -\sin\theta \\ 0 & \sin\theta & \cos\theta \end{bmatrix}.$$

A is an orthonormal matrix.

Note 5.1.3: The orthogonal $n \times n$ matrices $O(n)$ form a subgroup of $GL(n,R)$. The orthogonal matrices having determinant $+1$ form a subgroup called the special orthogonal group and is denoted by $SO(n)$.

$$SO(n) = \{A \in GL(n,R) \mid A^t A = I,\ \det A = 1\}.$$

(c) Reflection: A reflection in the plane is a function μ that carries each point of a fixed line l into itself and every point not on the line into the mirror image point straight across l and the same distance from l. Reflection r about the x_1-axis:

$$r(x) = \begin{bmatrix} 1 & 0 \\ 0 & -1 \end{bmatrix} \begin{bmatrix} x_1 \\ x_2 \end{bmatrix} = \begin{bmatrix} x_1 \\ -x_2 \end{bmatrix}.$$

(d) Glide reflection: Glide reflection is obtained by reflecting about a line l and then translating by a nonzero vector a parallel to l.

Note 5.1.4: Translation and rotation are orientation-preserving motions while reflection and glide reflection are orientation-reversing motions.

Note 5.1.5: A translation does not leave any point fixed. A rotation fixes exactly one point, that is, centre of rotation. Reflection fixes the point on the line of reflection. A glide has no fixed point.

Theorem 5.1.1. *Any rigid motion of the plane is a consequence of translation, a rotation, a reflection and glide reflection.*

The group S_3 given in Definition 5.1.5 has a geometric interpretation. Consider an equilateral triangle with vertices $1, 2, 3$ and consider the following transformations of the triangle to itself.

(a) The three rotations about the centre through $0°, 120°$ and $240°$ (counterclockwise). These correspond to the permutations

$$\rho_o = \begin{pmatrix} 1 & 2 & 3 \\ 1 & 2 & 3 \end{pmatrix}, \rho_1 = \begin{pmatrix} 1 & 2 & 3 \\ 2 & 3 & 1 \end{pmatrix}, \rho_2 = \begin{pmatrix} 1 & 2 & 3 \\ 3 & 1 & 2 \end{pmatrix}.$$

(b) The three reflections along the three bisectors. These correspond to the permutations

$$\mu_1 = \begin{pmatrix} 1 & 2 & 3 \\ 1 & 3 & 2 \end{pmatrix}, \mu_2 = \begin{pmatrix} 1 & 2 & 3 \\ 3 & 2 & 1 \end{pmatrix}, \mu_3 = \begin{pmatrix} 1 & 2 & 3 \\ 2 & 1 & 3 \end{pmatrix}.$$

Note 5.1.6: S_3 is also called the group D_3 of symmetries of an equilateral triangle.

Note 5.1.7: The n-th dihedral group D_n is the group of symmetries of the regular n-gon. D_n contains $2n$ elements.

5.1.4 Finite groups of motion

Theorem 5.1.2. *(A fixed point theorem)* *Let G be a finite subgroup of the group of motions M. There is a point p in the plane which is left fixed by every element of G. That is, there is a point p such that $g(p) = p$ for all $g \in G$.*

Let s be any point in the plane and let $S = \{s' : s' = g(s) \text{ for some } g \in G\}$. S is called the orbit of s. Let

$$S = \{s'_1, s'_2, \ldots, s'_n\}$$

then

$$p = \frac{1}{n}(s'_1 + s'_2 + \cdots + s'_n)$$

is the fixed point.

Theorem 5.1.3. *Let G be a finite subgroup of the group O of rigid motions which fix the origin. Then G is the cyclic group of order n generated by the rotation ρ_θ where $\theta = \frac{2\pi}{n}$, or G is the dihedral group D_n of order 2n generated by the rotation ρ_θ where $\theta = \frac{2\pi}{n}$ and a reflection r' about a line through the origin.*

Note 5.1.8: Let $\rho_\theta \left(\theta = \frac{2\pi}{n}\right) = x$ and $r' = y$. Then

$$D_n = \{x, y : x^n = y^2 = 1 \text{ and } yx = x^{-1}y\}$$
$$= \{x^i y^j \mid 0 \leq i < n, 0 \leq j < 2\}.$$

5.1.5 Discrete groups of motions

In this section we will discuss the symmetry groups of unbounded figures such as wall paper patterns. Such patterns do not admit arbitrarily small translations or rotations.

Definition 5.1.11. A subgroup G of the group of motions M is called discrete if it does not contain arbitrarily small translations or rotations. That is, a subgroup $G \subseteq M$ is discrete if there is some real number $\varepsilon > 0$ such that for any translation $t_a \in G$, $|a| \geq \varepsilon$ and for any $\rho_\theta \in G$ $(\theta \neq 0)$, $|\theta| \geq \varepsilon$.

Definition 5.1.12. Let G be a discrete group. The translation group of G denoted by L_G is defined as

$$L_G = \{a \in R^2 \mid t_a \in G\}.$$

Note 5.1.9: L_G is a subgroup of G.

Note 5.1.10: L_G contains no vector of length $< \varepsilon$, except for the null vector.

Proposition 5.1.1 Every discrete subgroup L of R^2 has one of these forms

(a) $L = \{0\}$
(b) L is generated as an additive group by one non-null vector a.

$$L = \{ma \mid m \in Z\}.$$

(c) L is generated by two linearly independent vectors a, b

$$L = \{ma + nb \mid m, n \in Z\}.$$

Discrete groups of type (c) are called plane lattices and the generating set (a, b) is called a lattice basis.

Note 5.1.11: If L_G is a lattice then G is called a two dimensional crystallographic group or a lattice group. These groups are the groups of symmetries of wall paper patterns and of two-dimensional crystals.

5.2 Lie Group and Special Functions

Definition 5.2.1. Let F be a field. A vector space V over F is a nonempty set V satisfying the following conditions:

(i) $(V, +)$ is an Abelian group
(ii) $\alpha \in F, v \in V \Rightarrow \alpha v \in V$
(iii) $(\alpha + \beta)v = \alpha v + \beta v$
(iv) $\alpha(v + w) = \alpha v + \alpha w$
(v) $\alpha(\beta v) = (\alpha \beta)v$
(vi) $1v = v$,

where $v, w \in V$ and $\alpha, \beta, 1 \in F$.

Proposition 5.2.1 Let V be a vector space (v.s) over F. Then

(i) $0v = O$
(ii) $\alpha O = O$
(iii) $(-1)v = -v$

where $\alpha, 0 \in F$ and $v, O \in V$.

Example 5.2.1. Let F be a field and

$$F^{(n)} = \{(x_1, x_2, \ldots, x_n) : x_i \in F\}.$$

$F^{(n)}$ is a vector space over F of dimension n having a basis

$$B = \{(1, 0, \ldots, 0), \ldots, (0, 0, \ldots, 1)\}.$$

Note 5.2.1:

$\dim(R, R) = 1$;
$\dim(\mathcal{C}, \mathcal{C}) = 1$;
$\dim(\mathcal{C}, R) = 2$.

Example 5.2.2. The set of all polynomials in x having their coefficients in F. That is,

$$F[x] = \{a_0 + a_1 x + \cdots + a_n x^n : a_1, a_2, \ldots, a_n \in F, \text{ and } n \in N\}$$

is an infinite dimensional vector space over F having a basis $B = \{1, x, x^2, \ldots\}$.

Example 5.2.3. The set of all polynomials in x of degree less than n having coefficients in F,

$$V_n = \{a_0 + a_1 x + \cdots + a_m x^m, 0 \le m \le n - 1, a_0, \ldots, a_m \in F\}$$

is an n-dimensional vector space over F.

5.2.1 Subspace of a vector space

Definition 5.2.2. Let V be a vector space over F. A nonempty subset W of V is said to be a subspace of V if $\alpha, \beta \in F$ and $v, w \in W \Rightarrow \alpha v + \beta w \in W$.

Example 5.2.4.

1. V_n is a subspace of $F[x]$.
2. F is a subspace of $F[x]$.
3. Subspaces of R^2
   (a) $\{O\}$
   (b) Any line passing through the origin, that is,

$$L = \{(x, mx) : x \in R\}$$

   is a subspace of R^2.
4. Subspaces of R^3
   (a) $\{O\}$
   (b) Any line passing through the origin.
   (c) Any plane passing through the origin.

Definition 5.2.3. Let U and V be vector spaces over F. A homomorphism T of U into V is a mapping $T : U \rightarrow V$ such that

$$T(\alpha u + \beta v) = \alpha T(u) + \beta T(v), \text{ for all } \alpha, \beta \in F \text{ and } u, v \in V.$$

If T is bijective then T is called an isomorphism of U onto V.

Example 5.2.5. V_n and $F^{(n)}$ are isomorphic. Define $T : V_n \rightarrow F^{(n)}$ by

$$T(\alpha_1 x + \cdots + \alpha_{n-1} x^{n-1}) = (\alpha_1, \alpha_2, \ldots, \alpha_{n-1}).$$

Then T is an isomorphism of V_n onto F^n.

Note 5.2.2: Two finite dimensional vector spaces over the same field F having the same dimension are always isomorphic.

Theorem 5.2.1. *Let U and V be vector spaces over the same field F and let $T : U \rightarrow V$ be an isomorphism of U onto V. Then T maps a basis of U onto a basis of V.*

Note 5.2.3: The set of all homomorphisms of U into V written as $\text{Hom}_F(U, V)$ is a vector space of dimension mn.

Definition 5.2.4. Let F be a field. An associative algebra A over F is a vector space over F such that

(a) A is an associative ring
(b) $\alpha(uv) = (\alpha u)v = u(\alpha v)$, $\alpha \in F$ and $u, v \in A$.

Example 5.2.6.

(1) $F[x]$ is an associative algebra over F.

(2) F_n, the set of all $n \times n$ matrices having their entries in F is an n^2 dimensional associative algebra over F.

(3) $\text{Hom}_F(V,V)$ is an n^2 dimensional associative algebra over F.

5.3 Lie Algebra

Definition 5.3.1. Let F be a field. A Lie algebra G over F is a vector space over F equipped with an operation

$$[,] : G \times G \to G$$

having the following properties:

(a) $[x,y] = -[y,x]$

(b) $[\alpha x + \beta y, z] = \alpha[x,z] + \beta[y,z]$

(c) $[[x,y],z] + [[y,z],x] + [[z,x],y] = O$

where $x,y,z \in G$ and $\alpha, \beta \in F$.

Note 5.3.1: Property a) is referred to as antisymmetry, properties a) and b) together are called bilinear property and c) is termed as Jacobi identity.

Note 5.3.2: $[,]$ is called a commutator or a bracket or a Lie product.

Example 5.3.1.

(1) R^3 is a Lie algebra over R. For, R^3 is a 3-dimensional vector space over R, let $\bar{x} = (x_1, x_2, x_3)$ and $\bar{y} = (y_1, y_2, y_3) \in R^3$. Define $[,] : R^3 \times R^3 \to R^3$ by

$$[\bar{x}, \bar{y}] = \bar{x} \times \bar{y}$$
$$= (x_2 y_3 - x_3 y_2, x_3 y_1 - x_1 y_3, x_1 y_2 - x_2 y_1).$$

Clearly $[,]$ satisfies all the properties $a), b)$ and $c)$. Therefore R^3 is a Lie algebra.

(2) $A_F(V) = \text{Hom}_F(V,V)$ is a Lie algebra over F.

$\text{Hom}_F(V,V)$ is a vector space, for, let $T_1, T_2 \in A_F(V)$. Define $[,]$:

$$A_F(V) \times A_F(V) \to A_F(V)$$

by

$$[T_1, T_2] = T_1 T_2 - T_2 T_1.$$

Then, $(A_F(V), [,])$ is a Lie algebra.

(3) $F_n = \{(a_{ij}) : a_{ij} \in F \text{ and } 1 \le i, j \le n\}$ is a Lie algebra.

F_n is an n-dimensional vector space over F. For $X, Y \in F_n$ define $[X, Y] = XY - YX$. $(F_n, [,])$ is a Lie algebra. This Lie algebra is usually denoted by $\mathrm{gl}(n, F)$ and is called general Lie algebra.

(4) Special Lie algebra $(\mathrm{sl}(n, F))$.

$\mathrm{sl}(n, F) = \{A = (a_{ij}) : 1 \le i, j \le n \text{ and trace of } A = 0\}$ is an $n^2 - 1$ dimensional Lie algebra.

For $n = 2$, $\mathrm{sl}(2, F)$ is a 3-dimensional Lie algebra having a basis

$$B = \{\mathcal{J}^0, \mathcal{J}^+, \mathcal{J}^-\}$$

where

$$\mathcal{J}^0 = \begin{bmatrix} \frac{1}{2} & 0 \\ 0 & -\frac{1}{2} \end{bmatrix}, \; \mathcal{J}^+ = \begin{bmatrix} 0 & -1 \\ 0 & 0 \end{bmatrix}, \; \mathcal{J}^- = \begin{bmatrix} 0 & 0 \\ -1 & 0 \end{bmatrix}.$$

Also

$$\begin{aligned}
[\mathcal{J}^0, \mathcal{J}^+] &= \mathcal{J}^0 \mathcal{J}^+ - \mathcal{J}^+ \mathcal{J}^0 \\
&= \begin{bmatrix} \frac{1}{2} & 0 \\ 0 & -\frac{1}{2} \end{bmatrix} \begin{bmatrix} 0 & -1 \\ 0 & 0 \end{bmatrix} - \begin{bmatrix} 0 & -1 \\ 0 & 0 \end{bmatrix} \begin{bmatrix} \frac{1}{2} & 0 \\ 0 & -\frac{1}{2} \end{bmatrix} \\
&= \begin{bmatrix} 0 & -\frac{1}{2} \\ 0 & 0 \end{bmatrix} - \begin{bmatrix} 0 & \frac{1}{2} \\ 0 & 0 \end{bmatrix} \\
&= \begin{bmatrix} 0 & -1 \\ 0 & 0 \end{bmatrix} \\
&= \mathcal{J}^+.
\end{aligned}$$

Similarly, $[\mathcal{J}^+, \mathcal{J}^-] = 2\mathcal{J}^0$ and $[\mathcal{J}^0, \mathcal{J}^-] = -\mathcal{J}^-$.

Proposition 5.3.1 *Let $B = \{J_1, J_2, \ldots\}$ be a basis of a vector space G. Then G is a Lie algebra if each commutator $[J_i, J_j]$ is a linear combination of the vectors in the basis B.*

Q.1 Prove that $e(2) = \left\{ \begin{bmatrix} 0 & -x_1 & 0 \\ x_1 & 0 & 0 \\ x_2 & x_3 & 0 \end{bmatrix} : x_1, x_2, x_3 \in \mathbb{C} \right\}$ is a 3-dimensional complex Lie algebra.

Clearly, $e(2)$ is a vector space. Consider the elements

$$M = \begin{bmatrix} 0 & -1 & 0 \\ 1 & 0 & 0 \\ 0 & 0 & 0 \end{bmatrix}, \; P_1 = \begin{bmatrix} 0 & 0 & 0 \\ 0 & 0 & 0 \\ 1 & 0 & 0 \end{bmatrix}, \; P_2 = \begin{bmatrix} 0 & 0 & 0 \\ 0 & 0 & 0 \\ 0 & 1 & 0 \end{bmatrix}.$$

M, P_1, P_2 are linearly independent of $e(2)$. Let

$$A = \begin{bmatrix} 0 & -x_1 & 0 \\ x_1 & 0 & 0 \\ x_2 & x_3 & 0 \end{bmatrix} \in e(2).$$

Clearly, $A = x_1 M + x_2 P_1 + x_3 P_2$ where $x_1, x_2, x_3 \in \mathbb{C}$. Therefore $\{M, P_1, P_2\}$ is a basis of e(2).

$$[M, P_1] = M P_1 - P_1 M = P_2.$$
$$[M, P_2] = M P_2 - P_2 M = -P_1.$$
$$[P_1, P_2] = P_1 P_2 - P_2 P_1 = O.$$

Hence $O \in V, O$ is a linear combination of the basis elements. By Proposition 5.3.1, it follows that e(2) is a Lie algebra.

Q.2 Prove that the differential operators

$$J^\circ = -u + z \frac{d}{dz}$$

$$J^+ = -2uz + z^2 \frac{d}{dz}$$

and

$$J^- = -\frac{d}{dz}$$

generate a 3-dimensional Lie algebra.

J°, J^+, J^- are linearly independent. For, $\alpha J^\circ + \beta J^+ + \gamma J^- = 0$

$$\Rightarrow \alpha \left(-u + z \frac{d}{dz} \right) + \beta \left(-2uz + z^2 \frac{d}{dz} \right) + \gamma \left(-\frac{d}{dz} \right)$$

$$\Rightarrow -u\alpha - 2u\beta z + (\alpha z + \beta z^2 - \gamma) \frac{d}{dz} = 0$$

$$\Rightarrow \alpha = \beta = \gamma = 0$$

$$\Rightarrow J^\circ, J^+, J^- \text{ are linearly independent.}$$

Therefore $B = \{J^\circ, J^+, J^-\}$ is a basis of a 3-dimensional vector space spanned by B.

$$[J^\circ, J^-] f(z) = (J^\circ J^+ - J^+ J^\circ) f(z)$$

$$= \left(-u + z \frac{d}{dz} \right) \left(-2uzf + z^2 \frac{df}{dz} \right) - \left(-2uz + z^2 \frac{d}{dz} \right) \times \left(-uf + z \frac{df}{dz} \right)$$

$$= 2u^2 zf - uz^2 \frac{df}{dz} - 2uz \frac{d(zf)}{dz} + z^3 \frac{d^2 f}{dz^2} + 2z^2 \frac{df}{dz} - 2u^2 zf$$

$$+ 2uz^2 \frac{df}{dz} + z^2 u \frac{df}{dz} - z^3 \frac{d^2 f}{dz^2} - z^2 \frac{df}{dz}$$

$$= -2uz \left(z \frac{df}{dz} + f \right) + 2uz^2 \frac{df}{dz} + 2z^2 \frac{df}{dz} - z^2 \frac{df}{dz}$$

$$= -2uz^2 \frac{df}{dz} - 2uzf + 2uz^2 \frac{df}{dz} + z^2 \frac{df}{dz}$$

$$= -2uzf + z^2 \frac{df}{dz}$$

$$= \left(-2uz + z^2 \frac{d}{dz} \right) f.$$

Therefore

$$[J^\circ, J^+]f(z) = J^+ f(z) \quad \forall f(z)$$
$$\Rightarrow [J^\circ, J^+] = J^+.$$

Similarly, $[J^\circ, J^-] = -J^-$ and $[J^+, J^-] = 2J^\circ$. By Proposition 5.3.1, $\{J^0, J^+, J^-\}$ generate a Lie algebra.

Q.3   Prove that the operators $\left\{ \frac{\partial}{\partial x}, \frac{\partial}{\partial y}, y\frac{\partial}{\partial x} - x\frac{\partial}{\partial y} \right\}$ generate a Lie algebra.

Let

$$P_1 = \frac{\partial}{\partial x}, \; P_2 = \frac{\partial}{\partial y}, \; M = y\frac{\partial}{\partial x} - x\frac{\partial}{\partial y}.$$

P_1, P_2, M are linearly independent elements. Therefore $B = \{P_1, P_2, M\}$ is a basis of a 3-dimensional vector space spanned by B.

$$[P_1, P_2]f(x, y) = (P_1 P_2 - P_2 P_1)f(x, y)$$
$$= \left(\frac{\partial}{\partial x}\frac{\partial}{\partial y} - \frac{\partial}{\partial y}\frac{\partial}{\partial x} \right) f(x, y)$$
$$= \frac{\partial}{\partial x}\left(\frac{\partial f}{\partial y} \right) - \frac{\partial}{\partial y}\left(\frac{\partial f}{\partial x} \right)$$
$$= \frac{\partial^2 f}{\partial x \partial y} - \frac{\partial^2 f}{\partial y \partial x}$$
$$= 0.$$
$$\Rightarrow [P_1, P_2]f(x, y) = 0f(x, y)$$
$$\Rightarrow [P_1 P_2] = 0.$$

Similarly

$$[M, P_1] = P_2 \text{ and } [M, P_2] = -P_1.$$

By Proposition 5.3.1, the given operators generate a 3-dimensional Lie algebra.

5.4 Representations of Lie Algebra

Definition 5.4.1. Let G and G' be two Lie algebras over F. A homomorphism T of G into G' is a mapping $T : G \rightarrow G'$ satisfying the following conditions:

(i) $T(\alpha x + \beta y) = \alpha T(x) + \beta T(y)$

(ii) $T[x, y] = [T(x), T(y)], x, y \in G$ and $\alpha, \beta \in F$.

If T is one-one then the homomorphism is called isomorphism. If T is an isomorphism of G onto G' then we say G and G' are isomorphic.

Theorem 5.4.1. *Let T be a vector space isomorphism of a Lie algebra G onto a Lie algebra G', both over a field F. Let $\{j_1, j_2, \ldots\}$ be a basis of G and $T(j_i) = J_i$. Then G' as Lie algebra is an isomorphic image of G if $T[j_i, j_j] = [T(j_i), T(j_j)]$.*

Q.4 Show that the Lie algebra $\mathrm{sl}(2, \mathbb{C})$ and the Lie algebra generated by $\{\mathbf{J}^\circ, \mathbf{J}^+, \mathbf{J}^-\}$ are isomorphic.

We know that $\mathrm{sl}(2, \mathbb{C})$ is a Lie algebra having basis $\{j^\circ, j^+, j^-\}$. Let G be the Lie algebra spanned by $\{\mathbf{J}^\circ, \mathbf{J}^+, \mathbf{J}^-\}$. $\dim \mathrm{sl}(2, \mathbb{C}) = \dim G$. Therefore, as vector spaces, they are isomorphic.

Let $T : \mathrm{sl}(2, \mathbb{C}) \to G$ be the isomorphism. Therefore,

$$T(j^\circ) = \mathbf{J}^\circ,\; T(j^+) = \mathbf{J}^+,\; T(j^-) = \mathbf{J}^-.$$

Also we have

$$[\mathbf{J}^\circ, \mathbf{J}^+] = \mathbf{J}^+,\; [\mathbf{J}^\circ, \mathbf{J}^-] = -\mathbf{J}^- \text{ and } [\mathbf{J}^+, \mathbf{J}^-] = 2\mathbf{J}^\circ.$$

Now,

$$T[j^\circ, j^+] = T[j^+] = \mathbf{J}^+ = [\mathbf{J}^\circ, \mathbf{J}^+] = [T(j^\circ), T(j^+)]$$
$$T[j^\circ, j^-] = T[-j^-] = -T[j^-] = -\mathbf{J}^- = [\mathbf{J}^\circ, \mathbf{J}^-] = [T(j^\circ), T(j^-)].$$

Similarly,

$$T[j^+, j^-] = [Tj^+, Tj^-].$$

Therefore, by Theorem 5.4.1,

$$\mathrm{sl}(2, \mathbb{C}) \cong G.$$

Q.5 Prove that the Lie algebra $e(2)$ is isomorphic to the matrix Lie algebra generated by $\{M, P_1, P_2\}$ where $P_1 = \frac{\partial}{\partial x}, P_2 = \frac{\partial}{\partial y}, M = y\frac{\partial}{\partial x} - x\frac{\partial}{\partial y}$.

Solution: $e(2)$ is the Lie algebra generated by $\{M, P_1, P_2\}$ where

$$M = \begin{bmatrix} 0 & -1 & 0 \\ 1 & 0 & 0 \\ 0 & 0 & 0 \end{bmatrix},\; P_1 = \begin{bmatrix} 0 & 0 & 0 \\ 0 & 0 & 0 \\ 1 & 0 & 0 \end{bmatrix},\; P_2 = \begin{bmatrix} 0 & 0 & 0 \\ 0 & 0 & 0 \\ 0 & 1 & 0 \end{bmatrix}.$$

Also,

$$[M, P_1] = P_2,\; [M, P_2] = -P_1 \text{ and } [P_1, P_2] = O.$$

Let G be the Lie algebra generated by $\{M, P_1, P_2\}$. $e(2)$ and G are finite dimensional vector spaces of the same dimension. Therefore they are isomorphic. Let $T : e(2) \to G$ be the isomorphism.

$$T(M) = \mathcal{M},\; T(P_1) = \mathcal{P}_1,\; T(P_2) = \mathcal{P}_2$$

$$T[M, P_1] = T(P_2) = \mathcal{P}_2 = [\mathcal{M}, \mathcal{P}_1] = [T(M), T(P_1)].$$

Similarly,

$$T[M, P_2] = [T(M), T(P_2)] \text{ and } T[P_1, P_2] = [T(P_1), T(P_2)].$$

By Theorem 5.4.1,
$$e(2) \cong G.$$

Definition 5.4.2. Let V be a vector space over F and G be a matrix Lie algebra over F. A representation of G on V is a homomorphism ρ of G into $\mathcal{L}(V)$. That is,

$$\rho : G \to \mathcal{L}(V)$$

such that

(i) $\rho(\alpha x + \beta y) = \alpha \rho(x) + \beta \rho(y)$

(ii) $\rho[x, y] = [\rho(x), \rho(y)]$

where $\alpha, \beta \in F$ and $x, y \in G$.

Theorem 5.4.2. *Let V be a vector space over F and let $\mathbf{J}_1, \mathbf{J}_2, \ldots$ be operators in $\mathcal{L}(V)$ spanning a Lie algebra G'. If G' is an isomorphic image of a matrix Lie algebra G then the isomorphism $\rho : G \to G'$ provides a representation of G on the representation space V.*

Q.6   Obtain a representation of $sl(2,\mathcal{C})$ on V_{2u+1} where V_{2u+1} is the complex vector space having a basis $B = \{1, z, \ldots, z^{2u}\}$, $u \in \mathbb{N}$.

$sl(2,\mathcal{C})$ is a Lie algebra having a basis $\{j^\circ, j^+, j^-\}$. Consider the operators $\mathbf{J}^\circ, \mathbf{J}^+, \mathbf{J}^-$ defined by

$$\mathbf{J}^\circ = -u + z\frac{d}{dz}, \mathbf{J}^+ = -2uz + z^2\frac{d}{dz} \text{ and } \mathbf{J}^- = -\frac{d}{dz}.$$

We shall show that these operators are linear transformations from V_{2u+1} to V_{2u+1}. $B = \{1, z, \ldots, z^{2u}\}$ is a basis of V_{2u+1}.

$$\mathbf{J}^\circ(z^k) = \left(-u + z\frac{d}{dz}\right)(z^k)$$
$$= -uz^k + zkz^{k-1}$$
$$= -uz^k + kz^k$$
$$= (-u+k)z^k \in V_{2u+1}.$$

$$\mathbf{J}^+(z^k) = \left(-2uz + z^2\frac{d}{dz}\right)(z^k)$$
$$= -2uz^{k+1} + z^2 kz^{k-1}$$
$$= (-2u+k)z^{k+1} \in V_{2u+1}.$$

Also,

$$\mathbf{J}^-(z^k) = \frac{-\mathrm{d}(z^k)}{\mathrm{d}z} = -kz^{k-1} \in V_{2u+1}.$$

Therefore, $\mathbf{J}^\circ, \mathbf{J}^+, \mathbf{J}^-$ are linear transformations from $V_{2u+1} \to V_{2u+1}$. Hence, $\mathbf{J}^\circ, \mathbf{J}^+, \mathbf{J}^- \in \mathcal{L}(V_{2u+1})$. Therefore $\{\mathbf{J}^\circ, \mathbf{J}^+, \mathbf{J}^-\}$ spans a Lie algebra which is a subalgebra of $\mathcal{L}(V_{2u+1})$. Define

$$\rho : \mathrm{sl}(2,\mathbb{C}) \to \mathcal{L}(V_{2u+1})$$

by

$$\rho(j^\circ) = \mathbf{J}^\circ, \rho(j^+) = \mathbf{J}^+, \rho(j^-) = \mathbf{J}^-$$
$$\rho[j^\circ, j^+] = \rho(j^+) = \mathbf{J}^+ = [\mathbf{J}^\circ, \mathbf{J}^+] = [\rho(j^\circ), \rho(j^+)].$$

Similarly,

$$\rho[j^\circ, j^-] = [\rho j^\circ, \rho j^-] \text{ and } \rho[j^+, j^-] = [\rho(j^+), \rho(j^-)].$$

Hence, ρ is a representation of $\mathrm{sl}(2, \mathbb{C})$ onto V_{2u+1}.

Exercises 5.4.

5.4.1. Let V be an infinite dimensional vector space having a basis $\{1, z, z^2, \ldots, z^n, \ldots\}$, and $U \in \mathbb{C}$ such that $2u$ is not a non-negative integer. Prove that the operators

$$\mathbf{J}^\circ = -u + z\frac{\mathrm{d}}{\mathrm{d}z}, \mathbf{J}^+ = -2uz + z^2\frac{\mathrm{d}}{\mathrm{d}z} \text{ and } \mathbf{J}^- = -\frac{\mathrm{d}}{\mathrm{d}z}$$

span an infinite dimensional representation of $\mathrm{sl}(2,\mathbb{C})$ on V.

5.4.2. Prove that the set of all matrices

$$\left\{ \begin{bmatrix} 0 & x_2 & x_4 & x_3 \\ 0 & x_3 & x_1 & 0 \\ 0 & 0 & 0 & 0 \\ 0 & 0 & 0 & 0 \end{bmatrix} ; x_1, x_2, x_3, x_4 \in \mathbf{C} \right\}$$

is a 4-dimensional Lie algebra having a basis

$$g^+ = \begin{bmatrix} 0 & 0 & 0 & 0 \\ 0 & 0 & 1 & 0 \\ 0 & 0 & 0 & 0 \\ 0 & 0 & 0 & 0 \end{bmatrix}, g^- = \begin{bmatrix} 0 & 1 & 0 & 0 \\ 0 & 0 & 0 & 0 \\ 0 & 0 & 0 & 0 \\ 0 & 0 & 0 & 0 \end{bmatrix},$$

$$g^\circ = \begin{bmatrix} 0 & 0 & 0 & 1 \\ 0 & 1 & 0 & 0 \\ 0 & 0 & 0 & 0 \\ 0 & 0 & 0 & 0 \end{bmatrix}, \xi = \begin{bmatrix} 0 & 0 & 1 & 0 \\ 0 & 0 & 0 & 0 \\ 0 & 0 & 0 & 0 \\ 0 & 0 & 0 & 0 \end{bmatrix},$$

satisfying

$$[g^\circ, g^+] = g^+, \ [g^\circ, \ g^-] = -g^-, \ [g^+, g^-] = -\xi,$$
$$[\xi, g^\circ] = [\xi, g^+] = [\xi, g^-] = O.$$

5.4.3. Prove that the set of all matrices

$$\left\{ \begin{bmatrix} 0 & 0 & 0 & x_3 \\ 0 & -x_3 & 0 & x_2 \\ 0 & 0 & x_3 & x_1 \\ 0 & 0 & 0 & 0 \end{bmatrix} ; x_1, x_2, x_3, \in \mathbf{C} \right\}$$

is a 3-dimensional Lie algebra having a basis

$$g^+ = \begin{bmatrix} 0 & 0 & 0 & 0 \\ 0 & 0 & 0 & 0 \\ 0 & 0 & 0 & 1 \\ 0 & 0 & 0 & 0 \end{bmatrix}, g^- = \begin{bmatrix} 0 & 0 & 0 & 0 \\ 0 & 0 & 0 & 1 \\ 0 & 0 & 0 & 0 \\ 0 & 0 & 0 & 0 \end{bmatrix}$$

$$g^\circ = \begin{bmatrix} 0 & 0 & 0 & 1 \\ 0 & -1 & 0 & 0 \\ 0 & 0 & 1 & 0 \\ 0 & 0 & 0 & 0 \end{bmatrix}$$

satisfying

$$[g^\circ, g^+] = g^+, \ [g^\circ, g^-] = -g^-, [g^+, \ g^-] = O.$$

5.4.4. Let

$$j^\circ = t\frac{\partial}{\partial t} + \frac{1}{2}(\alpha + 1), \ J^+ = t\left[x\frac{\partial}{\partial x} + t\frac{\partial}{\partial t} + (1 + \alpha - x)\right],$$
$$J^- = t^{-1}\left(x\frac{\partial}{\partial x} - t\frac{\partial}{\partial t}\right).$$

Prove that $\{J^\circ, J^+, J^-\}$ generate a 3-dimensional Lie algebra which is isomorphic image of sl$(2, \mathbb{C})$.

5.5 Special Functions

5.5.1 Gauss hypergeometric function

Consider a power series in z

$$c_0 + c_1 z + c_2 z^2 + \cdots + c_n z^n + \cdots \tag{5.5.1}$$

where $c_i \in \mathbb{C}$. For the convergence of the power series, we evaluate $\lim_{n \to \infty} |c_n|^{\frac{1}{n}} = \frac{1}{R}$.
We say that (5.5.1) is absolutely convergent if $|z| < R$ and R is called the radius of convergence. Let

$$c_0 = 1 \text{ and } \frac{c_{n+1}}{c_n} = \frac{(\alpha + x)(\beta + x)}{(1+x)(\gamma + x)}, \; n = 0, 1, 2, \ldots$$

where $\alpha, \beta \in \mathbb{C}$ and $\gamma \in \mathbb{C} - \{0, -1, -2, \ldots\}$. With this the power series (5.5.1) becomes

$$1 + \frac{\alpha \beta}{\gamma} z + \frac{\alpha(\alpha+1)\beta(\beta+1)}{\gamma(\gamma+1)} \frac{z^2}{2!} + \frac{\alpha(\alpha+1)(\alpha+2)\beta(\beta+1)(\beta+2)}{\gamma(\gamma+1)(\gamma+2)} \frac{z^3}{3!} + \cdots$$

$$= 1 + \frac{(\alpha)_1 (\beta)_1}{(\gamma)_1} \frac{z}{1!} + \frac{(\alpha)_2 (\beta)_2}{(\gamma)_2} \frac{z^2}{2!} + \frac{(\alpha)_3 (\beta)_3}{(\gamma)_3} \frac{z^3}{3!} + \cdots \qquad (5.5.2)$$

$$= 1 + \sum_{n=1}^{\infty} \frac{(\alpha)_n (\beta)_n}{(\gamma)_n} \frac{z^n}{n!}$$

$$= \sum_{n=0}^{\infty} \frac{(\alpha)_n (\beta)_n}{(\gamma)_n} \frac{z^n}{n!}$$

where

$$(\alpha)_n = \begin{cases} \alpha(\alpha+1) \cdots (\alpha + (n-1)), & n = 1, 2, \ldots, \alpha \neq 0 \\ 1, & n = 0. \end{cases}$$

Series (5.5.2) is called Gauss hypergeometric series and is denoted by $_2F_1(\alpha, \beta; \gamma; z)$. To find the radius of convergence let us write,

$$u_n = \frac{(\alpha)_n (\beta)_n}{(\gamma)_n n!}.$$

$$\frac{u_{n+1}}{u_n} = \frac{(\alpha)_{n+1}(\beta)_{n+1}}{(\gamma)_{n+1}(n+1)!} \frac{(\gamma)_n n!}{(\alpha)_n (\beta)_n}$$

$$= \frac{(\alpha+n)(\beta+n)}{(\gamma+n)(n+1)}.$$

$$\lim_{n \to \infty} \left| \frac{u_{n+1}}{u_n} \right| = \lim_{n \to \infty} \left| \frac{(\alpha+n)(\beta+n)}{(\gamma+n)(n+1)} \right|$$

$$= \lim_{n \to \infty} \left| \frac{(1+\frac{\alpha}{n})(1+\frac{\beta}{n})}{(1+\frac{\gamma}{n})(1+\frac{1}{n})} \right|$$

$$= 1.$$

Therefore

$$\frac{1}{R} = 1, \; R = 1.$$

5.5.2 Differential equation satisfied by $_2F_1$

Let $\theta = z\frac{d}{dz}$ and $w = {}_2F_1(\alpha, \beta; \gamma; z)$. Then $\theta z^n = z\frac{d(z^n)}{dz} = nz^n$.
Consider,

$$\theta(\theta + \gamma - 1)w = \theta(\theta + \gamma - 1) \sum_{n=0}^{\infty} \frac{(\alpha)_n(\beta)_n}{(\gamma)_n} \frac{z^n}{n!}$$

$$= \theta \left[\sum_{n=0}^{\infty} \frac{(\alpha)_n(\beta)_n}{(\gamma)_n} \frac{(\gamma + n - 1)z^n}{n!} \right]$$

$$= \sum_{n=1}^{\infty} \frac{(\alpha)_n(\beta)_n}{(\gamma)_{n-1}} \frac{z^n}{(n-1)!}$$

$$= \sum_{k=0}^{\infty} \frac{(\alpha)_{k+1}(\beta)_{k+1}}{(\gamma)_k} \frac{z^{k+1}}{k!}, \qquad n = k+1.$$

Therefore,

$$\theta(\theta + \gamma - 1)w = \sum_{n=0}^{\infty} \frac{(\alpha)_{n+1}(\beta)_{n+1}}{(\gamma)_n} \frac{z^{n+1}}{n!}. \tag{5.5.3}$$

$$z(\theta + \alpha)(\theta + \beta)w = z(\theta + \alpha) \sum_{n=0}^{\infty} \frac{(\alpha)_n(\beta)_n}{(\gamma)_n n!}(n + \beta)z^n$$

$$= z \sum_{n=0}^{\infty} \frac{(\alpha)_n(\beta)_n}{(\gamma)_n n!}(\alpha + n)(\beta + n)z^n$$

$$= \sum_{n=0}^{\infty} \frac{(\alpha)_{n+1}(\beta)_{n+1}}{(\gamma)_n} \frac{z^{n+1}}{n!}. \tag{5.5.4}$$

(5.5.3) and (5.5.4) give,

$$\theta(\theta + \gamma - 1)w - z(\theta + \alpha)(\theta + \beta)w = 0$$

$$\Rightarrow \quad [\theta(\theta + \gamma - 1) - z(\theta + \alpha)(\theta + \beta)]w = 0$$

$$\Rightarrow \quad \left[z\left(z\frac{d}{dz} + \alpha\right)\left(z\frac{d}{dz} + \beta\right) - z\frac{d}{dz}\left(z\frac{d}{dz} + \gamma - 1\right) \right]w = 0$$

$$\Rightarrow \quad z\left(z\frac{d}{dz} + \alpha\right)\left(z\frac{dw}{dz} + \beta w\right) - z\frac{d}{dz}\left(z\frac{dw}{dz} + \gamma w - w\right) = 0$$

$$\Rightarrow \quad z\left[z\frac{dw}{dz} + z^2\frac{d^2w}{dz^2} + \beta z\frac{dw}{dz} + \alpha z\frac{dw}{dz} + \alpha\beta w \right]$$

$$\qquad - \left[z\frac{dw}{dz} + z^2\frac{d^2w}{dz^2} + \gamma z\frac{dw}{dz} - z\frac{dw}{dz} \right] = 0$$

$$\Rightarrow \quad z^2\frac{dw}{dz} + z^3\frac{d^2w}{dz^2} + \beta z^2\frac{dw}{dz} + \alpha z^2\frac{dw}{dz} + \alpha\beta wz$$

$$\qquad - z\frac{dw}{dz} - z^2\frac{d^2w}{dz^2} - \gamma z\frac{dw}{dz} + z\frac{dw}{dz} = 0$$

$$\Rightarrow \quad z^2(\alpha+\beta+1)\frac{dw}{dz}+(z^3-z^2)\frac{d^2w}{dz^2}-\gamma z\frac{dw}{dz}+\alpha\beta wz=0$$

$$\Rightarrow \quad [z^2(\alpha+\beta+1)-z\gamma]\frac{dw}{dz}+z^2(z-1)\frac{d^2w}{dz^2}+\alpha\beta wz=0$$

$$\Rightarrow \quad z(1-z)\frac{d^2w}{dz^2}+[\gamma-(\alpha+\beta+1)z]\frac{dw}{dz}-\alpha\beta w=0,$$

is the differential equation satisfied by $w = {}_2F_1$.

Q.7 Prove that

$$\frac{d}{dz}[{}_2F_1(\alpha,\beta;\gamma;z)]=\frac{\alpha\beta}{\gamma}\,{}_2F_1(\alpha+1,\beta+1;\gamma+1;z).$$

Solution:

$$\frac{d}{dz}[{}_2F_1(\alpha,\beta;\gamma;z)]=\frac{d}{dz}\sum_{n=0}^{\infty}\frac{(\alpha)_n\,(\beta)_n}{(\gamma)_n}\frac{z^n}{n!}$$

$$=\sum_{n=0}^{\infty}\frac{(\alpha)_n\,(\beta)_n}{(\gamma)_n}\frac{n\,z^{n-1}}{n!}$$

$$=\sum_{n=1}^{\infty}\frac{(\alpha)_n\,(\beta)_n}{(\gamma)_n}\frac{z^{n-1}}{(n-1)!}$$

$$=\sum_{n=0}^{\infty}\frac{(\alpha)_{n+1}\,(\beta)_{n+1}}{(\gamma)_{n+1}}\frac{z^n}{n!}$$

$$=\frac{\alpha\beta}{\gamma}\sum_{n=0}^{\infty}\frac{(\alpha+1)_n\,(\beta+1)_n}{(\gamma+1)_n}\frac{z^n}{n!}$$

$$=\frac{\alpha\beta}{\gamma}\,{}_2F_1(\alpha+1,\beta+1;\gamma+1;z).$$

Q.8 Find the hypergeometric series corresponding to $\sin z$ and $\cos z$.

Solution:

$$\sin z = z-\frac{z^3}{3!}+\frac{z^5}{5!}+\cdots = \sum_{n=0}^{\infty}\frac{(-1)^n z^{2n+1}}{(2n+1)!}$$

$$=\sum_{n=0}^{\infty}\frac{(-1)^n z^{2n+1}}{1\,2\,3\cdots 2n\,(2n+1)}=z\sum_{n=0}^{\infty}\frac{(-z^2)^n}{2^n\,n!\,3\,5\cdots(2n+1)}$$

$$=z\sum_{n=0}^{\infty}\frac{(-z^2)^n}{2^n\,n!\,2^n\left\{\left(\frac{3}{2}\right)\left(\frac{5}{2}\right)\cdots\left(\frac{2n+1}{2}\right)\right\}}=z\sum_{n=0}^{\infty}\frac{(-z^2)^n}{4^n\,n!\,\left(\frac{3}{2}\right)_n}$$

$$=z\sum_{n=0}^{\infty}\frac{\left(\frac{(-z^2)}{4}\right)^n}{\left(\frac{3}{2}\right)_n\,n!}.$$

Therefore,

$$\sin z = z\,_0F_1\left(\ ;\frac{3}{2};-\frac{z^2}{4}\right).$$

$$\cos z = 1 - \frac{z^2}{2!} + \frac{z^4}{4!} - \frac{z^6}{6!} + \cdots$$

$$= \sum_{n=0}^{\infty} \frac{(-1)^n z^{2n}}{(2n)!} = \sum_{n=0}^{\infty} \frac{(-1)^n z^{2n}}{(1)(2)(3)\cdots(2n-1)2n}$$

$$= \sum_{n=0}^{\infty} \frac{(-z^2)^n}{(1)(3)(5)\cdots(2n-1)\,2^n\,n!}$$

$$= \sum_{n=0}^{\infty} \frac{(-z^2)^n}{2^n \left\{\frac{1}{2}\frac{3}{2}\frac{5}{2}\cdots\left(n-\frac{1}{2}\right)\right\} n!}$$

$$= \sum_{n=0}^{\infty} \frac{1}{\left(\frac{1}{2}\right)_n} \frac{\left(\frac{-z^2}{4}\right)^n}{n!}.$$

Therefore,

$$\cos z = \,_0F_1\left(\ ;\frac{1}{2};-\frac{z^2}{4}\right).$$

Q.9   Express $\ln(1+z)$ in terms of $_2F_1$.

$$\ln(1+z) = z - \frac{z^2}{2} + \frac{z^3}{3} - \cdots = \sum_{n=0}^{\infty} \frac{(-1)^n z^{n+1}}{n+1}$$

$$= z \sum_{n=0}^{\infty} \frac{n!\,(-z)^n}{n!\,(n+1)} = z \sum_{n=0}^{\infty} \frac{(1)_n\,(-z)^n}{n!\,(n+1)}$$

$$= z \sum_{n=0}^{\infty} \frac{(1)_n\,(1)_n}{(1)_n\,(n+1)} \frac{(-z)^n}{n!} = z\,_2F_1(1,1;2;-z).$$

5.5.3 *Integral representation of* $_pF_q\left(\begin{matrix}\alpha_1,\alpha_2,...,\alpha_p\\ \beta_1,\beta_2,...,\beta_q\end{matrix};z\right)$

$$\begin{aligned}&_pF_q(\alpha_1,\alpha_2,\ldots,\alpha_p;\ \beta_1,\beta_2,\ldots,\beta_q;z)\\ &= \frac{\Gamma(\beta_1)}{\Gamma(\alpha_1)\Gamma(\beta_1-\alpha_1)} \int_0^1 t^{\alpha_1-1}(1-t)^{\beta_1-\alpha_1-1}\\ &\quad \times\,_{p-1}F_{q-1}\left[\alpha_2,\alpha_3,\ldots,\alpha_p;\ \beta_2,\beta_3,\ldots,\beta_q;zt\right]dt,\end{aligned}$$

provided that integral exists.

Proof:

$$_pF_q\left(\begin{matrix}\alpha_1,\alpha_2,...,\alpha_p\\ \beta_1,\beta_2,...,\beta_q\end{matrix}; z\right) = \sum_{n=0}^{\infty} \frac{(\alpha_1)_n\,(\alpha_2)_n\cdots(\alpha_p)_n}{(\beta_1)_n\,(\beta_2)_n\cdots(\beta_q)_n}\,\frac{z^n}{n!}. \tag{5.5.5}$$

Consider

$$\frac{(\alpha_1)_n}{(\beta_1)_n} = \frac{\Gamma(\alpha_1+n)}{\Gamma(\alpha_1)}\,\frac{\Gamma(\beta_1)}{\Gamma(\beta_1+n)}$$

$$= \frac{\Gamma(\beta_1)}{\Gamma(\alpha_1)}\,\frac{\Gamma(\alpha_1+n)\Gamma(\beta_1-\alpha_1)}{\Gamma(\beta_1-\alpha_1)\Gamma(\beta_1+n)}$$

$$= \frac{\Gamma(\beta_1)}{\Gamma(\alpha_1)\Gamma(\beta_1-\alpha_1)}\,B(\alpha_1+n,\beta_1-\alpha_1)$$

$$= \frac{\Gamma(\beta_1)}{\Gamma(\alpha_1)\Gamma(\beta_1-\alpha_1)}\int_0^1 t^{\alpha_1+n-1}(1-t)^{\beta_1-\alpha_1-1}\mathrm{d}t.$$

Substituting in (5.5.5)

$$_pF_q\left(\begin{matrix}\alpha_1,\alpha_2,...,\alpha_p\\ \beta_1,\beta_2,...,\beta_q\end{matrix}; z\right) = \sum_{n=0}^{\infty} \frac{\Gamma(\beta_1)}{\Gamma(\alpha_1)\Gamma(\beta_1-\alpha_1)}\,\frac{(\alpha_2)_n\cdots(\alpha_p)_n}{(\beta_2)_n\cdots(\beta_q)_n}$$

$$\times \int_0^1 t^{\alpha_1+n-1}(1-t)^{\beta_1-\alpha_1-1}\mathrm{d}t\,\frac{z^n}{n!}$$

$$= \frac{\Gamma(\beta_1)}{\Gamma(\alpha_1)\Gamma(\beta_1-\alpha_1)}\int_0^1 t^{\alpha_1-1}(1-t)^{\beta_1-\alpha_1-1}$$

$$\times\left(\sum_{n=0}^{\infty} \frac{(\alpha_2)_n\cdots(\alpha_p)_n}{(\beta_2)_n\cdots(\beta_q)_n}\,\frac{(tz)^n}{n!}\right)\mathrm{d}t$$

$$= \frac{\Gamma(\beta_1)}{\Gamma(\alpha_1)\Gamma(\beta_1-\alpha_1)}\int_0^1 t^{\alpha_1-1}(1-t)^{\beta_1-\alpha_1-1}$$

$$\times\ _{p-1}F_{q-1}(\alpha_2,\ldots,\alpha_p;\,\beta_2,\ldots,\beta_q;\,tz)\mathrm{d}t$$

5.6 Laguerre Polynomial $L_n^{(\alpha)}(x)$

Definition 5.6.1. Laguerre polynomial $L_n^{(\alpha)}(x)$ is defined as

$$L_n^{(\alpha)}(x) = \frac{(1+\alpha)_n}{n!}\ _1F_1(-n;\,1+\alpha;\,x)\ n\in\mathbb{N}$$

Note 5.6.1: $L_n^{(\alpha)}(x)$ is a polynomial of degree n.

The Laguerre function $L_\delta^{(\alpha)}(x)$ is obtained by replacing $-n$ by δ. That is,

$$L_\delta^{(\alpha)}(x) = \frac{\Gamma(1+\alpha-\delta)}{\Gamma(1+\alpha)\Gamma(1-\delta)}\ _1F_1(\delta;\,1+\alpha;\,z).$$

5.6.1 Laguerre polynomial and Lie algebra

Consider the operator $\mathbf{J}^\circ, \mathbf{J}^+, \mathbf{J}^-$ defined by

$$\mathbf{J}^\circ = t\frac{\partial}{\partial t} + \frac{1}{2}(\alpha+1)$$

$$\mathbf{J}^+ = t\left[x\frac{\partial}{\partial x} + t\frac{\partial}{\partial t} + (1+\alpha - x)\right]$$

$$\mathbf{J}^- = t^{-1}\left[x\frac{\partial}{\partial x} - t\frac{\partial}{\partial t}\right].$$

$$\mathbf{J}^\circ\left[t^n \mathbf{L}_n^{(\alpha)}(x)\right] = a_n t^n \mathbf{L}_n^{(\alpha)}(x),$$

$$\mathbf{J}^+\left[t^n \mathbf{L}_n^{(\alpha)}(x)\right] = b_n t^{n+1} \mathbf{L}_{n+1}^{(\alpha)}(x),$$

$$\mathbf{J}^-\left[t^n \mathbf{L}_n^{(\alpha)}(x)\right] = c_n t^{n-1} \mathbf{L}_{n-1}^{(\alpha)}(x),$$

where a_n, b_n, c_n are constants. It is easy to check that

$$[\mathbf{J}^\circ, \mathbf{J}^+] = \mathbf{J}^+, \quad [\mathbf{J}^\circ, \mathbf{J}^-] = -\mathbf{J}^-, \quad [\mathbf{J}^+, \mathbf{J}^-] = 2\mathbf{J}^\circ.$$

Let V be a vector space having a basis $\left\{t^n \mathbf{L}_n^{(\alpha)}(x), n = 0, 1, 2, \ldots\right\}$. V is an infinite dimensional vector space. Operators $\mathbf{J}^\circ, \mathbf{J}^+, \mathbf{J}^- \in \mathcal{L}(V)$ span a Lie algebra, say G' which is an isomorphic image of $\mathrm{sl}(2, \mathbb{C})$. Hence, these operators provide a representation of $\mathrm{sl}(2, \mathbb{C})$ on the representation space V.

Exercises 5.6.

5.6.1. Evaluate

$$\int_{-1}^1 (1+x)^{p-1}(1-x)^{q-1}dx.$$

5.6.2. Show that for $0 \le k \le n$,

$$(\alpha)_{n-k} = \frac{(-1)^k (\alpha)_n}{(1-\alpha-n)_k}.$$

5.6.3. Prove that

$$\frac{d^n}{dx^n}\left[x^{\alpha-1+n} {}_2F_1(\alpha, \beta; \gamma; x)\right] = (\alpha)_n x^{\alpha-1} {}_2F_1(\alpha+n, \beta; \gamma; x).$$

5.6.4. Obtain the results:

(i) $\ln(1+x) = x \, {}_2F_1(1, 1; 2; -x),$

(ii) $\sin^{-1}x = x \, {}_2F_1\left(\frac{1}{2}, \frac{1}{2}; \frac{3}{2}; x^2\right),$

(iii) $\tan^{-1}x = x \, {}_2F_1\left(\frac{1}{2}, 1; \frac{3}{2}; -x^2\right).$

5.6.5. The complete elliptic integral of the first kind is

$$\mathbf{K} = \int_0^{\frac{\pi}{2}} \frac{dt}{\sqrt{1 - k^2 \sin^2 t}}.$$

Show that

$$\mathbf{K} = \frac{1}{2}\pi \, {}_2F_1\left(\frac{1}{2}, \frac{1}{2}; 1; k^2\right).$$

5.6.6. Prove that $w = {}_pF_q$ is a solution of the differential equation

$$\left[\theta \prod_{j=1}^{q}(\theta + \beta_j - 1) - z\prod_{j=1}^{p}(\theta + \alpha_j)\right] w = 0.$$

5.6.7. If $p \le q + 1$, $\Re(\beta_1) > \Re(\alpha_1) > 0$ and none of $\beta_1, \beta_2 \ldots, \beta_q$ is zero or a negative integer, and if $|z| < 1$, prove that

$$_pF_q(\alpha_1, \alpha_2, \ldots, \alpha_p; \beta_1, \beta_2, \ldots, \beta_q; z)$$

$$= \frac{\Gamma(\beta_1)}{\Gamma(\alpha_1)\Gamma(\beta_1 - \alpha_1)} \int_0^1 t^{\alpha_1 - 1}(1 - t)^{\beta_1 - \alpha_1 - 1}$$

$$\times {}_{p-1}F_{q-1}\left[\alpha_2, \alpha_3, \ldots, \alpha_p; \beta_2, \beta_3, \ldots, \beta_q; zt\right] dt.$$

If $p \le q$, the condition $|z| < 1$ may be omitted (See section 5.5.3).

5.6.8. Prove that

$$_2F_1(\gamma - \alpha, \gamma - \beta; \gamma; z) = (1 - z)^{\alpha + \beta - \gamma}{}_2F_1(\alpha, \beta; \gamma; z).$$

5.6.9. Let

$$\mathbf{L}_n^{(\alpha)}(x) = \frac{(1 + \alpha)_n}{n!} \, {}_1F_1[{}^{-n}_{1+\alpha}; x].$$

Using

$$\frac{d}{dx}\mathbf{L}_n^{(\alpha)}(x) = \frac{1}{x}\left[n\mathbf{L}_n^{(\alpha)}(x) - (\alpha + n)\mathbf{L}_{n-1}^{(\alpha)}(x)\right],$$

and

$$\frac{d}{dx}\mathbf{L}_n^{(\alpha)}(x) = \frac{1}{x}\left[(x - \alpha - n - 1)\mathbf{L}_n^{(\alpha)}(x) + (n + 1)\mathbf{L}_{n+1}^{(\alpha)}(x)\right],$$

prove that

$$\mathbf{J}^\circ\left[t^n\mathbf{L}_n^{(\alpha)}(x)\right] = a_n t^n \mathbf{L}_n^{(\alpha)}(x),$$

$$\mathbf{J}^+\left[t^n\mathbf{L}_n^{(\alpha)}(x)\right] = b_n t^{n+1} \mathbf{L}_{n+1}^{(\alpha)}(x),$$

$$\mathbf{J}^-\left[t^n\mathbf{L}_n^{(\alpha)}(x)\right] = c_n t^{n-1} \mathbf{L}_{n-1}^{(\alpha)}(x),$$

where a_n, b_n, c_n are expressions in n independent of x and y, and $\{\mathbf{J}^\circ, \mathbf{J}^+, \mathbf{J}^-\}$ are operators defined in Exercise 5.1.4.

5.6.10. Let V be a vector space having a basis $\left\{ \mathbf{L}_n^{(\alpha)}(x) : n = 1,2,3,\ldots \right\}$. Prove that the operators $\{\mathbf{J}^\circ, \mathbf{J}^+, \mathbf{J}^-\}$ of Exercise 5.1.4 provide a representation of $sl(2,\mathbb{C})$ on V.

5.7 Helmholtz Equation

Definition 5.7.1. The Helmholtz equation in two variables x and y is

$$(\partial_{xx} + \partial_{yy} + w^2)\Psi = 0, \ w > 0. \tag{5.7.1}$$

Let $\mathcal{F}$ be the complex vector space of all functions analytic in real x and y.

Note 5.7.1: A function $f(x)$ where x is real, is said to be analytic in x if it is expressible as a power series (or polynomial) in x.

Example 5.7.1.

1. $\sin x$ is analytic in x if x is real.
   For,

$$\sin x = x - \frac{x^3}{3!} + \frac{x^5}{5!} - \cdots \quad \text{(power series)}.$$

   But,

$$\sin(\sqrt{x}) = \sqrt{x} - \frac{x^{\frac{3}{2}}}{3!} + \frac{x^{\frac{5}{2}}}{5!} + \cdots$$

   is not analytic because it is not a power series.

2.

$$\cos x = 1 - \frac{x^2}{2!} + \frac{x^4}{4!} + \cdots$$

   and

$$\cos\sqrt{x} = 1 - \frac{x}{2!} + \frac{x^2}{2!} + \frac{x^4}{4!} - \cdots$$

   are both analytic functions.

Let $\mathcal{F}_\circ$ be the solution space of (5.7.1) containing solutions analytic in x and y. $\mathcal{F}_\circ$ is obviously a subspace of $\mathcal{F}$. Writing

$$Q = \partial_{xx} + \partial_{yy} + w^2,$$

we conclude that $Q : \mathcal{F} \to \mathcal{F}$ is a linear transformation having $\mathcal{F}_\circ$ as its kernel. $[\ker Q = \{T \in \mathcal{F} : Q(T) = O\} = \mathcal{F}_\circ]$.

Definition 5.7.2. A first order linear differential operator

$$L = X(x,y)\partial_x + Y(x,y)\partial_y + Z(x,y),$$

where $X, Y, Z \in \mathcal{F}$ is called a symmetry operator of (5.7.1) if the commutator of L and Q, that is $[L,Q] = R(x,y)Q$ for some $R \in \mathcal{F}$.

Theorem 5.7.1. $\Psi \in \mathcal{F}_o \Rightarrow L\Psi \in \mathcal{F}_o.$

Proof:

$$[L,Q] = RQ$$
$$\Rightarrow LQ - QL = RQ$$
$$\Rightarrow QL = LQ - RQ$$
$$\Rightarrow Q(L\Psi) = (QL)\Psi$$
$$= (LQ - RQ)\Psi$$
$$= L(Q\Psi) - R(Q\Psi)$$
$$= L(O) - R(O), \quad \text{since } \Psi \text{ is a solution of } Q$$
$$= O.$$

So $L\Psi \in \mathcal{F}_o.$

Theorem 5.7.2. *The set of all symmetry operators of* (5.7.1) *is a Lie algebra, that is, if* L_1 *and* L_2 *are symmetric operators then*

$$(i) \quad \alpha L_1 + \beta L_2$$

is a symmetry operator

$$(ii) \quad [L_1, L_2]$$

is a symmetry operator.

Proof: Let G be the set of all symmetry operators of (5.7.1) and let $L_1, L_2 \in G$. $[L_1, Q] = R_1 Q$ and $[L_2, Q] = R_2 Q$ for some $R_1, R_2 \in \mathbb{F}$. Consider,

$$[\alpha L_1, + \beta L_2, Q] = \alpha[L_1, Q] + \beta[L_2, Q]$$
$$= \alpha(R_1 Q) + \beta(R_2 Q)$$
$$= (\alpha R_1 + \beta R_2)Q$$
$$= R_3 Q$$

where

$$R_3 = \alpha R_1 + \beta R_2 \in \mathbb{F}.$$
$$\Rightarrow \alpha L_1 + \beta L_2 \in G.$$

Hence (i). We can write,

$$L_1 = X_1 \partial_x + Y_1 \partial_y + Z_1$$

and

$$L_2 = X_2 \partial_x + Y_2 \partial_y + Z_2.$$

It is easy to check that

$$[L_1 L_2 - L_2 L_1, Q] = [(L_1 + Z_2)R_2 - (L_2 + Z_1)R_1]Q.$$

Therefore

$$L_1 L_2 - L_2 L_1 \in G \Rightarrow [L_1, L_2] \in G.$$

Hence (ii) and hence the theorem.

Since L is a symmetry operator, we have

$$LQ - QL = RQ \tag{5.7.2}$$
$$L = X\partial_x + Y\partial_y + Z$$

and

$$Q = \partial_{xx} + \partial_{yy} + w^2.$$

(5.7.2) gives,

$$[X\partial_x + Y\partial_y + Z][\partial_{xx} + \partial_{yy} + w^2]f - [\partial_{xx} + \partial_{yy} + w^2][X\partial_x + Y\partial_y + Z]f$$
$$= R(\partial_{xx} + \partial_{yy} + w^2)f. \tag{5.7.3}$$

Left-side of (5.7.3) gives

$$\big[X\partial_{xxx}f + X\partial_{xyy}f + Xw^2\partial_x f + Y\partial_{yxx}f + Y\partial_{yyy}f + w^2 Y\partial_y f + Z\partial_{xx}f + Z\partial_{yy}f + Zw^2 f\big]$$
$$- \big[X_{xx}\partial_x f + 2X_x\partial_{xx} + X\partial_{xxx}f + Y_{xx}\partial_y f + 2Y_x\partial_{xy} + Y\partial_{xxy}f + Z_{xx}f + 2Z_x\partial_{xx}f$$
$$+ Z\partial_{xx}f + X_{yy}\partial_x f + 2X_y\partial_{xy}f + X\partial_{yyx}f + Y_{yy}\partial_y f + 2Y_y\partial_{yy}f + Y\partial_{yyy}f$$
$$+ Z_{yy}f + 2Z_y\partial_y f + Z\partial_{yy}f + w^2 X\partial_x f + w^2 Y\partial_y f + w^2 Z\big]\,f$$
$$= \big[-2X_x\partial_{xx} - 2(X_y + Y_x)\partial_{xy} - 2Y_y\partial_{yy} - (X_{xx} + X_{yy} + 2Z_x)\partial_x - (Y_{xx} + Y_{yy} + 2Z_y)\partial_y$$
$$- (Z_{xx} + Z_{yy})\big]\,f$$

From (5.7.3),

$$2X_x\partial_{xx} + 2(X_y + Y_x)\partial_{xy} + 2Y_y\partial_{yy} + (X_{xx} + X_{yy} + 2Z_x)\partial_x$$
$$+ (Y_{xx} + Y_{yy} + 2Z_y)\partial_y + (Z_{xx} + Z_{yy}) = -R(\partial_{xx} + \partial_{yy} + w^2).$$

$$\Rightarrow \quad (a) \qquad 2X_x = -R = 2Y_y$$
$$X_y + Y_x = 0 \Rightarrow X_y = -Y_x, \tag{5.7.4}$$
$$(b) \quad X_{xx} + X_{yy} + 2Z_x = 0,\ Y_{xx} + Y_{yy} + 2Z_y = 0, \tag{5.7.5}$$
$$(c) \qquad Z_{xx} + Z_{yy} = -Rw^2. \tag{5.7.6}$$

From (a)

$$X_{xx} + X_{yy} = 0.$$
$$Y_{xx} + Y_{yy} = 0. \tag{5.7.7}$$

Putting (5.7.7) in (5.7.5) we have

$$Z_x = 0 \text{ and } Z_y = 0.$$

Hence

$$Z = \delta \text{ (constant)}; (c) \Rightarrow R = 0. \qquad (5.7.8)$$

Therefore (5.7.4) gives

$$X_x = 0 = Y_y$$
$$\Rightarrow \quad X = f(y) \text{ and } Y = g(x)$$
$$\Rightarrow \quad X_y = f'(y) \text{ and } Y_x = g'(x).$$

Hence

$$X_y = -Y_x$$
$$\Rightarrow f'(y) = -g'(x).$$

Therefore

$$X = \alpha + \gamma y \text{ and } Y = \beta - \gamma x; \ \alpha, \beta, \gamma \in \mathbb{C}.$$

Now,

$$L = X\partial_x + Y\partial_y + Z$$
$$= \alpha\partial x + \beta\partial y + Z + \gamma(y\partial x - x\partial y),$$

which is a linear combination of independent vectors $\partial x, \partial y, y\partial x - x\partial y, 1$.

Hence

$$B = \{\partial x, \partial y, y\partial x - x\partial y, 1\}$$

is a basis of the Lie algebra G.

This Lie algebra having a basis $\{P_1, P_2, M, E\}$ where $P_1 = \partial x, P_2 = \partial y, M = y\partial x - x\partial y, E = 1$ is called the symmetry algebra of the given Helmholtz equation. The elements are called symmetry operators.

Let G be a matrix Lie algebra having a basis $\{\mathbb{P}_1, \mathbb{P}_2, \mathbb{M}, \xi\}$ where,

$$\mathbb{P}_1 = \begin{bmatrix} 0 & 0 & 0 \\ 0 & 0 & 0 \\ 1 & 0 & 0 \end{bmatrix}, \mathbb{P}_2 = \begin{bmatrix} 0 & 0 & 0 \\ 0 & 0 & 0 \\ 0 & 1 & 0 \end{bmatrix}, \mathbb{M} = \begin{bmatrix} 0 & -1 & 1 \\ 1 & 0 & 0 \\ 0 & 0 & 0 \end{bmatrix}, \xi = \begin{bmatrix} 1 & 0 & 0 \\ 0 & 1 & 0 \\ 0 & 0 & 1 \end{bmatrix},$$

having commutation relation

$$[\mathbb{P}_1, \mathbb{M}] = \mathbb{P}_2, \ [\mathbb{P}_2, \mathbb{M}] = -\mathbb{P}_1, \ [\mathbb{P}_1, \mathbb{P}_2] = O$$

and

$$[\xi, \mathbb{P}_1] = [\xi, \mathbb{P}_2] = [\xi, \mathbb{M}] = O.$$

Hence the symmetry algebra of Helmholtz equation (5.7.4) provides a representation of matrix Lie algebra on the representation space $\mathbb{F}_0$.

5.7.1 Helmholtz equation in three variables

$$Q\Psi = (\partial_{x_1 x_1} + \partial_{x_2 x_2} + \partial_{x_3 x_3} + w^2)\Psi(x_1, x_2, x_3) = 0. \qquad (5.7.9)$$

Definition 5.7.3. A linear differential operator

$$L = X\partial x_1 + Y\partial x_2 + U\partial x_3 + Z, \quad X, Y, U, Z \in \mathbb{F}$$

is called a symmetry operator if

$$[L, Q] = RQ \text{ for some } R \in \mathbb{F}.$$

Proceeding as in the case of two variables, we arrive at

$$L = \alpha_1 \partial x_1 + \alpha_2 \partial x_2 + \alpha_3 \partial x_3 + \alpha_4(x_3 \partial x_2 - x_2 \partial x_3)$$
$$+ \alpha_5(x_1 \partial x_3 - x_3 \partial x_1) + \alpha_6(x_2 \partial x_1 - x_2 \partial x_1).$$

Take

$$P_1 = \partial x_1, P_2 = \partial x_2, P_3 = \partial x_3,$$
$$J_1 = x_3 \partial x_2 - x_2 \partial x_3, J_2 = x_1 \partial x_3 - x_3 \partial x_1, J_3 = x_2 \partial x_1 - x_1 \partial x_2.$$

$P_1, P_2, P_3, J_1, J_2, J_3$ are linearly independent and so they generate a six dimensional vector space.

$$[P_l, P_m] = O$$
$$[J_l, J_m] = \sum_{n=1}^{3} e_{lmn} P_n$$
$$[J_l, P_m] = \sum_{n=1}^{3} e_{lmn} P_n$$

where

$$e_{123} = e_{231} = e_{312} = 1$$
$$e_{213} = e_{321} = e_{132} = -1$$

while all other e's are zero.

$$[J_1, J_2] = e_{121}J_1 + e_{122}J_2 + e_{123}J_3$$
$$= O + O + 1J_3 = J_3$$
$$[J_1, P_2] = e_{121}P_1 + e_{122}P_2 + e_{123}P_3$$
$$= O + O + 1P_3 = P_3$$

and so on. These six operators generate a 15 dimensional Lie algebra.

Consider the matrix Lie algebra $e(3)$ having a basis

$$\mathbb{P}_1 = \begin{bmatrix} 0 & 0 & 0 & 0 \\ 0 & 0 & 0 & 0 \\ 0 & 0 & 0 & 0 \\ 1 & 0 & 0 & 0 \end{bmatrix}, \quad \mathbb{P}_2 = \begin{bmatrix} 0 & 0 & 0 & 0 \\ 0 & 0 & 0 & 0 \\ 0 & 0 & 0 & 0 \\ 0 & 1 & 0 & 0 \end{bmatrix}$$

$$\mathbb{P}_3 = \begin{bmatrix} 0 & 0 & 0 & 0 \\ 0 & 0 & 0 & 0 \\ 0 & 0 & 1 & 0 \\ 0 & 0 & 0 & 0 \end{bmatrix}, \quad \mathbb{J}_1 = \begin{bmatrix} 0 & 0 & 0 & 0 \\ 0 & 0 & -1 & 0 \\ 0 & 1 & 0 & 0 \\ 1 & 0 & 0 & 0 \end{bmatrix}$$

$$\mathbb{J}_2 = \begin{bmatrix} 0 & 0 & 1 & 0 \\ 0 & 0 & 0 & 0 \\ -1 & 0 & 0 & 0 \\ 0 & 0 & 0 & 0 \end{bmatrix}, \quad \mathbb{J}_3 = \begin{bmatrix} 0 & -1 & 0 & 0 \\ 1 & 0 & 0 & 0 \\ 0 & 0 & 0 & 0 \\ 1 & 0 & 0 & 0 \end{bmatrix}.$$

$\{\mathbb{P}_1, \mathbb{P}_2, \mathbb{P}_3, \mathbb{J}_1, \mathbb{J}_2, \mathbb{J}_3\}$ is a basis of $e(3)$, and both Lie algebra G and matrix Lie algebra $e(3)$ are isomorphic.

Note 5.7.2: Gauss' hypergeometric function will give rise to a symmetric algebra which is isomorphic to $sl(4, \mathbb{C})$.

Note 5.7.3:

$$w(z) = {}_2F_1(\alpha, \beta; \gamma; z)$$

is a solution of $[z(\theta + \alpha)(\theta + \beta) - \theta(\theta + \gamma - 1)]w = 0$

$$\Rightarrow z\left[\left(z\frac{d}{dz} + \alpha\right)\left(z\frac{d}{dz} + \beta\right) - \frac{d}{dz}\left(z\frac{d}{dz} + \gamma - 1\right)\right]w(z) = 0$$

$$\Leftrightarrow w(s, u, t, z) = s^\alpha u^\beta t^{\gamma-1} {}_2F_1(\alpha, \beta; \gamma; z)$$

is a solution of

$$\left[\left\{s\left(z\frac{\partial}{\partial z} + s\frac{\partial}{\partial s}\right)\right\}\left\{u\left(z\frac{\partial}{\partial z} + u\frac{\partial}{\partial u}\right)\right\} - \left(sut\frac{\partial}{\partial z}\right)\left\{t^{-1}\left(z\frac{\partial}{\partial z} + t\frac{\partial}{\partial t}\right)\right\}\right]w(s, u, t, z) = 0 \qquad (5.7.10)$$

(The significance of writing this is to get a wave equation). For example, consider

$$\left(z\frac{d}{dz} + \alpha\right)f(z) = 0 \Leftrightarrow \left(z\frac{\partial}{\partial z} + s\frac{\partial}{\partial s}\right)s^\alpha f(z) = 0$$

$$\Leftrightarrow s^\alpha z\frac{\partial f(z)}{\partial z} + \alpha s^\alpha f(z) = 0 \Leftrightarrow z\frac{df(z)}{dz} + \alpha f(z) = 0$$

$$\Leftrightarrow \left[z\frac{d}{dz} + \alpha\right]f(z) = 0.$$

Let

$$L^\alpha = s\left(z\frac{\partial}{\partial z} + s\frac{\partial}{\partial s}\right), \; L^\beta = u\left(z\frac{\partial}{\partial z} + u\frac{\partial}{\partial u}\right)$$

$$L_\gamma = t^{-1}\left(z\frac{\partial}{\partial z} + t\frac{\partial}{\partial t}\right), \; L^{\alpha\beta\gamma} = sut\frac{\partial}{\partial z}.$$

Equation (5.7.10) becomes

$$\left(L^\alpha L^\beta - L^{\alpha\beta\gamma}L_\gamma\right) w(s,u,t,z) = 0.$$

Four variable wave equation is

$$\left(\frac{\partial^2}{\partial v_1 \partial v_2} - \frac{\partial^2}{\partial v_3 \partial v_4}\right)\Psi = 0.$$

The operators $L^\alpha, L^\beta, L^{\alpha\beta\gamma}, L_\gamma$ generate a 4-dimensional Abelian Lie algebra. For,

$$\left[L^\alpha, L^\beta\right] = O \Rightarrow L^\alpha L^\beta - L^\beta L^\alpha = O$$

$$\Rightarrow L^\alpha L^\beta = L^\beta L^\alpha \text{ (commute, therefore Abelian).}$$

Since $L^\alpha, L^\beta, L^{\alpha\beta\gamma}, L_\gamma$ generate an Abelian Lie algebra, we introduce another coordinate system (v_1, v_2, v_3, v_4) such that

$$L^\alpha = \frac{\partial}{\partial v_1}, \; L^\beta = \frac{\partial}{\partial v_2}, \; L^{\alpha\beta\gamma} = \frac{\partial}{\partial v_4} \text{ and } L_\gamma = \frac{\partial}{\partial v_3}.$$

The relationship between (s,t,u,z) and (v_1, v_2, v_3, v_4) is

$$s = s(v_1, v_2, v_3, v_4), u = u(v_1, v_2, v_3, v_4)$$

$$t = t(v_1, v_2, v_3, v_4), z = z(v_1, v_2, v_3, v_4).$$

Using chain rule,

$$sz\frac{\partial}{\partial z} + s^2\frac{\partial}{\partial s} = \frac{\partial}{\partial v_1} = \frac{\partial}{\partial s}\frac{\partial s}{\partial v_1} + \frac{\partial}{\partial u}\frac{\partial u}{\partial v_1} + \frac{\partial}{\partial t}\frac{\partial t}{\partial v_1} + \frac{\partial}{\partial z}\frac{\partial z}{\partial v_1}.$$

$$\frac{\partial s}{\partial v_1} = s, \; \frac{\partial u}{\partial v_1} = O, \; \frac{\partial t}{\partial v_1} = O, \; \frac{\partial z}{\partial v_1} = sz$$

$$\frac{\partial s}{\partial v_1} = s^2 \Rightarrow s = -\frac{1}{v_1}$$

$$\frac{\partial z}{\partial v_1} = sz \Rightarrow \frac{\partial z}{\partial v_1} = -\frac{z}{v_1}$$

$$uz\frac{\partial}{\partial z} + u^2\frac{\partial}{\partial u} = \frac{\partial}{\partial v_2} = \frac{\partial}{\partial s}\frac{\partial s}{\partial v_2} + \frac{\partial}{\partial u}\frac{\partial u}{\partial v_2} + \frac{\partial}{\partial t}\frac{\partial t}{\partial v_2} + \frac{\partial}{\partial z}\frac{\partial z}{\partial v_2}$$

$$\Rightarrow u = -\frac{1}{v_2} \text{ and } \frac{\partial z}{\partial v_2} = -\frac{z}{v_2}.$$

$$t^{-1}z\frac{\partial}{\partial z} + \frac{\partial}{\partial t} = \frac{\partial}{\partial v_3} = \frac{\partial}{\partial s}\frac{\partial s}{\partial v_3} + \frac{\partial}{\partial u}\frac{\partial u}{\partial v_3} + \frac{\partial}{\partial t}\frac{\partial t}{\partial v_3} + \frac{\partial}{\partial z}\frac{\partial z}{\partial v_3}.$$

Therefore

$$\frac{\partial s}{\partial v_3} = 0 = \frac{\partial u}{\partial v_3}, \frac{\partial t}{\partial v_3} = 1, \frac{\partial z}{\partial v_3} = \frac{z}{t} = \frac{z}{v_3}.$$

That is $t = v_3$. But

$$\frac{\partial}{\partial z} = \frac{\partial}{\partial v_4} = \frac{\partial}{\partial s}\frac{\partial s}{\partial v_4} + \frac{\partial}{\partial u}\frac{\partial u}{\partial v_4} + \frac{\partial}{\partial t}\frac{\partial t}{\partial v_4} + \frac{\partial}{\partial z}\frac{\partial z}{\partial v_4}$$

$$\Rightarrow \frac{\partial s}{\partial v_4} = \frac{\partial u}{\partial v_4} = \frac{\partial t}{\partial v_4} = 0 \text{ and } \frac{\partial z}{\partial v_4} = sut.$$

Therefore

$$z = sut\, v_4 = \left[-\frac{1}{v_1}\right] \times \left[-\frac{1}{v_2}\right] \times v_3\, v_4.$$

That is

$$z = \frac{v_3 v_4}{v_1 v_2}.$$

$$w(v_1, v_2, v_3, v_4) = (v_1)^{-\alpha}(v_2)^{-\beta}(v_3)^{\gamma-1}{}_2F_1\left(\alpha, \beta; \gamma; \frac{v_3 v_4}{v_1 v_2}\right)$$

is a solution of a 4-variable wave equation

$$(\partial_{v_1 v_2} - \partial_{v_3 v_4})w(v_1, v_2, v_3, v_4) = 0.$$

Let V be a vector space having a basis

$$B = \{s^{\alpha+n} u^{\beta+n} t^{\gamma+p-1}{}_2F_1\left[\alpha+n, \beta+n; \gamma+p; z\right], m, n, p \in \mathbb{I}\}.$$

The symmetry algebra G provides an infinite dimensional representation of $sl(4,\mathbb{C})$ on the representation space V.

5.8 Lie Group

Definition 5.8.1. A Lie group is a nonempty set G having the following properties:
   (i) G is a group, say, with respect to multiplication,
   (ii) G is an analytic manifold,
   (iii) Group multiplication $\circ : G \times G \to G$ and group inversion $\lambda : G \to G$ are analytic with respect to the manifold structure.

Example 5.8.1.

$$E(2) = \left\{ \begin{bmatrix} \cos\theta & -\sin\theta & 0 \\ \sin\theta & \cos\theta & 0 \\ a & b & 1 \end{bmatrix} : a, b \in \mathbb{R}, 0 \leq \theta < 2\pi \right\}.$$

Clearly $E(2)$ is a group with respect to multiplication. Also, $E(2) = \{(a,b,\theta) : a, b \in \mathbb{R} \text{ and } 0 \leq \theta < 2\pi\}$. Since the parameters a, b, θ are independent, $E(2)$ is a 3-dimensional manifold. $E(2)$ satisfies (iii). Therefore $E(2)$ is a Lie group.

Theorem 5.8.1. *Let $G(t_1, t_2, \ldots, t_m)$ be an m-dimensional Lie group such that the parameters $t_1, t_2, \ldots, t_m$ are independent. Then the Lie algebra corresponding to the Lie group G has a basis*

$$\left\{ \frac{\partial A(t_1, t_2, \ldots, t_m)}{\partial t_k} \Big|_{t_1 = t_2 = \cdots = t_m = 0} : A \in G, k = 1, 2, \ldots m \right\}.$$

Example 5.8.2. Basis of the Lie algebra of $E(2)$. Consider

$$E(2) = \left\{ \begin{bmatrix} \cos\theta & -\sin\theta & 0 \\ \sin\theta & \cos\theta & 0 \\ a & b & 1 \end{bmatrix} : a, b \in \mathbb{R}, 0 \le \theta < 2\pi \right\}$$

$$= \{A(a, b, \theta) : a, b \in \mathbb{R} \text{ and } 0 \le \theta < 2\pi\}.$$

$$\frac{\partial A}{\partial a}\Big|_{\theta=a=b=0} = \begin{bmatrix} 0 & 0 & 0 \\ 0 & 0 & 0 \\ 1 & 0 & 0 \end{bmatrix} = P_1,$$

$$\frac{\partial A}{\partial b}\Big|_{\theta=a=b=0} = \begin{bmatrix} 0 & 0 & 0 \\ 0 & 0 & 0 \\ 0 & 1 & 0 \end{bmatrix} = P_2,$$

$$\frac{\partial A}{\partial \theta}\Big|_{\theta=a=b=0} = \begin{bmatrix} -\sin\theta & -\cos\theta & 0 \\ \cos\theta & -\sin\theta & 0 \\ 0 & 0 & 0 \end{bmatrix}\Big|_{\theta=0}$$

$$= \begin{bmatrix} 0 & -1 & 0 \\ 1 & 0 & 0 \\ 0 & 0 & 0 \end{bmatrix} = M.$$

Therefore, by Theorem 5.8.1, $\{P_1, P_2, M\}$ is a basis.

5.8.1 Basis of the Lie algebra of the Lie group $SL(2)$

$$SL(2) = \left\{ \begin{bmatrix} a+1 & b \\ c & d+1 \end{bmatrix} : (a+1)(d+1) - bc = 1 \right\}$$

$$= \left\{ \begin{bmatrix} a+1 & b \\ c & \frac{1+bc}{a+1} \end{bmatrix} : a \ne -1 \right\}$$

$$= \{A(a, b, c) : |A| = 1\}$$

$$\frac{\partial A}{\partial a}\Big|_{a=b=c=0} = \begin{bmatrix} 1 & 0 \\ 0 & -1 \end{bmatrix}$$

$$\frac{\partial A}{\partial b}\Big|_{a=b=c=0} = \begin{bmatrix} 0 & 1 \\ 0 & 0 \end{bmatrix}$$

$$\frac{\partial A}{\partial c}\Big|_{a=b=c=0} = \begin{bmatrix} 0 & 0 \\ 1 & 0 \end{bmatrix}.$$

Therefore

$$\left\{ \begin{bmatrix} 1 & 0 \\ 0 & -1 \end{bmatrix}, \begin{bmatrix} 0 & 1 \\ 0 & 0 \end{bmatrix}, \begin{bmatrix} 0 & 0 \\ 1 & 0 \end{bmatrix} \right\}$$

is a basis.

Note 5.8.1: Trace of basis elements $=0$.

Note 5.8.2: For special functions Miller takes the basis as

$$\begin{bmatrix} \frac{1}{2} & 0 \\ 0 & -\frac{1}{2} \end{bmatrix}, \begin{bmatrix} 0 & -1 \\ 0 & 0 \end{bmatrix} \text{ and } \begin{bmatrix} 0 & 0 \\ -1 & 0 \end{bmatrix}.$$

Example 5.8.3. General linear Lie group $GL(2,F)$

$$GL(2,F) = \left\{ \begin{bmatrix} a+1 & b \\ c & d+1 \end{bmatrix} : (a+1)(d+1) - bc \neq 0 \right\}$$
$$= \{A(a,b,c,d) : |A| \neq 0\}.$$

Then

$$\frac{\partial A}{\partial a}|_{a=b=c=d=0} = \begin{bmatrix} 1 & 0 \\ 0 & 0 \end{bmatrix}, \frac{\partial A}{\partial b}|_{a=b=c=d=0} = \begin{bmatrix} 0 & 1 \\ 0 & 0 \end{bmatrix}$$

$$\frac{\partial A}{\partial c}|_{a=b=c=d=0} = \begin{bmatrix} 0 & 0 \\ 1 & 0 \end{bmatrix}, \frac{\partial A}{\partial a}|_{a=b=c=d=0} = \begin{bmatrix} 0 & 0 \\ 0 & 1 \end{bmatrix}.$$

Therefore by Theorem 5.8.1,

$$\left\{ \begin{bmatrix} 1 & 0 \\ 0 & 0 \end{bmatrix}, \begin{bmatrix} 0 & 1 \\ 0 & 0 \end{bmatrix}, \begin{bmatrix} 0 & 0 \\ 1 & 0 \end{bmatrix}, \begin{bmatrix} 0 & 0 \\ 0 & 1 \end{bmatrix} \right\}$$

is a basis.

Exercises 5.8.

5.8.1. Prove that the solution space $\mathbb{F}_o$ of the Helmholtz equation

$$(\Delta_2 + w^2)\Psi(x,y) = 0$$

has basis

$$\left\{ \exp\{i[kx + (w^2 - k^2)^{\frac{1}{2}}y]\} : k \in \mathbb{R} \right\}.$$

5.8.2. Prove that the operators $\{P_1, P_2, M\}$ given by Proposition 5.3.1, Question 1 provide a representation of the Lie algebra $e(2)$ on the solution space of $\mathbb{F}_o$ of the Helmholtz equation.

5.8.3. Prove that the symmetry algebra of the three-dimensional Helmholtz equation

$$(\Delta_3 + w^3)\Psi(x_1, x_2, x_3) = 0, \Delta_3 = \partial_{x_1 x_1} + \partial_{x_2 x_2} + \partial_{x_3 x_3}$$

is a six-dimensional Lie algebra having a basis

$$P_1 = \partial_{x_1}, \ P_2 = \partial_{x_2}, \ P_3 = \partial_{x_3}, \ J_1 = x_3\partial_{x_2} - x_2\partial_{x_3}$$
$$J_2 = x_1\partial_{x_3} - x_3\partial_{x_1}, \ J_3 = x_2\partial_{x_1} - x_1\partial_{x_2}$$

satisfying the commutation relations

$$[J_l, J_m] = \sum_n e_{lmn}J_n, \ [J_l, P_m] = \sum_n e_{lmn}J_n, \ [P_l, P_m] = O, \ e,m,n = 1,2,3$$

where e_{lmn} is the tensor such that

$$e_{123} = e_{231} = e_{312} = 1, \ e_{132} = e_{321} = e_{213} = -1$$

and all other components zero.

5.8.4. Prove that the symmetry algebra of Exercise 5.7.1 is an isomorphic image of the Lie algebra $e(3)$ having a basis

$$\mathcal{P}_1' = \begin{bmatrix} 0 & 0 & 0 & 0 \\ 0 & 0 & 0 & 0 \\ 0 & 0 & 0 & 0 \\ 1 & 0 & 0 & 0 \end{bmatrix}, \quad \mathcal{P}_2' = \begin{bmatrix} 0 & 0 & 0 & 0 \\ 0 & 0 & 0 & 0 \\ 0 & 0 & 0 & 0 \\ 0 & 1 & 0 & 0 \end{bmatrix}, \quad \mathcal{P}_3' = \begin{bmatrix} 0 & 0 & 0 & 0 \\ 0 & 0 & 0 & 0 \\ 0 & 0 & 0 & 0 \\ 0 & 0 & 1 & 0 \end{bmatrix}$$

$$\mathcal{J}_1' = \begin{bmatrix} 0 & 0 & 0 & 0 \\ 0 & 0 & -1 & 0 \\ 0 & 1 & 0 & 0 \\ 0 & 0 & 0 & 0 \end{bmatrix}, \quad \mathcal{J}_2' = \begin{bmatrix} 0 & 0 & 1 & 0 \\ 0 & 0 & 0 & 0 \\ -1 & 0 & 0 & 0 \\ 0 & 0 & 0 & 0 \end{bmatrix}, \quad \mathcal{J}_3' = \begin{bmatrix} 0 & -1 & 0 & 0 \\ 1 & 0 & 0 & 0 \\ 0 & 0 & 0 & 0 \\ 0 & 0 & 0 & 0 \end{bmatrix}$$

satisfying commutation relations as in Exercise 5.7.1.

5.8.5. Prove that the set of numbers of the form $a + b\sqrt{2}$, where a and b are rational numbers, is a field.

5.8.6. Let V be a vector space of dimension n over F. Prove that V contains a subspace of dimension r, $0 \leq r \leq n$.

5.8.7. Prove that the set $\{a + ib : a, b \in F_3\}$ forms a field of 9 elements.

TEST
on Lie Theory and Special Functions

Time: $1\frac{1}{2}$ hours

5.1. Prove by mathematical induction that

$$\frac{d^n}{dx^n} \, {}_3F_2\left[\begin{matrix} \alpha_1, \alpha_2, \alpha_3 \\ \beta_1, \beta_2 \end{matrix}; x \right] = \frac{(\alpha_1)_n(\alpha_2)_n(\alpha_3)_n}{(\beta_1)_n(\beta_2)_n} \, {}_3F_2\left[\begin{matrix} \alpha_1+n, \alpha_2+n, \alpha_3+n \\ \beta_1+n, \beta_2+n \end{matrix}; x \right], \ n \in \mathbb{N}. \quad (8)$$

5.2. Prove that for $\Re(\gamma - \alpha - \beta) > 0$,

$$_2F_1[\alpha, \beta; \gamma; 1] = \frac{\Gamma(\gamma)\Gamma(\gamma - \alpha - \beta)}{\Gamma(\gamma - \alpha)\Gamma(\gamma - \beta)}.$$

What is the result when $\alpha = -n$, where n is a +ve integer? (8)

5.3. Using the operators

$$J^\circ = -vz + z\frac{d}{dz}, \quad J^+ = z\left(-2v + z\frac{d}{dz}\right), \quad J^- = -\frac{d}{dz},$$

show that there exists a vector space V so that

1. the representation of $sl(2, \mathbb{C})$ on V is finite dimensional.
2. the representation of $sl(2, \mathbb{C})$ on V is infinite dimensional. (12)

5.4. Prove that the symmetry algebra of the Helmholtz equation

$$(\partial_{xx} + \partial_{yy} + w^2)\Psi(x, y) = 0$$

is an isomorphic image of some matrix Lie algebra. (14)

5.5. Find the Lie algebra corresponding to the Lie Group E. (8)

(H. L. Manocha)

Chapter 6
Applications to Stochastic Process and Time Series

[This chapter is based on the lectures of Dr. K.K. Jose, Department of Statistics, St. Thomas College, Pala, Mahatma Gandhi University, Kerala, India.]

6.0 Introduction

In this chapter we discuss some elementary theory of Stochastic Processes and Time Series Modeling. Stochastic processes are introduced in Section 6.1. Some modern concepts in distribution theory which are of frequent use in this chapter are discussed in Section 6.2. Section 6.3 deals with stationary time series models. In Section 6.4, we consider a structural relationship and some new autoregressive models. Section 6.5 deals with tailed processes. In section 6.6, semi-Weibull time series models with minification structure are discussed.

6.1 Stochastic Processes

The theory of stochastic processes is generally regarded as the dynamic part of probability theory, in which one studies a collection of random variables indexed by a parameter. One is observing a stochastic process whenever one examines a system developing in time in a manner controlled by probabilistic laws. In other words, a stochastic process can be regarded as an empirical abstraction of a phenomenon developing in nature according to some probabilistic rules.

If a scientist is to take account of the probabilistic nature of the phenomenon with which he is dealing, he should undoubtedly make use of the theory of stochastic processes. The scientist making measurements in his laboratory, the meteorologist attempting to forecast weather, the control systems engineer designing a servomechanism, the electrical engineer designing a communication system, the hardware engineer developing a computer network, the economist studying price fluctuations

247

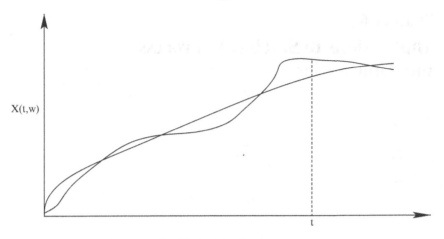

Fig. 6.1 A stochastic process

and business cycles, the seismologist studying earthquake vibrations, the neurosurgeon studying brainwave records, the cardiologist studying the electro cardiogram etc are encountering problems to which the theory of stochastic processes can be applied. Financial modeling and insurance mathematics are emerging areas where the theory of stochastic processes is widely used.

Examples of stochastic processes are provided by the generation sizes of populations such as a bacterial colony, life length of items under different renewals, service times in a queuing system, waiting times in front of a service counter, displacement of a particle executing Brownian motion, number of events during a particular time interval, number of deaths in a hospital on different days, voltage in an electrical system during different time instants, maximum temperature in a particular place on different days, deviation of an artificial satellite from its stipulated path at each instant of time after its launch, the quantity purchased of a particular inventory on different days etc. Suppose that a scientist is observing the trajectory of a satellite after its launch. At random time intervals, the scientist is observing whether it is deviating from the designed path or not and also the magnitude of the deviation.

Let $x(t,w)$ denote the altitude of the satellite from sea-level at time t where w is the outcome associated with the random experiment. Here the random experiment is noting the weather conditions with regard to temperature, pressure, wind velocity, humidity etc. These outcomes may vary continuously. Hence $\{x(t,w); t \in T; w \in \Omega\}$ gives rise to a stochastic process.

Thus a stochastic process is a family of random variables indexed by a parameter t taking values from a set T called the index set or parameter space. It may be denoted by $\{x(t,w); t \in T, w \in \Omega\}$. A more precise definition may be given as follows.

Definition 6.1.1. A stochastic process is a family of indexed random variables $\{x(t,w); t \in T; w \in \Omega\}$ defined on a probability space (Ω, β, P) where T is an arbitrary set.

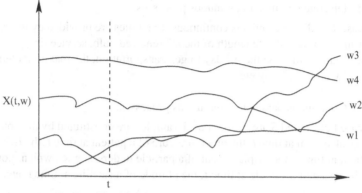

$X(t,w)$

w3
w4
w2
w1

Fig. 6.2 Different stochastic processes

There are many ways of visualizing a stochastic process.

(i) For each choice of $t \in T, x(t,w)$ is a random variable.
(ii) For each choice of $\omega \in \Omega, x(t,w)$ is a function of t.
(iii) For each choice of w and $t, x(t,w)$ is a number.
(iv) In general it is an ensemble (family) of functions $x(t,w)$ where t and w can take different possible values.

Hereafter we shall use the notation $x(t)$ to represent a stochastic process, omitting w, as in the case of random variables. It is convention to use x_n and $x(t)$ according as the indexing parameter is discrete or continuous.

The values assumed by the r.v. (random variable) $x(t)$ are called states and the set of all possible values of $x(t)$, is called the state space of the process and is denoted by S. The state space can be discrete or continuous. When S is discrete, by a proper labeling, we can take the state-space as the set of natural numbers namely $N = \{1, 2, \cdots\}$. It may be finite or infinite.

The main elements distinguishing stochastic processes are the nature of the state space S and parameter space T, and the dependence relations among the random variables $x(t)$. Accordingly there are four types of processes.

Type 1: Discrete parameter discrete processes:

In this case both S and T are discrete. Examples are provided by the number of customers reported in a bank counter on the nth day, the nth generation size of a population, the number of births in a hospital on the nth day etc. There may be multidimensional processes also. For example consider the process (x_n, y_n) where x_n and y_n are the number of births and deaths in a municipality on the nth day.

Type 2: Continuous parameter discrete processes:

In this case T is continuous and S is discrete. Examples constitute the number of persons in a queue at time t, the number of telephone calls during $(0,t)$, the number of vehicles passing through a specific junction during $(0,t)$ etc.

Type 3: Discrete parameter continuous processes:

In this case T is discrete and S is continuous. Examples are provided by the renewal time for the nth renewal, life length of the nth renewed bulb, service time for the nth customer, waiting time on the nth day to get transportation, the maximum temperature in a city on the nth day etc.

Type 4: Continuous parameter continuous processes:

In this case both T and S are continuous. Examples are constituted by the voltage in an electrical system at time t, the blood pressure of a patient at time t, the ECG level of a patient at time t, the displacement of a particle undergoing Brownian motion at time t, the speed of a vehicle at time t, the altitude of a satellite at time t, etc.

For more details see Karlin and Taylor (2002), Papoulis (2000), Medhi (2004, 2006). Feller (1966) gives a good account of infinite divisible distributions. Ross (2002) gives a good description of stochastic processes and their applications. Medhi (2004) gives a good introduction to the theory and application of stochastic processes.

Consider a computer system with jobs arriving at random points in time, queuing for service, and departing from the system after service completion. Let N_k be the number of jobs in the system at the time of departure of the kth customer (after service completion). The stochastic process $\{N_k; k = 1, 2, \cdots\}$ is a discrete-parameter, discrete-state process. A realization of this process is shown in Figure 6.3

Next let $x(t)$ be the number of jobs in the system at time t. Then $\{x(t); t \in T\}$ is a continuous parameter discrete-state process. A realization is given in Figure 6.4.

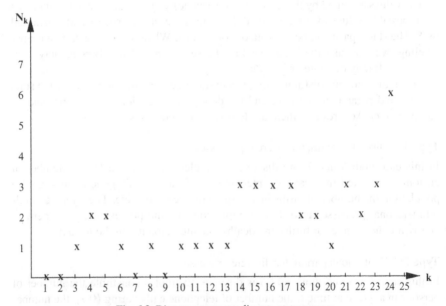

Fig. 6.3 Discrete parameter discrete state process

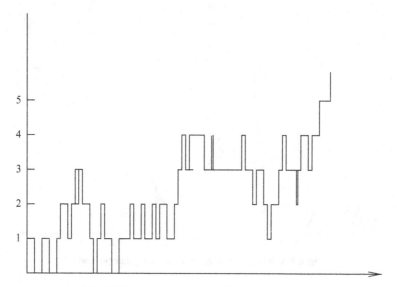

Fig. 6.4 Continuous parameter discrete state process

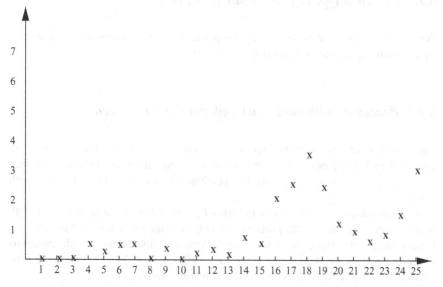

Fig. 6.5 Discrete parameter continuous state process

Let w_k be the time that the kth customer has to wait in the system before receiving service. Then $\{w_k; k \in T\}$ is a discrete-parameter, continuous-state process, see Figure 6.5

Finally, let $y(t)$ be the cumulative service requirement of all jobs in the system at time t. Then $\{y(t)\}$ is a continuous parameter continuous-state process, see Figure 6.6.

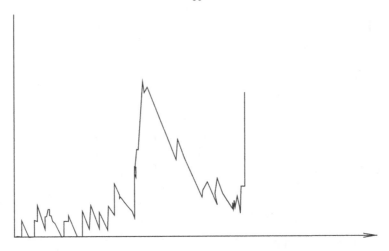

Fig. 6.6 Continuous parameter continuous state process

6.1.1 Classical types of stochastic processes

We now describe some of the classical types of stochastic processes characterized by different dependence relationships among $x(t)$.

6.1.2 Processes with stationary independent increments

Consider a stochastic process $\{x(t); t \in T\}$ where $T = [0, \infty)$. Then the process $\{x(t)\}$ is called a process with independent increments if the random variables $x_{t_1} - x_{t_0}, x_{t_2} - x_{t_1}, \cdots, x_{t_n} - x_{t_{n-1}}$ are independent for all choices of $t_0, t_1, \cdots, t_n$ such that $t_0 < t_1 < \cdots < t_n$.

If the distribution of the increments $x(t_i + h) - x(t_i)$ depends only on h, the length of the interval, and not on the particular time t_i, then the process is said to have stationary increments. Hence for a process with stationary increments, the distributions of the increments $x(t_0 + h) - x(t_0), x(t_1 + h) - x(t_1), x(t_2 + h) - x(t_2), \cdots$ etc are the same and depend only on h, irrespective of the time points $t_0, t_1, \cdots$

If a process $\{x(t)\}$ has both independent and stationary increments, then it is called a process with stationary independent increments.

Result: If a process $\{x_t; t \in T\}$ has stationary independent increments and has finite mean value, then $E(x_t) = m_0 + m_1 t$ where $m_0 = E(x_0)$ and $m_1 = E(x_1) - m_0, E$ denoting the expected value.

6.1.3 Stationary processes

A stochastic process $\{x_t\}$ is said to be stationary in the strict sense (SSS) if the joint distribution function of the families of the random variables $[x_{t_1+h}, \cdots x_{t_n+h}]$ and $[x_{t_1}, x_{t_2}, \cdots, x_{t_n}]$ are the same for all $h > 0$ and arbitrary selections $t_1, t_2, \cdots, t_n$ from T. This condition asserts that the process is in probabilistic equilibrium and that the particular times at which one examines the process are of no relevance. In particular the distribution of x_t is the same for each t.

Thus stationarity of a process implies that the probabilistic structure of the process is invariant under translation of the time axis. Many processes encountered in practice exhibit such a characteristic. So, stationary processes are appropriate models for describing many phenomena that occur in communication theory, astronomy, biology, economics etc.

However strict sense stationarity is seldom observed in practice. Moreover, many important questions relating to a stochastic process can be adequately answered in terms of the first two moments of the process. Therefore we relax the condition of strict sense stationarity to describe weak sense stationarity (WSS), also known as wide sense stationarity.

A Stochastic process $\{x_t\}$ is said to be wide sense stationary if its first two moments (mean function and variance function) are finite and independent of t and the covariance function $\text{Cov}(x_t, x_{t+s})$ is a function only of s, the time difference, for all t. Such processes are also known as covariance stationary or second order stationary processes. A process, which is not stationary, in any sense, is said to be evolutionary.

6.1.4 Gaussian processes and stationarity

If a process $\{x_t\}$ is such that the joint distribution of $(x_{t_1}, x_{t_2}, x_{t_n})$ for all $t_1, t_2, \cdots, t_n$ is multivariate normal, then $\{x_t\}$ is called a Gaussian (normal) process. For a Gaussian process weak sense stationarity and strict sense stationarity are identical. This follows from the fact that a multivariate normal distribution is completely determined by its mean vector and variance-covariance matrix. Here we need only mean, variance and covariance functions. In other words, if a Gaussian process $\{x_t\}$ is covariance stationary, then it is strictly stationary and vice versa.

Example 6.1.1. Let $\{x_n; n \geq 1\}$ be uncorrelated random variables with mean value 0 and variance 1. Then

$$\text{Cov}(x_n, x_m) = \begin{cases} 0 & \text{if } n \neq m \\ 1 & \text{if } n = m. \end{cases}$$

Hence $\text{Cov}(x_n, x_m)$ is a function of $n - m$ and so the process is covariance stationary. If x_n are identically distributed also, then the process is strictly stationary.

Example 6.1.2. Consider the Poisson process $\{x(t)\}$ where

$$P[x(t) = n] = \frac{(\lambda t)^n}{n!} e^{-\lambda t}; n = 0, 1, \cdots.$$

Clearly,

$$E[x(t)] = \lambda t$$
$$\text{Var}[x(t)] = \lambda t \text{ which depends on } t$$

Therefore the process is not stationary. It is evolutionary.

Example 6.1.3. Consider the process $\{x(t)\}$ where $x(t) = A_1 + A_2 t$ where A_1, A_2 are independent r.v.'s with $E(A_i) = a_i$, $\text{Var}(A_i) = \sigma_i^2, i = 1, 2$. Obviously

$$E[x(t)] = a_1 + a_2 t$$
$$\text{Var}[\mathrm{x(t)}] = \sigma_1^2 + \sigma_2^2 t^2$$
$$\text{Cov}[\mathrm{x(s)}, \mathrm{x(t)}] = \sigma_1^2 + st\sigma_2^2.$$

These are functions of t and hence the process is evolutionary.

Example 6.1.4. Consider the process $\{x(t)\}$ where $x(t) = A\cos wt + B\sin wt$, where A and B are uncorrelated r.v.'s with mean value 0 and variance 1, and w is a positive constant.

In this case $E[x(t)] = 0$ and $\text{Var}[\mathrm{x(t)}] = 1$, $\text{Cov}[\mathrm{x(t)}, \mathrm{x(t+h)}] = \cos(hw)$, where E denotes the expected value. Hence the above process is covariance stationary. This process is called a sinusoidal process.

Example 6.1.5. Consider the process $\{x(t)\}$ such that

$$P[x(t) = n] = \begin{cases} \frac{(at)^{n-1}}{(1+at)^{n+1}}; n = 1, 2, \cdots \\ \\ \frac{(at)}{(1+at)}; n = 0. \end{cases}$$

Obviously

$$E[x(t)] = \sum_{n=0}^{\infty} nP[x(t) = n]$$

$$= \frac{1}{(1+at)^2} \sum_{n=1}^{\infty} n\frac{(at)^{n-1}}{(1+at)^{n-1}} = 1$$

$$E[x^2(t)] = \sum_{n=1}^{\infty} n^2 \frac{(at)^{n-1}}{(1+at)^{n-1}}$$

$$= \frac{at}{(1+at)^3} \left\{ \sum_{n=2}^{\infty} n(n-1)\frac{(at)^{n-2}}{(1+at)^{n-2}} \right\} + \sum_{n=1}^{\infty} n\frac{(at)^{n-1}}{(1+at)^{n+1}}$$

$$= 2at + 1,$$

which is a function of t. Hence the process is not stationary. It is an evolutionary process.

Example 6.1.6. Consider the Bernoulli process: Consider a sequence of independent Bernoulli trials with outcomes as success and failure. Let

$$x_n = \begin{cases} 1, & \text{if the outcome is a success} \\ 0, & \text{otherwise.} \end{cases}$$

Then the process $\{x_n; n \geq 1\}$ has states 0 and 1 and the process is called a Bernoulli process. Let us define $\{y_n\}$ by $y_n = 0$ for $n = 0$ and $y_n = x_1 + \cdots + x_n, n \geq 1$. Then the process $\{y_n; n \geq 0\}$ has the set of non-negative integers as the state space. This y_n is binomially distributed with

$$P[y_n = k] = \binom{n}{k} p^k (1-p)^{n-k}; k = 0, 1, 2, \cdots n; \ 0 < p < 1$$

where p is the probability of success in a trial.

Example 6.1.7. (The random telegraph signal process): Let $\{N(t), t \geq 0\}$ denote a Poisson process, and let x_0 be independent of this process, and be such that $P(x_0 = 1) = P(x_0 = -1) = \frac{1}{2}$. Define $x(t) = x_0(-1)^{N(t)}$. Then $\{x(t); t \geq 0\}$ is called a random telegraph signal process. In this case $P[N(t) = k] = e^{-\lambda t} \frac{(\lambda t)^k}{k!}$ for $k = 0, 1, 2, \cdots$. Clearly

$$E[x(t)] = E[x_0(-1)^{N(t)}]$$
$$= E[x_0]E[(-1)^{N(t)}] = 0$$
$$\text{Cov}[x(t), x(t+s)] = E[x(t)x(t+s)]$$
$$= E[x_0^2(-1)^{N(t)+N(t+s)}]$$
$$= E[x_0^2]E[(-1)^{2N(t)+N(t+s)-N(t)}]$$
$$= 1E[(-1)^{2N(t)}(-1)^{N(t+s)-N(t)}]$$
$$= E[(-1)^{2N(t)}]E[(-1)^{N(t+s)}]$$
$$= 1E[(-1)^{N(s)}]$$
$$= \sum_{k=0}^{\infty} (-1)^k \frac{e^{-\lambda s}(\lambda s)^k}{k!}$$
$$= e^{-2\lambda s}; s \geq 0.$$

Also

$$\text{Var}[x(t)] = 1 < \infty.$$

Hence the above process is covariance stationary.

For an application of the above random telegraph signal, consider a particle moving at a constant unit velocity along a straight line and suppose that collisions involving this particle occur at Poisson rate λ. Also suppose that each time when the particle suffers from a collision it reverses direction. If x_0 represents the initial velocity of the particle, then its velocity at time t is given by $x(t) = x_0(-1)^{N(t)}$. If we take $D(t) = \int_0^t x(s)ds$, then $D(t)$ represents the displacement of the particle during $(0,t)$. It can be shown that $\{D(t); t \geq 0\}$ is also a weakly stationary process.

Example 6.1.8. Consider an Autoregressive Process $\{x_n\}$ where $x_0 = z_0$ and $x_n = \rho x_{n-1} + z_n; n \geq 1, |\rho| < 1$ where $z_0, z_1, z_2, \cdots$ are uncorrelated random variables with $E(z_n) = 0; n \geq 0$ and

$$\text{Var}(z_n) = \begin{cases} \sigma^2; & n \geq 1 \\ \frac{\sigma^2}{1-\rho^2}; & n = 0. \end{cases}$$

Then

$$x_n = \rho x_{n-1} + z_n$$
$$= \rho(\rho x_{n-2} + z_{n-1}) + z_n$$
$$= \rho^2 x_{n-2} + \rho z_{n-1} + z_n$$
$$= \sum_{i=0}^{n} \rho^{n-i} z_i.$$

Therefore

$$E(x_n) = 0$$

$$\text{Cov}[x_n, x_{n+m}] = \text{Cov}\left[\sum_{i=0}^{n} \rho^{n-i} z_i, \sum_{i=0}^{n+m} \rho^{n+m-i} z_i\right]$$

$$= \sum_{i=0}^{n} \rho^{n-i} \rho^{n+m-i} \text{Cov}(z_i, z_i)$$

$$= \sigma^2 \rho^{2n+m}\left[\frac{1}{1-\rho^2} + \sum_{i=1}^{n} \rho^{-2i}\right]$$

$$= \frac{\sigma^2 \rho^m}{1-\rho^2}, \quad n \to \infty.$$

Therefore this process is also covariance stationary.

Now we consider a special type of Gaussian Process, which is stationary in both senses and has a wide range of applications.

6.1.5 Brownian processes

We consider a symmetric random walk in which, in each time unit, there is a chance for one unit step forward or backward. Now suppose that we speed up this process by taking smaller and smaller steps in smaller and smaller time intervals. In the limit we obtain the Brownian motion process. It is also known as the Wiener process, after Wiener who developed this concept in a series of papers from 1918 onwards. Actually it originated in Physics, as the notion associated with the random movements of a small particle immersed in a liquid or gas. This was first discovered by the British botanist Robert Brown. The process can be more precisely developed as follows: Suppose that, in the random walk, in each time interval of duration Δt we take a step of size Δx either to the left or to the right with equal probabilities. If we let $x(t)$ denote the position at time t, then

$$x(t) = \Delta x \left[x_1 + \cdots + x_{\left[\frac{t}{\Delta t}\right]} \right]$$

where

$$x_i = \begin{cases} +1 & \text{if the } i^{th} \text{ step is to the right} \\ -1 & \text{if it is to the left} \end{cases}$$

and $\left[\frac{t}{\Delta t}\right]$ is the integer part of $\frac{t}{\Delta t}$. We assume that x_i's are independent with $P(x_i = 1) = P(x_i = -1) = \frac{1}{2}$. Since $E(x_i) = 0$, $\text{Var}(x_i) = 1$ we have $E[x(t)] = 0$, $\text{Var}[x(t)] = (\Delta x)^2 \left[\frac{t}{\Delta t}\right]$

Now we consider the case when $\Delta x \to 0$ and $\Delta t \to 0$ in such a way that $E[x(t)] = 0$ and $\text{Var}[x(t)] \to \sigma^2 t$. The resulting process $\{x(t)\}$ is such that $x(t)$ is normally distributed with mean 0 and variance $\sigma^2 t$, and has independent, stationary increments. This leads us to the formal definition of a Brownian motion process.

Definition 6.1.2. A stochastic process $\{x(t); t \geq 0\}$ is said to be a Brownian motion process if (i) $x(0) = 0$ (ii) $\{x(t)\}$ has stationary independent increments (iii) for every $t > 0$, $x(t)$ is normally distributed with mean value 0 and variance $\sigma^2 t$.

When $\sigma = 1$, the process is called a standard Brownian motion. Any Brownian motion $x(t)$ can be converted to a standard Brownian motion by taking $B(t) = \frac{x(t)}{\sigma}$. If $\{B(t)\}$ is a standard Brownian motion and $x(t) = \sigma B(t) + \mu t$, then $x(t)$ is normally distributed with mean value μt and variance $t\sigma^2$. Then $\{x(t); t \geq 0\}$ is called a Brownian motion with drift coefficient μ. If $\{x(t); t \geq 0\}$ is a Brownian motion process with drift coefficient μ and variance parameter $\sigma^2 t$, then $\{y(t); t \geq 0\}$ where $y(t) = \exp[x(t)]$ is a called a geometric Brownian motion process. It is useful in modeling of stock prices over time when the percentage changes are independent and identically distributed.

If $\{x(t); t \geq 0\}$ is a Brownian motion process then each of $x(t_1), x(t_2), \cdots$ can be expressed as a linear combination of the independent normal random variables $x(t_1), x(t_2) - x(t_1), x(t_3) - x(t_2), \cdots x(t_n) - x(t_{n-1})$. Hence it follows that a Brownian motion is a Gaussian process.

Since a multivariate normal distribution is completely determined by the marginal mean values and covariance values, it follows that a standard Brownian motion could also be defined as a Gaussian process having $E[x(t)] = 0$ and for $s \geq t$,

$$\begin{aligned}
\text{Cov}[x(s), x(t)] &= \text{Cov}[x(s), x(s) + x(t) - x(s)] \\
&= \text{Cov}[x(s), x(s)] + \text{Cov}[x(s), x(t) - x(s)] \\
&= \text{Var}[x(s)] \\
&= s\sigma^2.
\end{aligned}$$

Let $\{x(t); t \geq 0\}$ be a standard Brownian motion process and consider the process values between 0 and 1 conditional on $x(1) = 0$. Consider the conditional stochastic process, $\{x(t); 0 \leq t \leq 1 | x(1) = 0\}$. Since this conditional distribution is also multivariate normal it follows that this conditional process is a Gaussian process. This conditional process is known as the Brownian bridge.

Brownian motion theory is a major topic in fluid dynamics and has applications in aeronautical engineering in the designing of aeroplanes, submarines, satellites, space crafts etc. It also finds applications in financial modeling.

6.1.6 Markov chains

An elementary form of dependence between values of x_n in successive transitions, was introduced by the celebrated Russian probabilist A.A. Markov, and is known as Markov dependence. Markov dependence is a form of dependence which states that x_{n+1} depends only on x_n when it is known and is independent of $x_{n-1}, x_{n-2}, \cdots, x_0$. This implies that the future of the process depends only on the present, irrespective of the past. This property is known as Markov property. In probabilistic terms, the Markov property can be stated as

$$\begin{aligned}
P[x_{n+1}, &= i_{n+1} | x_0 = i_0, x_1 = i_1, \cdots, x_{n-1} = i_{n-1}, x_n = i_n] \\
&= P[x_{n+1} = i_{n+1} | x_n = i_n]
\end{aligned}$$

for all states $i_0, i_1, \cdots, i_{n+1}$ and for all n. This is called Markov dependence of the first order.

A stochastic process $\{x_n\}$ with discrete state space and discrete parameter space is called a Markov chain if for all states $i, j, i_0, i_1, \cdots, i_{n-1}$ we have

$$\begin{aligned}
P[x_{n+1} &= j | x_0 = i_0, x_1 = i_1, \cdots, x_{n-1} = i_{n-1}, x_n = i] \\
&= P\{x_{n+1} = j | x_n = i\} \qquad \text{for all } n.
\end{aligned}$$

The probability that the system is in state j at the end of $(n+1)$ transitions given that the system was in state i at the end of n transitions is denoted by $p_{ij}^{(1)}$ and is called a one-step transition probability. In general this probability depends on i, j and n. If these probabilities are independent of n, we say that the Markov chain is homogeneous and has stationary transition probabilities. Here we consider only such chains. Thus

$$p_{ij}^{(1)} = P[x_{n+1} = j | x_n = i].$$

In a similar manner we can consider m-step transition probabilities denoted by $p_{ij}^{(m)}$ where

$$p_{ij}^{(m)} = P[x_{n+m} = j | x_n = i].$$

If the state space of a Markov chain consists of only a finite number of states, it is called a finite Markov chain. Otherwise we call it an infinite Markov chain.

The square matirx P consisting of the elements $p_{ij}^{(1)}$ for all possible states i and j is called one-step transition probability matrix of the chain. Therefore

$$P = [p_{ij}^{(1)}].$$

Similarly the square matrix $P^{(m)}$ consisting of the elements $p_{ij}^{(m)}$ for all possible values of the states i and j is called the m-step transition matrix of the chain. Hence

$$P^{(m)} = [p_{ij}^{(m)}].$$

Obviously we have $P^{(1)} = P$ and

$$p_{ij}^{(m)} \geq 0 \qquad \text{and} \qquad \sum_j p_{ij}^{(m)} = 1.$$

Now we consider $p_j^{(0)} = P[x_0 = j]$. It may be noted that $p_j^{(0)}$ describes the probability distribution of x_0. The vector $p^{(0)} = (p_0^{(0)}, p_1^{(0)}, \cdots p_j^{(0)} \cdots)$ is called the initial probability vector. Similarly $p_j^{(n)} = P[x_n = j]$ gives the probability distribution of x_n. The vector $p^{(n)} = (p_0^{(n)}, p_1^{(n)}, \cdots p_j^{(n)} \cdots)$ is called the n-step absolute probability vector.

Theorem 6.1.1. *A Markov chain is completely defined by its one-step transition probability matrix and the initial probability vector.*

Proof 6.1.1: Consider

$$P[x_0 = i | x_1 = j, x_2 = k, \cdots, x_{n-1} = r, x_n = s]$$
$$= P[x_0 = i] \, P[x_1 = j | x_0 = i] \, P[x_2 = k | x_0 = i, x_1 = j]$$
$$\cdots P\{x_n = s | x_0 = i, \cdots, x_{n-1} = r\}$$
$$= P(x_0 = i) P(x_1 = j | x_0 = i) \, P(x_2 = k | x_1 = j) \cdots P(x_n = s | x_{n-1} = r)$$
$$= p_i^{(0)} p_{ij}^{(1)} \cdots p_{rs}^{(1)}.$$

This shows that any finite dimensional joint distribution for the chain can be obtained in terms of the initial probabilities and one-step transition probabilities, and this establishes the theorem.

Theorem 6.1.2. *(Chapman-Kolmogorov Equations): The transition probabilities of Markov chains satisfy the equation*

$$p_{ij}^{(m+n)} = \sum_k p_{ik}^{(m)} p_{kj}^{(n)} \qquad \text{where } p_{ij}^{(0)} = \begin{cases} 1, & i=j \\ 0, & i \neq j \end{cases}$$

or equivalently,

$$P^{(n)} = P^n \quad \text{and} \quad P^{(m+n)} = P^{(m)} P^{(n)}$$

(i) Computation of absolute probabilities

Consider

$$p_j^{(n)} = P[x_n = j]$$
$$P(x_n = j) = \sum_i P(x_n = j, x_0 = i)$$
$$= \sum_i P(x_0 = i) P(x_n = j | x_0 = i).$$

Therefore

$$p_j^{(n)} = \sum_i p_i^{(0)} p_{ij}^{(n)}$$

(ii) Inverse transition probabilities

The n-step inverse transition probabilities denoted by $q_{ij}^{(n)}$ is defined as

$$q_{ij}^{(n)} = P(x_m = j | x_{n+m} = i)$$

for $m \geq 0, n \geq 0$. They describe the past behaviour of the process when the present is given. But transition probabilities describe the future behaviour of the process when the present is given.

Now

$$P(x_m = j | x_{n+m} = i) P(x_{n+m} = i)$$
$$= P(x_{n+m} = i | x_n = j) P(x_n = j).$$

Therefore

$$P(x_m = j | x_{n+m} = i) = \frac{P(x_{n+m} = i | x_m = j) P(x_m = j)}{P(x_{n+m} = i)}.$$

Hence

$$q_j^{(n)} = \frac{p_{ij}^{(n)} p_j^{(m)}}{P_i^{(n+m)}}, \qquad m \geq 0$$

whenever the denominator is nonzero.

(iii) **Taboo probabilities**

In this case the movement of the system to some specified states is prohibited. For example consider

$$P(x_2 = j, x_1 \neq k | x_0 = i) = P$$

(the system reaches state j at the end of 2 transitions without visiting state k given that the system started from state i). This is usually denoted by $_kP_{ij}^{(2)}$. Using the Chapman-Kolmogorov equations, we have

$$_kP_{ij}^{(2)} = \sum_{m \neq k} P_{im}^{(1)} P_{mj}^{(1)}$$

It may be noted that $_kP_{ij}^{(2)}$ is different from $P[x_2 = j | x_0 = i, x_1 \neq k]$ which is equal to $P[x_2 = j | x_1 \neq k]$. Similarly $P[x_2 = j, x_1 \neq k, m | x_0 = i]$ may be denoted by $_{m,k}P_{ij}^{(2)}$. Obviously,

$$_{m,k}P_{ij}^{(2)} = \sum_{v \neq k,m} P_{iv}^{(1)} P_{vj}^{(1)}$$

Problems relating to taboo probabilities can be solved as shown above.

Exercises 6.1.

6.1.1. Give two examples each of the four types of stochastic processes.

6.1.2. Define a stochastic process with stationary independent increments.

6.1.3. For a process with stationary independent increments show that $E(x_t) = m_0 + m_1 t$ where $m_0 = E(x_0)$ and $m_1 = E(x_1) - m_0$.

6.1.4. What is a Poisson process ? Show that it is evolutionary.

6.1.5. Give an example of a strictly stationary process.

6.1.6. Give an example of a covariance stationary process.

6.1.7. Let $\{x_n\}$ be uncorrelated r.v.'s with $E(x_n) = 0$, $\text{Var}(x_n) = 1$. Show that $\{x_n\}$ is strictly stationary.

6.1.8. Consider a Poisson process $\{x(t)\}$ where $p[x(t) = n] = \frac{e^{-\lambda t}(\lambda t)^n}{n!}$; $n = 0, 1, \cdots$. Find $E(x(t))$ and $\text{Var}(x(t))$. Is the process stationary ?

6.1.9. Consider a Poisson process $\{x(t)\}$ as above. Let x_0 be independent of $x(t)$ such that $p(x_0 = 1) = p(x_0 = -1) = \frac{1}{2}$. Define $N(t) = x_0(-1)^{N(t)}$. Find $E(N(t))$ and $\text{Cov}(N(t), N(t+s))$.

6.1.10. Define a Brownian process and show that it is an approximation of the random walk process.

6.1.11. Obtain an expression for the covariance function of a Brownian motion process.

6.1.12. What is geometric Brownian motion process? Discuss its uses ?

6.1.13. Consider the numbers $1,2,3,4,5$. We select one number out of these at random and note it as x_1. Then select a number at random from $1,2\cdots x_1$ and denote it as x_2. The process is continued. Write down the one step and two step transition matrices of the chain $\{x_n\}$.

6.1.14. 4 white and 4 red balls are randomly distributed in two urns so that each urn contains 4 balls. At each step one ball is selected at random from each urn and the two balls are interchanged. Let x_n denote the number of white balls in the first urn at the end of the nth interchange. Then write down the one-step transition matrix and the initial distribution. Also find

(i) $P[x_3 = 4|x_1 = 4]$;　　(ii) $P[x_2 = 3]$;

(iii) $P[x_1 = 4, x_2 = 3, x_3 = 2, x_4 = 1]$;　　(iv) $P[x_1 = 3|x_3 = 4]$.

6.1.15. If x_n denotes the maximum face value observed in n tosses of a balanced die with faces marked $1,2,3,4,5,6$ write down the state space and parameter space of the process $\{x_n\}$. Also obtain the transition matrix.

6.2 Modern Concepts in Distribution Theory

6.2.1 Introduction

In this section we discuss some modern concepts in distribution theory which will be of frequent use in this chapter.

Definition 6.2.1.　Infinite divisibility.
  A random variable x is said to be infinitely divisible if for every $n \in \mathbb{N}$, there exists independently and identically distributed random variables $y_{1n}, y_{2n}, \ldots, y_{nn}$ such that $x \overset{d}{=} y_{1n} + y_{2n} + \cdots + y_{nn}$, where $\overset{d}{=}$ denotes equality in distributions. In terms of distribution functions, a distribution function F is said to be infinitely divisible if for every positive integer n, there exists a distribution function F_n such that $F = \underbrace{F_n \star F_n \star \cdots \star F_n}_{n \text{ times}}$, where $\star$ denotes convolution. This is equivalent to the existence of a characteristic function $\varphi_n(t)$ for every $n \in \mathbb{N}$ such that $\varphi(t) = [\varphi_n(t)]^n$ where $\varphi(t)$ is the characteristic function of x.

Infinitely divisible distributions occur in various contexts in the modeling of many real phenomena. For instance when modeling the amount of rain x that falls in a period of length T, one can divide x into more general independent parts from the same family. That is,

$$x \stackrel{d}{=} x_{t_1} + x_{t_2 - t_1} + \cdots + x_{T - t_{n-1}}.$$

Similarly, the amount of money x paid by an insurance company during a year is expressible as the sum of the corresponding amounts $x_1, x_2, \ldots, x_{52}$ in each week. That is,

$$x \stackrel{d}{=} x_1 + x_2 + \cdots + x_{52}.$$

A large number of distributions such as normal, exponential, Weibull, gamma, Cauchy, Laplace, logistic, lognormal, Pareto, geometric, Poisson, etc., are infinitely divisible. Various properties and applications of infinitely divisible distributions can be found in Laha and Rohatgi (1979) and Steutel (1979).

6.2.2 Geometric infinite divisibility

The concept of geometric infinite divisibility (g.i.d.) was introduced by Klebanov et al. (1985). A random variable y is said to be g.i.d. if for every $p \in (0,1)$, there exists a sequence of independently and identically distributed random variables $x_1^{(p)}, x_2^{(p)}, \ldots$ such that

$$y \stackrel{d}{=} \sum_{j=1}^{N(p)} x_j^{(p)} \tag{6.2.1}$$

and

$$P\{N(p) = k\} = p(1-p)^{k-1}, \qquad k = 1, 2, \cdots$$

where $y, N(p)$ and $x_j^{(p)}$, $j = 1, 2, \ldots$ are independent. The relation (6.2.1) is equivalent to

$$\varphi(t) = \sum_{j=1}^{\infty} [g(t)]^j \, p(1-p)^{j-1}$$

$$= \frac{pg(t)}{1 - (1-p)g(t)}$$

where $\varphi(t)$ and $g(t)$ are the characteristic functions of y and $x_j^{(p)}$ respectively.

The class of g.i.d. distributions is a proper subclass of infinitely divisible distributions. The g.i.d. distributions play the same role in 'geometric summation' as infinitely divisible distributions play in the usual summation of independent random variables. Klebanov et al. (1985) established that a distribution function F with characteristic function $\varphi(t)$ is g.i.d. if and only if $\exp\left\{1 - \frac{1}{\varphi(t)}\right\}$ is infinitely divisible. Exponential and Laplace distributions are obvious examples of g.i.d. distributions. Pillai (1990b), Mohan et al. (1993) discuss properties of g.i.d. distributions. It may be noted that normal distribution is not geometrically infinitely divisible.

6.2.3 Bernstein functions

A C^∞-function f from $(0, \infty)$ to R is said to be completely monotone if $(-1)^n \frac{d^n f}{dx^n} \geq 0$ for all integers $n \geq 0$. A C^∞-function f from $(0, \infty)$ to R is said to be a Bernstein function, if $f(x) \geq 0$, $x > 0$ and $(-1)^n \frac{d^n f}{dx^n} \leq 0$ for all integers $n \geq 1$. Then f is also referred to as a function with complete monotone derivative (c.m.d.). A completely monotone function is positive, decreasing and convex while a Bernstein function is positive, increasing and concave (see Berg and Forst (1975)).

Fujita (1993) established that a cumulative distribution function G with $G(0) = 0$ is geometrically infinitely divisible, if and only if G can be expressed as

$$G(x) = \sum_{n=1}^{\infty} (-1)^{n+1} W^{n*}([0,x]), \qquad x > 0$$

where $W^{n*}(dx)$ is the n-fold convolution measure of a unique positive measure $W(dx)$ such that

$$\frac{1}{f(x)} = \int_0^\infty e^{-sx} W(ds), \qquad x > 0$$

where $f(x)$ is a Bernstein function, satisfying the conditions

$$\lim_{x \downarrow 0} f(x) = 0 \text{ and } \lim_{x \to \infty} f(x) = \infty.$$

A distribution is said to have complete monotone derivative if its distribution function $F(x)$ is Bernstein. Pillai and Sandhya (1990) proved that the class of distributions having complete monotone derivative is a proper subclass of g.i.d. distributions. This implies that all distributions with complete monotone densities are geometrically infinitely divisible. It is easier to verify the complete monotone criterion and using this approach we can establish the geometric infinite divisibility of many distributions such as Pareto, gamma and Weibull.

The class of non-degenerate generalized gamma convolutions with densities of the form given by

$$f_1(x) = c\, x^{\beta-1} \prod_{j=1}^{M} (1 + c_j x)^{-r_j}, \qquad x > 0$$

is geometrically infinitely divisible for $0 < \beta \leq 1$. Similarly distributions having densities of the form

$$f_2(x) = cx^{\beta-1} \exp(-bx^\alpha); \qquad 0 < \alpha \leq 1$$

is g.i.d. for $0 < \beta \leq 1$. Also the Bondesson family of distributions with densities of the form

$$f_3(x) = cx^{\beta-1} \prod_{j=1}^{M} \left[1 + \sum_{k=1}^{N_j} c_{jk} x^{\alpha_{jk}} \right]^{-r_j}$$

is g.i.d. for $0 \leq \beta \leq 1$, $\alpha_{jk} \leq 1$ provided all parameters are strictly positive (see Bondesson(1992)).

6.2.4 Self-decomposability

Let $\{x_n;\ n \geq 1\}$ be a sequence of independent random variables, and let $\{b_n\}$ be a sequence of positive real numbers such that

$$\lim_{n \to \infty} \max_{1 \leq k \leq n} P\{|x_k| \geq b_n \epsilon\} = 0 \text{ for every } \epsilon > 0 .$$

Let $s_n = \sum_{k=1}^{n} x_k$ for $n \geq 1$. Then the class of distributions which are the weak limits of the distributions of the sums $b_n^{-1} s_n - a_n;\ n \geq 1$ where a_n and $b_n > 0$ are suitably chosen constants, is said to constitute class L. Such distributions are called self-decomposable.

A distribution F with characteristic function $\varphi(t)$ is called self-decomposable, if and only if, for every $\alpha \in (0,1)$, there exists a characteristic function $\varphi_\alpha(t)$ such that $\varphi(t) = \varphi(\alpha t)\varphi_\alpha(t)$, for $t \in R$.

Clearly, apart from $x \equiv 0$, no lattice random variable can be self–decomposable. All non-degenerate self–decomposable distributions are absolutely continuous.

A discrete analogue of self–decomposability was introduced by Steutel and Van Harn (1979). A distribution on $N_0 \equiv \{0,1,2,\ldots\}$ with probability generating function (p.g.f.) $P(z)$ is called discrete self-decomposable if and only if $P(z) = P(1 - \alpha + \alpha z)P_\alpha(z);\ |z| \leq 1,\ \alpha \in (0,1)$ where $P_\alpha(z)$ is a p.g.f.

If we define $G(z) = P(1 - z)$, then $G(z)$ is called the alternate probability generating function (a.p.g.f.). Then it follows that a distribution is discrete self–decomposable if and only if $G(z) = G(\alpha z)G_\alpha(z);\ |z| \leq 1,\ \alpha \in (0,1)$ where $G_\alpha(z)$ is some a.p.g.f.

6.2.5 Stable distributions

A distribution function F with characteristic function $\varphi(t)$ is stable if for every pair of positive real numbers b_1 and b_2, there exist finite constants a and $b > 0$ such that $\varphi(b_1 t)\varphi(b_2 t) = \varphi(bt)e^{iat}$ where $i = \sqrt{-1}$.

Clearly, stable distributions are in class L with the additional condition that the random variables $x_n;\ n \geq 1$ in Subsection 6.2.4 are identically distributed also. F is stable if and only if its characteristic function can be expressed as

$$\ln \varphi(t) = i\alpha t - c|t|^\beta [1 + i\gamma\omega(t,\beta)\text{sgn } t]$$

where α, β, γ are constants with $c \geq 0,\ 0 < \beta \leq 2,\ |\gamma| \leq 1$ and

$$\omega(t,\beta) = \begin{cases} \tan \frac{\pi\beta}{2}; & \beta \neq 1 \\ \frac{2}{\pi} \ln |t|; & \beta = 1. \end{cases}$$

The value $c = 0$ corresponds to the degenerate distribution, and $\beta = 2$ to the normal distribution. The case $\gamma = 0,\ \beta = 1$ corresponds to the Cauchy law (see Laha and Rohatgi (1979)).

6.2.6 Geometrically strictly stable distributions

A random variable y is said to be geometrically strictly stable (g.s.s.) if for any $p \in (0,1)$ there exists a constant $c = c(p) > 0$ and a sequence of independent and identically distributed random variables $y_1, y_2, \ldots$ such that

$$y \overset{d}{=} c(p) \sum_{j=1}^{N(p)} y_j$$

where $P\{N(p) = k\} = p(1-p)^{k-1}$, $k = 1, 2, \ldots$ and $y, N(p)$ and y_j, $j = 1, 2, \ldots$ are independent.

If $\varphi(t)$ is the characteristic function of y, then it implies that

$$\varphi(t) = \frac{p\varphi(ct)}{1 - (1-p)\varphi(ct)}, \qquad p \in (0,1).$$

Among the geometrically strictly stable distributions, the Laplace distribution and exponential distribution possess all moments. A geometrically strictly stable random variable is clearly geometrically infinitely divisible.

A non–degenerate random variable y is geometrically strictly stable if and only if its characteristic function is of the form

$$\varphi(t) = \frac{1}{\left[1 + \lambda |t|^\alpha \exp\left(-i\frac{\pi}{2}\theta \alpha \operatorname{sgn} t\right)\right]}$$

where $0 < \alpha \le 2$, $\lambda > 0$, $|\theta| \le \min[1, \frac{2}{\alpha} - 1]$. When $\alpha = 2$, it corresponds to the Laplace distribution. Thus it is apparent that when ordinary summation of random variables is replaced by geometric summation, the Laplace distribution plays the role of the normal distribution, and exponential distribution replaces the degenerate distribution (see Klebanov *et al.* (1985)).

6.2.7 Mittag-Leffler distribution

The Mittag-Leffler distribution was introduced by Pillai (1990a) and has cumulative distribution function given by

$$F_\alpha(x) = \sum_{k=1}^{\infty} \frac{(-1)^{k-1} x^{k\alpha}}{\Gamma(1 + k\alpha)}, \qquad 0 < \alpha \le 1, x > 0.$$

Its Laplace transform is given by $\phi(t) = \frac{1}{1+t^\alpha}$, $0 < \alpha \le 1, t \ge 0$, and the distribution may be denoted by $ML(\alpha)$. Here α is called the exponent. It can be regarded as a generalization of the exponential distribution in the sense that $\alpha = 1$ corresponds to the exponential distribution. The Mittag-Leffler distribution is geometrically infinitely divisible and belongs to class L. It is normally attracted to the stable law with exponent α.

If u is exponential with unit mean and y is positive stable with exponent α, then $x = u^{1/\alpha}y$ is distributed as Mittag-Leffler (α) when u and y are independent. If u is Mittag-Leffler (α) and v is exponential and u and v are independent, then $w = \frac{u}{v}$ is distributed as Pareto type III with survival function $\bar{F}_w(x) = P(w > x) = \frac{1}{1+x^\alpha}$; $0 < \alpha \le 1$.

For the Mittag-Leffler distribution, $E(x^\delta)$ exists for $0 \le \delta < \alpha$ and is given by

$$E(x^\delta) = \frac{\Gamma\left(1 - \frac{\delta}{\alpha}\right)\Gamma\left(1 + \frac{\delta}{\alpha}\right)}{\Gamma(1 - \delta)}.$$

A two parameter Mittag-Leffler distribution can also be defined with the corresponding Laplace transform $\phi(t) = \frac{\lambda^\alpha}{\lambda^\alpha + t^\alpha}$, $0 < \alpha \le 1$. It may be denoted by $ML(\alpha, \lambda)$.

Jayakumar and Pillai (1993) considered a more general class called semi–Mittag-Leffler distribution which included the Mittag-Leffler distribution as a special case. A random variable x with positive support is said to have a semi–Mittag-Leffler distribution if its Laplace transform is given by

$$\phi(t) = \frac{1}{1 + \eta(t)}$$

where $\eta(t)$ satisfies the functional equation $\eta(t) = a\eta(bt)$ where $0 < b < 1$ and α is the unique solution of $ab^\alpha = 1$. It may be denoted by $SML(\alpha)$. Then it follows that $\eta(bt) = b^\alpha h(t)$ where $h(t)$ is a periodic function in t with period $\frac{-\ln b}{2\pi\alpha}$. When $h(t)$ is a constant, the distribution reduces to the Mittag-Leffler distribution. The semi–Mittag-Leffler distribution is also geometrically infinitely divisible and belongs to class L.

6.2.8 α–Laplace distribution

The α–Laplace distribution has characteristic function given by $\varphi(t) = \frac{1}{1+|t|^\alpha}$; $0 < \alpha \le 2$, $-\infty < t < \infty$. This is also called Linnik's distribution. Pillai (1985) refers to it as the α–Laplace distribution since $\alpha = 2$ corresponds to the Laplace distribution. It is unimodal, geometrically strictly stable and belongs to class L. It is normally attracted to the symmetric stable law with exponent α. Also

$$E(|x|^\delta) = \frac{2^\delta \ \Gamma\left(1 + \frac{\delta}{\alpha}\right)\Gamma\left(1 - \frac{\delta}{\alpha}\right)\Gamma\left(\frac{(1+\delta)}{2}\right)}{\sqrt{\pi} \ \Gamma\left(1 - \frac{\delta}{2}\right)}$$

where $0 < \delta < \alpha$; $0 < \alpha \le 2$.

If u and v are independent random variables where u is exponential with unit mean and v is symmetric stable with exponent α, then $x = u^{1/\alpha}v$ is distributed as

α–Laplace. Using this result, Devroye (1990) develops an algorithm for generating random variables having α–Laplace distribution.

Pillai (1985) introduced a larger class of distributions called semi–α–Laplace distribution, with characteristic function given by

$$\varphi(t) = \frac{1}{1 + \eta(t)}$$

where $\eta(t)$ satisfies the functional equation $\eta(t) = a\eta(bt)$ for $0 < b < 1$ and a is the unique solution of $ab^\alpha = 1, 0 < \alpha \leq 2$. Here b is called the order and α is called the exponent of the distribution. If b_1 and b_2 are the orders of the distribution such that $\frac{\ln b_1}{\ln b_2}$ is irrational, then $\eta(t) = c|t|^\alpha$, where c is some constant. Pillai (1985) established that, for a semi–α–Laplace distribution with exponent α, $E|x|^\delta$ exists for $0 \leq \delta < \alpha$. It can be shown that

$$\varphi(t) = \frac{1}{1 + |t|^\alpha [1 - A\cos(k \ln |t|)]}$$

where $k = \frac{2\pi}{\ln b}, 0 < b < 1$ is the characteristic function of a semi–α–Laplace distribution for suitable choices of A and $\alpha < 1$.

The semi–α–Laplace distribution is also geometrically infinitely divisible and belongs to class L. It is useful in modeling household income data. Mohan *et al.*(1993) refer to it as a geometrically right semi–stable law.

6.2.9 Semi–Pareto distribution

The semi–Pareto distribution was introduced by Pillai (1991). A random variable x with positive support has semi–Pareto distribution $SP(\alpha, p)$ if its survival function is given by $\bar{F}_x(x_0) = P(x > x_0) = \frac{1}{1 + \psi(x_0)}$ where $\psi(x_0)$ satisfies the functional equation $p\psi(x_0) = \psi(p^{1/\alpha} x_0); 0 < p < 1, \alpha > 0$.

The above definition is analogous to that of the semi–stable law defined by Levy (see Pillai (1971)). It can be shown that $\psi(x) = x^\alpha h(x)$ where $h(x)$ is periodic in $\ln x$ with period $\frac{-2\pi\alpha}{\ln p}$. For example if $h(x) = \exp[\beta \cos(\alpha \ln x)]$, then it satisfies the above functional equation with $p = \exp(-2\pi)$ and $\psi(x)$ is monotone increasing with $0 < \beta < 1$. The semi–Pareto distribution can be viewed as a more general class which includes the Pareto type III distribution when $\psi(x) = cx^\alpha$, where c is a constant.

Exercises 6.2.

6.2.1. Examine whether the following distributions are infinitely divisible.

(i) normal   (ii) exponential   (iii) Laplace   (iv) Cauchy
(v) binomial   (vi) Poisson   (vii) geometric   (viii) negative binomial

6.2.2. Show that exponential distribution is geometric infinite divisible and self-decomposable.

6.2.3. Examine whether Cauchy distribution is self-decomposable.

6.2.4. Show that (i) Mittag-Leffler distribution is g.i.d and belongs to class L. (ii) α-Laplace distribution is g.i.d and self-decomposable.

6.2.5. Give a distribution which is infinitely divisible but not g.i.d.

6.2.6. Show that AR(1) structure $x_n = ax_{n-1} + \varepsilon_n, a \in (0,1)$ is stationary Markovian if and only if $\{x_n\}$ is self-decomposable.

6.2.7. Show that geometric and negative binomial distributions are discrete self-decomposable.

6.2.8. Consider the symmetric stable distribution with characteristic function $\varphi(t) = e^{-|t|^\alpha}$. Is it self-decomposable?

6.3 Stationary Time Series Models

6.3.1 Introduction

A time series is a realization of a stochastic process. In other words, a time series, $\{x_t\}$, is a family of real–valued random variables indexed by $t \in \mathbb{Z}$, where $\mathbb{Z}$ denotes the set of integers. More specifically, it is referred to as a discrete parameter time series. The time series $\{x_t\}$ is said to be stationary if, for any $t_1, t_2, \ldots, t_n \in \mathbb{Z}$, any $k \in \mathbb{Z}$, and $n = 1, 2, \ldots,$

$$F_{x_{t_1}, x_{t_2}, \ldots, x_{t_n}}(x_1, x_2, \ldots, x_n) = F_{x_{t_1+k}, x_{t_2+k}, \ldots, x_{t_n+k}}(x_1, x_2, \ldots, x_n)$$

where F denotes the joint distribution function of the set of random variables which appear as suffices. This is called stationarity in the strict sense. Less stringently, we say a process $\{x_n\}$ is weakly stationary if the mean and variance of x_t remain constant over time and the covariance between any two values x_t and x_s depends only on the time difference and not on their individual time points. $\{x_t\}$ is called a Gaussian process if, for all $t_n; n \geq 1$ the set of random variables $\{x_{t_1}, x_{t_2}, \ldots, x_{t_n}\}$ has a multivariate normal distribution. Since a multivariate normal distribution is completely specified by its mean vector and covariance matrix, it follows that for a Gaussian process weak stationarity implies complete stationarity. But for non–Gaussian processes, this may not hold.

6.3.2 Autoregressive models

The era of linear time series models began with autoregressive models first introduced by Yule in 1927. The standard form of an autoregressive model of order p, denoted by AR(p), is given by

$$x_t = \sum_{j=1}^{p} a_j x_{t-j} + \epsilon_t; \qquad t = 0, \pm 1, \pm 2, \ldots$$

where $\{\epsilon_t\}$ are independent and identically distributed random variables called innovations and a_j, p are fixed parameters, with $a_p \neq 0$.

Another kind of model of great practical importance in the representation of observed time series is the moving average model. The standard form of a moving average model of order q, denoted by $MA(q)$, is given by $x_t = \sum_{j=1}^{q} b_j \epsilon_{t-j} + \epsilon_t, t \in \mathbb{Z}$ where b_j, q are fixed parameters, with $b_q \neq 0$. In order to achieve greater flexibility in the fitting of actually observed time series, it is more advantageous to include both autoregressive and moving average terms in the model. Such models, called autoregressive–moving average models, denoted by ARMA (p,q), have the form

$$x_t = \sum_{j=1}^{p} a_j x_{t-j} + \sum_{k=1}^{q} b_k \epsilon_{t-k} + \epsilon_t, \qquad t \in \mathbb{Z}$$

where $\{a_j\}_{j=1}^{p}$ and $\{b_k\}_{k=1}^{q}$ are real constants called parameters of the model. It can be seen that an AR(p) model is the same as an ARMA(p,o) model and a MA(q) model is the same as an ARMA(o,q) model.

With the introduction of various non–Gaussian and non–linear models, the standard form of autoregression was widened in several respects. A more general definition of autoregression of order p is given in terms of the linear conditional expectation requirement that

$$E(x_t | x_{t-1}, x_{t-2}, \ldots) = \sum_{j=1}^{p} a_j x_{t-j}.$$

This definition could apply to models which are not of the linear form (see Lawrance (1991)).

6.3.3 A general solution

We consider a first order autoregressive model with innovation given by the structural relationship

$$x_n = \epsilon_n + \begin{cases} 0 & \text{with probability } p \\ x_{n-1} & \text{with probability } 1-p \end{cases} \qquad (6.3.1)$$

where $p \in (0,1)$ and $\{\epsilon_n\}$ is a sequence of independent and identically distributed (i.i.d.) random variables selected in such a way that $\{x_n\}$ is stationary Markovian with a given marginal distribution function F.

Let $\phi_x(t) = E[e^{-tx}]$ be the Laplace–Stieltjes transform of x. Then (6.3.1) gives

$$\phi_{x_n}(t) = \phi_{\epsilon_n}(t)[p + (1-p)\phi_{x_{n-1}}(t)].$$

If we assume stationarity, this simplifies to

$$\phi_\epsilon(t) = \frac{\phi_x(t)}{p + (1-p)\phi_x(t)} \tag{6.3.2}$$

or equivalently

$$\phi_x(t) = \frac{p\phi_\epsilon(t)}{1 - (1-p)\phi_\epsilon(t)}. \tag{6.3.3}$$

When $\{x_n\}$ is marginally distributed as exponential, it is easy to see that (6.3.1) gives the TEAR(1) model. We note that $\phi_\epsilon(t)$ in (6.3.2) does not represent a Laplace transform always. In order that the process given by (6.3.1) is properly defined, there should exist an innovation distribution such that $\phi_\epsilon(t)$ is a Laplace transform for all $p \in (0,1)$. For establishing the main results we need the following lemmas.

Lemma 6.3.1: (Pillai (1990b)): Let F be a distribution with positive support and $\phi(t)$ be its Laplace transform. Then F is geometrically infinitely divisible if and only if

$$\phi(t) = \frac{1}{1 + \psi(t)}$$

where $\psi(t)$ is Bernstein with $\psi(0) = 0$.

Now we consider the following definition from Pillai (1990b).

Definition 6.3.1. For any non–vanishing Laplace transform $\phi(t)$, the function

$$\psi(t) = \frac{1}{\phi(t)} - 1$$

is called the third characteristic.

Lemma 6.3.2: Let $\psi(t)$ be the third characteristic of $\phi(t)$. Then $p\psi(t)$ is a third characteristic for all $p \in (0,1)$ if and only if $\psi(t)$ has complete monotone derivative and $\psi(0) = 0$.

Thus we have the following theorem.

Theorem 6.3.1. $\phi_\epsilon(t)$ in (6.3.2) *represents a Laplace transform for all* $p \in (0,1)$ *if and only if* $\phi_x(t)$ *is the Laplace transform of a geometrically infinitely divisible distribution.*

This leads to the following theorem which brings out the role of geometrically infinitely divisible distributions in defining the new first order autoregressive model given by (6.3.1).

Theorem 6.3.2. *The innovation sequence $\{\epsilon_n\}$ defining the first order autoregressive model given by*

$$x_n = \epsilon_n + \begin{cases} 0 & \text{with probability } p \\ x_{n-1} & \text{with probability } 1-p \end{cases}$$

where $p \in (0,1)$, exists if and only if the stationary marginal distribution of x_n is geometrically infinitely divisible. Then the innovation distribution is also geometrically infinitely divisible.

Proof 6.3.1: Suppose that an innovation sequence $\{\epsilon_n\}$, such that the model (6.3.1) is properly defined, exists. This implies that $\phi_\epsilon(t)$ in (6.3.3) is a Laplace transform for all $p \in (0,1)$. Then from (6.3.3)

$$\phi_x(t) = p\phi_\epsilon(t)[1 - (1-p)\phi_\epsilon(t)]^{-1}$$
$$= \sum_{n=1}^{\infty} p(1-p)^{n-1}[\phi_\epsilon(t)]^n$$

showing that the stationary marginal distribution of x_n is geometrically infinitely divisible. Conversely, if x_n has a stationary marginal distribution which is geometrically infinitely divisible, then $\phi_x(t) = \frac{1}{1+\psi(t)}$ where $\psi(t)$ has complete monotone derivative and $\psi(0) = 0$. Then from (6.3.2) we get $\phi_\epsilon(t) = \frac{1}{1+p\psi(t)}$, which establishes the existence of an innovation distribution, which is geometrically infinitely divisible.

6.3.4 Extension to a k-th order autoregressive model

In this section we consider an extension of the model given by (6.3.1) to the k-th order. The structure of this model is given by

$$x_n = \epsilon_n + \begin{cases} 0 & \text{with probability } p_0 \\ x_{n-1} & \text{with probability } p_1 \\ \vdots & \\ x_{n-k} & \text{with probability } p_k \end{cases} \tag{6.3.4}$$

where $p_i \in (0,1)$ for $i = 0, 1, \ldots, k$ and $p_0 + p_1 + \cdots + p_k = 1$. Taking Laplace transforms on both sides of (6.3.4) we get

$$\phi_{x_n}(t) = \phi_{\epsilon_n}(t)\left[p_0 + \sum_{i=1}^{k} p_i \phi_{x_{n-i}}(t)\right].$$

Assuming stationarity, it simplifies to

$$\phi_x(t) = \phi_\epsilon(t)\left[p_0 + \sum_{i=1}^{k} p_i \phi_x(t)\right]$$

$$= \phi_\epsilon(t)[p_0 + (1-p_0)\phi_x(t)].$$

This yields

$$\phi_\epsilon(t) = \frac{\phi_x(t)}{p_0 + (1-p_0)\phi_x(t)} \qquad (6.3.5)$$

which is analogous to the expression (6.3.2).

It may be noted that $k = 1$ corresponds to the first order model with $p = p_0$. From (6.3.5) it follows that the results obtained in Section 6.1.2 hold good for the k-th order model given by (6.3.4). This establishes the importance of geometrically infinitely divisible distributions in autoregressive modeling.

6.3.5 Mittag-Leffler autoregressive structure

The Mittag-Leffler distribution was introduced by Pillai (1990a) and has Laplace transform $\phi(t) = \frac{1}{1+t^\alpha}, 0 < \alpha \leq 1$. When $\alpha = 1$, this corresponds to the exponential distribution with unit mean. Jayakumar and Pillai (1993) considered the semi–Mittag-Leffler distribution with exponent α. Its Laplace transform is of the form $\frac{1}{1+\eta(t)}$ where $\eta(t)$ satisfies the functional equation

$$\eta(t) = a\eta(bt), \qquad 0 < b < 1 \qquad (6.3.6)$$

and a is the unique solution of $ab^\alpha = 1$ where $0 < \alpha \leq 1$. Then by Lemma 6.3.1 of Jayakumar and Pillai (1993), the solution of the functional equation (6.3.6) is $\eta(t) = t^\alpha h(t)$ where $h(t)$ is periodic in $\ln t$ with period $-\frac{2\pi\alpha}{\ln b}$. When $h(t) = 1$, $\eta(t) = t^\alpha$ and hence the Mittag-Leffler distribution is a special case of the semi-Mittag-Leffler distribution. It is obvious that the semi–Mittag-Leffler distribution is geometrically infinitely divisible.

Now we bring out the importance of the semi-Mittag-Leffler distribution in the context of the new autoregressive structure given by (6.3.1). The following theorem establishes this.

Theorem 6.3.3. *For a positive valued first order autoregressive process $\{x_n\}$ satisfying (6.3.1) the stationary marginal distribution of x_n and ϵ_n are identical except for a scale change if and only if x_n's are marginally distributed as semi-Mittag-Leffler.*

Proof 6.3.2: Suppose that the stationary marginal distributions of x_n and ϵ_n are identical. This implies $\phi_\epsilon(t) = \phi_x(ct)$ where c is a constant. Then from (6.3.2) we get

$$\phi_x(ct) = \frac{\phi_x(t)}{p + (1-p)\phi_x(t)}. \tag{6.3.7}$$

Writing $\phi_x(t) = \dfrac{1}{1+\eta(t)}$ in (6.3.7) we get

$$\frac{1}{1+\eta(ct)} = \frac{1}{1+p\eta(t)}$$

so that

$$\eta(ct) = p\eta(t).$$

By choosing $c = p^{1/\alpha}$, it follows that x_n is distributed as semi–Mittag-Leffler with exponent α.

Conversely, we assume that the stationary marginal distribution of x_n is semi-Mittag-Leffler. Then from (6.3.2)

$$\phi_\epsilon(t) = \frac{1}{1+p\eta(t)} = \frac{1}{1+\eta(p^{1/\alpha}t)}.$$

This establishes that $\epsilon_n \overset{d}{=} p^{1/\alpha} M_n$ where $\{M_n\}$ are independently and identically distributed as semi–Mittag-Leffler. It can be easily seen that the above result is true in the case of the k-th order autoregressive model given by (6.3.4) also.

Exercises 6.3.

6.3.1. Define an AR(1) process and obtain the stationary solution for the distribution of $\{\epsilon_n\}$ when $\{x_n\}$ are exponentially distributed.

6.3.2. Show that an AR(1) model can be expressed as $MA(\infty)$ model.

6.3.3. Establish a new AR(1) model with exponential innovations.

6.3.4. Examine whether two-parameter gamma distribution is g.i.d., giving conditions if any.

6.3.5. Show that the exponential distribution is a special case of Mittag-Leffler distribution.

6.3.6. Obtain the stationary distribution of $\{\epsilon_n\}$ in the AR(1) structure $x_n = ax_{n-1} + \epsilon_n; a \in (0,1)$ when $\{x_n\}$ follows exponential distribution. Generalize it to the case of Mittag-Leffler random variables.

6.3.7. Obtain the structure of the innovation distribution if $\{x_n\}$ follows α-Laplace distribution where $x_n = ax_{n-1} + \epsilon_n$. Deduce the case when $\alpha = 2$.

6.3.8. Show that if $\{x_n\}$ follows Cauchy distribution then $\{\epsilon_n\}$ also follows a Cauchy distribution in the AR(1) equation $x_n = ax_{n-1} + \epsilon_n$.

6.4 A Structural Relationship and New Processes

In this section we obtain the specific structural relationship between the stationary marginal distributions of x_n and ϵ_n in the new autoregressive model. Fujita (1993) generalized the results on Mittag-Leffler distributions and obtained a new characterization of geometrically infinitely divisible distributions with positive support using Bernstein functions. It was established that a distribution function G with $G(0) = 0$ is geometrically infinitely divisible if and only if G can be expressed in the form.

$$G(x) = \sum_{n=1}^{\infty} (-1)^{n+1} \lambda^n W^{n*}([0,x]); \qquad x > 0, \lambda > 0 \qquad (6.4.1)$$

where $W^{n*}(dx)$ is the n-fold convolution measure of a unique positive measure $W(dx)$ on $[0, \infty)$ such that

$$\frac{1}{f(x)} = \int_0^{\infty} e^{-sx} W(ds); \qquad x > 0 \qquad (6.4.2)$$

for some Bernstein function f such that $\lim_{x \downarrow 0}(x) = 0$ and $\lim_{x \to \infty} f(x) = \infty$. Then the Laplace transform of $G(x)$, with parameter t, is $\dfrac{\lambda}{\lambda + f(t)}$. Using this result we get the following theorem.

Theorem 6.4.1. *The k-th order autoregressive equation given by (6.3.4) defines a stationary process with a given marginal distribution function $F_x(x)$ for x_n if and only if $F_x(x)$ can be expressed in the form*

$$F_x(x) = \sum_{n=1}^{\infty} (-1)^{n+1} \lambda^n W^{n*}([0,x]); \qquad x > 0, \lambda > 0. \qquad (6.4.3)$$

Then the innovations $\{\epsilon_n\}$ have a distribution function $F_\epsilon(x)$ given by

$$F_\epsilon(x) = \sum_{n=1}^{\infty} (-1)^{n+1} (\lambda/p_0)^n W^{n*}([0,x]); \qquad x > 0, \lambda > 0, \qquad (6.4.4)$$

where $p_0 \in (0,1)$ and W^{n*} is as in (6.4.1).

Proof 6.4.1: We have from Theorem 6.3.1 that $F_x(x)$ is geometrically infinitely divisible. Then (6.4.3) follows directly from Fujita (1993). Now by substituting $\phi_x(t) = \dfrac{\lambda}{\lambda + f(t)}$ in (6.3.2) we get

$$\phi_\epsilon(t) = \frac{\lambda}{\lambda + p_0 f(t)} = \frac{(\lambda/p_0)}{(\lambda/p_0) + f(t)}$$

which leads to (6.4.4). This completes the proof.

The above theorem can be used to construct various autoregressive models under different stationary marginal distributions for x_n. For example, the TEAR(1) model of Lawrance and Lewis (1981) can be obtained by taking $f(t) = t$. Then $W^{n*}([0,x]) = \frac{x^n}{n!}$ so that $F_x(x) = 1 - e^{-\lambda x}$ and $F_\epsilon(x) = 1 - e^{-(\lambda/p)x}$. If we take $\lambda = 1$ and $f(t) = t^\alpha; 0 < \alpha \le 1$ we can obtain an easily tractable first order autoregressive Mittag-Leffler process denoted by TMLAR(1). In this case $W^{n*}([0,x]) = \frac{x^{n\alpha}}{\Gamma(1+n\alpha)}$. In a similar manner by taking $\lambda = 1$ and $f(t)$ satisfying the functional equation $f(t) = af(bt)$ where $a = b^{-\alpha}; 0 < b < 1, 0 < \alpha \le 1$, we can obtain an easily tractable first order autoregressive semi–Mittag-Leffler process denoted by TSMLAR(1).

6.4.1 The TMLAR(1) process

An easily tractable form of a first order autoregressive Mittag-Leffler process, called TMLAR(1), is constituted by $\{x_n\}$ having a structure of the form

$$x_n = p^{1/\alpha}M_n + \begin{cases} 0 & \text{with probability p} \\ x_{n-1} & \text{with probability 1-p} \end{cases} \tag{6.4.5}$$

where $p \in (0,1); 0 < \alpha \le 1$ and $\{M_n\}$ are independently and identically distributed as Mittag-Leffler with exponent α and $x_0 \overset{d}{=} M_1$. The model (6.4.5) can be rewritten in the form

$$x_n = p^{1/\alpha}M_n + I_n x_{n-1} \tag{6.4.6}$$

where $\{I_n\}$ is a Bernoulli sequence such that $P(I_n = 0) = p$ and $P(I_n = 1) = 1 - p$.

If in the structural form (6.4.5), we assume that $\{M_n\}$ are distributed as semi–Mittag-Leffler with exponent α, then $\{x_n\}$ constitute a tractable semi–Mittag-Leffler autoregressive process of order 1, called TSMLAR(1). Both models are Markovian and stationary. It can be seen that the TMLAR(1) process is a special case of the TSMLAR(1) process since the Mittag-Leffler distribution is a special case of the semi–Mittag-Leffler distribution.

Now we shall consider the TSMLAR(1) process and establish that it is strictly stationary and Markovian, provided x_0 is distributed as semi–Mittag-Leffler. In order to prove this we use the method of induction. Suppose that x_{n-1} is distributed as semi–Mittag-Leffler (α). Then by taking Laplace transforms on both sides of (6.4.5), we get

$$\begin{aligned} \phi_{x_n}(t) &= \phi_{M_n}(p^{1/\alpha}t)[p + (1-p)\phi_{x_{n-1}}(t)] \\ &= \frac{1}{1+\eta(p^{1/\alpha}t)}\left[p + (1-p)\frac{1}{1+\eta(t)}\right] \\ &= \left[\frac{1}{1+p\eta(t)}\right]\left[\frac{1+p\eta(t)}{1+\eta(t)}\right] \\ &= \frac{1}{1+\eta(t)}. \end{aligned}$$

Hence x_n is distributed as semi–Mittag-Leffler with exponent α. If x_0 is arbitrary, then also it is easy to establish that $\{x_n\}$ is asymptotically stationary. Thus we have the following theorem.

Theorem 6.4.2. *The first order autoregressive equation*

$$x_n = p^{1/\alpha}M_n + I_n x_{n-1}, \qquad n = 1,2,\ldots, \ p \in (0,1)$$

where $\{I_n\}$ are independent Bernoulli random variables such that $P(I_n = 0) = p = 1 - P(I_n = 1)$ defines a positive valued strictly stationary first order autoregressive process if and only if $\{M_n\}$ are independently and identically distributed as semi–Mittag-Leffler with exponent α and $x_0 \overset{d}{=} M_1$.

Remark 6.4.1: If we consider characteristic functions instead of Laplace transforms, the results can be applied to real valued autoregressive processes. Then the role of semi–Mittag-Leffler distributions is played by semi–α–Laplace distributions introduced by Pillai (1985).

6.4.2 The NEAR(1) model

In this section we consider a generalized form of the first order autoregressive equation. The new structure is given by

$$x_n = \epsilon_n + \begin{cases} 0 & \text{with probability } p \\ ax_{n-1} & \text{with probability } 1-p \end{cases} \qquad (6.4.7)$$

where $0 \le p \le 1; 0 \le a \le 1$ and $\{\epsilon_n\}$ is a sequence of independent and identically distributed random variables such that $\{x_n\}$ have a given stationary marginal distribution. Let $\phi_x(t) = E[e^{-tx}]$ be the Laplace–Stieltjes transform of x. Then (6.4.7) gives

$$\phi_{x_n}(t) = \phi_{\epsilon_n}(t)[p + (1-p)\phi_{x_{n-1}}(at)].$$

Assuming stationarity, it simplifies to

$$\phi_\epsilon(t) = \frac{\phi_x(t)}{p + (1-p)\phi_x(at)}. \qquad (6.4.8)$$

When $p = 0$ and $0 < a < 1$, the model (6.4.7) is the standard first order autoregressive model. Then the model is properly defined if and only if the stationary marginal distribution of x_n is self–decomposable. When $a = 1, 0 < p < 1$ the model is the same as the model (6.4.1), which is properly defined if and only if the stationary marginal distribution of x_n is geometrically infinitely divisible. When $a = 0$ or $p = 1$, x_n and ϵ_n are identically distributed.

Now we consider the case when $a \in (0, 1]$ and $p \in (0, 1]$, but not simultaneously equal to 1. Lawrance and Lewis (1981) developed a NEAR(1) model with exponential (λ) marginal distribution for x_n. Then $\phi_x(t) = \dfrac{\lambda}{\lambda + t}$ and substitution in (6.4.8) gives

$$\phi_\epsilon(t) = \left[\frac{\lambda + at}{\lambda + t}\right]\left[\frac{\lambda}{\lambda + pat}\right] \tag{6.4.9}$$

which can be rewritten as

$$\phi_\epsilon(t) = \left(\frac{1-a}{1-pa}\right)\left(\frac{\lambda}{\lambda + t}\right) + \left[\frac{(1-p)a}{1-pa}\right]\left[\frac{\lambda}{\lambda + pat}\right].$$

Hence ϵ_n can be regarded as a convex exponential mixture of the form

$$\epsilon_n = \begin{cases} E_n & \text{with probability } \frac{1-a}{1-pa} \\ paE_n & \text{with probability } \frac{(1-p)a}{1-pa} \end{cases} \tag{6.4.10}$$

where $\{E_n\}$, $n = 1, 2, \ldots$ are independent and identically distributed as exponential (λ) random variables. Another representation for ϵ_n can be obtained from (6.4.9) by writing

$$\phi_\epsilon(t) = \left[a + (1-a)\frac{\lambda}{\lambda + t}\right]\left[\frac{\lambda}{\lambda + pat}\right]. \tag{6.4.11}$$

Then writing w.p. for 'with probability' ϵ_n can be regarded as the sum of two independent random variables u_n and v_n where

$$u_n = \begin{cases} 0 & \text{w. p. } a \\ e_n & \text{w. p. } 1 - a \text{ and} \end{cases} \tag{6.4.12}$$

$$v_n = pae_n$$

where $\{e_n\}$, $n = 1, 2, \ldots$ are exponential (λ). It may be noted that when $p = 0$, the model is identical with the EAR(1) process, of Gaver and Lewis (1980). Thus the new representation of ϵ_n seems to be more appropriate, when NEAR(1) process is regarded as a generalization of the EAR(1) process.

6.4.3 New Mittag-Leffler autoregressive models

Now we construct a new first order autoregressive process with Mittag-Leffler marginal distribution, called the NMLAR(1) model. The structure of the model is as in (6.4.7) and the innovations can be derived by substituting $\phi_x(t) = \dfrac{1}{1 + t^\alpha}$; $0 < \alpha \le 1$ in (6.4.8). This gives

$$\phi_\epsilon(t) = \left[\frac{1 + a^\alpha t^\alpha}{1 + t^\alpha}\right]\left[\frac{1}{1 + a^\alpha pt^\alpha}\right].$$

Hence the innovations ϵ_n can be given in the form

$$\epsilon_n = \begin{cases} M_n & \text{w.p. } \frac{1-a^\alpha}{1-pa^\alpha} \\ pa^\alpha M_n & \text{w.p. } \frac{(1-p)a^\alpha}{1-pa^\alpha} \end{cases} \qquad (6.4.13)$$

where $\{M_n\}$ are Mittag-Leffler (α) random variables.

An alternate representation of ϵ_n is $\epsilon_n = u_n + v_n$ where u_n and v_n are independent random variables such that

$$u_n = \begin{cases} 0 & \text{w.p. } a^\alpha \\ M_n & \text{w.p. } 1-a^\alpha \text{ and} \end{cases} \qquad (6.4.14)$$

$$v_n = ap^{1/\alpha}M_n$$

where $\{M_n\}, n = 1,2,\ldots$ are independent Mittag-Leffler (α) random variables.

It can be shown that the process is strictly stationary and Markovian. This gives us the following theorem.

Theorem 6.4.3. *The first order autoregressive equation given by* $(6.4.7)$ *defines a strictly stationary* AR(1) *process with a Mittag-Leffler* (α) *marginal distribution for* x_n *if and only if the innovations are of the form* $\epsilon_n = u_n + v_n$ *where* u_n *and* v_n *are as in* $(6.4.14)$ *with* x_0 *distributed as Mittag-Leffler* (α).

Proof 6.4.2: We prove this by induction. We assume that x_{n-1} is Mittag-Leffler (α). Then by taking Laplace transforms, we get

$$\phi_{x_n}(t) = [\phi_{M_n}(ap^{1/\alpha}t)][a^\alpha + (1-a^\alpha)\phi_{M_n}(t)]$$
$$\times [p + (1-p)\phi_{x_{n-1}}(at)]$$
$$= \left[\frac{1}{1+a^\alpha pt^\alpha}\right]\left[a^\alpha + (1-a^\alpha)\frac{1}{1+t^\alpha}\right]$$
$$\times \left[p + (1-p)\frac{1}{1+a^\alpha t^\alpha}\right]$$
$$= \left[\frac{1}{1+a^\alpha pt^\alpha}\right]\left[\frac{1+a^\alpha t^\alpha}{1+t^\alpha}\right]\left[\frac{1+pa^\alpha t^\alpha}{1+a^\alpha t^\alpha}\right]$$
$$= \frac{1}{1+t^\alpha}.$$

This shows that x_n is distributed as Mittag-Leffler (α), and this establishes the sufficiency part. The necessary part is obvious from the derivation of the innovation sequence. This completes the proof.

The joint distribution of (x_n, x_{n-1}) is of interest in describing the process and matching it with data. Therefore, we shall obtain the joint distribution with the use of Laplace-Stieltjes transforms. The bivariate Laplace transform is given by

$$\phi_{x_n,x_{n-1}}(s,t) = E\{\exp(-sx_n - tx_{n-1})\}$$

$$= \phi_\epsilon(s)\{p\phi_x(t) + (1-p)\phi_x(as+t)\}$$

$$= \left[\frac{1+a^\alpha s^\alpha}{1+s^\alpha}\right]\left[\frac{1}{1+pa^\alpha s^\alpha}\right]\left\{\frac{p}{1+t^\alpha} + \frac{1-p}{1+(as+t)^\alpha}\right\}.$$

It is possible to obtain the joint distribution by inverting this expression.

6.4.4 The NSMLAR(1) process

Now we extend the NMLAR(1) process to a wider class to construct a new semi–Mittag-Leffler first order autoregressive process. The process has the structure

$$x_n = \epsilon_n + \begin{cases} 0 & \text{w.p. } p \\ ax_{n-1} & \text{w.p. } 1-p \end{cases}$$

where $\{\epsilon_n\}$ are independently and identically distributed as the sum of two independent random variables u_n and v_n where

$$u_n = \begin{cases} 0 & \text{w.p. } a^\alpha \\ M_n & \text{w.p. } 1-a^\alpha \text{ and} \end{cases} \tag{6.4.15}$$

$$v_n = ap^{1/\alpha}M_n$$

where $\{M_n\}$, $n = 1,2,\ldots$ are independently and identically distributed as semi–Mittag-Leffler (α). This process is also clearly strictly stationary and Markovian provided x_0 is semi–Mittag-Leffler (α). This follows by induction. In terms of Laplace transforms we have

$$\phi_{x_n}(t) = [p + (1-p)\phi_{x_{n-1}}(at)][a^\alpha + (1-a^\alpha)\phi_{M_n}(t)]$$

$$\times [\phi_{M_n}(ap^{1/\alpha}t)]$$

$$= \left[p + \frac{(1-p)}{1+\eta(at)}\right]\left[a^\alpha + \frac{(1-a^\alpha)}{1+\eta(t)}\right]$$

$$\times \frac{1}{1+\eta(ap^{1/\alpha}t)}$$

$$= \left[p + (1-p)\frac{1}{1+a^\alpha\eta(t)}\right]\left[\frac{1+a^\alpha\eta(t)}{1+\eta(t)}\right]$$

$$\times \frac{1}{1+a^\alpha p\eta(t)}$$

$$= \frac{1}{1+\eta(t)}.$$

Thus we have established the following theorem.

Theorem 6.4.4. *The first order autoregressive equation*

$$x_n = aI_n x_{n-1} + \epsilon_n, \quad n = 1, 2, \ldots$$

where $\{I_n\}$ are independent Bernoulli sequences such that $P(I_n = 0) = p$ and $P(I_n = 1) = 1 - p;\ p \in (0,1),\ a \in (0,1)$ is a strictly stationary AR(1) process with semi–Mittag-Leffler (α) marginal distribution if and only if $\{\epsilon_n\}$ are independently and identically distributed as the sum of two independent random variables u_n and v_n as in (6.4.15) and x_0 is distributed as semi–Mittag-Leffler (α).

When $\eta(t) = t^\alpha$, the NSMLAR(1) model becomes the NMLAR(1) model.

Remark 6.4.2: If we consider characteristic functions instead of Laplace transforms, the results can be applied to real valued autoregressive processes. Then the role of semi-Mittag-Leffler distributions is played by semi-α-Laplace distributions introduced by Pillai (1985). As special cases we get Laplace and α-Laplace processes.

Exercises 6.4.

6.4.1. If $f(t) = t$, find $W^{n*}([0,x])$.

6.4.2. If $f(t) = t^\alpha$, find $F_\epsilon(x)$.

6.4.3. State any three distributions belonging to the semi-Mittag-Leffler family.

6.4.4. Show that the stationary solution of equation (6.4.7) is a family consisting of g.i.d. and class L distributions.

6.4.5. Obtain the innovation structure of the NEAR(1) model.

6.4.6. Obtain the innovation structure of the NMLAR(1) model.

6.5 Tailed Processes

In an attempt to develop autoregressive models for time series with exact zeros Littlejohn (1993) formulated an autoregressive process with exponential tailed marginal distribution, after the new exponential autoregressive process (NEAR(1)) of Lawrance and Lewis (1981). However, the primary aim of Littlejohn was to extend the time reversibility theorem of Chernick *et al.*(1988) and hence the model was not studied in detail. Hence we intend to make a detailed study of this process. Here the tail of a non–negative random variable refers to the positive part of the sample space, excluding only the point zero.

Definition 6.5.1. A random variable E is said to have the exponential tailed distribution denoted by $ET(\lambda, \theta)$ if $P(E = 0) = \theta$ and $P(E > x) = (1 - \theta)e^{-\lambda x}$, $x > 0$ where $\lambda > 0$ and $0 \le \theta < 1$. Then the Laplace-Stieltjes transform of E is given by

$$\phi_E(t) = \theta + (1 - \theta)\frac{\lambda}{\lambda + t} = \frac{\lambda + \theta t}{\lambda + t}.$$

6.5.1 The exponential tailed autoregressive process [ETAR(1)]

It is evident that the exponential tailed distribution is not self–decomposable and so it cannot be marginal to the autoregressive structure of Gaver and Lewis (1980). But an autoregressive process satisfying the NEAR(1) structure given by (6.4.7) can be constructed as follows: We have from (6.4.8), by substituting $\phi_x(t) = \frac{\lambda + \theta t}{\lambda + t}$, the Laplace transform of the innovation ϵ_n in the stationary case as

$$\phi_\epsilon(t) = \left[\frac{\lambda + \theta t}{\lambda + t}\right]\left[\frac{\lambda + at}{\lambda + a[p + (1 - p)\theta]t}\right]$$

$$= \left[\frac{\lambda + at}{\lambda + t}\right]\left[\frac{\lambda + \theta t}{\lambda + bt}\right]$$

where $b = a[p + (1 - p)\theta]$.

$$\phi_\epsilon(t) = \left[a + (1 - a)\frac{\lambda}{\lambda + t}\right]\left[\frac{\theta}{b} + \left(1 - \frac{\theta}{b}\right)\frac{(\lambda/b)}{(\lambda/b) + t}\right]$$

so that the innovations $\{\epsilon_n\}$ can be represented as the sum of two independent exponential tailed random variables u_n and v_n where

$$u_n \stackrel{d}{=} ET(\lambda, a) \qquad \text{and} \quad v_n \stackrel{d}{=} ET(\lambda', \theta') \tag{6.5.1}$$

where $\lambda' = \lambda/b$ and $\theta' = \theta/b$, provided $\theta \le b$. Since $p \le 1$, we require that $\theta \le a$. Thus the ETAR(1) process can be defined as a sequence $\{x_n\}$ satisfying (6.4.7) where $\{\epsilon_n\}$ is a sequence of independent and identically distributed random variables such that $\epsilon_n = u_n + v_n$ where u_n and v_n are as in (6.5.1).

It can be easily shown that the process is strictly stationary and Markovian provided x_0 is distributed as $ET(\lambda, \theta)$. This follows by mathematical induction since

$$\phi_{x_n}(t) = \phi_{\epsilon_n}(t)[p + (1 - p)\phi_{x_{n-1}}(at)]$$

$$= \left[\frac{\lambda + at}{\lambda + t}\right]\left[\frac{\lambda + \theta t}{\lambda + bt}\right]\left[p + (1 - p)\left(\frac{\lambda + \theta at}{\lambda + at}\right)\right]$$

$$= \left[\frac{\lambda + at}{\lambda + t}\right]\left[\frac{\lambda + \theta t}{\lambda + bt}\right]\left[\frac{\lambda + bt}{\lambda + at}\right]$$

$$= \frac{\lambda + \theta t}{\lambda + t}.$$

When $\theta = 0$, the $ET(\lambda, \theta)$ distribution reduces to the exponential (λ) distribution and the ETAR(1) model then becomes the NEAR(1) model.

6.5.2 The Mittag-Leffler tailed autoregressive process [MLTAR(1)]

The Mittag-Leffler tailed distribution has Laplace transform given by

$$\phi_x(t) = \theta + \frac{(1-\theta)}{1+t^\alpha}$$
$$= \frac{1+\theta t^\alpha}{1+t^\alpha}, \qquad 0 < \alpha \le 1$$

and the distribution shall be denoted by MLT (α, θ). Similarly for a two-parameter Mittag-Leffler random variable $ML(\alpha, \lambda)$ the Laplace transform of the tailed Mittag-Leffler distribution is given by $\phi_x(t) = \theta + (1-\theta)\frac{\lambda^\alpha}{\lambda^\alpha+t^\alpha} = \frac{\lambda^\alpha+\theta t^\alpha}{\lambda^\alpha+t^\alpha}$. This shall be denoted by $MLT(\alpha, \lambda, \theta)$. The MLTAR(1) process has the general structure given by the equation (6.4.7). The innovation structure can be derived as follows.

$$\phi_\epsilon(t) = \left[\frac{1+\theta t^\alpha}{1+t^\alpha}\right]\left[\frac{1+a^\alpha t^\alpha}{1+a^\alpha[p+(1-p)\theta]t^\alpha}\right]$$
$$= \left[\frac{1+a^\alpha t^\alpha}{1+t^\alpha}\right]\left[\frac{1+\theta t^\alpha}{1+ct^\alpha}\right]$$

where $c = a^\alpha[p+(1-p)\theta]$. Therefore

$$\phi_\epsilon(t) = \left[a^\alpha + (1-a^\alpha)\frac{1}{1+t^\alpha}\right]\left[\frac{\frac{1}{c}+\frac{\theta}{c}t^\alpha}{\frac{1}{c}+t^\alpha}\right].$$

Hence the innovation $\{\epsilon_n\}$ can be viewed as the sum of two independently distributed random variables u_n and v_n where

$$u_n \overset{d}{=} MLT(\alpha, a^\alpha)$$

and

$$v_n \overset{d}{=} MLT(\alpha, \lambda', \theta')$$

where $\lambda' = 1/c^{1/\alpha}$ and $\theta' = \theta/c$ provided $\theta \le c$. This holds when $\theta \le a^\alpha$.

The model can be extended to the class of semi–Mittag-Leffler distributions. Here we consider a semi–Mittag-Leffler distribution with Laplace transform

$$\phi_x(t) = \frac{\lambda^\alpha}{\lambda^\alpha + \eta(t)}$$

where $\eta(t)$ satisfies the functional equation

$$\eta(mt) = m^\alpha \eta(t); \qquad 0 < m < 1; \quad 0 < \alpha \le 1.$$

This is denoted by $SML(\alpha, \lambda)$. Then the semi–Mittag-Leffler tailed distribution denoted by $SMLT(\alpha, \lambda, \theta)$ has Laplace transform

$$\phi_x(t) = \frac{\lambda^\alpha + \theta\eta(t)}{\lambda^\alpha + \eta(t)}.$$

The first order semi-Mittag-Leffler tailed autoregressive $(SMLTAR(1))$ process has innovations whose Laplace transform is given by

$$\phi_\epsilon(t) = \left[\frac{\lambda^\alpha + \theta\eta(t)}{\lambda^\alpha + \eta(t)}\right] \left[\frac{\lambda^\alpha + a^\alpha\eta(t)}{\lambda^\alpha + c\eta(t)}\right]$$

where $c = a^\alpha[p + (1-p)\theta]$. Therefore

$$\phi_\epsilon(t) = \left[\frac{\lambda^\alpha + a^\alpha\eta(t)}{\lambda^\alpha + \eta(t)}\right] \left[\frac{\lambda^\alpha + \theta\eta(t)}{\lambda^\alpha + c\eta(t)}\right]$$

$$= \left[a^\alpha + (1 - a^\alpha)\frac{\lambda^\alpha}{\lambda^\alpha + \eta(t)}\right] \left[\frac{\theta}{c} + \left(1 - \frac{\theta}{c}\right)\frac{\lambda^\alpha/c}{\lambda^\alpha/c + \eta(t)}\right].$$

Therefore, the innovations $\{\epsilon_n\}$ can be represented as the sum of two independent semi–Mittag-Leffler tailed random variables u_n and v_n where

$$u_n \overset{d}{=} SMLT(\alpha, \lambda, a^\alpha) \text{ and } v_n \overset{d}{=} SMLT(\alpha, \lambda', \theta') \tag{6.5.2}$$

where $\lambda' = \lambda/c^{1/\alpha}, \theta' = \theta/c$. Then we have the following theorem which gives the stationary solution of the $SMLTAR(1)$ model.

Theorem 6.5.1. *For $0 < p < 1$, $0 < a < 1$ the stationary Markov process $\{x_n\}$ defined by (6.4.7) has a semi–Mittag-Leffler tailed $SMLT(\alpha, \lambda, \theta)$ marginal distribution if and only if the innovation sequence $\{\epsilon_n\}$ are independent and identically distributed as the sum of two independent semi–Mittag-Leffler tailed random variables as in (6.5.2), provided $x_0 \overset{d}{=} SMLT(\alpha, \lambda, \theta)$.*

The stationarity of the process can be easily established, as given below.

$$\phi_{x_n}(t) = \phi_{\epsilon_n}(t)[p + (1-p)\phi_{x_{n-1}}(at)]$$

$$= \left[\frac{\lambda^\alpha + a^\alpha\eta(t)}{\lambda^\alpha + \eta(t)}\right] \left[\frac{\lambda^\alpha + \theta\eta(t)}{\lambda^\alpha + c\eta(t)}\right]$$

$$\times \left[p + (1-p)\frac{\lambda^\alpha + \theta\eta(at)}{\lambda^\alpha + \eta(at)}\right]$$

$$= \left[\frac{\lambda^\alpha + a^\alpha\eta(t)}{\lambda^\alpha + \eta(t)}\right] \left[\frac{\lambda^\alpha + \theta\eta(t)}{\lambda^\alpha + c\eta(t)}\right] \left[\frac{\lambda^\alpha + c\eta(t)}{\lambda^\alpha + \eta(at)}\right]$$

$$= \frac{\lambda^\alpha + \theta\eta(t)}{\lambda^\alpha + \eta(t)} \quad \text{since } \eta(at) = a^\alpha\eta(t).$$

Hence x_n is distributed as $SMLT(\alpha, \lambda, \theta)$. The necessity part follows easily from the derivation of the structure of the innovation sequence. Now we consider the following theorem.

Theorem 6.5.2. *In a positive valued stationary Markov process $\{x_n\}$ satisfying the first order autoregressive equation $x_n = ax_{n-1} + \epsilon_n$, $0 < a < 1$ the innovations $\{\epsilon_n\}$ are independently and identically distributed as a tailed distribution of the same type as that of $\{x_n\}$ if and only if $\{x_n\}$ are distributed as semi–Mittag-Leffler .*

Proof 6.5.1: We have, assuming stationarity,

$$\phi_x(t) = \phi_x(at)\phi_\epsilon(t).$$

Suppose

$$\phi_\epsilon(t) = \theta + (1-\theta)\phi_x(t) \text{ where } 0 \le \theta < 1.$$

Then

$$\phi_x(t) = \phi_x(at)[\theta + (1-\theta)\phi_x(t)].$$

Writing

$$\phi_x(t) = \frac{1}{1+\eta(t)}, \qquad \text{we get}$$

$$\frac{1}{1+\eta(t)} = \frac{1}{1+\eta(at)}\left[\theta + (1-\theta)\frac{1}{1+\eta(t)}\right]$$

$$= \left[\frac{1}{1+\eta(at)}\right]\left[\frac{1+\theta\eta(t)}{1+\eta(t)}\right].$$

This implies that $\eta(at) = \theta\eta(t)$. By taking $\theta = a^\alpha$ it follows that the distribution of x_n is semi–Mittag-Leffler . Conversely, if $\{x_n\}$ are semi–Mittag-Leffler, we get

$$\phi_\epsilon(t) = \frac{\phi_x(t)}{\phi_x(at)} = \frac{1+\eta(at)}{1+\eta(t)}$$

$$= \frac{1+a^\alpha\eta(t)}{1+\eta(t)}$$

$$= a^\alpha + (1-a^\alpha)\frac{1}{1+\eta(t)}.$$

Hence $\{\epsilon_n\}$ is distributed as $SMLT(\alpha, a^\alpha)$.

The SMLTAR(1) process can be regarded as generalizations of the EAR(1), NEAR(1), MLAR(1), NMLAR(1), TEAR(1), ETAR(1) and MLTAR(1) processes. These processes are useful to model non-negative time series data which exhibit zeros, as in the case of stream flow data of rivers that are dry during part of the year. They are useful for modeling life times of devices which have some probability for damage immediately after putting it to use. In a similar manner, the models can be extended to the semi-α-Laplace case and its special cases. Also geometric Mittag-Leffler and geometric alpha-Laplace distributions and time series models can be developed.

Exercises 6.5.

6.5.1. Derive the Laplace transform of the exponential tailed distribution.

6.5.2. Derive the innovation structure of the Mittag-Leffler tailed autoregressive process.

6.5.3. Examine whether the Mittag-Leffler tailed distribution is self-decomposable.

6.5.4. Give a real life example where the exponential tailed distribution can be used for modeling.

6.5.5. Show that Laplace distribution belongs to the semi-α-Laplace family.

6.5.6. Define a geometric exponential distribution similar to the geometric stable distribution.

6.5.7. Try to develop a generalized Laplacian model, with characteristic function $\varphi_x(t) = \left(\frac{1}{1-\beta^2 t^2}\right)^\alpha$.

6.5.8. Develop the concept in geometric infinite divisibility by replacing addition by minimum in the case of g.i.d.

6.5.9. Develop an autoregressive minification structure by replacing addition by minimum in the standard AR(1) equation.

6.6 Marshall-Olkin Weibull Time Series Models

6.6.1 Introduction

The need for developing time series models having non-Gaussian marginal distributions has been long felt from the fact that many naturally occurring time series are non-Gaussian with Markovian structure. In recent years Tavares (1980), Yeh et al. (1988), Arnold and Robertson (1989), Pillai (1991), Alice and Jose (2004, 2005) and others have developed various autoregressive models with minification structure. The Weibull distribution, including exponential distribution plays a central role in the modeling of survival or lifetime data and time series data of nonnegative random variables such as hydrological data and wind velocity magnitudes. Lewis and McKenzie (1991), Brown et al (1984) note that although studies have shown that Weibull marginal distributions have been found adequate for wind velocity magnitudes, unfortunately 'no time series models have been rigorously developed for random variables possessing a Weibull distribution'. Wind power data are even more likely to need very long tailed marginal distributions. Again in reliability studies, sequences of times between failures are correlated and models with non-constant marginal hazard rate are needed to model them adequately.

6.6.2 Marshall-Olkin semi-Weibull distribution

We say that a random variable x with positive support has a semi-Weibull distribution and write $x \overset{d}{=} SW(\beta, \rho)$ if its survival function is given by

$$\bar{F}_x(x_o) = P(x > x_o) = \exp(-\Psi(x_o)) \tag{6.6.1}$$

where $\Psi(x)$ satisfies the functional equation,

$$\rho\Psi(x) = \Psi(\rho^{\frac{1}{\beta}}x), \beta > 0, 0 < \rho < 1. \tag{6.6.2}$$

Equation (6.6.2) will give on iteration

$$\rho^n\Psi(x) = \Psi(\rho^{\frac{n}{\beta}}x).$$

On solving (6.6.2) we obtain $\Psi(x) = x^\beta h(x)$, where $h(x)$ is periodic in $\ln x$ with period $\left(\frac{-2\pi\beta}{\ln \rho}\right)$. For details see Jose (1994, 2005).

We consider a new family of distributions introduced by Marshall and Olkin (1997). Considering a survival function $\bar{F}$, we get the one-parameter family of survival functions

$$\bar{G}(x; \alpha) = \frac{\alpha\bar{F}(x)}{[1 - (1 - \alpha)\bar{F}(x)]}; -\infty < x < \infty, 0 < \alpha < \infty. \tag{6.6.3}$$

It can be easily seen that when $\alpha = 1$, $\bar{G} = \bar{F}$. Whenever F has a density, the family of survival functions given by $\bar{G}(x; \alpha)$ in (6.6.3) has easily computed densities. In particular, if F has a density f and hazard rate r_F, then G has the density g given by

$$g(x; \alpha) = \frac{\alpha f(x)}{\{1 - (1 - \alpha)\bar{F}(x)\}^2} \tag{6.6.4}$$

and hazard rate

$$r(x; \alpha) = \frac{r_F(x)}{(1 - (1 - \alpha)\bar{F}(x)}, -\infty < x < \infty. \tag{6.6.5}$$

Substituting (6.6.1) in (6.6.3) we get a new family of distributions, which we shall refer to as the survival function of Marshall-Olkin semi-Weibull [MOSW (α, β, ρ)] family, whose survival function is given by

$$\bar{G}(x; \alpha) = \frac{\alpha}{e^{\Psi(x)} - (1 - \alpha)}, x > 0, \alpha > 0.$$

The probability density function corresponding to G is given by

$$g(x; \alpha) = \frac{\alpha e^{\Psi(x)}\Psi'(x)}{[e^{\Psi(x)} - (1 - \alpha)]^2}, x > 0, \alpha > 0.$$

The hazard rate is given by

$$r(x;\alpha) = \frac{\Psi'(x)}{1-(1-\alpha)e^{-\Psi(x)}}, x > 0, \alpha > 0.$$

Now we establish the following properties:

Theorem 6.6.1. *Let N be an integer valued random variable independent of the x_n's such that $P[N \geq 2] = 1$ where $\{x_n\}$ is a sequence of independent and identically distributed MOSW random variables. Then $y = \left(\frac{N}{\alpha}\right)^{\frac{1}{\beta}} min\,(x_1, x_2, ..., x_N); N > \alpha, N > 1$ is distributed as semi-Weibull.*

Proof 6.6.1: We have

$$\bar{F}_y(x) = P[y > x]$$

$$= \sum_{n=2}^{\infty} P[N=n]P[y > x | N = n]$$

$$= \sum_{n=2}^{\infty} P[N=n]\left[\bar{F}_x\left(\left(\frac{n}{\alpha}\right)^{-1/\beta}x\right)\right]^n$$

$$= \sum_{n=2}^{\infty} P[N=n]\left[\frac{1}{1+(\frac{1}{\alpha})\Psi((\frac{n}{\alpha})^{-1/\beta}x)}\right]^n = e^{-\Psi(x)}.$$

Hence Y is distributed as semi-Weibull.

Theorem 6.6.2. *If $\{x_1, x_2, \cdots, x_n\}$ are independently and identically distributed as MOSP (α, β, p), then $z_n = \left(\frac{n}{\alpha}\right)^{\frac{1}{\beta}} min(x_1, x_2, \cdots, x_n), \alpha, \beta > 0, n > 1, n > \alpha,$ is asymptotically distributed as semi-Weibull.*

Proof 6.6.2: If x is distributed as Marshall-Olkin semi-Pareto, MOSP (α, β, p), then

$$\bar{F}(x;\alpha,\beta,p) = \frac{1}{1+\frac{1}{\alpha}\psi(x)}$$

where

$$\psi(x) = \psi(p^{\frac{1}{\beta}}x).$$

Hence

$$\bar{F}_{z_n}(x) = P\left[\left(\frac{n}{\alpha}\right)^{\frac{1}{\beta}} min(x_1, x_2, \cdots, x_n) > x\right]$$

$$= \left[\bar{F}_x\left(\left(\frac{n}{\alpha}\right)^{-\frac{1}{\beta}}x\right)\right]^n$$

$$= \left[\frac{1}{1+\frac{\psi(x)}{n}}\right]^n \to e^{-\psi(x)}$$

as n tends to infinity.

Similar results can be obtained in the case of Marshall-Olkin Pareto and Weibull distributions as a special cases.

Theorem 6.6.3. *Let $\{x_i, i \geq 1\}$ be a sequence of independent and identically distributed random variables with common survival function $\bar{F}(x)$ and N be a geometric random variable with parameter p and $P(N = n) = pq^{n-1}, n = 1, 2, ..., 0 < p < 1$, $q = 1 - p$, which is independent of $\{x_i\}$ for all $i \geq 1$. Let $u_N = \min_{1 \leq i \leq N} x_i$. Then $\{u_N\}$ is distributed as MOSW if and only if $\{x_i\}$ is distributed as semi-Weibull.*

Proof 6.6.3:

$$\bar{H}(x) = P(u_N > x)$$

$$= \sum_{n=1}^{\infty} [\bar{F}(x)]^n pq^{n-1}$$

$$= \frac{p\bar{F}(x)}{1 - (1 - p)\bar{F}(x)}.$$

Suppose

$$\bar{F}(x) = \exp(-\Psi(x)).$$

Then

$$\bar{H}(x) = \frac{1}{1 + \left(\frac{1}{p}\right)(e^{\Psi(x)} - 1)},$$

which is the survival function of MOSW. This proves the sufficiency part of the theorem. Conversely, suppose

$$\bar{H}(x) = \frac{1}{1 + \left(\frac{1}{p}\right)(e^{\Psi(x)} - 1)}.$$

Then we get

$$\bar{F}(x) = \exp(-\Psi(x)),$$

which is the survival function of semi-Weibull.

6.6.3 An AR (1) model with MOSW marginal distribution

In this section we consider a first order autoregressive model.

Theorem 6.6.4. *Consider an AR (1) structure given by*

$$x_n = \begin{cases} \varepsilon_n & w.p. \ p \\ \min(x_{n-1}, \varepsilon_n) & w.p. \ (1 - p) \end{cases}. \qquad (6.6.6)$$

where $\{\varepsilon_n\}$ is a sequence of independently and identically distributed random variables independent of x_n, then $\{x_n\}$ is a stationary Markovian AR(1) process with MOSW marginals if and only if $\{\varepsilon_n\}$ is distributed as semi-Weibull distribution.

Proof 6.6.4: From (6.6.6) it follows that

$$\bar{F}_{x_n}(x) = p\bar{F}_{\varepsilon_n}(x) + (1-p)\bar{F}_{x_{n-1}}(x)\bar{F}_{\varepsilon_n}(x). \tag{6.6.7}$$

Under stationary equilibrium,

$$\bar{F}_x(x) = \frac{p\bar{F}_\epsilon(x)}{[1-(1-p)\bar{F}_\epsilon(x)]}.$$

If we take $\bar{F}_\epsilon(x) = e^{-\Psi(x)}$, then it easily follows that

$$\bar{F}_x(x) = \frac{p}{e^{\Psi(x)} - (1-p)},$$

which is the survival function of MOSW. Conversely, if we take,

$$\bar{F}_{x_n}(x) = \frac{p}{e^{\Psi(x)} - (1-p)},$$

it is easy to show that $F_{\varepsilon_n}(x)$ is distributed as semi-Weibull and the process is station-ary. In order to establish stationarity we proceed as follows: Assume $x_{n-1} \overset{d}{=} MOSW$ and $\varepsilon_n \overset{d}{=}$ semi-Weibull. Then

$$\bar{F}_{x_n}(x) = \frac{pe^{-\Psi(x)}}{1-(1-p)e^{-\Psi(x)}}.$$

This establishes that $\{x_n\}$ is distributed as $MOSW$. Even if x_0 is arbitrary, it is easy to establish that $\{x_n\}$ is stationary and is asymptotically marginally distributed as $MOSW$.

The following theorem is regarding a kth order autoregressive model.

Theorem 6.6.5. *Consider an autoregressive model of order k as follows:*

$$x_n = \begin{cases} \epsilon_n & w.p. \ p_0 \\ \min(x_{n-1}, \epsilon_n) & w.p. \ p_1 \\ \min(x_{n-2}, \epsilon_n) & w.p. \ p_2 \\ \vdots & \\ \min(x_{n-k}, \epsilon_n) & w.p. \ p_k \end{cases} \tag{6.6.8}$$

where $0 < p_i < 1, (p_1 + p_2 + \cdots + p_k) = 1 - p_0$. Then $\{x_n\}$ has stationary marginal distribution as MOSW if and only if $\{\varepsilon_n\}$ is distributed as semi-Weibull.

The proof follows from the following facts.

$$\bar{F}_{x_n}(x) = p_0 \bar{F}_{\epsilon_n}(x) + p_1 \bar{F}_{x_{n-1}}(x)\bar{F}_{\epsilon_n}(x) + \cdots + p_k \bar{F}_{x_{n-k}}(x)\bar{F}_{\epsilon_n}(x).$$

Under stationary equilibrium,

$$\bar{F}_x(x) = p_0 \, \bar{F}_\in(x) + p_1 \, \bar{F}_x(x) \, \bar{F}_\in(x) + \cdots + p_k \, \bar{F}_x(x) \, \bar{F}_\in(x).$$

This reduces to

$$\bar{F}_x(x) = \frac{p_0 \, \bar{F}_\in(x)}{[1 - (1 - p_0) \, \bar{F}_\in(x)]}.$$

It can be seen that the semi-Weibull distribution is a more general class of distributions which includes Weibull distribution in the sense that for $h(x) = 1$, we have $\Psi(x) = x^\beta$.

6.6.4 Marshall-Olkin generalized Weibull distribution

Consider the two-parameter Weibull distribution with survival function

$$\bar{F}(x) = \exp(-(\lambda x)^\beta), x > 0, \lambda > 0, \beta > 0.$$

Then substituting in (6.3.2) we get a new family of distributions, which we shall refer to as the Marshall-Olkin Generalized Weibull (MOGW) family, whose survival function is given by

$$\bar{G}(x; \alpha, \lambda, \beta) = \frac{\alpha \exp[-(\lambda x)^\beta]}{1 - (1 - \alpha) \exp[-(\lambda x)^\beta]}, x > 0, \lambda, \beta, \alpha > 0.$$

The probability density function corresponding to G is given by

$$g(x; \alpha, \lambda, \beta) = \frac{\alpha \beta \lambda^\beta x^{\beta-1} \exp(\lambda x)^\beta}{\exp[(\lambda x)^\beta - (1 - \alpha)]^2}, x > 0, p, \beta, \alpha > 0.$$

The hazard rate is given by

$$r(x; p, \alpha, \beta) = \frac{\lambda^\beta \beta(x)^{\beta-1} \exp(\lambda x)^\beta}{\{\exp(\lambda x)^\beta - (1 - \alpha)\}}, x > 0, \lambda, \beta, \alpha > 0.$$

We also explore the nature of the hazard rate $r(x)$. It is increasing if $\alpha \geq 1$, $\beta \geq 1$ and decreasing if $\alpha < 1$, $\beta < 1$. If $\beta > 1$, then r(x) is initially increasing and eventually increasing, but there may be an interval where it is decreasing. Similarly if $\beta < 1$, then r(x) is initially decreasing and eventually decreasing but there is an interval where it is increasing. When $\alpha = 1$, it coincides with the Weibull distribution. This points out the wide applicability of the *MOGW* distribution for modeling various types of reliability data. Theorem 6.6.4 and Theorem 6.6.5 can be extended in this case also.

6.6.5 An AR (1) model with MOGW marginal distribution

Theorem 6.6.6. *Consider the AR (1) structure given by*

$$x_n = \begin{cases} \varepsilon_n & w.p.\ p \\ \min(x_{n-1}, \varepsilon_n) & w.p.\ (1-p) \end{cases} \tag{6.6.9}$$

where $\{\in_n\}$ is a sequence of independently and identically distributed random variables, independent of x_n, then $\{x_n\}$ is a stationary Markovian AR (1) process with MOGW (p, λ, β) marginals if and only if $\{\in_n\}$ is distributed as Weibull distribution with parameters λ and β.

Proof 6.6.5: Proceeding as in the case of Theorem 6.6.4 if we take

$$\bar{F}_\in(x) = \exp(-\lambda x)^\beta,$$

then it easily follows that

$$\begin{aligned} \bar{F}_x(x) &= \frac{p \exp(-\lambda x)^\beta}{[1-(1-p)\exp(-\lambda x)^\beta]} \\ &= \frac{p}{[\exp(\lambda x)^\beta - (1-p)]} \end{aligned}$$

which is the survival function of MOGW (p, λ, β). Conversely, if we take,

$$\bar{F}_{x_n}(x) = \frac{p}{[\exp(\lambda x)^\beta - (1-p)]},$$

it is easy to show that $\bar{F}_{\in_n}(x)$ is distributed as Weibull with parameters λ, β and the process is stationary. In order to establish stationarity we proceed as follows: Assume $x_{n-1} \overset{d}{=} MOGW(p, \lambda, \beta)$ and $\{\in_n\} \overset{d}{=}$ Weibull (λ, β). Then

$$\bar{F}_{x_n}(x) = \frac{p[\exp(-\lambda x)^\beta]}{\{1-(1-p)\exp(-\lambda x)^\beta\}}.$$

This establishes that $\{x_n\}$ is distributed as $MOGW(p, \lambda, \beta)$. Even if x_0 is arbitrary, it is easy to establish that $\{x_n\}$ is stationary and is asymptotically marginally distributed as $MOGW(p, \lambda, \beta)$.

Theorem 6.6.7. *Consider an autoregressive model x_n of order k with structure (6.6.8). Then $\{x_n\}$ has stationary marginal distribution as MOGW if and only if $\{\varepsilon_n\}$ is distributed as Weibull.*

Proof is similar to that of Theorem 6.6.6

Table 1 shows $P(x_n < x_{n-1})$, which are obtained through a Monte Carlo simulation procedure. Sequences of 100, 300, 500, 700, 900 observations from MOGWAR (1) process are generated repeatedly for ten times and for each sequence the probability is estimated. A table of such probabilities is provided with the average from ten trials along with an estimate of standard error in brackets. (see Table 1).

Table 6.1 $P(x_n < x_{n-1})$ for the MOGWAR(1) process where $\lambda = 1, \beta = 5$.

	Sample size				
p/n	200	400	600	800	1000
0.1	0.7705171	0.7523951	0.7573636	0.745249	0.7589128
	(0.002001358)	(0.003661458)	(0.2899318)	(0.002795277)	(0.002238917)
0.2	0.6740114	0.6597776	0.6564568	0.6621319	0.6654536
	(0.005064845)	(0.003381987)	(0.003264189)	(0.002337129)	(0.001765257)
0.3	0.6253248	0.5887756	0.5930888	0.5941391	0.5913183
	(0.0551385)	(0.002164813)	(0.0259571)	(0.002407355)	(0.003051416)
0.4	0.5027496	0.4992585	0.5230972	0.5237219	0.5153204
	(0.005699338)	(0.004394413)	(0.003465077)	(0.001724348)	(0.003295175)
0.5	0.4097892	0.4390949	0.4486389	0.4265652	0.4378709
	(0.003480096)	(0.004124351)	(0.003437423)	(0.003071082)	(0.001833982)
0.6	0.3228981	0.3585286	0.3523585	0.3440362	0.357975
	(0.005857569)	(0.005020905)	(0.003365158)	(0.002421289)	(0.003009052)
0.7	0.2695292	0.2676458	0.2646377	0.2773244	0.2731059
	(0.005099691)	(0.003165013)	(0.003349412)	(0.003459127)	(0.003940234)
0.8	0.1873956	0.207557	0.1761628	0.1863542	0.2003661
	(0.005808668)	(0.006569641)	(0.002695993)	(0.0033855295)	(0.003308759)
0.9	0.1175264	0.1194025	0.1119517	0.1012818	0.1130184
	(0.007946155)	(0.006232362)	(0.005194928)	(0.002785095)	(0.003507806)

6.6.6 Case study

In this section, we illustrate the application of the MOGWAR(1) process in modeling a hydrology data as a case study. The data consists of total daily weighted discharge (in mm^3) of Neyyar river in Kerala, India, at the location Amaravilla (near Amaravilla bridge) during 1993. Neyyar is one of the west flowing rivers in Kerala, located in the Southern most part of Kerala. It originates from Agasthyamala at an elevation of about 1,860 m. above mean sea level. From there it flows down rapidly along steep slopes in its higher reaches and then winds its way through flat country in the lower reaches. In the initial stages the course is in a southwestern direction but at Ottasekharamangalam the river turns and flows west. It again takes a southwestern course from Valappallikanam upto its fall. The Neyyar is 56 Km. long and has a total drainage area of 497 sq. Km. It is a main source of irrigation in southern Kerala and the Neyyar Dam is a main source of hydroelectric power in that region.

The arithmetic mean of the given data is 0.81. The estimates are obtained as $p = 0.5$ and $\beta = 0.7$. The calculated value of χ^2 is 0.626, which is significantly less than the tabled value. Hence MOGW distribution is found to be a good fit in this situation. It is found that the simulated MOGWAR (1) process has close resemblance to the actual data.

Exercises 6.6.

6.6.1. Define a minification process of order 1.

6.6.2. Obtain the class of distributions for which a stationary minification process is defined.

6.6.3. Develop a minification process with Pareto marginals.

6.6.4. Develop a semi-Weibull minification process.

6.6.5. Obtain the relationship between semi-Weibull and semi-Pareto distributions.

6.6.6. Obtain the innovation structure of a general Marshall-Olkin minification process.

6.6.7. Develop a bivariate Pareto minification process.

6.6.8. Develop a bivariate exponential minification process of order α.

6.6.9. Derive the hazard rate function of a Marshall-Olkin exponential distribution.

6.6.10. Derive the stationary solution of a kth order minification process.

References

Alice, T. and Jose, K. K. (2004). Bivariate semi-Pareto minification processes. *Metrika,* **59**, 305–313.

Alice, T. and Jose, K. K. (2005). Marshall-Olkin Semi-Weibull Minification Processes. *Recent Advances in Statistical Theory and Applications*, **1**, 6–17.

Anderson, D. N. and Arnold, B. C. (1993). Linnik distributions and processes, *J. Appl. Prob.,* **30**, 330–340.

Arnold, B. C. and Robertson, C. A. (1989). Autoregressive logistic processes. *J. Appl. Prob.,* **26**,524–531.

Berg, C. and Forst, G. (1975). *Potential Theory on Locally Compact Abelian Groups*, Springer, Berlin.

Bondesson, L. (1992). *Generalized Gamma Convolutions and Related Classes of Distributions and Densities*, Lecture Notes in Statistics **76**, Springer–Verlag, New York.

Brown, B. G.; Katz, R. W. and Murphy, A. H. (1984). Time series models to simulate and forecast wind speed and wind power. *J. Climate Appl. Meteorol.* **23**, 1184–1195.

Chernick, M. R., Daley, D. J. and Littlejohn, R. P. (1988). A time–reversibility relationship between two Markov chains with stationary exponential distributions, *J. Appl. Prob.,* **25**, 418–422.

Devroye, L. (1990). A note on Linnik's distribution, *Statist. Prob. Letters.* **9**, 305–306.

Feller, W. (1966). *An Introduction to Probability Theory and Its Applications*, Vol. II, Wiley, New York.

Fujita, Y. (1993). A generalization of the results of Pillai, *Ann. Inst. Statist. Math.,* **45**, 2, 361–365.

Gaver, D. P. and Lewis, P. A. W. (1980). First order autoregressive gamma sequences and point processes, *Adv. Appl. Prob.,* **12**, 727–745.

Jayakumar, K. and Pillai, R. N. (1993). The first order autoregressive Mittag-Leffler processes, *J. Appl. Prob.,* **30**, 462–466.

Jose, K. K. (1994). *Some Aspects of Non-Gaussian Autoregressive Time Series Modeling*, Unpublished Ph.D. Thesis submitted to University of Kerala.

Jose, K. K. (2005). Autoregressive models for time series data with exact zeros, (preprint).

Karlin, S. and Taylor, E. (2002) *A First Course in Stochastic Processes,* Academic press, London.

Klebanov, L. B., Maniya, G. M. and Melamed, I. A. (1985). A problem of Zolotarev and analogues of infinitely divisible and stable distributions in a scheme for summing a random number of random variables, *Theory Prob. Appl.,* **29**, 791–794.

Laha, R. G. and Rohatgi, V. K. (1979). *Probability Theory*, John Wiley and Sons, New York.

Lawrance, A. J. (1991). Directionality and reversibility in time series. *Int. Stat. Review*, **59(1)**, 67–79.

Lawrance, A. J. and Lewis, P. A. W. (1981). A new autoregressive time series model in exponential variables (NEAR(1)), *Adv. Appl. Prob.*, **13**, 826–845.

Lewis, P. A. W. and McKenzie, E. (1991). Minification processes and their transformations. *J. Appl. Prob.*, **28**, 45–57.

Littlejohn, R. P. (1993). A reversibility relation for two Markovian time series models with stationary exponential tailed distribution, *J. Appl. Prob.*, (Preprint).

Marshall, A. W. and Olkin, I. (1997). A new method for adding a parameter to a family of distributions with application to the exponential and Weibull families. *Biometrika*, **84**, 3, 641–652.

Medhi, A. (2006) *Stochastic Models in Queueing Theory*, Academic Press, California.

Medhi, A. (2004) *Stochastic Processes*, New Age Publishers, New Delhi.

Mohan, N. R., Vasudeva, R. and Hebbar, H. V. (1993). On geometrically infinitely divisible laws and geometric domains of attraction, *Sankhya B.* **55** A, 2, 171–179.

Papoulis, E. (2000). *Probability, Random Variables and Stochastic Processes*, McGraw-Hill, New York.

Pillai, R. N. (1971). Semi-stable laws as limit distributions, *Ann. Math. Statist.*, **42**, 2, 780–783.

Pillai, R. N. (1985). Semi–α–Laplace distributions, *Commun. Statist.–Theor. Meth.*, **14(4)**, 991–1000.

Pillai, R. N. (1990a). On Mittag-Leffler functions and related distributions, *Ann. Inst. Statist. Math.*, **42**, 1, 157–161.

Pillai, R. N. (1990b). Harmonic mixtures and geometric infinite divisibility, *J. Indian Statist. Assoc.*, **28**, 87–98.

Pillai, R. N. (1991). Semi–Pareto processes, *J. Appl. Prob.*, **28**, 461–465.

Pillai, R. N. and Sandhya, E. (1990). Distributions with complete monotone derivative and geometric infinite divisibility, *Adv. Appl. Prob.*, **22**, 751–754.

Rachev, S. T. and SenGupta, A. (1992). Geometric stable distributions and Laplace–Weibull mixtures, *Statistics and Decisions*, **10**, 251–271.

Ross, S. M. (2002). *Probability Models in Stochastics*, Academic Press, New Delhi.

Steutel, F.W. (1979). Infinite divisibility in theory and practice, *Scand. J. Statist.*, **6**, 57–64.

Steutel, F. W. and Van Harn, K. (1979). Discrete analogues of self–decomposability and stability, *Ann. Prob.*, **7**, 893–899.

Tavares, L. V. (1980). An exponential Markovian stationary process. *J. Appl. Prob.*, **17**, 1117–1120.

Yeh, H. C.; Arnold, B. C. and Robertson, C. A. (1988). Pareto processes. *J. Appl. Prob.*, **25**, 291–301.

Chapter 7
Applications to Density Estimation

[*This chapter is based on the lectures of Professor Serge B. Provost of the Department of Statistical and Actuarial Sciences, The University of Western Ontario, Canada.*]

7.0 Density Estimation and Orthogonal Polynomials

It is often the case that the exact moments of a continuous distribution can be explicitly determined, while its exact density function either does not lend itself to numerical evaluation or proves to be mathematically intractable. The density approximants proposed in this article are entirely specified by the first few moments of a given distribution. First, it is shown that the density functions of random variables confined to closed intervals can be approximated in terms of linear combinations of Legendre polynomials. In an application, the density function of a mixture of two beta random variables is approximated. It is also explained that the density functions of many statistics whose support is the positive half-line can be approximated by means of sums involving Laguerre polynomials; this approach is applied to a mixture of three gamma random variables. It is then shown that density approximants that are based on orthogonal polynomials such as Legendre, Laguerre, Jacobi and Hermite polynomials can be equivalently obtained by solving a linear system involving the moments of a so-called base density function.

7.1 Introduction

This lecture is concerned with the problem of approximating a density function from the theoretical moments (or cumulants) of the corresponding distribution. Approximants of this type can be obtained for instance by making use of Pearson or Johnson curves [Solomon and Stephens (1978); Elderton and Johnson (1969)], or saddlepoint approximations [Reid (1988)]. These methodologies can provide adequate approximations in a variety of applications involving unimodal distributions.

297

However, they may prove difficult to implement and their applicability can be subject to restrictive conditions. The approximants proposed here are expressed in terms of relatively simple formulae and apply to a very wide array of distributions; moreover, their accuracy can be improved by making use of additional moments. Interestingly, another technique called the inverse Mellin transform, which is based on the complex moments of certain distributions, provides representations of their exact density functions in terms of generalized hypergeometric functions; for theoretical considerations as well as various applications, the reader is referred to [Mathai and Saxena (1978)] and [Provost and Rudiuk (1995)].

First, it should be noted that the hth moment of a statistic, $u(x_1,\ldots,x_n)$, whose exact density is unknown, can be determined exactly or numerically by integrating the product $u(x_1,\ldots,x_n)^h g(x_1,\ldots,x_n)$ over the range of integration of the $x_i's$ where $g(x_1,\ldots,x_n)$ denotes the joint density of the $x_i's$, $n = 1,2,\ldots$. Alternatively, the moments of a random variable x can be obtained from the derivatives of its moment-generating function or by making use of a relationship between the moments and the cumulants when the latter are known, see [Mathai (1993), Smith (1995)]. Moments can also be derived recursively as for instance is the case in connection with certain queuing models. Once the moments of a statistic are available, one can often approximate its density function in terms of sums involving orthogonal polynomials. The approximant obtained for nonnegative random variables depends on two parameters that are determined so as to produce the best initial gamma approximation on the basis of the first two moments of the distribution. Furthermore, it was determined that for commonly encountered unimodal distributions, twelve moments usually suffice to produce reasonably accurate approximations.

The approximant proposed for distributions defined on semi-infinite intervals applies to a wide class of statistics which includes those whose asymptotic distribution is chi-square, such as $-2\ln\lambda$ where λ denotes a likelihood ratio statistic, as well as those that are distributed as quadratic forms in normal variables, such as the sample serial covariance. It should be noted that an indefinite quadratic form can be expressed as the difference of two independent nonnegative definite quadratic forms whose cumulants, incidentally, are well-known. As for distributions having compact supports, one has for example the Durbin-Watson statistic, the sample correlation coefficient, as well as many other useful statistics that can be expressed as the ratio of two quadratic forms, as discussed for instance in [Provost and Cheong (2000)].

In Section 7.4, we propose a unified approach for approximating density functions, which turns out to be mathematically equivalent to making use of orthogonal polynomials. This semiparametric methodology is also based on the moments of a distribution and only requires solving a linear system involving the moments of a so-called base density function.

Several illustrative examples are presented. For comparison purposes, each of them involves a distribution whose exact density function can be determined. First, the distribution of a mixture of two beta distributions is considered. The approximation technique presented in Section 7.3 is applied to a mixture of three gamma random variables. A mixture of three Gaussian random variables is considered in Section 7.4.

For results on the convergence of approximating sums that are expressed in terms of orthogonal polynomials, the reader is referred to Sansone (1959), Alexits (1961), Devroye and Györfi (1985) and Jones and Ranga (1998). Since the proposed methodology allows for the use of a large number of theoretical moments and the functions being approximated are nonnegative, the approximants can be regarded as nearly exact bona fide density functions, and quantiles can thereupon easily be estimated with great accuracy. As well, the polynomial representations of the approximants make them easy to report and amenable to complex calculations.

Up to now, orthogonal polynomials have been scarcely discussed in the statistical literature in connection with the approximation of distributions. This state of affairs might be due to difficulties encountered in deriving moments of high orders or in obtaining accurate results from high degree polynomials. In any case, given the powerful computational resources that are widely available these days, such complications can hardly any longer be viewed as impediments. It should be pointed out that the simple semiparametric technique proposed in Section 7.4 eliminates some of the complications associated with the use of orthogonal polynomials while yielding identical density approximants.

7.2 Approximants Based on Legendre Polynomials

A polynomial density approximation formula is obtained in this section for distributions having compact supports. This approximant is derived from an analytical result stated in Alexits (1961), which is couched below in statistical nomenclature. It should be pointed out that no a priori restrictions on the shape of the distribution need to be made in this case.

The density function of a random variable x that is defined on the interval $[-1, 1]$ can be expressed as follows:

$$f_x[x] = \sum_{k=0}^{\infty} \lambda_k P_k[x] \tag{7.2.1}$$

where $P_k[x]$ is a Legendre polynomial of degree k, that is,

$$P_k[x] = \frac{1}{2^k k!} \frac{\partial^k}{\partial x^k} \left(-1 + x^2\right)^k = \sum_{i=0}^{\text{Floor}[k/2]} \frac{(-1)^i x^{-2i+k} (-2i+2k)!}{2^k i! (-2i+k)! (-i+k)!}, \tag{7.2.2}$$

$Floor[k/2]$ denoting the largest integer less than or equal to $k/2$, and

$$\lambda_k = \frac{1+2k}{2} \sum_{i=0}^{\text{Floor}[k/2]} \frac{(-1)^i (-2i+2k)! \mu_x(-2i+k)}{2^k i! (-2i+k)! (-i+k)!}$$

$$= \frac{1+2k}{2} P^*_k(\omega) \tag{7.2.3}$$

with $P^*_k(x) = P_k(x)$ wherein x^{k-2i} is replaced by the $(k-2i)$th moment of x:

$$\mu_x[-2i+k] = E(x^{-2i+k}) = \int_{-1}^{1} x^{k-2i} f(x)dx, \qquad (7.2.4)$$

see also Devroye (1989). Legendre polynomials can also be derived by means of a recurrence relationship, available for example in Sansone (1959), p. 178. Given the first n moments of $x, \mu[1], \ldots, \mu[n]$, and setting $\mu[0] = 1$, the following truncated sum denoted by $f_{x_n}(x)$ can be used as a polynomial approximation to $f_x(x)$:

$$f_{x_n}[x] = \sum_{k=0}^{n} \lambda_k P_k(x). \qquad (7.2.5)$$

As explained in Burden and Faires (1993), Chapter 8, this polynomial turns out to be the least-squares approximating polynomial that minimizes $\int_{-1}^{1} (f_x(x) - f_{x_n}(x))^2 dx$, the integrated squared error. As stated in Rao (1965), p. 106, the moments of any continuous random variable whose support is a closed interval, uniquely determine its distribution, and as shown by Alexits (1961), p. 304, the rate of convergence of the supremum of the absolute error, $|f_x(x) - f_{x_n}(x)|$, depends on $f_x(x)$ and n, the degree of $f_{x_n}(x)$, via a continuity modulus. Therefore, more accurate approximants can always be obtained by making use of higher degree polynomials.

We now turn our attention to the more general case of a continuous random variable y defined on the closed interval [a,b], whose kth moment is denoted by

$$\mu_y[k] = E(y^k) = \int_a^b y^k f_y(y)dy, \ k = 0, 1, \ldots, \qquad (7.2.6)$$

where $f_y(y)$ denotes the density function of y. As pointed out in Section 7.1, there exist several alternative methods for evaluating the moments of a distribution when the exact density is unknown. On mapping y onto x by means of the linear transformation

$$x = \frac{2y - (a+b)}{b-a}, \qquad (7.2.7)$$

one has the desired range for x, that is, the interval $[-1, 1]$. The jth moment of x, expressed as the expected value of the binomial expansion of $((2y - (a+b))/(b-a))^j$ is then given by

$$\mu_x[j] = \frac{1}{(b-a)^j} \sum_{k=0}^{j} \binom{j}{k} 2^k \mu_y[k](-1)^{j-k}(a+b)^{j-k} \qquad (7.2.8)$$

and (7.2.5) can then be used to provide an approximant to the density function of x. On transforming x back to y with the affine change of variables specified in (7.2.7) and noting that $dx/dy = 2/(b-a)$, one obtains the following polynomial approximation for the density function of y:

$$f_{y_n}[y] = [2/(b-a)] \sum_{k=0}^{n} \lambda_k P_k \left(\frac{2y - (a+b)}{b-a} \right). \qquad (7.2.9)$$

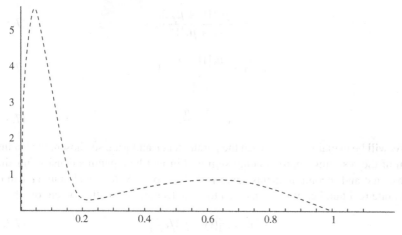

Fig. 7.1 Exact and approximate (dashed) PDF's

Example 7.2.1. **Approximate density of a mixture of beta random variables**

Consider a mixture of two equally weighted beta distributions with parameters (3,2) and (2,30), respectively. A fifteenth-degree polynomial approximation was obtained from (7.2.9). The exact density function of this mixture and its approximant, both plotted in Figure 7.1, are manifestly in close agreement. Obviously, approaches that are based on a few moments would fail to provide satisfactory approximations in this case.

As will be mentioned in Section 7.4, beta-shaped density functions defined on closed intervals can be approximated in terms of Jacobi polynomials. However, it should be pointed out that approximants expressed in terms of Legendre polynomials can accommodate a much wider class of distributions defined on closed intervals.

7.3 Approximants Based on Laguerre Polynomials

As mentioned in Section 7.1, the density functions of numerous statistics distributed on the positive half-line can be approximated from their exact moments by means of sums involving Laguerre polynomials. It should be pointed out that such an approximant should only be used when the underlying distribution possesses the tail behaviour of a gamma random variable; thankfully, this is often the case for statistics whose support is semi-infinite. Note that for other types of distributions whose support is the positive half-line, such as the lognormal, the moments may not uniquely determine the distribution; see for instance Rao (1965), p. 106 for conditions ensuring that they do.

Consider a random variable y defined on the interval (a, ∞), whose jth moment is denoted by $\mu_y[j]$, $j = 0, 1, 2, \ldots$, and let

$$c = \frac{-\mu_y[1]^2 + \mu_y[2]}{-a + \mu_y[1]} \tag{7.3.1}$$

$$v = \frac{\mu_y[1] - a}{c} - 1 \tag{7.3.2}$$

and

$$x = \frac{y - a}{c}. \tag{7.3.3}$$

As will be explained later, when the parameters c and v are so chosen, the leading term of the resulting approximating sum will in fact be a gamma density function whose first and second moments agree with those of y. Note that although a can be any finite real number, it is often equal to zero. Denoting the jth moment of x by

$$\mu_x[j] = E[((y - \mu)/c)^j], \tag{7.3.4}$$

the density function of the random variable x defined on the interval $(0, \infty)$ can be expressed as

$$f_x[x] = x^v e^{-x} \sum_{j=0}^{\infty} \delta_j L_j(v, x) \tag{7.3.5}$$

where

$$L_j[v, x] = \sum_{k=0}^{j} \frac{(-1)^k \Gamma(1 + j + v) x^{j-k}}{\Gamma(1 + j - k + v)(j - k)! k!} \tag{7.3.6}$$

is a Laguerre polynomial of order j with parameter v and

$$\delta_j = \sum_{k=0}^{j} \frac{(-1)^k j! \mu_x[j - k]}{\Gamma(1 + j - k + v)(j - k)! k!} \tag{7.3.7}$$

which also can be represented by $j!/\Gamma(v + j + 1)$ times $L_j[v, x]$ wherein x^k is replaced with $\mu_x[k]$, see for example Szegö (1959) and Devroye (1989). Then, on truncating the series given in (7.3.5) and making the change of variables $y = cx + a$, one obtains the following density approximant for y:

$$f_{y_n}[y_-] = \frac{(y - a)^v e^{-(y-a)/c}}{c^{v+1}} \sum_{j=0}^{n} \delta_j L_j(v, (y - a)/c).$$

Remark 7.3.1. On observing that $f_{y_0}(y)$ is a shifted gamma density function with parameters $\alpha \equiv v + 1 = (\mu[1] - a)^2/(\mu[2] - \mu[1]^2)$ and $\beta \equiv c = (\mu[2] - \mu[1]^2)/(\mu[1] - a)$, one can express $f_{y_n}(x)$ as the product of an initial shifted gamma density approximation whose first two central moments, $\alpha\beta + a = \mu[[1]]$ and $\alpha\beta^2 = \mu[[2]] - \mu[[1]]^2$, match those of y, times a polynomial adjustment; that is,

$$f_{y_n}(y) = \frac{e^{\frac{a-y}{\beta}} (-a + y)^{-1+\alpha}}{\beta^\alpha \Gamma[\alpha]} \sum_{j=0}^{n} a_j L_j \Gamma(\alpha)(\alpha - 1, (y - a)/c). \tag{7.3.8}$$

The following example is relevant as nonnegative definite quadratic forms in normal variables which happen to be ubiquitous in Statistics can be expressed as mixtures of chi-square random variables, see for instance Mathai and Provost (1992), Chapters 2 and 7.

Example 7.3.1. Approximate density of a mixture of gamma random variables

Let the random variable y be a mixture of three equally weighted shifted gamma random variables with parameters $(\alpha_1 = 8, \beta_1 = 1), (\alpha_2 = 16, \beta_2 = 1)$ and $(\alpha_3 = 64, \beta_3 = 1/2)$, all defined on the interval $(5, \infty)$. The hth moment of this distribution is determined by evaluating the hth derivative of its moment-generating function, $M_x(t)$, with respect to t at $t = 0$.

Figure 7.2 shows the exact density function of the mixture as well as the initial gamma density approximation given by $f_{y_0}[y]$. Clearly, traditional approximants such as those mentioned in Section 7.1, could not capture adequately all the distinctive features of this particular distribution.

The exact density function, $f_y[y]$ and its polynomial approximant, $f_{y_{60}}[y]$, are plotted in Figure 7.3. (Once such an approximant is obtained, one could for instance approximate it with a spline composed of third-degree polynomial arcs, in order to reduce the degree of precision required in further calculations.)

This example illustrates that the proposed methodology can also accomodate multimodal distributions and that calculations involving high order Laguerre polynomials will readily produce remarkably accurate approximations when performed in an advanced computing environment such as that provided by Mathematica.

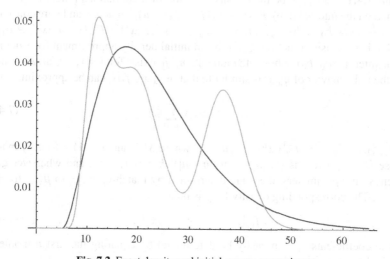

Fig. 7.2 Exact density and initial gamma approximant

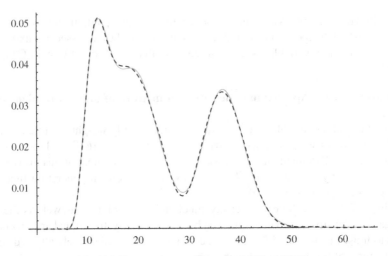

Fig. 7.3 Exact and approximate (dashed line) PDF's

7.4 A Unified Methodology

The remark made in the previous section suggests the following general semiparametric approach to density approximation, which consists of approximating the density function of a distribution whose first n moments are known by means of a base density function whose parameters are determined by matching moments, times a polynomial of degree n whose coefficients are also obtained by matching moments.

Result 7.4.1 Let $f_y[y]$ be the density function of a continuous random variable y defined on the interval (a,b), $E(y^j) \equiv \mu_y[j], x = (y-u)/s$ where u and s are constants, $a_0 = (a-u)/s, b_0 = (b-u)/s, f_x[x] = |s| f_y[u+sx], E(x^j) = E[((y-u)/s)^j] \equiv \mu_x[j]$, and the base density function, $\psi_x[x]$, be an initial density approximant for x defined on the interval (a_0, b_0), whose jth moment, $m_x[j]$, exists for $j = 1, \ldots, 2n$. Assuming that the tail behavior of $\psi_x[x]$ is similar to that of $f_x[x]$, $f_x[x]$ can be approximated by

$$f_{x_n}[x] = \psi_x[x] \sum_{i=0}^{n} \xi_i x^i \qquad (7.4.1)$$

with $(\xi_0, \ldots, \xi_n)' = M^{-1}(\mu[0], \ldots, \mu_x[n])'$ where M is an $(n+1) \times (n+1)$ matrix whose $(h+1)$th row is $(m_x[h], \ldots, m_x[h+n])', h = 0, 1, \ldots, n$, and whenever $\xi_x[x]$ depends on r parameters, these are determined by matching $m_x[j]$ to $\mu_x[j]$ for $j = 1, \ldots, r$. The corresponding density for y is then

$$f_{y_n}^*[y] = f_{x_n}[(y-u)/s]/s. \qquad (7.4.2)$$

The coefficients ξ_i, can easily be determined by equating the first n moments obtained from $f_{x_n}[x]$ to those of x :

$$\int_{a_0}^{b_0} x^h \psi_x[x] \sum_{i=0}^{n} \xi_i x^i dx = \int_{a_0}^{b_0} x^h f[x] dx, h = 0, 1, \ldots, n, \qquad (7.4.3)$$

which is equivalent to

$$(m_x[h], \ldots, m_x[h+n]).(\xi_0, \ldots, \xi_n) = \mu_x[h], \, h = 0, 1, \ldots, n; \tag{7.4.4}$$

this linear system can be represented in matrix form as

$$M(\xi_0, \ldots, \xi_n)' = (\mu_x[0], \ldots, \mu_x[n])'$$

where M is as defined in Result 7.4.1.

7.5 Approximants Expressed in Terms of Orthogonal Polynomials

By making use of the same notation, we now show that the unified approach described above provides approximants that are mathematically equivalent to those obtained from orthogonal polynomials whose weights are proportional to a certain base density function.

Let $\{T_i[x] = \sum_{k=0}^{i} \delta_{ik} x^k, \, i = 0, 1, \ldots, n\}$ be a set of orthogonal polynomials on the interval (a_0, b_0) such that

$$\int_{a_0}^{b_0} w[x] T_i[x] T_h[x] \mathrm{d}x = \theta_h \text{ when } i = h, \, h = 0, 1, \ldots, n, \text{ and zero otherwise,} \tag{7.5.1}$$

where $w[x]$ is a weight function, and let c_T be a normalizing constant such that $c_T w[x] \equiv \psi_x[x]$ integrates to one over the interval (a_0, b_0). On noting that the orthogonal polynomials T_i are linearly independent [Burden and Faires (1993), Corollary 8.8], one can write (7.4.1) as

$$f_{x_n}[x] = c_T w[x] \sum_{i=0}^{n} \eta_i T_i p[x] \tag{7.5.2}$$

where the η_i's are obtained from equating $\int_{a_0}^{b_0} T_h[x] f_{x_n}[x] \mathrm{d}x$ to $\int_{a_0}^{b_0} T_h[x] f[x] \mathrm{d}x$ for $h = 0, 1, \ldots, n$, which yields the following linear system:

$$c_T \int_{a_0}^{b_0} T_h[x] w[x] \sum_{i=0}^{n} \eta_i T_i[x] \mathrm{d}x = \int_{a_0}^{b_0} T_h[x] f[x] \mathrm{d}x, \, h = 0, 1, \ldots, n, \tag{7.5.3}$$

which is equivalent to

$$\sum_{i=0}^{n} \eta_i c_T \int_{a_0}^{b_0} w[x] T_i[x] T_h[x] \mathrm{d}x = \sum_{k=0}^{h} \delta_{hk} \mu_x[k], \, h = 0, 1, \ldots, n, \tag{7.5.4}$$

where δ_{hk} is the coefficient of x^k in T_h. Thus, by virtue of the orthogonality property given in (7.5.1), one has

$$\eta_h = \frac{1}{c_T \theta_h} \sum_{k=0}^{h} \delta_{hk} \mu_x[k], \, h = 0, 1, \ldots, n, \tag{7.5.5}$$

and

$$f_{x_n}[x] = \psi[x] \sum_{i=0}^{n} \left(\frac{1}{c_T \theta_i} \sum_{k=0}^{i} \delta_{ik} \mu_x[k] \right) T_i[x]. \tag{7.5.6}$$

Now, letting $y = u + sx$, $a = u + sa_0$, $b = u + sb_0$, and denoting the density functions of y and y_n corresponding to those of x and x_n by $f_y[y]$ and $f_{y_n}[y]$, respectively, $f_y[y]$ whose support is the interval (a,b) can be approximated by

$$f_{y_n}[y] = w[(y-u)/s] \sum_{i=0}^{n} \left(\frac{1}{s\theta_i} \sum_{k=0}^{i} \delta_{ik} \mu_x[k] \right) T_i[(y-u)/s]. \tag{7.5.7}$$

It is apparent that several complications associated with the use of orthogonal polynomials can be avoided by resorting to the direct approach described in Result 7.4.1. Density approximants expressed in terms of Laguerre, Legendre, Jacobi and Hermite polynomials are discussed below.

7.5.1 Approximants expressed in terms of Laguerre polynomials

Consider the approximants based on Laguerre polynomials discussed in Section 7.3. In that case, $y = cx + a$, so that $u = a, s = c, a_0 = 0, b_0 = \infty, w[x] = x^\nu e^{-x}, T_i[x]$ is the Laguerre $L_i[\nu, x]$ orthogonal polynomial and $\theta_h = \Gamma[\nu + h + 1]/h!$. It is easily seen that the density expressions given in (7.5.7) and (7.3.8) coincide.

In this case, the base density function, $\psi_x[x]$ is that of a gamma random variable with parameters $\nu + 1$ and 1. Note that after the transformation, our base density is a shifted gamma distribution with parameters $\nu + 1$ and c, whose support is the interval (a, ∞).

Alternatively, one can obtain an identical density approximant by making use of Result 7.4.1 where in $\psi_x[x]$ is a $Gamma(\lambda + 1, 1)$ density function whose jth moment, $m_x[j]$, which is needed to determine the ξ's, is given by $\Gamma[\nu + 1 + j]/\Gamma[\nu + 1]$, $j = 0, 1, \ldots, 2n$.

7.5.2 Approximants expressed in terms of Legendre polynomials

First, we note that whenever the finite interval (a, b) is mapped onto the interval (a_0, b_0), the requisite affine transformation is

$$x = \frac{y - u}{s} \tag{7.5.8}$$

with $u = (ab_0 - a_0 b)/(b_0 - a_0)$ and $s = (b - a)/(b_0 - a_0)$. Consider the approximants based on Legendre polynomials discussed in Section 7.2, which are defined on the interval $(-1,1)$. In that case, $u = \frac{(a+b)}{2}$, $s = (b-a)/2$, $w[x] = 1$, $T_i[x]$ is the Legendre orthogonal polynomial $P_i[x]$ and $\theta_h = \frac{2}{(2h+1)}$. It is easily seen that $f_{y_n}[y]$ given in (7.5.7) yields $f_{y_n}[y]$ of (7.2.9).

7.5.3 Approximants expressed in terms of Jacobi polynomials

In order to approximate densities for which a beta type density is suitable as a base density, we shall make use of the following alternative form of the Jacobi polynomials

$$G_n[\alpha,\beta,x] = n!\frac{\Gamma[n+\alpha]}{\Gamma[2n+\alpha]}JacobiP[n,\alpha-\beta,\beta-1,2x-1] \qquad (7.5.9)$$

defined on the interval (0, 1), where $JacobiP[n,a_1,b_1,z]$ denotes a standard Jacobi polynomial of order n in z with parameters a_1 and b_1. In this case, the weight function is $x^\alpha(1-x)^\beta$ and the base density is that of a $Beta(\alpha+1,\beta+1)$ random variable, that is,

$$\psi_x[x] = \frac{1}{B[\alpha+1,\beta+1]}x^\alpha(1-x)^\beta, 0 < x < 1, \qquad (7.5.10)$$

whose jth moment is given by

$$m_x[j] = \frac{\Gamma[\alpha+\beta+2]\Gamma[\alpha+1+j]}{\Gamma[\alpha+1]\Gamma[\alpha+j+\beta+2]}. \qquad (7.5.11)$$

The parameters α and β can be determined as follows:

$$\begin{aligned}
\alpha &= \mu_x[1](\mu_x[1]-\mu_x[2])/(\mu_x[2]-\mu_x[1]^2)-1, \\
\beta &= (1-\mu_x[1])(\alpha+1)/(\mu_x[1]-1),
\end{aligned} \qquad (7.5.12)$$

see Johnson and Kotz (1970). Moreover, in this case,

$$\theta_k^{-1} = \frac{(2k+a+b+1)\Gamma[2k+a+b+1]^2}{k!\Gamma[k+a+1]\Gamma[k+a+b+1]\Gamma[k+b+1]}. \qquad (7.5.13)$$

7.5.4 Approximants expressed in terms of Hermite polynomials

Densities of random variables for which a normal density can provide a reasonable initial approximation can be expressed in terms of the modified Hermite polynomials given by

$$H_k^*[x] = (-1)^k 2^{-k/2} HermiteH[k,x\sqrt{2}] \qquad (7.5.14)$$

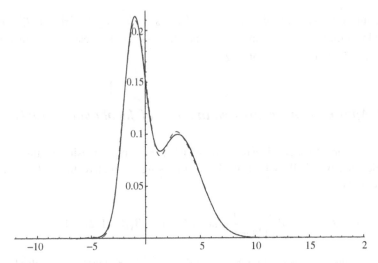

Fig. 7.4 Exact and approximate (dashed line) PDF's

where $HermiteH[k,z]$ denotes a standard Hermite polynomial of order k in z. $H_k^*[x]$ is also defined on the interval $(-\infty, \infty)$, and its associated weight function is $w[x] = e^{-x^2/2}$. Clearly, $x = (y - u)/s$ with $u = \mu_y[1]$ and $s = \sqrt{\mu_y[2] - \mu_y[1]^2}$. In this case, the base density is that of a standard normal random variable, that is,

$$\psi_x[x] = \frac{1}{\sqrt{2\pi}} e^{-x^2/2}, \qquad -\infty < x < \infty, \tag{7.5.15}$$

whose jth moment is given by

$$m_x[j] = \frac{2^{\frac{1}{2}(-1+j)}(1 + (-1)^j)\Gamma[\frac{1+j}{2}]}{\sqrt{2\pi}}, \quad j = 0, 1, \ldots, \tag{7.5.16}$$

and

$$\theta_k = \sqrt{2\pi} k!. \tag{7.5.17}$$

Example 7.5.1. Consider an equally weighted mixture of a $N(\mu, \sigma^2) = N(3, 4)$ and a $N(1, 1)$ distributions. The exact and approximate densities are shown in Figure 7.4.

References

Alexits, G. (1961). *Convergence Problems of Orthogonal Series*, Pergamon Press, New York.
Barndorff-Nielsen, O. and Cox, D. R. (1979). Edgeworth and Saddlepoint Approximations with Statistical Applications, *Journal of the Royal Statistical Society, Series B*, **41**, 279-312.
Burden, R. L. and Faires, J. D. (1993). *Numerical Analysis*, 5th Ed. Brooks/ Cole, New York.

Devroye, L. (1989). On random variate generation when only moments or Fourier coefficients are known, *Mathematics and Computers in Simulation*, **31,** 71-89.

Devroye, L. and Györfi, L. (1985). *Nonparametric Density Estimation, The L_1 View*, Wiley, New York.

Elderton, W. P. and Johnson, N. L. (1969). *Systems of Frequency Curves*, Cambridge University Press.

Johnson, N. L. and Kotz, S. (1970). *Distributions in Statistics: Continuous Univariate Distributions-2*, Wiley, New York

Jones, W. B. and Ranga, A. S., Eds. (1998). *Orthogonal Functions, Moment Theory, and Continued Fractions, Theory and Applications*, Marcel Dekker, New York.

Mathai, A. M. and Provost, S. B. (1992). *Quadratic Forms in Random Variables, Theory and Applications*, Marcel Dekker, New York.

Mathai, A. M. and Saxena, R. K. (1978). *The H-function with Applications in Statistics and Other Disciplines*, Wiley Halsted, New York.

Provost, S. B. and Rudiuk, E. M. (1995). Moments and densities of tests statistics for covariance structures, *International Journal of Mathematical and Statistical Sciences*, **4,** 85-104.

Provost, S. B. and Cheong, Y.-H. (2000). On the distribution of linear combinations of the components of a Dirichlet random vector, *The Canadian Journal of Statistics*, **28,** 417-425.

Rao, C. R. (1965). *Linear Statistical Inference and Its Applications*, Wiley, New York.

Reid, N. (1988). Saddlepoint methods and statistical inference, *Statistical Science*, **3,** 213-227.

Sansone, G. (1959). *Orthogonal Functions*, Interscience Publishers, New York.

Smith, P. J. (1995). A recursive formulation of the old problem of obtaining moments from cumulants and vice versa, *The American Statistician*, **49,** 217-219.

Solomon, H. and Stephens, M.A. (1978). Approximations to density functions using Pearson curves, *Journal of the American Statistical Association*, **73,** 153-160.

Szegö, G. (1959). *Orthogonal Polynomials*, American Mathematical Society, New York.

Chapter 8
Applications to Order Statistics

[*This chapter is based on the lectures of Dr. Yageen Thomas, Kerala University, Kariavattom, Kerala, India, at the SERC Schools.*]

8.0 Introduction

Order statistics deal with the properties and applications of ordered random variables and their functions. In the study of many natural problems related to flood, longevity, breaking strength, atmospheric temperature, atmospheric pressure, wind etc., the future possibilities in the recurrence of extreme situations are of much importance and accordingly the problem of interest in these cases reduces to that of the extreme observations. Let $z_1, z_2, \ldots, z_n$ be n random variables. Define $z_{1:n} = \min(z_1, z_2, \ldots, z_n)$, $z_{2:n} =$ second smallest of $(z_1, z_2, \ldots, z_n)$, $\ldots, z_{i:n} = i$-th smallest of $(z_1, z_2, \ldots, z_n)$, $\ldots, z_{n:n} = n$-th smallest (the largest) of $(z_1, z_2, \ldots, z_n) = \max(z_1, z_2, \ldots, z_n)$. Then we have $z_{1:n} \leq z_{2:n} \leq \cdots \leq z_{n:n}$ and these ordered random variables $z_{1:n}, z_{2:n}, \ldots, z_{n:n}$ are known as the order statistics of the given set of random variables. In particular $z_{i:n}$ is called the i-th order statistic. In most of the applications of order statistics in statistical inference problems, our interest is with the order statistics of a random sample of size n drawn from a parent population. We may write $z_{1:n}, z_{2:n}, \ldots, z_{n:n}$ to denote the order statistics of a random sample of size n drawn from an arbitrary distribution. When one considers problems similar to contamination of a random sample by outliers, order statistics of independent but not identically distributed random variables are to be considered.

8.1 Distribution Function

Let $x_{1:n}, x_{2:n}, \ldots, x_{n:n}$ be the order statistics of n independently and identically distributed $(i.i.d)$ random variables. Then for any real number x, the distribution function of $x_{r:n}$ is denoted by $F_{r:n}(x)$ and is given by

$$F_{r:n}(x) = P(x_{r:n} \le x)$$

$$= P\{r \text{ or more of the } n, x_j\text{'s are } \le x\}$$

$$= \sum_{i=r}^{n} P\{i \text{ of the } n, x_j\text{'s are } \le x\}$$

$$= \sum_{i=r}^{n} \binom{n}{i} [F(x)]^i [1 - F(x)]^{n-i}, \quad x \in R. \tag{8.1.1}$$

From the well-known relation between binomial sums and incomplete beta functions we may also write (8.1.1) as

$$F_{r:n}(x) = \int_0^{F(x)} \frac{\Gamma(n+1)}{\Gamma(r)\Gamma(n-r+1)} t^{r-1} (1-t)^{(n-r+1)-1} dt$$

$$= I_{F(x)}(r, n-r+1), \quad -\infty < x < \infty. \tag{8.1.2}$$

Example 8.1.1. Let $x_{1:n}, x_{2:n}, \ldots, x_{n:n}$ be the order statistics of a random sample of size n drawn from the uniform distribution over $[0, 1]$. Obtain $F_{r:n}(x)$.

Solution 8.1.1: A random variable x is said to have a uniform distribution over $[0, 1]$ if its probability density function $f(x)$ is given by

$$f(x) = \begin{cases} 1, & 0 \le x \le 1 \\ 0, & \text{elsewhere.} \end{cases}$$

The distribution function $F(x)$ of the uniform distribution is then given by

$$F(x) = \begin{cases} 0, & x < 0 \\ x, & 0 \le x \le 1 \\ 1, & x \ge 1. \end{cases}$$

Then by using (8.1.1), the distribution function $F_{r:n}(x)$ of $x_{r:n}$ is given by

$$F_{r:n}(x) = \sum_{i=r}^{n} \binom{n}{i} x^i (1-x)^{n-i}. \tag{8.1.3}$$

As a consequence of (8.1.2), we can write the following alternate expression for $F_{r:n}(x)$:

$$F_{r:n}(x) = \int_0^x \frac{\Gamma(n+1)}{\Gamma(r)\Gamma(n+1)} u^{r-1} (1-u)^{(n-r+1)-1} du. \tag{8.1.4}$$

Note 8.1.1: It may be noted that if $x_1, x_2, \ldots, x_n$ are the observations of a random sample of size n drawn from a population then the order statistics $x_{1:n}, x_{2:n}, \ldots, x_{n:n}$ can be considered as the order statistics of $i.i.d.$ random variables.

8.1.1 Density of the r-th order statistic

Suppose $x_{1:n}, x_{2:n}, \ldots, x_{n:n}$ are the order statistics of n i.i.d. random variables which are distributed identically with distribution function $F(x)$ and is absolutely continuous and having the density function $f(x)$, then the pdf (probability density function) $f_{r:n}(x)$ of $x_{r:n}$ can be obtained from (8.1.1) by differentiation. However $f_{r:n}(x)$ also can be obtained in the following manner: The event $x_{r:n} \in (x, x+\Delta x]$ is the same as $x_i \leq x$ for $r-1$ of the x_i's, $x < x_i \leq x+\Delta x$ for exactly one of the x_i's and $x_i > x+\Delta x$ for the remaining $n-r$ of the x_i's. As $\Delta x \to 0$ the probability that an observation lies in the above mutually exclusive classes is $F(x)$, $f(x)dx$ and $1-F(x)$ respectively. Then as $\Delta x \to 0$, by the multinomial probability law we have

$$f_{r:n}(x) = \lim_{\Delta x \to 0} \frac{P(x < x_{r:n} \leq x+\Delta x)}{\Delta x}$$

$$= \frac{n!}{(r-1)!\,(n-r)!}[F(x)]^{r-1}\,[1-F(x)]^{n-r}\,f(x), \qquad (8.1.5)$$

$$r = 1, 2, \ldots, n; \ -\infty < x < \infty.$$

Example 8.1.2. Let $x_{r:n}$ be the r-th order statistic of a random sample of size n drawn from the uniform distribution over $[0,1]$. Obtain the pdf of $x_{r:n}$

Solution 8.1.2: Clearly

$$f_{r:n}(x) = \frac{n!}{(r-1)!\,(n-r)!}x^{r-1}(1-x)^{n-r}, \ 0 \leq x \leq 1.$$

Example 8.1.3. Let $x_{1:n}, x_{2:n}, \ldots, x_{n:n}$ be the order statistics of a random sample of size n from the exponential distribution with the following pdf,

$$f(x) = \begin{cases} e^{-x}, & x > 0 \\ 0, & \text{elsewhere.} \end{cases}$$

Obtain the pdf of $x_{r:n}$.

Solution 8.1.3: Clearly,

$$F(x) = 1 - e^{-x}.$$

Therefore

$$f_{r:n} = \frac{n!}{(r-1)!\,(n-r)!}(1-e^{-x})^{r-1}\,e^{-(n-r+1)x}, \ x \geq 0.$$

In particular,

$$f_{1:n}(x) = n e^{-nx}, \ x > 0.$$

8.1.2 Joint distribution function of two order statistics

If x and y are two random variables and $(u,v) \in R^2$ then $F(u,v) = P(x \le u, y \le v)$ is called the joint distribution function of x and y. It may be noted that if $F_1(u) = P(x \le u)$ and $F_2(v) = P(y \le v)$ then $F_1(u)$ and $F_2(v)$ are called the marginal distribution functions of x and y respectively. Further,

$$F_1(u) = \lim_{v \to \infty} F(u,v) \text{ and } F_2(v) = \lim_{u \to \infty} F(u,v).$$

If $x_{r:n}$ and $x_{s:n}$ are the r-th and s-th order statistics for $1 \le r < s \le n$ of a random sample of size n arising from a population with distribution function $F(x)$ and if $F_{r,s:n}(x,y)$ is the joint distribution function of $x_{r:n}$ and $x_{s:n}$ then it is clear that

$$F_{r,s:n}(x,y) = \lim_{x \to \infty} F_{r,s:n}(x,y) \text{ for every } x > y.$$

That is,

$$F_{r,s:n}(x,y) = F_{s:n}(y) \text{ for all } x \ge y. \tag{8.1.6}$$

However for $x < y$ we have,

$$
\begin{aligned}
F_{r,s:n}(x,y) &= P(x_{r:n} \le x, x_{s:n} \le y) \\
&= P\{\text{at least } r \text{ of the } x_j\text{'s are at most } x \text{ and at least } s \\
&\quad \text{ of the } x_j\text{'s are at most } y\} \\
&= \sum_{j=s}^{n} \sum_{i=r}^{j} \frac{n!}{i!\,(j-i)!\,(n-j)!} \\
&\quad \times [F(x)]^i \, [F(y) - F(x)]^{j-i} \, [1 - F(y)]^{n-j}.
\end{aligned}
\tag{8.1.7}
$$

(8.1.6) and (8.1.7) define the distribution function $F_{r,s:n}(x,y)$.

8.1.3 Joint density of two order statistics

Suppose $x_{r:n}$ and $x_{s:n}$ for $1 \le r < s \le n$ be the r-th and s-th order statistics of a random sample of size n arising from an arbitrary continuous distribution with distribution function $F(x)$ and pdf $f(x)$. Then for $x \le y$, the event $(x_{r:n} \in (x, x+\Delta x], x_{s:n} \in (y, y+\Delta y])$ is the same as $x_i \le x$ for $(r-1)$ of the x_i's, $x < x_i \le x+\Delta x$ for exactly one of the x_i's, $x+\Delta x < x_i \le y$ for exactly $s-r-1$ of the x_i's, $y < x_i \le y+\Delta y$ for exactly one of the x_i's and $x_i > y+\Delta y$ for exactly $n-s$ of the x_i's. Then by using the multinomial probability law we have

$$
\begin{aligned}
f_{r,s:n}(x,y) &= \lim_{\Delta x \to 0, \Delta y \to 0} \frac{P(x < x_{r:n} \le x+\Delta x, y < x_{s:n} \le y+\Delta y)}{\Delta x \Delta y} \\
&= \frac{n!}{(r-1)!\,(s-r-1)!\,(n-s)!} [F(x)]^{r-1} \, [F(y) - F(x)]^{s-r-1} \\
&\quad \times [1 - F(y)]^{n-s} f(x) f(y), \quad -\infty < x < y < \infty.
\end{aligned}
\tag{8.1.8}
$$

In particular, if we put $r = 1$ and $s = n$ in (8.1.8) then we obtain the joint pdf of the most extreme order statistics and is given by

$$f_{1,n:n}(x, y) = n(n-1)[F(y) - F(x)]^{n-2} f(x) f(y), \quad -\infty < x < y < \infty.$$

Example 8.1.4. Let $x_{r:n}, x_{s:n}, 1 \leq r < s \leq n$ be the r-th and s-th order statistics of a random sample of size n arising from the uniform distribution over $[0, 1]$. Then the joint pdf of $x_{r:n}$ and $x_{s:n}$ is given by

$$f_{r,s:n}(x, y) = \frac{n!}{(r-1)!\,(s-r-1)!\,(n-s)!}$$
$$\times x^{r-1} (y-x)^{s-r-1} (1-y)^{n-s}, \quad 0 \leq x < y \leq 1.$$

8.1.4 Moments of order statistics

Let $x_{1:n}, x_{2:n}, \ldots, x_{n:n}$ be the order statistics of a random sample of size n drawn from a population with distribution function $F(x)$ and pdf $f(x)$. Then the k-th moment of the r-th order statistic is denoted by $\mu_{r:n}^{(k)}$, and is given by

$$\mu_{r:n}^{(k)} = \int_{-\infty}^{\infty} x f_{r:n}(x) dx$$
$$= \frac{n!}{(r-1)!\,(n-r)!} \tag{8.1.9}$$
$$\times \int_{-\infty}^{\infty} x [F(x)]^{r-1} [1 - F(x)]^{n-r} f(x)\, dx, \quad k = 1, 2, \ldots$$

The variance of $x_{r:n}$ is then given by

$$\text{Var}(x_{r:n}) = E(x_{r:n}^2) - [E(x_{r:n})]^2$$
$$= \mu_{r:n}^2 - [\mu_{r:n}]^2.$$

The product moment of $x_{r:n}$ and $x_{s:n}$ for $1 \leq r < s \leq n$ is given by

$$\mu_{r,s:n} = E(x_{r:n} x_{s:n})$$
$$= \int\int_{-\infty < x < y < \infty} xy\, f_{r,s:n}(x, y) dx \wedge dy$$
$$= C(r, s : n) \int_{-\infty}^{\infty} \int_{-\infty}^{y} xy [F(x)]^{r-1} [F(y) - F(x)]^{s-r-1}$$
$$\times [1 - F(y)]^{n-s} f(x) f(y) dx \wedge dy$$

where

$$C(r, s : n) = \frac{n!}{(r-1)!\,(s-r-1)!\,(n-s)!}.$$

Then the covariance between $x_{r:n}$ and $x_{s:n}$ is given by

$$\text{Cov}(x_{r:n}, x_{s:n}) = \mu_{r,s:n} - \mu_{r:n} \mu_{s:n}.$$

Example 8.1.5. Let $x_{1:n}, x_{2:n}, \ldots, x_{n:n}$ be the order statistics of a random sample of size n drawn from the uniform distribution over $[0,1]$. Then obtain $\mu_{r:n}^{(k)}$, $\mathrm{Var}(x_{r:n})$, $\mathrm{Cov}(x_{r:n}, x_{s:n})$ where $1 \leq r < s \leq n$.

Solution 8.1.4:

$$\mu_{r:n}^{(k)} = \frac{n!}{(r-1)!\,(n-r)!} \int_0^1 x^k x^{r-1}(1-x)^{n-r}dx$$

$$= \frac{n!}{(r-1)!\,(n-r)!}\, \frac{\Gamma(k+r)\Gamma(n-r+1)}{\Gamma(n+k+1)}$$

$$= \frac{n!\,(k+r-1)!}{(r-1)!\,(n+k)!}.$$

When we put $k = 1$ and $k = 2$ in the above expression we get

$$\mu_{r:n} = \frac{r}{n+1}$$

$$\mu_{r:n}^{(2)} = \frac{(r+1)r}{(n+1)(n+2)}.$$

Therefore

$$\mathrm{Var}(x_{r:n}) = \mu_{r:n}^{(2)} - [\mu_{r:n}]^2$$

$$= \frac{r(r+1)}{(n+1)(n+2)} - \frac{r^2}{(n+1)^2}$$

$$= \frac{r(n-r+1)}{(n+1)^2(n+2)}.$$

$$\mu_{r,s:n} = \frac{n!}{(r-1)!\,(s-r-1)!\,(n-s)!}$$

$$\times \int_0^1 \int_0^y xy x^{r-1}(y-x)^{s-r-1}(1-y)^{n-s}dx \wedge dy$$

$$= \frac{n!}{(r-1)!\,(s-r-1)!\,(n-s)!}$$

$$\times \int_0^1 (1-y)^{n-s}y^{s+1}dy \int_0^1 t^r(1-t)^{s-r-1}dt,$$

on putting $x = yt$.

$$= \frac{n!}{(r-1)!\,(s-r-1)!\,(n-s)!}\, \frac{\Gamma(n-s+1)\Gamma(s+2)}{\Gamma(n+3)}$$

$$\times \frac{\Gamma(r+1)\Gamma(s-r)}{\Gamma(s+1)}$$

$$= \frac{r(s+1)}{(n+1)(n+2)}.$$

Hence,

$$\begin{aligned}
\text{Cov}(x_{r:n}, x_{s:n}) &= \mu_{r,s:n} - \mu_{r:n}\mu_{s:n} \\
&= \frac{r(s+1)}{(n+1)(n+2)} - \frac{rs}{(n+1)^2} \\
&= \frac{r(n-s+1)}{(n+1)^2(n+2)}.
\end{aligned}$$

8.1.5 Recurrence relations for moments

The following is a basic recurrence relation which is very helpful in the evaluation of the moments of order statistics.

Theorem 8.1.1. *For an arbitrary distribution*

$$r\mu_{r+1:n}^{(k)} + (n-r)\mu_{r:n}^{(k)} = n\,\mu_{r:n-1}^{(k)} \tag{8.1.10}$$

for $n \geq 2$, $1 \leq r \leq n-1$ and $k = 1, 2, \ldots$

Proof 8.1.1: Suppose the parent distribution admits the moment of order k. We have

$$\begin{aligned}
\mu_{r:n-1}^{(k)} &= \frac{(n-1)!}{(r-1)!(n-r-1)!} \int_{-\infty}^{\infty} x^k [F(x)]^{r-1} [1-F(x)]^{n-r-1} f(x)\,dx \\
&= \frac{(n-1)!}{(r-1)!(n-r-1)!} \\
&\quad \times \int_{-\infty}^{\infty} x^k [F(x) + (1-F(x))] [F(x)]^{r-1} [1-F(x)]^{n-r-1} f(x)\,dx \\
&= \frac{(n-1)!}{(r-1)!(n-r-1)!} \left\{ \int_{-\infty}^{\infty} x^k [F(x)]^{r} [1-F(x)]^{n-r-1} f(x)\,dx \right. \\
&\quad \left. + \int_{-\infty}^{\infty} x^k [F(x)]^{r-1} [1-F(x)]^{n-r} f(x)\,dx \right\} \\
&= \frac{r}{n} \int_{-\infty}^{\infty} x^k f_{r+1:n}(x)\,dx + \frac{n-r}{n} \int_{-\infty}^{\infty} x^k f_{r:n}(x)\,dx.
\end{aligned}$$

That is,

$$\mu_{r:n-1}^{(k)} = \frac{r}{n}\mu_{r+1:n}^{(k)} + \frac{(n-r)}{n}\mu_{r:n}^{(k)}.$$

From the above equation we get the required result.

Note 8.1.2: In several applications of order statistics, the moments of order statistics are to be evaluated. The recurrence relation (8.1.10) is very helpful in the evaluation of these moments, when explicit expressions for the moments of order statistics fail to exist. It is to be noted that by using (8.1.10) one can evaluate $\mu_{r:n}^{(k)}, r = 1, 2, \ldots, n$ provided one integral among all integrals for the above moments is evaluated directly and moments of order statistics of lower sample sizes are also known.

8.1.6 Recurrence relations on the product moments

The following theorem provides a basic recurrence relation which has been established by Govindarajulu and is very helpful in the evaluation of the product moments of order statistics.

Theorem 8.1.2. *For an arbitrary distribution with finite mean value,*

$$(r-1)\mu_{r,s:n} + (s-r)\mu_{r-1,s:n} + (n-s+1)\mu_{r-1,s-1:n}$$
$$= n\,\mu_{r-1,s-1:n-1} \qquad (8.1.11)$$

for $n \geq 3$ and $2 \leq r < s \leq n$.

Proof 8.1.2:

$$\mu_{r-1,s-1:n-1} = \frac{(n-1)!}{(r-2)!(s-r-1)!(n-s)!}$$
$$\times \int_{-\infty}^{\infty} \int_{-\infty}^{y} xy[F(x)]^{r-2}[F(x) - F(y)]^{s-r-1}[1 - F(y)]^{n-s}$$
$$\times f(x)f(y)dx \wedge dy.$$

Introducing $1 = F(x) + (F(y) - F(x)) + (1 - F(y))$ within the above integral and expanding it into three integrals we obtain the following:

$$\mu_{r-1,s-1:n} = \frac{(n-1)!}{(r-2)!(s-r-1)!(n-s)!}$$
$$\times \left\{ \int_{-\infty}^{\infty} \int_{-\infty}^{y} xy[F(x)]^{r-1}[F(x) - F(y)]^{s-r-1}[1 - F(y)]^{n-s} \right.$$
$$\times f(x)f(y)dx \wedge dy$$
$$+ \int_{-\infty}^{\infty} \int_{-\infty}^{y} xy[F(x)]^{r-2}[F(x) - F(y)]^{s-r}[1 - F(y)]^{n-s}$$
$$\times f(x)f(y)dx \wedge dy$$
$$+ \int_{-\infty}^{\infty} \int_{-\infty}^{y} xy[F(x)]^{r-2}[F(x) - F(y)]^{s-r}[1 - F(y)]^{n-s+1}$$
$$\left. \times f(x)f(y)dx \wedge dy \right\}.$$

That is,

$$\mu_{r-1,s-1:n} = \frac{r-1}{n}\mu_{r,s:n} + \frac{s-r}{n}\mu_{r-1,s:n} + \frac{n-s+1}{n}\mu_{r-1,s-1:n}.$$

Now the required result (8.1.11) is obtained from the above relation.

Note 8.1.3: The above recurrence relation helps us to compute all product moments $\mu_{r,s:n}$, $1 \leq r < s \leq n$ if we know $n-1$ suitably chosen moments (for example, the knowledge of the $n-1$ immediate upper diagonal product moments $\mu_{r,r+1:n}$, $1 \leq r \leq n-1$ is enough for this purpose) provided the product moments of order statistics of lower sample size are also known.

The above recurrence relation can be further modified and for such results see, Arnold, Balakrishnan and Nagaraja (1998).

8.1.7 Order statistics from symmetric distributions

Suppose a random variable x has the distribution which is continuous and symmetric about 0. Then if $F(x)$ is the distribution function and $f(x)$ the pdf of x, then $f(x) = f(-x)$ and $F(-x) = 1 - F(x)$. Further in this case x and $-x$ are identically distributed and if $x_{1:n}, x_{2:n}, \ldots, x_{n:n}$ are the order statistics of a random sample of size n from this distribution then $(x_{1:n}, x_{2:n}, \ldots, x_{n-1:n}, x_{n:n})$ is distributed identically as $(-x_{n:n}, -x_{n-1:n}, \ldots, -x_{2:n}, -x_{1:n})$ as we have $(x_{1:n} \leq x_{2:n} \leq \ldots \leq x_{n-1:n} \leq x_{n:n})$ iff $(-x_{n:n} \leq -x_{n-1:n} \leq \ldots \leq -x_{2:n} \leq -x_{1:n})$. In particular, we have

$$x_{r:n} \overset{d}{=} x_{n-r+1:n}$$

(that is, $x_{r:n}$ distributed identically as $-x_{n-r+1:n}$) and

$$(x_{r:n}, x_{s:n}) \overset{d}{=} (-x_{n-s+1:n}, -x_{n-r+1:n}).$$

As a result of the above properties we have

$$\mu_{r:n}^{(k)} = (-1)^k \mu_{n-r+1:n}^{(k)}, \qquad 1 \leq r \leq n;$$
$$\mu_{r,s:n} = \mu_{n-s+1,n-r+1:n}, \qquad 1 \leq r < s \leq n.$$

8.2 Discrete Order Statistics

If the distribution of a random variable is discrete then the order statistics of a random sample of size n arising from such a distribution are known as the discrete order statistics of a random sample. In general, if $x_1, x_2, \ldots, x_n$ are discrete random variables then the ordered random variables $x_{1:n}, x_{2:n}, \ldots, x_{n:n}$ are known as discrete order statistics. If the distribution function of the parent distribution is $F(x)$ and $x_{r:n}$ is the r-th order statistic of a sample of size n arising from $F(x)$ then the distribution function $F_{r:n}(x)$ of $x_{r:n}$ is

$$F_{r:n}(x) = \sum_{i=r}^{n} \binom{n}{i} [F(x)]^i [1 - F(x)]^{n-i}, \; r = 1, 2, \ldots, n \text{ and } x \in R.$$

Example 8.2.1. If the parent random variable is discrete uniform with support $S = \{1, 2, \ldots, N\}$ with probability mass function given by $f(x) = 1/N, x = 1, 2, \ldots, N$ then

$$F(x) = \begin{cases} 0, & x < 1 \\ \frac{x}{N}, & x \in S \\ \frac{[x]}{N}, & x \in R, \text{and } x \notin S, 1 < x < N \\ 1, & x > N \end{cases}$$

where $[x]$ denotes the greatest integer contained in x. Then for $x \in S$, we have

$$F_{r:n}(x) = \sum_{i=r}^{n} \binom{n}{r} \left(\frac{x}{N}\right)^i \left(1 - \frac{x}{N}\right)^{n-i}.$$

Similarly for other possible values of x also we can obtain $F_{r:n}(x)$.

8.2.1 Probability function of discrete order statistics

First method: Let $F_{r:n}(x)$ be the distribution function of $x_{r:n}$, the r-th order statistic of a random sample of size n drawn from a discrete distribution. Then the probability mass function $f_{r:n}(x)$ of $x_{r:n}$ is given by

$$P(x_{r:n} = x) = f_{r:n}(x) = F_{r:n}(x) - F_{r:n}(x-). \tag{8.2.1}$$

Also from the relation between binomial sums and incomplete beta function (8.2.1) can be also written as

$$f_{r:n}(x) = \frac{\Gamma(n+1)}{\Gamma(r)\Gamma(n-r+1)} \int_{F(x-)}^{F(x)} t^{r-1}(1-t)^{n-r}dt. \tag{8.2.2}$$

Second method: For each observation x, we can associate a multinomial trial with outcomes $\{x < u\}, \{x = u\}$ and $\{x > u\}$ with corresponding probabilities $F(u-), f(u)$ and $1 - F(x)$ respectively. Now, the event $\{x_{r:n} = x\}$ can be related in $r(n - r + 1)$ distinct and mutually exclusive ways as $(r - 1 - i)$ observations are less than $x, (n - r - j)$ observations exceed x and the rest equal to x, where $i = 0, 1, \ldots, r-1$ and $j = 0, 1, \ldots, n-r$. Thus we have

$$f_{r:n} = \sum_{i=0}^{r-1} \sum_{j=0}^{n-r} \frac{n![F(x-)]^{r-1-i}[1 - F(x)]^{n-r-j}[f(x)]^{i+j+1}}{(r-1-i)!(n-r-j)!(i+j+1)!}.$$

8.2.2 Joint probability function of two order statistics

The joint distribution function of $x_{r:n}$ and $x_{s:n}$, denoted by $F_{r,s:n}(x,y)$ for $x \leq y$, is given in (8.1.7). Then the probability mass function $f_{r,s:n}(x,y)$ for $x \leq y$ is given by

$$f_{r,s:n}(x,y) = F_{r,s:n}(x,y) - F_{r,s:n}(x-1,y) - F_{r,s:n}(x,y-1) + F_{r,s:n}(x-1,y-1).$$

8.2.3 Bernoulli order statistics

Let $x_{1:n}, x_{2:n}, \ldots, x_{n:n}$ be the order statistics of a random sample of size n arising from the Bernoulli population with probability mass function (pmf)

$$P(x = i) = \begin{cases} \pi, & \text{if } i = 1 \\ 1 - \pi, & \text{if } i = 0 \\ 0, & \text{otherwise}, 0 < \pi < 1. \end{cases}$$

Then,

$$P(x_{r:n} = 1) = P(\text{at least } n - r + 1 \text{ of the } x_j\text{'s are taking the value 1})$$
$$= \sum_{i=n-r+1}^{n} \binom{n}{i} \pi^i (1 - \pi)^{n-i}$$
$$= \pi_r^* \text{ (say)}.$$

Then $P(x_{r:n} = 0) = 1 - \pi_r^*$. Hence it follows that the r-th order statistic $x_{r:n}$ from a Bernoulli distribution is also a Bernoulli random variable with probability of success π_r^* and hence we have

$$E(x_{r:n}) = \pi_r^*$$
$$\text{Var}(x_{r:n}) = \pi_r^* (1 - \pi_r^*).$$

The joint probability mass function $f_{r,s:n}(x,y)$ for $1 \leq r < s \leq n$ is described below.

$$P(x_{r:n} = 0, x_{s:n} = 0) = P(x_{s:n} = 0) = 1 - \pi_s^*$$
$$P(x_{r:n} = 0, x_{s:n} = 1) = P(\text{at least } n - s + 1 \text{ and at most } n - r$$
$$\text{of the } x_j\text{'s take the value 1})$$
$$= \sum_{i=n-s+1}^{n-r} \binom{n}{i} \pi^i (1 - \pi)^{n-i}$$
$$= \pi_s^* - \pi_r^*.$$
$$P(x_{r:n} = 1, x_{s:n} = 1) = P(x_{r:n} = 1) = \pi_r^*.$$

Trivially we have $P(x_{r:n} = 1, x_{s:n} = 0) = 0$. Then from the above relations we have

$$\mu_{r,s:n} = \pi_r^*$$

and

$$\text{Cov}(x_{r:n}, x_{s:n}) = \pi_r^* - \pi_r^* \pi_s^*$$
$$= \pi_r^*(1 - \pi_s^*).$$

Note 8.2.1: From the above results we may take $x_{r:n}$ as an unbiased estimator of the parametric function $\pi_r^* = \sum_{i=n-r+1}^{n} \binom{n}{i} \pi^i (1-\pi)^{n-i}$ with variance equal to $\pi_r^*(1 - \pi_r^*)$.

8.3 Independent Random Variables

Let $x_1, x_2, \ldots, x_n$ be n random variables with distribution functions $F_1, F_2, \ldots, F_n$ respectively. Let the distribution functions be continuous with F_i having pdf f_i, $i = 1, 2, \ldots, n$. Let $x_{1:n}, x_{2:n}, \ldots, x_{n:n}$ be the order statistics of the given random variables. Now we will obtain the pdf of $x_{r:n}$ and the joint pdf of $x_{r:n}$ and $x_{s:n}$. In the subsequent sections we will show that these densities are expressed as functions of matrix arguments. Now we define the following:

Definition 8.3.1. *Permanent of a square matrix* Let A be a square matrix then we write $|A|^+$ to denote the permanent of the square matrix which is defined by an expansion just like that of the determinant of A with all terms positive. For example

$$\begin{vmatrix} 1 & 2 & 1 \\ 0 & 3 & 1 \\ 1 & 2 & 3 \end{vmatrix}^+ = 1 \begin{vmatrix} 3 & 1 \\ 2 & 3 \end{vmatrix}^+ + 2 \begin{vmatrix} 0 & 1 \\ 1 & 3 \end{vmatrix}^+ + 1 \begin{vmatrix} 0 & 3 \\ 1 & 2 \end{vmatrix}^+$$
$$= 11 + 2 + 3 = 16.$$

8.3.1 Distribution of a single order statistic

Consider the case in which $x_{1:3}, x_{2:3} x_{3:3}$ are the order statistics of the random variables x_1, x_2, x_3 which are independently distributed with x_i having the distribution function F_i and density f_i, $i = 1, 2, 3$. Now consider the event $x_{1:3} \in (x, x + \Delta x]$. Then

$$P(x_{1:3}) = P(x_1 \in (x, x + \Delta x]; x_2, X_3 \in (x + \Delta x, \infty))$$
$$+ P(x_2 \in (x, x + \Delta x]; x_1, x_3 \in (x + \Delta x, \infty))$$
$$+ P(x_3 \in (x, x + \Delta x]; x_1, x_2 \in (x + \Delta x, \infty)).$$

That is,

$$f_{1:3}(x) = \lim_{\Delta x \to 0} \frac{P(x_{1:3} \in (x, x + \Delta x))}{\Delta x}$$
$$= f_1(x)(1 - F_2(x))(1 - F_3(x))$$
$$+ f_2(x)(1 - F_1(x))(1 - F_3(x))$$
$$+ f_3(x)(1 - F_1(x))(1 - F_2(x)), \quad -\infty < x < \infty. \qquad (8.3.1)$$

Then we can easily show that the expression given in (8.3.1) is also given by

$$f_{1:3}(x) = \frac{1}{2!} \begin{vmatrix} f_1(x) & 1 - F_1(x) & 1 - F_1(x) \\ f_2(x) & 1 - F_2(x) & 1 - F_2(x) \\ f_3(x) & 1 - F_3(x) & 1 - F_3(x) \end{vmatrix}^+, \quad -\infty < x < \infty.$$

Similarly one can further prove that

$$f_{2:3}(x) = \frac{1}{1!} \begin{vmatrix} F_1(x) & f_1(x) & 1 - F_1(x) \\ F_2(x) & f_2(x) & 1 - F_2(x) \\ F_3(x) & f_3(x) & 1 - F_3(x) \end{vmatrix}^+, \quad -\infty < x < \infty;$$

$$f_{3:3}(x) = \frac{1}{2!} \begin{vmatrix} F_1(x) & F_1(x) & f_1(x) \\ F_2(x) & F_2(x) & f_2(x) \\ F_3(x) & F_3(x) & f_3(x) \end{vmatrix}^+, \quad -\infty < x < \infty.$$

This method of obtaining the pdf $f_{1:3}$, can be generalized in this manner and thus one has the following pdf $f_{r:n}(x)$ of $x_{r:n}$:

$$f_{r:n}(x) = \frac{1}{(r-1)!(n-r)!}$$
$$\times \begin{vmatrix} F_1(x) & \cdots & F_1(x) & f_1(x) & 1 - F_1(x) & \cdots & 1 - F_1(x) \\ F_2(x) & \cdots & F_2(x) & f_2(x) & 1 - F_2(x) & \cdots & 1 - F_2(x) \\ \vdots & \ddots & \vdots & \vdots & \vdots & \ddots & \vdots \\ F_n(x) & \cdots & F_n(x) & f_n(x) & 1 - F_n(x) & \cdots & 1 - F_n(x) \end{vmatrix}^+ \qquad (8.3.2)$$

where $1 \le r \le n$ and $-\infty < x < \infty$.

Example 8.3.1. Let x_i have an exponential distribution with pdf

$$f_i(x) = \begin{cases} a_i e^{-a_i x}, & x > 0, a_i > 0 \\ 0, & \text{elsewhere}, \end{cases}$$

where $i = 1, 2, 3$ and x_1, x_2, x_3 are independently distributed. Then obtain the pdf of $x_{1:3}$ and hence obtain $E(x_{1:3})$ and $\text{Var}(x_{1:3})$.

Solution 8.3.1: Clearly $F_i(x) = 1 - e^{-a_i x}$, $i = 1, 2, 3$

$$f_{1:3}(x) = \frac{1}{2!} \begin{vmatrix} a_1 e^{-a_1 x} & e^{-a_1 x} & e^{-a_1 x} \\ a_2 e^{-a_2 x} & e^{-a_2 x} & e^{-a_2 x} \\ a_3 e^{-a_3 x} & e^{-a_3 x} & e^{-a_3 x} \end{vmatrix}^+ , x > 0$$

$$= (a_1 + a_2 + a_3) e^{-(a_1 + a_2 + a_3)x}, \; x > 0. \tag{8.3.3}$$

That is, $x_{1:3}$ is exponentially distributed with parameter $a_1 + a_2 + a_3$ and hence $E(x_{1:3})$ and $\text{Var}(x_{1:3})$ are easily available.

8.3.2 Joint distribution of two order statistics

Let x_i have distribution function $F_i(x)$ and density $f_i(x)$, $i = 1, 2, 3$ and be independent. Then consider the event $\{x_{1:3} \in (x, x + \Delta x], x_{2:3} \in (y, y + \Delta y]\}$, for $x < y$. Let $M(x) = P\{x_{1:3} \in (x, x + \Delta x], x_{2:3} \in (y, y + \Delta y]\}$. Then

$$\begin{aligned}
M(x) = {} & P\{x_1 \in (x, x + \Delta x]; x_2 \in (y, y + \Delta y]; x_3 \in (y + \Delta y, \infty)\} \\
& + P\{x_1 \in (x, x + \Delta x]; x_3 \in (y, y + \Delta y]; x_2 \in (y + \Delta y, \infty)\} \\
& + P\{x_2 \in (x, x + \Delta x]; x_1 \in (y, y + \Delta y]; x_3 \in (y + \Delta y, \infty)\} \\
& + P\{x_2 \in (x, x + \Delta x]; x_3 \in (y, y + \Delta y]; x_1 \in (y + \Delta y, \infty)\} \\
& + P\{x_3 \in (x, x + \Delta x]; x_1 \in (y, y + \Delta y]; x_2 \in (y + \Delta y, \infty)\} \\
& + P\{x_3 \in (x, x + \Delta x]; x_2 \in (y, y + \Delta y]; x_1 \in (y + \Delta y, \infty)\}.
\end{aligned}$$

Therefore

$$f_{1,2:3}(x, y) = \lim_{\substack{\Delta x \to 0 \\ \Delta y \to 0}} \frac{P\{x_{1:3} \in (x, x + \Delta x], x_{2:3} \in (y, y + \Delta y)\}}{\Delta x \Delta y}$$

$$\begin{aligned}
= {} & f_1(x)\{f_2(y)(1 - F_3(y)) + f_3(y)(1 - F_2(y))\} \\
& + f_2(x)\{f_1(y)(1 - F_3(y)) + f_3(y)(1 - F_1(y))\} \\
& + f_3(x)\{f_1(y)(1 - F_2(y)) + f_2(y)(1 - F_1(y))\}, \; -\infty < x < y < \infty.
\end{aligned} \tag{8.3.4}$$

One can easily verify that (8.3.4) is also equal to

$$f_{1,2:3}(x) = \frac{1}{1!} \begin{vmatrix} f_1(x) & f_1(y) & 1 - F_1(y) \\ f_2(x) & f_2(y) & 1 - F_2(y) \\ f_3(x) & f_3(y) & 1 - F_3(y) \end{vmatrix}^+ , \; -\infty < x < y < \infty.$$

Similarly we can show that

$$f_{1,3:3}(x) = \frac{1}{1!} \begin{vmatrix} f_1(x) & F_1(y) - F_1(x) & f_1(y) \\ f_2(x) & F_2(y) - F_2(x) & f_2(y) \\ f_3(x) & F_3(y) - F_3(x) & f_3(y) \end{vmatrix}^+ , \; -\infty < x < y < \infty$$

and

$$f_{2,3:3}(x) = \frac{1}{1!} \begin{vmatrix} F_1(x) & f_1(x) & f_1(y) \\ F_2(x) & f_2(x) & f_2(y) \\ F_3(x) & f_3(x) & f_3(y) \end{vmatrix}^+, \quad -\infty < x < y < \infty.$$

In general we can have a straight forward generalization of the above approach to obtain an expression for $f_{r,s:n}(x,y)$.

TEST

on Order Statistics

Time: 1 hour

Three fifth of the paper carries full marks. All questions carry equal marks.

8.1. Define order statistics. If $x_{1:2}$ and $x_{2:2}$ are the order statistics of a random sample of size two drawn from a continuous distribution with probability density function $f(x)$, obtain the pdf of $x_{1:2}$ and $x_{2:2}$.

8.2. Obtain the expression for the distribution function (df) $F_{r:n}(x)$ of the r-th order statistic $x_{r:n}$ of a random sample of size n arising from a distribution with df $F(x)$. Give the connection between $F_{r:n}$ and the incomplete beta function.

8.3. Let $x_{1:n}, x_{2:n}, \ldots, x_{n:n}$ be the order statistics of a random sample of size n arising from the uniform distribution over $(0, 1)$. Obtain the pdf of $x_{r:n}$ and $x_{s:n}$ and the expressions for $E(x_{r:n}^k)$ and $\text{Var}(x_{r:n})$.

8.4. Let $\mu_{r:n}^{(k)}$ be the k-th moment of the order statistic $x_{r:n}$ of a random sample arising from a distribution with pdf $f(x)$. Then show that

$$n\, \mu_{r-1:n-1}^{(k)} = (r-1)\mu_{r:n}^{(k)} + (n-r+1)\mu_{r-1:n}^{(k)}, \quad n \geq 2, r \geq 2.$$

Describe the importance of this relation.

8.5. Let x_i be an observation from $f_i(x)$, $i = 1, 2, 3$ where

$$f_1(x) = ae^{-ax}, x > 0, \quad f_2(x) = be^{-bx}, x > 0, \quad f_3(x) = ce^{-cx}, x > 0$$

where $a > 0, b > 0, c > 0$ Derive the pdf $x_{1:3}$ and obtain its mean and variance.

(Yageen Thomas)

8.4 On Concomitants of Order Statistics

8.4.1 Application of concomitants of order statistics

Since there is no direct extension of order concept to multivariate random variables, the extension of procedure based on order statistics to such situations is inapplicable. But however from a random sample arising from a bivariate distribution, ordering of the values recorded on the first variable x generates a set of random variables associated with the corresponding y variate. These random variables obtained due to the ordering of the x's are known as the concomitants of order statistics. Let (x,y) be a random vector with joint cumulative distribution function $(cdf), F(x,y)$ and joint probability density function $(pdf), f(x,y)$. Let (x_i, y_i), $i = 1, 2, ..., n$ be a random sample drawn from the distribution of (x,y). Let $x_{i:n}$ be the ith order statistic of the x observation, then the y variate associated with the $x_{i:n}$ is called the concomitant of the ith order statistic and is denoted by $y_{[i:n]}$. It may be noted that Bhattacharya (1974) has independently developed the above concept of concomitants of order statistics and he called them as induced order statistics.

Applications of concomitants of order statistics arises in several problems of study. The most important use of concomitants of order statistics arises in selection procedures when $k(< n)$ individuals are chosen on the basis of their x-values. Then the corresponding y-values represent the performance on an associated characteristic. For example, if the top k out of n bulls, as judged by their genetic make up, are selected for breeding, then $y_{[n-k+1:n]}, \cdots, y_{[n:n]}$ might represent the average milk yield of their female offspring. As another example, x might be the score on a screening test and y the score on a latter test. In this example only the top k performers in the screening test are selected for further training and their scores on a second test generates the concomitants of order statistics. These concomitants of order statistics help one to reduce the complexity of identifying the best performers among a group of individuals.

Suppose the parent bivariate distribution is defined with cdf $F(x,y)$ and pdf $f(x,y)$, then the pdf of the rth concomitant $y_{[r:n]}$ for $1 \leq r \leq n$ is given by (see, David and Nagaraja, 2003, p.144),

$$g_{[r:n]}(y) = \int_x f(y|x) f_{r:n}(x) dx, \tag{8.4.1}$$

where $f_{r:n}(x)$ is the pdf of the rth order statistic $x_{r:n}$ of the x variate and $f(y|x)$ is the conditional pdf of y at a given x. The joint pdf of $y_{[r:n]}$ and $y_{[s:n]}$ for $1 \leq r < s \leq n$ is given by (see, David and Nagaraja, 2003, p.144),

$$g_{[r,s:n]}(y_1, y_2) = \int_{-\infty}^{\infty} \int_{-\infty}^{x_2} f(y_1|x_1) f(y_2|x_2) f_{r,s:n}(x_1, x_2) dx_1 dx_2, \tag{8.4.2}$$

where $f_{r,s:n}(x_1, x_2)$ is the joint pdf of $x_{r:n}$ and $x_{s:n}$. From Yang (1977) we get the expressions for, $E(y_{[r:n]})$, $\text{Var}(y_{[r:n]})$, for $1 \leq r \leq n$ and $\text{Cov}(y_{[r:n]}, y_{[s:n]})$ for $1 \leq r < s \leq n$ and are given below.

$$E(y_{[r:n]}) = E[E(y|x_{r:n})], \qquad (8.4.3)$$

$$\text{Var}(y_{[r:n]}) = \text{Var}[E(y|x_{r:n})] + E[\text{Var}(y|x_{r:n})] \qquad (8.4.4)$$

and

$$\text{Cov}(y_{[r:n]}, y_{[s:n]}) = \text{Cov}[E(y|x_{r:n}), E(y|x_{s:n})]. \qquad (8.4.5)$$

There is extensive literature available on the application of concomitants of order statistics such as in biological selection problem (see, Yeo and David,1984), ocean engineering (see, Castillo,1988), development of structural designs (see, Coles and Tawn,1994) and so on. Concomitants of order statistics have been used by several authors in estimating the parameters of bivariate distributions. Harrell and Sen (1979) and Gill et al. (1990) have used concomitants of order statistics to estimate the parameters of a bivariate normal distribution. Spruill and Gastwirth (1982) have considered another interesting use of concomitants in estimating the correlation coefficient between two random variables x and y. Barnett et al. (1976) have considered different estimators for the correlation coefficient of a bivariate normal distribution based on concomitants of order statistics. The distribution theory of concomitants in the bivariate Weibull distribution of Marshall and Olkin is discussed in Begum and Khan (2000a). Begum and Khan (2000b) have also developed the distribution theory of concomitants of order statistics from Gumbel's bivariate logistic distribution. In Section 5, we consider an application of concomitants of order statistics in estimating a parameter of Morgenstern type bivariate uniform distribution.

Let $(x_1,y_1), (x_2,y_2), \cdots$ be a sequence of independent and identically distributed random variables with cdf $F(x,y), (x,y) \in R \times R$. Let $F_x(x)$ and $F_y(y)$ be the marginal $cdfs$ of x and y respectively. Let $\{R_n, n \geq 1\}$ be the sequence of upper record values (see, Arnold et al., 1998, p.8) in the sequence of x's as defined by,

$$R_n = x_{t_n}, \quad n = 1,2,\cdots$$

where $t_1 = 1$ and $t_n = \min\{j : x_j > x_{t_{n-1}}\}$ for $n \geq 2$. Then the y-variate associated with the x-value, which qualified as the nth record will be called the concomitant of the nth record and will be denoted by $R_{[n]}$. Suppose in an experiment, individuals are measured based on an inexpensive test and only those individuals whose measurement breaks the previous records are retained for the measurement based on an expensive test; then the resulting data involves concomitants of record values. For a detailed discussion on the distribution theory of concomitants of record values see, Arnold et al. (1998) and Ahsanullah and Nevzorov (2000).

The pdf of nth ($n \geq 1$) record value is given by,

$$g_{R_n}(x) = \frac{1}{(n-1)!}[-log(1 - F_x(x))]^{n-1} f_x(x) \qquad (8.4.6)$$

and the joint pdf of mth and n^{th} record values for $m < n$ is given by,

$$g_{R_m,R_n}(x_1,x_2) = \frac{[-log(1 - F_x(x_1))]^{m-1}}{(m-1)!} \frac{[-log(1 - F_x(x_2)) + log(1 - F_x(x_1))]^{n-m-1}}{(n-m-1)!}$$

$$\times \frac{f_x(x_1)f_x(x_2)}{1 - F_x(x_1)}. \qquad (8.4.7)$$

Thus the pdf of the concomitant of nth record value is given by

$$f_{R_{[n]}}(y) = \int_{-\infty}^{\infty} f(y|x) g_{R_n}(x) dx,$$

where $g_{R_n}(x)$ is as defined in (8.4.6) and $f(y|x)$ is the conditional pdf of y at a given value of x of the parent bivariate distribution.

The joint pdf of concomitants of mth and n^{th} record values is given by (see, Ahsanullah and Nevzorov, 2000),

$$g_{R_{[m]},R_{[n]}}(y_1,y_2) = \int_{-\infty}^{\infty} \int_{-\infty}^{x_2} f(y_1|x_1) f(y_2|x_2) g_{R_m,R_n}(x_1,x_2) dx_1 dx_2,$$

where $g_{R_m,R_n}(x_1,x_2)$ is defined by (8.4.7). Some properties of concomitants of record values were discussed in Houchens (1984), Ahsanullah and Nevzorov (2000) and Arnold et al. (1998). However, not much work is seen done in the distribution theory and applications of concomitants of records in statistical inference problems. In Subsection 8.4.2, we provide an application of concomitants of record values in estimating some parameters of Morgenstern type bivariate logistic distribution.

8.4.2 Application in estimation

Scaria and Nair (1999) have discussed the distribution theory of concomitants of order statistics arising from Morgenstern family of distributions (MFD) with cdf defined by (see, Kotz et al., 2000, P.52),

$$F(x,y) = F_x(x)F_y(y)\{1+\alpha(1-F_x(x))(1-F_y(y))\}, \quad -1 \le \alpha \le 1. \qquad (8.4.8)$$

An important member of the MFD is Morgenstern type bivariate uniform distribution with pdf given by,

$$F(x,y) = \frac{xy}{\theta_1 \theta_2}\left\{1+\alpha\left(1-\frac{x}{\theta_1}\right)\left(1-\frac{y}{\theta_2}\right)\right\}, \qquad (8.4.9)$$
$$0 < x < \theta_1, \ 0 < y < \theta_2; \ -1 \le \alpha \le 1.$$

Now we derive the Best Linear Unbiased Estimator (BLUE) of the parameter θ_2 involved in (8.4.9) using concomitants of order statistics (see, Chacko and Thomas, 2004).

Let $y_{[r:n]}$, $r = 1,2,\cdots,n$ be the concomitants of order statistics of a random sample of size n drawn from (8.4.9). Then the pdf of $y_{[r:n]}$ and the joint pdf of $y_{[r:n]}$ and $y_{[s:n]}$ are obtained as,

$$g_{[r:n]}(y) = \frac{1}{\theta_2}\left[1+\alpha\frac{n-2r+1}{n+1}\left(1-\frac{2y}{\theta_2}\right)\right], \ 1 \le r \le n. \qquad (8.4.10)$$

and

$$g_{[r,s:n]}(y_1,y_2) = \frac{1}{\theta_2^2}\left[1 + \alpha\frac{n-2r+1}{n+1}\left(1 - \frac{2y_1}{\theta_2}\right)\right.$$

$$+ \alpha\frac{n-2s+1}{n+1}\left(1 - \frac{2y_2}{\theta_2}\right) + \alpha^2\left(\frac{n-2s+1}{n+1} - \frac{2r(n-2s)}{(n+1)(n+2)}\right)$$

$$\left. \times \left(1 - \frac{2y_1}{\theta_2}\right)\left(1 - \frac{2y_2}{\theta_2}\right)\right], \quad 1 \le r < s \le n. \tag{8.4.11}$$

From (8.4.10) and (8.4.11) we get the means, variances and covariances of concomitants of order statistics as follows:

$$E[y_{[r:n]}] = \theta_2\left[\frac{1}{2} - \alpha\frac{n-2r+1}{6(n+1)}\right]$$

$$= \theta_2\xi_{r:n}, \tag{8.4.12}$$

where

$$\xi_{r:n} = \frac{1}{2} - \alpha\frac{n-2r+1}{6(n+1)}.$$

$$\mathrm{Var}[y_{[r:n]}] = \theta_2^2\left[\frac{1}{12} - \frac{\alpha^2(n-2r+1)^2}{36(n+1)^2}\right]$$

$$= \theta_2^2\rho_{r,r:n}, \tag{8.4.13}$$

where

$$\rho_{r,r:n} = \frac{1}{12} - \frac{\alpha^2(n-2r+1)^2}{36(n+1)^2}$$

and

$$\mathrm{Cov}[y_{[r:n]},y_{[s:n]}] = \theta_2^2\frac{\alpha^2}{36}\left[\frac{(n-2s+1)}{(n+1)} - \frac{2r(n-2s)}{(n+2)(n+1)}\right.$$

$$\left[-\frac{(n-2r+1)(n-2s+1)}{(n+1)^2}\right]$$

$$= \theta_2^2\rho_{r,s:n}, \tag{8.4.14}$$

where

$$\rho_{r,s:n} = \frac{\alpha^2}{36}\left[\frac{(n-2s+1)}{(n+1)} - \frac{2r(n-2s)}{(n+2)(n+1)} - \frac{(n-2r+1)(n-2s+1)}{(n+1)^2}\right].$$

Let $\mathbf{y}_{[n]} = [y_{[1:n]}, \cdots, y_{[n:n]}]'$ be the vector of concomitants. Then from (8.4.12) we can write

$$E(\mathbf{y}_{[n]}) = \theta_2 \xi,$$

where

$$\xi = [\xi_{[1:n]}, \cdots, \xi_{[n:n]}]'.$$

Then from (8.4.13) and (8.4.14), the variance-covariance matrix of $\mathbf{y}_{[n]}$ is given by

$$D(\mathbf{y}_{[n]}) = G\theta_2^2,$$

where

$$G = ((\rho_{r,s:n})).$$

If α is known then $(\mathbf{y}_{[n]}, \theta_2\xi, \theta_2^2 G)$ is a generalized Gauss-Markov setup and hence the BLUE ($\hat{\theta}_2$ of θ_2) is given by,

$$\hat{\theta}_2 = (\xi'G^{-1}\xi)^{-1}\xi'G^{-1}\mathbf{y}_{[n]}$$

and the variance of $\hat{\theta}_2$ is given by,

$$\mathrm{Var}(\hat{\theta}_2) = (\xi'G^{-1}\xi)^{-1}\theta_2^2.$$

It is clear that $\hat{\theta}_2$ is a linear function of the concomitants $y_{[r:n]}$ $r = 1, 2, \cdots, n$. Hence we can write $\hat{\theta}_2 = \sum_{r=1}^{n} a_r y_{[r:n]}$, where a_r, $r = 1, 2, \cdots, n$ are constants. It is to be noted that the possible values of α are in the interval $[-1, 1]$. If the estimate $\hat{\theta}_2$ of θ_2 for a given $\alpha = \alpha_0 \in [-1, 1]$ is evaluated, then one need not consider the estimate for θ_2 for $\alpha = -\alpha_0$ as the coefficients of the estimate in this case can be obtained from the coefficients of $\hat{\theta}_2$ for $\alpha = \alpha_0$. This property can be easily observed from the following theorem:

Theorem 8.4.1. *Let $y_{[r:n]}$, $r = 1, 2, \cdots, n$ be the concomitants of order statistics of a random sample $(x_i, y_i), i = 1, 2, \cdots, n$ arising from (8.4.9) for a given $\alpha = \alpha_0 \in [-1, 1]$. Let the BLUE $\hat{\theta}_2(\alpha_0)$ of θ_2 for given α_0 based on the concomitants $Y_{[r:n]}, r = 1, 2, \cdots, n$ be written as $\hat{\theta}_2(\alpha_0) = \sum_{r=1}^{n} a_r Y_{[r:n]}$. Then the BLUE of $\hat{\theta}_2(-\alpha_0)$ of θ_2 when $\alpha = -\alpha_0$ is given by*

$$\hat{\theta}_2(-\alpha_0) = \sum_{r=1}^{n} a_{n-r+1} Y_{[r:n]} \ \text{with } \mathrm{Var}[\hat{\theta}_2(-\alpha_0)] = \mathrm{Var}[\hat{\theta}_2(\alpha_0)].$$

Proof 8.4.1: From (8.4.13) and (8.4.14) for $1 \le r \le n$ we have $\rho_{r,r:n} = \rho_{n-r+1,n-r+1:n}$ and for $1 \le r < s \le n$, we have $\rho_{r,s:n} = \rho_{n-s+1,n-r+1:n}$. Moreover G is symmetric. Therefore we can write for any $\alpha \in [-1, 1]$, $G = JGJ$, where J is an an $n \times n$ matrix given by,

$$J = \begin{bmatrix} 0 & \cdots & 0 & 1 \\ 0 & \cdots & 1 & 0 \\ \vdots & \ddots & & \vdots \\ 1 & \cdots & 0 & 0 \end{bmatrix}$$

Again from (8.4.12) we have for any $\alpha_0 \in [-1, 1]$,

$$\xi_{r:n}(\alpha_0) = \frac{1}{2} - \alpha_0 \frac{n - 2r + 1}{6(n+1)}$$

$$= \xi_{n-r+1:n}(-\alpha_0).$$

Thus

$$\xi(-\alpha_0) = J\xi(-\alpha_0).$$

Therefore, if $\alpha = \alpha_0$ is changed to $\alpha = -\alpha_0$ then the estimate $\hat{\theta}_2(-\alpha_0)$ is given by,

$$\hat{\theta}_2(-\alpha_0) = (\xi'(-\alpha_0)G^{-1}\xi(-\alpha_0))^{-1}\xi'(-\alpha_0)G^{-1}Y_{[n]}$$

$$= (\xi'(\alpha_0)JG^{-1}J\xi(\alpha_0))^{-1}\xi'(\alpha_0)JG^{-1}Y_{[n]}(-\alpha_0).$$

Since $JJ = I$ and $JGJ = G$, we get,

$$\hat{\theta}_2(-\alpha_0) = (\xi'(\alpha_0)G^{-1}\xi(\alpha_0))^{-1}\xi'(\alpha_0)G^{-1}JY_{[n]}$$

$$= \sum_{r=1}^{n} a_r Y_{[n-r+1:n]}.$$

That is the coefficient of $y_{[r:n]}$ in $\hat{\theta}_2$ for $\alpha = \alpha_0$ is the same as the coefficient of $y_{[n-r+1:n]}$ in $\hat{\theta}_2$ for $\alpha = -\alpha_0$. Similarly we get

$$\text{Var}[\hat{\theta}_2(-\alpha_0)] = \text{Var}[\hat{\theta}_2(\alpha_0)].$$

Thus the theorem is proved.

We have evaluated the coefficients a_r of $y_{[r:n]}, 1 \le r \le n$ in $\hat{\theta}_2$ and $\text{Var}(\hat{\theta}_2)$ for $n = 2(1)10$ and $\alpha = 0.25(0.25)0.75$ and are given in Table 8.4.1. In order to obtain the efficiency of our estimate $\hat{\theta}_2$, we introduce a simple unbiased estimate of θ_2 as,

$$\tilde{\theta}_2 = y_{[1:n]} + y_{[n:n]},$$

with variance given by,

$$\text{Var}(\tilde{\theta}_2) = \theta_2^2 \left[\frac{1}{6} + \frac{\alpha^2}{18} + \left(\frac{2n}{(n+1)(n+2)} - \frac{n-1}{n+1} \right) \right].$$

We have obtained the ratio $\frac{\text{Var}(\hat{\theta}_2)}{\text{Var}(\tilde{\theta}_2)}$ as a measure of the efficiency $e_1 = e(\hat{\theta}_2|\tilde{\theta}_2)$ of our estimator $\hat{\theta}_2$ relative to the unbiased estimator $\tilde{\theta}_2$ for $n = 2(1)10$ and $\alpha = 0.25(0.25)0.75$. It can be seen that the efficiency of our estimator $\hat{\theta}_2$ of θ_2 is relatively very high when compared with $\tilde{\theta}_2$. An advantage of the above method of obtaining the BLUE of θ_2 is that with the expressions for $E[y_{[r:n]}]$ and $\text{Cov}[y_{[r:n]}, y_{[s:n]}]$ one can also obtain without any difficulty the BLUE of θ_2 even if a censored sample alone is available.

Table 8.4.1 The coefficients $a_i, i = 1 \cdots$ in the BLUE $\hat{\theta}_2 = i f_1^n \; a_i y_{[in]}, \; v_1 = \mathrm{Var}(\hat{\theta}_2)/\theta_2^2, \; v_2 = \mathrm{Var}(\hat{\hat{\theta}}_2)/\theta_2^2, \; e_1 = v_2/v_1$

n	α	a_1	a_2	a_3	a_4	a_5	a_6	a_7	a_8	a_9	a_{10}	v_1	v_2	e_1
	0.25	0.9713	1.0271									0.1665	0.1667	1.0012
2	0.50	0.9404	1.0533									0.1661	0.1667	1.0036
	0.75	0.9065	1.0791									0.1655	0.1667	1.0073
	0.25	0.6394	0.6631	0.6952								0.1110	0.1660	1.4955
3	0.50	0.6124	0.6523	0.7258								0.1106	0.1639	1.4819
	0.75	0.5850	0.6337	0.7595								0.1098	0.1604	1.4608
	0.25	0.4762	0.4889	0.5055	0.5266							0.0832	0.1655	1.9892
4	0.50	0.4543	0.4720	0.5053	0.5570							0.0828	0.1620	1.9565
	0.75	0.4333	0.4490	0.4989	0.5924							0.0821	0.1563	1.9038
	0.25	0.3794	0.3871	0.3970	0.4093	0.4242						0.0666	0.1652	2.4805
5	0.50	0.3614	0.3703	0.3878	0.4148	0.4530						0.0662	0.1607	2.4275
	0.75	0.3451	0.3492	0.3722	0.4161	0.4881						0.0655	0.1533	2.3405
	0.25	0.3153	0.3204	0.3268	0.3348	0.3442	0.3553					0.0555	0.1649	2.7712
6	0.50	0.3001	0.3050	0.3152	0.3309	0.3529	0.3824					0.0552	0.1597	2.8931
	0.75	0.2872	0.2867	0.2980	0.3214	0.3590	0.4162					0.0545	0.1510	2.7706
	0.25	0.2697	0.2733	0.2778	0.2832	0.2897	0.2972	0.3057				0.0475	0.1647	3.4674
7	0.50	0.2567	0.2595	0.2658	0.2757	0.2894	0.3076	0.3310				0.0473	0.1590	3.3615
	0.75	0.2462	0.2438	0.2494	0.2627	0.2845	0.3168	0.3633				0.0467	0.1493	3.1970
	0.25	0.2356	0.2383	0.2416	0.2455	0.2501	0.2554	0.2615	0.2684			0.0416	0.1646	3.9567
8	0.50	0.2244	0.2259	0.2300	0.2366	0.2457	0.2577	0.2729	0.2920			0.0413	0.1583	3.8329
	0.75	0.2157	0.2124	0.2150	0.2229	0.2364	0.2563	0.2841	0.3226			0.0408	0.1479	3.6250
	0.25	0.2092	0.2112	0.2137	0.2167	0.2201	0.2241	0.2285	0.2336	0.2392		0.0370	0.1645	4.4459
9	0.50	0.1993	0.2002	0.2029	0.2074	0.2137	0.2221	0.2325	0.2455	0.2613		0.0367	0.1578	4.2997
	0.75	0.1919	0.1885	0.1893	0.1941	0.2028	0.2158	0.2337	0.2579	0.2904		0.0362	0.1468	4.0552
	0.25	0.1881	0.1897	0.1917	0.1939	0.1966	0.1996	0.2030	0.2068	0.2110	0.2157	0.0333	0.1644	4.9369
10	0.50	0.1793	0.1797	0.1816	0.1848	0.1893	0.1953	0.2028	0.2120	0.2231	0.2364	0.0330	0.1574	4.7697
	0.75	0.1730	0.1695	0.1694	0.1723	0.1780	0.1868	0.1990	0.2151	0.2363	0.2641	0.0326	0.1458	4.4724

8.4.3 Concomitants of record values and estimation problems

In this section we (see, Chacko and Thomas, 2005) consider the concomitants of record values arising from Morgenstern family of distributions with cdf given in (8.4.1). We further derive the joint pdf of concomitants of mth and n^{th} $(m < n)$ record values arising from MFD. Based on these expressions we also derive the explicit expression for the product moments of concomitants of record values.

An important member of the MFD is the Morgenstern Type bivariate logistic distribution $(MTBLD)$ and its cdf is given by,

$$F_{X,Y}(x,y) = \left[1 + \exp\left\{-\frac{x-\theta_1}{\sigma_1}\right\}\right]^{-1} \left[1 + \exp\left\{-\frac{y-\theta_2}{\sigma_2}\right\}\right]^{-1}$$

$$\times \left[1 + \alpha\left\{1 - \left[1 + \exp\left\{-\frac{x-\theta_1}{\sigma_1}\right\}\right]^{-1}\right\}\right.$$

$$\times \left.\left\{1 - \left[1 + \exp\left\{-\frac{y-\theta_2}{\sigma_2}\right\}\right]^{-1}\right\}\right], \qquad (8.4.15)$$

$$(x,y) \in R^2; \quad (\theta_1, \theta_2) \in R^2; \quad \sigma_1 > 0, \ \sigma_2 > 0, \ -1 < \alpha < 1.$$

Suppose in certain complicated experiments significance is attributed to the values of the secondary measurement made by an accurate expensive test on individuals having record values with respect to the measurement made preliminarily on them by an inexpensive test. Now we derive (see, Chacko and Thomas 2005) the BLUE's of θ_2 and σ_2 involved in the $MTBLD$ defined by (8.4.15) when α is known and also obtain the BLUE of θ_2 when σ_2 and α are known based on concomitants of first n record values.

The joint cdf of the standard $MTBLD$ is obtained by making the transformation $u = \frac{x-\theta_1}{\sigma_1}$ and $v = \frac{y-\theta_2}{\sigma_2}$ in (8.4.15) and is given by,

$$F_{U,V}(u,v) = [1 + \exp(-u)]^{-1} [1 + \exp(-v)]^{-1}$$

$$\times \left\{1 + \alpha \frac{\exp(-u-v)}{[1+\exp(-u)][1+\exp(-v)]}\right\}. \qquad (8.4.16)$$

Let $(u_i, v_i), i = 1, 2, \cdots$ be a sequence of independent observations drawn from (8.4.16). Let $R^*_{[n]}$ be the concomitant of the nth record value $R_{[n]}$ arising from (8.4.16). Then the pdf $f^*_{[n]}(v)$ of $R^*_{[n]}$ and the joint pdf $f^*_{[m,n]}(v_1, v_2)$ of $R^*_{[m]}$ and $R^*_{[n]}$ for $m < n$ are given below,

$$f^*_{[n]}(v) = [1 + \exp(-v)]^{-2} \exp(-v) \left\{1 + \alpha(1 - 2^{1-n})\left[\frac{1-\exp(-v)}{1+\exp(-v)}\right]\right\}, \qquad (8.4.17)$$

and for $m < n$,

$$f^*_{[m,n]}(v_1, v_2) = [1 + \exp(-v_1)]^{-2} [1 + \exp(-v_2)]^{-2} \exp(-v_1 - v_2)$$
$$\times \left[1 + \alpha \{2I_1(m,n) - 1\} \left(\frac{1 - \exp(-v_1)}{1 + \exp(-v_1)} \right) \right.$$
$$+ \alpha \{2I_2(m,n) - 1\} \left(\frac{1 - \exp(-v_2)}{1 + \exp(-v_2)} \right)$$
$$+ \alpha^2 \{4I_3(m,n) - 2I_1(m,n) - 2I_2(m,n) + 1\}$$
$$\left. \times \left(\frac{1 - \exp(-v_1)}{1 + \exp(-v_1)} \right) \left(\frac{1 - \exp(-v_2)}{1 + \exp(-v_2)} \right) \right],$$

where,

$$I_1(m,n) = \frac{1}{(m-1)!(n-m-1)!} \sum_{r=0}^{n-m-1} (-1)^{n-m-r-1} \binom{n-m-1}{r}$$
$$\times \left[\frac{(n-1)!}{n-r-1} - (n-r-2)!r! + (n-r-2)! \sum_{s=0}^{n-r-2} \frac{1}{s!} \frac{(r+s)!}{2^{r+s+1}} \right],$$
$$(8.4.18)$$

$$I_2(m,n) = \frac{(n-1)!}{(m-1)!(n-m-1)!} \left(1 - \frac{1}{2^n} \right)$$
$$\times \sum_{r=0}^{n-m-1} (-1)^{n-m-r-1} \binom{n-m-1}{r} \frac{1}{n-r-1} \qquad (8.4.19)$$

and

$$I_3(m,n) = \frac{1}{(m-1)!(n-m-1)!} \sum_{r=0}^{n-m-1} (-1)^{n-m-r-1} \binom{n-m-1}{r} \left[\frac{(n-1)!}{n-r-1} \left(1 - \frac{1}{2^n} \right) \right.$$
$$\left. - (n-r-2) \left(r! \left(1 - \frac{1}{2^{r+1}} \right) - \sum_{s=0}^{n-r-2} \frac{(r+s)!}{s!} \left(\frac{1}{2^{r+s+1}} - \frac{1}{3^{r+s+1}} \right) \right) \right].$$
$$(8.4.20)$$

Thus the means, variances and covariances of concomitants of first n record values (for $n \geq 1$) arising from (8.4.16) are given by,

$$E[R^*_{[n]}] = \alpha(1 - 2^{1-n}) = \mu_n \text{ (say)}, \qquad (8.4.21)$$

$$\text{Var}[R^*_{[n]}] = \frac{\pi^2}{3} - \alpha^2(1 - 2^{1-n})^2 = v_{n,n} \text{ (say)} \qquad (8.4.22)$$

and for $m < n$,

$$\text{Cov}[R^*_{[m]}, R^*_{[n]}] = \alpha^2[\{4I_3(m,n) - 2I_1(m,n) - 2I_2(m,n) + 1\} - (1 - 2^{1-m})(1 - 2^{1-n})]$$
$$= V_{m,n} \text{ (say)}, \qquad (8.4.23)$$

Let (x_i, y_i) $i = 1, 2, \cdots$ be a sequence of independent observations drawn from a population with cdf defined by (8.4.15). If we write $u = \frac{x-\theta_1}{\sigma_1}$ and $v = \frac{y-\theta_2}{\sigma_2}$ then we have $x_i = \theta_1 + \sigma_1 u_i$ and $y_i = \theta_2 + \sigma_2 v_i$ for $i = 1, 2, \cdots$. Then by using (8.4.21), (8.4.22) and (8.4.23) we have for $n \geq 1$,

$$E[R_{[n]}] = \theta_2 + \sigma_2 \mu_n, \tag{8.4.24}$$

$$\text{Var}[R_{[n]}] = \sigma_2^2 v_{n,n} \tag{8.4.25}$$

and for $m < n$,

$$\text{Cov}[R_{[m]}, R_{[n]}] = \sigma_2^2 v_{m,n}. \tag{8.4.26}$$

Clearly from (8.4.20), (8.4.21) and (8.4.22) it follows that $\mu_n, v_{n,n}$ and $v_{m,n}$ are known constants provided α is known. Suppose $\mathbf{R}_{[n]} = (R_{[1]}, R_{[2]}, \cdots, R_{[n]})$ denote the vector of concomitants of first n record values. Then from (8.4.24) to (8.4.26), we can write

$$E[\mathbf{R}_{[n]}] = \theta_2 \mathbf{1} + \sigma_2 \mu, \tag{8.4.27}$$

where 1 is a column vector of n ones and $\mu = (\mu_1, \cdots, \mu_n)'$. Then the variance-covariance matrix of $\mathbf{R}_{[n]}$ is given by,

$$D[\mathbf{R}_{[n]}] = H\sigma_2^2, \tag{8.4.28}$$

where $H = ((v_{i,j}))$. If α involved in μ and H are known, then (8.4.27) and (8.4.28) together define a generalized Gauss-Markov setup and then (proceeding as in David and Nagaraja 2003, p. 185) the BLUE's of θ_2 and σ_2 are given by

$$\hat{\theta}_2 = \frac{\mu' H^{-1}(\mu \mathbf{1}' - \mathbf{1}\mu')H^{-1}}{\Delta} \mathbf{R}_{[n]} \tag{8.4.29}$$

and

$$\hat{\sigma}_2 = \frac{\mathbf{1}' H^{-1}(\mathbf{1}\mu' - \mu \mathbf{1}')H^{-1}}{\Delta} \mathbf{R}_{[n]}, \tag{8.4.30}$$

where

$$\Delta = (\mu' H^{-1}\mu)(\mathbf{1}'H^{-1}\mathbf{1}) - (\mu'H^{-1}\mathbf{1})^2.$$

The variances of the above estimators are given by

$$\text{Var}(\hat{\theta}_2) = \left(\frac{\mu' H^{-1}\mu}{\Delta}\right)\sigma_2^2, \tag{8.4.31}$$

and

$$\text{Var}(\hat{\sigma}_2) = \left(\frac{\mathbf{1}' H^{-1}\mathbf{1}}{\Delta}\right)\sigma_2^2. \tag{8.4.32}$$

Clearly $\hat{\theta}_2$ and $\hat{\sigma}_2$ can be written as $\hat{\theta}_2 = \sum_{i=1}^n b_i R_{[i]}$ and $\hat{\sigma}_2 = \sum_{i=1}^n c_i R_{[i]}$ where b_i and c_i, $i = 1, 2, \cdots, n$ are constants.

We have evaluated, the coefficients b_i and c_i of $R_{[i]}$, $1 \le i \le n$ in $\hat{\theta}_2$ and $\hat{\sigma}_2$; $\text{Var}(\hat{\theta}_2)$ and $\text{Var}(\hat{\sigma}_2)$ for $n = 2(1)10$ and $\alpha = 0.25(0.25)0.75$ and are given in Table 8.4.2 and Table 8.4.3 respectively. In order to compare the efficiencies of our estimators $\hat{\theta}_2$ and $\hat{\sigma}_2$ we introduce two simple unbiased estimators of θ_2 and σ_2 based on the concomitants of the first and n^{th} records as given below,

$$\tilde{\theta}_2 = R_{[1]}$$

and

$$\tilde{\sigma}_2 = \frac{R_{[n]} - R_{[1]}}{\alpha(1 - 2^{1-n})}.$$

Clearly from (8.4.23) it follows that $\tilde{\theta}_2$ is unbiased for θ_2 and $\tilde{\sigma}_2$ is unbiased for σ_2. By using (8.4.24), (8.4.25) and (8.4.26), we get the variances of $\tilde{\theta}_2$ and $\tilde{\sigma}_2$ as,

$$\text{Var}[\tilde{\theta}_2] = \frac{\pi^2}{3}\sigma_2^2$$

and

$$\text{Var}[\tilde{\sigma}_2] = \frac{1}{\alpha^2(1 - 2^{1-n})^2}\left[\left(\frac{2\pi^2}{3}\right) - \alpha^2(1 - 2^{1-n})^2 \right.$$
$$\left. - 2\alpha^2\{4I_3(1,n) - 2I_1(1,n) - 2I_2(1,n) + 1\} \right].$$

We have obtained the variance of $\tilde{\theta}_2$, the relative efficiency $\frac{\text{Var}(\tilde{\theta}_2)}{\text{Var}(\hat{\theta}_2)}$ of $\hat{\theta}_2$ relative to $\tilde{\theta}_2$ for $n = 2(1)10; \alpha = 0.25(0.25)0.75$ and are provided in Table 8.4.2. Again we have obtained the variance of $\tilde{\sigma}_2$, the relative efficiency $\frac{\text{Var}(\tilde{\sigma}_2)}{\text{Var}(\hat{\sigma}_2)}$ of $\hat{\sigma}_2$ relative to $\tilde{\sigma}_2$ for $n = 2(1)10; \alpha = 0.25(0.25)0.75$ are provided in Table 8.4.3.

Remark 8.4.1: We can see that the BLUE $\hat{\theta}_2$ of θ_2 does not depend much on the association parameter α but the BLUE $\hat{\sigma}_2$ of σ_2 depends very much on α and our assumption is that α is known. Therefore in the situation where α is unknown we introduce a rough estimator for α as follows, in order to make our estimators $\hat{\theta}_2$ and $\hat{\sigma}_2$ useful for the α unknown situation.

For $MTBLD$ the correlation coefficient between the two variates is given by $\rho = \frac{3}{\pi^2}\alpha$. If r is the simple correlation coefficient between R_i and $R_{[i]}, i = 1, 2, 3, \cdots$ then a rough moment type estimator for α is obtained by equating r with the population correlation coefficient ρ and is obtained as,

Table 8.4.2 The coefficients $a_i, i = 1, \cdots$ in the BLUE $\hat{t}_2 = {}_{ij} \int_1^m a_i R_{[ij]}$, $v_1 = \mathrm{Var}(\hat{t}_2)/\zeta_2^2$, $v_2 = \mathrm{Var}(\tilde{t}_2)/\zeta_2^2$, $e_1 = v_2/v_1$

| n | α | Coefficients | | | | | | | | | | v_1 | v_2 | e_1 |
		a_1	a_2	a_3	a_4	a_5	a_6	a_7	a_8	a_9	a_{10}			
2	0.25	-8.0000	8.0000									424.1031	425.1031	1.0024
	0.50	-4.0000	4.0000									108.2758	109.2758	1.0092
	0.75	-2.6667	2.6667									49.7892	50.7892	1.0201
3	0.25	-5.7103	1.1309	4.5794								179.3723	185.8606	1.0362
	0.50	-2.8489	0.5468	2.3021								44.0154	45.4929	1.0336
	0.75	-1.8919	0.3423	1.5496								18.9457	19.4989	1.0292
4	0.25	-4.7255	-0.2884	1.9434	3.0705							116.4690	136.3942	1.1711
	0.50	-2.3550	-0.1559	0.9645	1.5504							28.5842	33.2669	1.1638
	0.75	-1.5607	-0.1253	0.6338	1.0521							12.3052	14.1693	1.1515
5	0.25	-4.2101	-0.7821	0.9393	1.8077	2.2451						90.0503	118.7330	1.3185
	0.50	-2.0973	-0.4032	0.4586	0.9058	1.1361						22.1283	28.8977	1.3059
	0.75	-1.3889	-0.2831	0.2919	0.6058	0.7743						9.5466	12.2615	1.2844
6	0.25	-3.9072	-1.0069	0.4477	1.1809	0.5500	1.7354					76.2298	111.1551	1.4582
	0.50	-1.9464	-0.5129	0.2124	0.5876	0.7805	0.8787					18.7641	27.0221	1.4401
	0.75	-1.2888	-0.3530	0.1275	0.3878	0.5269	0.5997					8.1193	11.4419	1.4092
7	0.25	-3.7145	-1.1286	0.1669	0.8196	1.1480	1.3129	1.3956				67.9899	107.6336	1.5831
	0.50	-1.8506	-0.5718	0.0726	0.4051	0.5757	0.6626	0.7065				16.7655	26.1503	1.5598
	0.75	-1.2258	-0.3900	0.0350	0.2639	0.3857	0.4493	0.4819				7.2770	11.0608	1.5200
8	0.25	-3.5841	-1.2031	-0.0111	0.5892	0.8911	1.0427	1.1187	1.1567			62.6239	105.9349	1.6916
	0.50	-1.7861	-0.6076	-0.0156	0.2891	0.4454	0.5248	0.5649	0.5851			15.4681	25.7298	1.6634
	0.75	-1.1835	-0.4121	-0.0228	0.1857	0.2964	0.3541	0.3836	0.3986			6.7335	10.8770	1.6154
9	0.25	-3.4916	-1.2528	-0.1327	0.4312	0.7147	0.8570	0.9283	0.9640	0.9819		58.8975	105.1005	1.7845
	0.50	-1.7404	-0.6314	-0.0757	0.2099	0.3561	0.4304	0.4680	0.4868	0.4963		14.5695	25.5231	1.7518
	0.75	-1.1537	-0.4267	-0.0620	0.1326	0.2357	0.2892	0.3167	0.3306	0.3376		6.3589	10.7866	1.6963
10	0.25	-3.4233	-1.2282	-0.2205	0.3167	0.5869	0.7224	0.7904	0.8244	0.8414	0.8499	56.1794	104.6869	1.8634
	0.50	-1.7067	-0.6483	-0.1190	0.1526	0.2916	0.3622	0.3978	0.4157	0.4247	0.4292	13.9155	25.4707	1.8268
	0.75	-1.1318	-0.4370	-0.0901	0.0944	0.1920	0.2426	0.2685	0.2816	0.2882	0.2916	6.0874	10.7418	1.7646

Table 8.4.3 The coefficients $a_i, i = 1, \cdots, n$ in the BLUE $\hat{\sigma}_2 = \sum_{i=1}^n a_i R_{[i]}$, $v_1 = \text{Var}(\hat{\sigma}_2)/\sigma_2^2$, $v_2 = \text{Var}(\tilde{\sigma}_2)/\sigma_2^2$, $e_1 = v_2/v_1$

n	Δ	a_1	a_2	a_3	a_4	a_5	a_6	a_7	a_8	a_9	a_{10}	v_1	v_2	e_1
2	0.25	1.0000	0.0000									3.2899	3.2899	1.0000
	0.50	1.0000	0.0000									3.2899	3.2899	1.0000
	0.75	1.0000	0.0000									3.2899	3.2899	1.0000
3	0.25	0.9284	0.2148	-0.1432								3.0558	3.2899	1.0766
	0.50	0.9279	0.2163	-0.1442								3.0587	3.2899	1.0756
	0.75	0.9270	0.2190	-0.1460								3.0634	3.2899	1.0739
4	0.25	0.8780	0.2874	-0.0083	-0.1572							2.8911	3.2899	1.1379
	0.50	0.8772	0.2888	-0.0069	-0.1591							2.8961	3.2899	1.1360
	0.75	0.8758	0.2914	-0.0043	-0.1628							2.7044	3.2899	1.1327
5	0.25	0.8451	0.3189	0.0558	-0.0766	-0.1432						2.7836	3.2899	1.1819
	0.50	0.8442	0.3200	0.0579	-0.0766	-0.1454						2.7903	3.2899	1.1790
	0.75	0.8426	0.3219	0.0618	-0.0766	-0.1496						2.8014	3.2899	1.1744
6	0.25	0.8234	0.3350	0.0910	-0.0317	-0.0934	-0.1243					2.7127	3.2899	1.2128
	0.50	0.8225	0.3358	0.0933	-0.0309	-0.0943	-0.1264					2.7207	3.2899	1.2092
	0.75	0.8208	0.3371	0.0975	-0.0292	-0.0958	-0.1304					2.7339	3.2899	1.2034
7	0.25	0.8087	0.3443	0.1125	-0.0041	-0.0626	-0.0920	-0.1067				2.6645	3.2899	1.2347
	0.50	0.8078	0.3448	0.1147	-0.0028	-0.0628	-0.0932	-0.1085				2.6745	3.2899	1.2347
	0.75	0.8062	0.3456	0.1190	-0.0004	-0.0630	-0.0954	-0.1120				2.6885	3.2899	1.2237
8	0.25	0.7984	0.3502	0.1266	0.0142	-0.0422	-0.0706	-0.0847	-0.0918			2.6307	3.2899	1.2506
	0.50	0.7975	0.3505	0.1288	0.0157	-0.0420	-0.0712	-0.0859	-0.0933			2.6405	3.2899	1.2459
	0.75	0.7959	0.3510	0.1330	0.0185	-0.0415	-0.0724	-0.0882	-0.0962			2.6568	3.2899	1.2383
9	0.25	0.7909	0.3543	0.1365	0.0270	-0.0279	-0.0555	-0.0693	-0.0762	-0.0797		2.6061	3.2899	1.2624
	0.50	0.7901	0.3544	0.1386	0.0286	-0.0275	-0.0558	-0.0701	-0.0773	-0.0809		2.6167	3.2899	1.2573
	0.75	0.7886	0.3546	0.1426	0.0316	-0.0265	-0.0564	-0.0717	-0.0794	-0.0833		2.6340	3.2899	1.2490
10	0.25	0.7852	0.3572	0.1437	0.0364	-0.0174	-0.0444	-0.0580	-0.0647	-0.0681	-0.0698	2.5878	3.2899	1.2713
	0.50	0.7845	0.3572	0.1458	0.0380	-0.0168	-0.0446	-0.0586	-0.0656	-0.0691	-0.0709	2.5988	3.2899	1.2659
	0.75	0.7831	0.3572	0.1496	0.0411	-0.0156	-0.0448	-0.0597	-0.0672	-0.0710	-0.0729	2.6170	3.2899	1.2571

$$\hat{\alpha} = \begin{cases} -1, & \text{if } r \le -\frac{3}{\pi^2} \\ 1, & \text{if } r \ge \frac{3}{\pi^2} \\ r\frac{\pi^2}{3}, & \text{otherwise.} \end{cases}$$

Remark 8.4.2: From the tables we can see that the efficiency of the BLUE of θ_2, the location parameter ranges from 1 to 1.25 and the efficiency of the BLUE of σ_2 the scale parameter ranges from 1 to 1.75. It is clear that the efficiency of the BLUE of σ_2 is better than the efficiency of the BLUE of θ_2. However, one should keep in mind that competitors are naive estimators because those are the only available estimators to obtain the relative efficiency of our estimators in this situation.

References

Ahsanullah, M. and Nevzorov, V.B. (2000). Some distributions of induced record values, *Biometrical Journal*, **42**, 1069-1081.

Arnold, B.C., Balakrishnan, N. and Nagaraja, H.N. (1998). *Records*, Wiley, New York.

Balakrishnan, N., Ahsanullah M. and Chan, P. S. (1995). On the logistic record values and associated inference, *Journal of Applied Statistical Science*, **2**, 233-248.

Balasubramanian, K. and Beg, M.I. (1997). Concomitants of order statistics in Morgenstern type bivariate exponential distribution, *Journal of Applied Statistical Science*, **5**, 233-245.

Barnett, Y., Green, P.J. and Robinson, A. (1976). Concomitants and correlation estimates, *Biometrika*, **63**, 323-328.

Begum, A.A. and Khan, A.H. (2000a). Concomitants of order statistics from Marshall and Olkin's bivariate Weibull distribution, *Calcutta Statistical Association Bulletin*, **50**, 65-70.

Begum, A.A. and Khan, A.H. (2000b). Concomitants of order statistics from Gumbel's bivariate logistic distribution, *Journal of the Indian Society for Probability and Statistics*, **5**, 51-64.

Bhattacharya, P.K. (1974). Convergence of sample paths of normalized sums of induced order statistics, *Annals of Statistics*, **2**, 1034-1039.

Castillo, E. (1988). *Extreme Value Theory in Engineering*, Academic Press, New York.

Chacko, M. and Thomas, P.Y. (2004). Estimation of a parameter of Morgenstern type bivariate uniform distribution based on concomitants of order statistics and concomitants of record values, *Journal of the Kerala Statistical Association*, **15**, 13-26.

Chacko, M. and Thomas, P. Y. (2005). Concomitants of record values arising from Morgenstern type bivariate logistic distribution and some of their applications in parameter estimation, *Metrika* (to appear).

Coles, S.G. and Tawn, A. J. (1994). Statistical methods for multivariate extremes: An application to structural design, *Applied Statistics*, **43**, 1-48.

David, H. A.(1981): *Order Statistics*, Second edition, Wiley, New York.

David, H.A. and Nagaraja, H. N. (2003). *Order Statistics*, 3rd edition, John Wiley, New York.

Gill, P.S., Tiku, M.L. and Vaughan, D.C. (1990). Inference problems in life testing under multivariate normality, *Journal of Applied Statistics*, **17**, 133-149.

Harrel, F.E. and Sen., P.K. (1979). Statistical inference for censored bivariate normal distributions based on induced order statistics, *Biometrika*, **66**, 293-298.

Houchens, R. L. (1984). *Record Value Theory and Inference*, Ph. D. Dissertation, University of California, Riverside, California.

Kotz, S., Balakrishnan, N. and Johnson, N. L. (2000). *Continuous Multivariate Distributions*, John Wiley, New York.

Scaria, J. and Nair, N.U. (1999). On concomitants of order statistics from Morgenstern family, *Biometrical Journal*, **41**, 483-489.

Spruill, N.L. and Gastwirth, J. (1982). On the estimation of correlation coefficient from grouped data, *Journal of American Statistical Association*, **77**, 614-620.

Yang, S.S. (1977). General distribution theory of the concomitants of order statistics, *Annals of Statistics*, **5**, 996-1002.

Yeo, W.B. and David, H.A. (1984). Selection through an associated characteristic with application to the random effects model, *Journal of American Statistical Association*, **79**, 399-405.

Chapter 9
Applications to Astrophysics Problems

[This chapter is based on the lectures of Professor Dr. Hans Haubold of the Office of Outer Space Affairs, United Nations, Vienna, Austria.]

9.0 Introduction

Understanding Nature Through Reaction and Diffusion

2005: Albert Einstein 1879-1955: International Year of Physics (IYP), annus mirabilis of Einstein 1905. Two papers on statistical mechanics (Avogadro number and size of molecules, fluctuations), two papers on special relativity (velocity of light, $E = mc^2$), one paper on quantum mechanics (photoelectric effect). All of Einstein's papers start with a reference to experiments and subsequently develop theory that may explain the experiments and allow predictions.

2006: Ludwig Boltzmann 1844-1906: Discovers microphysical basis (statistical mechanics, entropy) of macrophysical theory (thermodynamics, entropy) and explains second law of thermodynamics with laws of statistical mechanics.

2007: International Heliophysical Year (IHY): Can science of IHY contribute to fundamental physics? Prigogine's quest for probabilistic foundation of classical and quantum mechanics? Haken's synergetics based on slaving principle: In general just a few collective modes become unstable and serve as "ordering parameters" which describe the macroscopic pattern. At the same time the macroscopic variables, i.e., the ordering parameters, govern the behavior of the microscopic parts by the "slaving principle". In this way, the occurrence of order parameters and their ability to enslave allows the system to find its own structure.

341

9.1 Entropy: Boltzmann, Planck, and Einstein on W

9.1.1 Entropic functional

- Clausius entropy (second law of thermodynamics)

$$\frac{dS}{dt} \geq 0. \tag{9.1.1}$$

- Boltzmann entropy (Boltzmann's principle)

$$S = k \ln W. \tag{9.1.2}$$

- Boltzmann-Gibbs statistical mechanics and Maxwell-Boltzmann distribution function for gases.
- Planck's law, Boltzmann's entropy, and the black-body radiation law
- Prigogine's strict formulation of second law of thermodynamics.

$$dS = d_{ext}S + d_{int}S; \ d_{ext}S = {}^{\geq}_{\leq} 0; \ d_{int}S \geq 0.$$

- Tsallis entropy

$$S_q = k \frac{W^{1-q} - 1}{1 - q}. \tag{9.1.3}$$

Nonextensive statistical mechanics and Tsallis distribution function.

9.1.2 Entropy and probability

- Boltzmann's first definition (ok for Einstein)

$$S_i, \tau_i, \ \frac{\tau_i}{\tau}, \ \tau \to \infty. \tag{9.1.4}$$

- Boltzmann's second definition (criticized by Einstein)

$$w = \frac{N!}{\Pi_i n_i!} \ \text{fine grained.} \tag{9.1.5}$$

- Boltzmann's complexions (gas, W?)

$$W = N! \Pi_A \frac{\omega_A^{N_A}}{N_A!} \ \text{coarse grained.} \tag{9.1.6}$$

- Planck's complexions (radiation, W?)

$$W = \Pi_s \frac{(N_s + P_s - 1)!}{N_s!(P_s - 1)!}. \tag{9.1.7}$$

- Einstein's definition of statistical probability

$$\frac{W^a}{W^b} = \left(\frac{V}{V_0}\right)^n. \tag{9.1.8}$$

Note 9.1.1: In systems far from thermal equilibrium, Shannon information plays the same role as entropy in systems in thermal equilibrium or close to it, namely as the cause of processes. However, the question remains whether the maximization of information (or entropy) is indeed the fundamental law which drives systems in a unique way. The question is whether evolution and development are governed by extremal principles, especially extremal principles connected with a single function, such as entropy or information (Haken).

9.1.3 Boltzmann-Gibbs

What can be added to thermodynamics by knowing something of the structure of matter? This question lead to the development of kinetic theory and statistical mechanics. In statistical mechanics, properties of matter are deduced by applying statistics to large numbers of molecules. In information theory, the information-carrying capacity of communications systems is deduced by applying statistics to large numbers of messages. Fundamental equation of information theory(S) = equation for entropy in statistical mechanics(S). Individual molecule $\rightarrow$ very large collection of macroscopically identical systems = ensemble. Total internal energy of ensemble (first law of thermodynamics):

$$E = \sum_q n_q E_q = \text{constant.} \tag{9.1.9}$$

Total of n members in the ensemble

$$n = \sum_q n_q = \text{constant.} \tag{9.1.10}$$

Number of different ways an ensemble of n members can be arranged for q states:

$$\omega = \frac{n!}{\Pi_q n_q!}. \tag{9.1.11}$$

Total entropy of the ensemble is some function of ω,

$$S = k \ln \omega \tag{9.1.12}$$

$$\ln \omega = \ln n! - \sum_q \ln n_q!. \tag{9.1.13}$$

Stirling's formula: From the asymptotic formula for gamma functions (see Chapter 1)

$$x! = \Gamma(x+1) \approx \sqrt{2\pi} x^{x+\frac{1}{2}} e^{-x}. \tag{9.1.14}$$

Hence

$$\ln x! \approx \frac{1}{2} \ln(2\pi) + \frac{1}{2} \ln x + x \ln x - x.$$

If we omit $\frac{1}{2}\ln(2\pi) + \frac{1}{2}\ln x$ then for very large x we have

$$\ln x! \approx x\ln x - x. \tag{9.1.15}$$

For very large x we will take it as equal to the right side. Then

$$\ln \omega = n\ln n - n - \sum_q (n_q \ln n_q) + \sum_q n_q \tag{9.1.16}$$

when all n_q's are large.

$$\ln \omega = n\ln n - \sum_q (n_q \ln n_q) = n\left[\ln n - \frac{1}{n}\sum_q (n_q \ln n_q)\right] \tag{9.1.17}$$

since $\sum_q n_q/n = 1$,

$$\ln \omega = n\left[\sum_q \left(\frac{n_q}{n}\ln n\right) - \sum_q \left(\frac{n_q}{n}\ln n_q\right)\right] \tag{9.1.18}$$

$$= -n\sum_q \left[\frac{n_q}{n}(\ln n_q - \ln n)\right] = -n\sum_q \left(\frac{n_q}{n}\ln \frac{n_q}{n}\right). \tag{9.1.19}$$

Estimation of the probability of quantum state q:

$$P_q = \frac{n_q}{n}; \quad \ln \omega = -n\sum_q (P_q \ln P_q) \tag{9.1.20}$$

$$S = -kn\sum_q (P_q \ln P_q). \tag{9.1.21}$$

Two constraints $\rightarrow$ maximum $\rightarrow$ Lagrange method:

$$\sum_q (n_q E_q) = n\sum_q \left(\frac{n_q}{n}E_q\right) = n\sum_q (P_q E_q) = U. \tag{9.1.22}$$

Let us maximize S subject to the conditions $\sum_q P_q = 1$ and $\sum_q P_q E_q = U$ where U is fixed. Let λ_1 and λ_2 be Langrangian multipliers. Let

$$f(P_1, P_2, \cdots) = -nk\sum_q (P_q \ln P_q) + \lambda_1 \left(\sum_q P_q E_q - U\right) + \lambda_2 \left(\sum_q P_q - 1\right).$$

Then for each i, $\frac{\partial f}{\partial P_i} = 0$, $i = 1, 2, \cdots$. That is,

$$\frac{\partial f}{\partial P_i} = 0 \Rightarrow -nk[\ln P_i + 1] + \lambda_1 E_i + \lambda_2 = 0 \Rightarrow$$

$$\ln P_i = A - \beta E_i \Rightarrow P_i = e^A e^{-\beta P_i} \tag{9.1.23}$$

where A and β are some constants. We can determine A by using the fact that

$$\sum_q P_q = e^A \sum_q e^{-\beta E_q} = 1 \tag{9.1.24}$$

$$e^A = \frac{1}{\sum_q e^{-\beta E_q}}. \tag{9.1.25}$$

Distribution of probabilities for the possible quantum states of ensemble = Maxwell-Boltzmann distribution:

$$P_q = \frac{e^{-\beta E_q}}{\sum_q e^{-\beta E_q}} = \frac{e^{-\beta E_q}}{Z}. \tag{9.1.26}$$

Definition 9.1.1. Partition function

$$Z = \sum_q e^{-\beta E_q}.$$

Note 9.1.2: Stirling's approximation, coming from an asymptotic expansion of a gamma function is the following: For $|z| \to \infty$ and α bounded

$$\Gamma(z+\alpha) \approx \sqrt{2\pi} z^{z+\alpha-\frac{1}{2}} e^{-z}. \tag{9.1.27}$$

Stirling's formula given in equation (9.1.14) holds for very large x and the remaining steps hold when each of $n_1, n_2, \cdots$ is very large.

9.2 Gravitationally Stabilized Fusion Reactor: The Sun

The Sun is a spherically symmetric gas sphere in hydrostatic equilibrium. Rotation and magnetic fields can be neglected. The innermost region of the Sun (solar core) is a gravitationally stabilized fusion reactor. Energy is being produced by thermonuclear reactions generating photons (surface source) and neutrinos (volume source). The evolution of the Sun proceeds through the change of chemical abundances (kinetic equations).

9.2.1 Internal solar structure

Solar structure is determined by conditions of mass conservation, momentum conservation, energy conservation, and the mode of energy transport. One can derive succinctly the equations of solar structure and develop a model in hydrostatic equilibrium as a model of the Sun in order to illustrate important physical requirements. Then by arguing physically that the density gradient can be matched to a simple function, one can derive a complete analytic representation of the solar interior in terms of a one-parameter family of models. Two different conditions can be used to select the appropriate value of the parameter specifying the best model within the family: (i) the solar luminosity is equated to the thermonuclear power generated near the center and/or (ii) the solar luminosity is equated to the radiative diffusion of energy from a central region. The central conditions of the Sun are well calculated by these analytic formulas. The model yields a good description of the solar center to be found by methods of differential and integral calculus, rendering it an excellent laboratory for applied calculus and special functions.

In the following we are concerned with the hydrostatic equilibrium of the purely gaseous spherical central region of the Sun generating energy by nuclear reactions at a certain rate. For this gaseous sphere we assume that the matter density varies non-linearly from the center outward, depending on two parameters δ and γ,

$$\rho(x) = \rho_c \, f_D(x), \tag{9.2.1}$$

$$f_D(x) = [1 - x^\delta]^\gamma, \tag{9.2.2}$$

where x denotes the dimensionless distance variable, $x = r/R_\odot, 0 \leq x \leq 1, R_\odot$ is the solar radius, $\delta > 0, \gamma > 0$, and γ is kept a positive integer in the following considerations. The choice of the density distribution in (9.2.1) and (9.2.2) reveals immediately that $\rho(x = 0) = \rho_c$ is the central density of the configuration and $\rho(x = 1) = 0$ is a boundary condition for hydrostatic equilibrium of the gaseous configuration. For the range $0 \leq x \leq 0.3$ the density distribution in (9.2.1) and (9.2.2) can be fit numerically to computed data for solar models by choosing $\delta = 1.28$ and $\gamma = 10$. The choice of restricting x to $x \leq 0.3$ is justified by looking at a Standard Solar Model which shows that $x \leq 0.3$ comprises what is considered to be the gravitationally stabilized solar fusion reactor. More precisely, 95% of the solar luminosity is produced within the region $x < 0.2$ ($M < 0.3 M_\odot$). The half-peak value for the matter density occurs at $x = 0.1$ and the half-peak value for the temperature occurs at $x = 0.25$. The region $x \leq 0.3$ is also the place where the solar neutrino fluxes are generated. As we are concerned with a spherically symmetrical distribution of matter, the mass $M(x)$ within the radius x having the density distribution given in (9.2.1) and (9.2.2) is the following:

$$\frac{\mathrm{d}M(x)}{\mathrm{d}x} = R_\odot^3 4\pi x^2 \rho(x) \tag{9.2.3}$$

which means

$$M(x) = R_\odot^3 4\pi \rho_c \int_0^x t^2 \left[1 - t^\delta\right]^\gamma \mathrm{d}t. \tag{9.2.4}$$

Put $u = t^\delta$ and for positive integer γ expand the binomial part to obtain

$$M(x) = R_\odot^3 \frac{4\pi \rho_c}{\delta} \sum_{m=0}^\gamma \frac{(-\gamma)_m}{m!} \int_0^{x^\delta} u^{\frac{3}{\delta} - 1 + m} \mathrm{d}u.$$

Writing $\frac{1}{(\frac{3}{\delta}+m)} = \frac{1}{(3/\delta)} \frac{(\frac{3}{\delta})_m}{(\frac{3}{\delta}+1)_m}$ we have

$$M(x) = \frac{4}{3}\pi R_\odot^3 \rho_c x^3 {}_2F_1\left(-\gamma, \frac{3}{\delta}; \frac{3}{\delta}+1; x^\delta\right)$$

$$= M_\odot \rho_c x^3 {}_2F_1\left(-\gamma, \frac{3}{\delta}; \frac{3}{\delta}+1; x^\delta\right) = M_\odot f_M(x), \qquad (9.2.5)$$

where $f_M(x)$ will be given below, $M_\odot$ denotes the solar mass and ${}_2F_1(.)$ is Gauss' hypergeometric function, $M_\odot = \frac{4}{3}\pi R_\odot^3$. Equations (9.2.5) is satisfying the boundary condition $M(x=0) = 0$ and determines the central value ρ_c of the matter density through the boundary condition $M(x=1) = M_\odot$, where ρ_c depends then only on δ and γ of the chosen density distribution in (9.2.1) and (9.2.2). Then

$$\rho_c = \frac{1}{{}_2F_1(-\gamma, \frac{3}{\delta}; \frac{3}{\delta}+1; 1)} = \frac{\Gamma(\frac{3}{\delta}+1+\gamma)\Gamma(1)}{\Gamma(\frac{3}{\delta}+1)\Gamma(\gamma+1)} \qquad (9.2.6)$$

by evaluating the ${}_2F_1$ at $x=1$, see Chapter 1,

$$\rho_c = \frac{(\frac{3}{\delta}+1)(\frac{3}{\delta}+2)...(\frac{3}{\delta}+\gamma)}{\gamma!}. \qquad (9.2.7)$$

Therefore

$$f_M(x) = \left[\frac{1}{\gamma!}\prod_{i=1}^{\gamma}\left(\frac{3}{\delta}+i\right)\right] x^3 {}_2F_1\left(-\gamma, \frac{3}{\delta}; \frac{3}{\delta}+1; x^\delta\right). \qquad (9.2.8)$$

For computing pressure and temperature use the following equations and then follow through the above procedure to obtain the results given below.

$$\frac{dP(r)}{dr} = -G\frac{M(r)\rho(r)}{r^2} \qquad (9.2.9)$$

$$P(r) = P(0) - \int_0^r \frac{G\,M(t)\rho(t)}{t^2}\,dt \qquad (9.2.10)$$

and

$$T(r) = \frac{\mu}{kN_A}\frac{P(r)}{\rho(r)} \qquad (9.2.11)$$

at an arbitrary distance r from the center. Converting to $x = \frac{r}{R_\odot}$ we have the expressions for $P(x)$ and $T(x)$ under the $\rho(x)$ in (9.2.2).

For hydrostatic equilibrium of the gaseous configuration the internal pressure needs to balance the gravitational attraction. The pressure distribution follows by integration of the respective differential equation for hydrostatic equilibrium, making use of the density distribution in (9.2.1) and the mass distribution in (9.2.5), that is

$$P(x) = \frac{9}{4\pi}G\frac{M_\odot^2}{R_\odot^4}f_P(x), \tag{9.2.12}$$

where

$$f_P(x) = \left[\frac{1}{\gamma!}\prod_{i=1}^{\gamma}\left(\frac{3}{\delta}+i\right)\right]^2 \frac{1}{\delta^2}\sum_{m=0}^{\gamma}\frac{(-\gamma)_m}{m!(\frac{3}{\delta}+m)(\frac{2}{\delta}+m)}$$

$$\times\left[\frac{\gamma!}{(\frac{2}{\delta}+m+1)_\gamma} - x^{\delta m+2}{}_2F_1\left(-\gamma,\frac{2}{\delta}+m;\frac{2}{\delta}+m+1;x^\delta\right)\right], \tag{9.2.13}$$

where G is Newton's constant and ${}_2F_1(.)$ denotes again Gauss' hypergeometric function.

The Pochhammer symbol $(\frac{2}{\delta}+m+1)_\gamma = \Gamma(\frac{2}{\delta}+m+1+\gamma)/\Gamma(\frac{2}{\delta}+m+1)$, $\gamma = 1,2$ often appears in series expansions for hypergeometric functions. Equations (9.2.12) and (9.2.13) give the value of the pressure P_c at the centre of the gaseous configuration and satisfy the condition $P(x = 1) = 0$.

It should be noted that $P(x)$ in (9.2.12) denotes the total pressure of the gaseous configuration, that is the sum of the gas kinetic pressure and the radiation pressure (according to Stefan-Boltzmann's law). However, the radiation pressure, although the ratio of radiation pressure to gas pressure increases towards the center of the Sun, remains negligibly small in comparison to the gas kinetic pressure. Thus, (9.2.12) can be considered to represent the run of the gas pressure through the configuration under consideration. Further, the matter density is so low that at the temperatures involved the material follows the equation of state of the perfect gas. Therefore, the temperature distribution throughout the gaseous configuration is given by

$$T(x) = \frac{\mu}{k\,N_A}\frac{P(x)}{\rho(x)} \tag{9.2.14}$$

$$= 3\frac{\mu}{kN_A}G\frac{M_\odot}{R_\odot}f_T(x), \tag{9.2.15}$$

where

$$f_T(x) = \left[\frac{1}{\gamma!}\prod_{i=1}^{\gamma}\left(\frac{3}{\delta}+i\right)\right]\frac{1}{\delta^2}\frac{1}{[1-x^\delta]^\gamma}\sum_{m=0}^{\gamma}\frac{(-\gamma)_m}{m!(\frac{3}{\delta}+m)(\frac{2}{\delta}+m)}$$

$$\times\left[\frac{\gamma!}{(\frac{2}{\delta}+m+1)_\gamma} - x^{\delta m+2}{}_2F_1\left(-\gamma,\frac{2}{\delta}+m;\frac{2}{\delta}+m+1;x^\delta\right)\right], \tag{9.2.16}$$

where k is the Boltzmann constant, N_A Avogadro's number, μ the mean molecular weight, and ${}_2F_1(.)$ Gauss' hypergeometric function. Equations (9.2.15) and (9.2.16) reveal the central temperature $T(x = 0) = T_c$ and satisfy the boundary condition

$T(x = 1) = 0$. Since the gas in the central region of the Sun can be treated as completely ionized, the mean molecular weight μ is given by $\mu = (2X + \frac{3}{4}Y + \frac{1}{2}Z)^{-1}$, where X, Y, Z are relative abundances by mass of hydrogen, helium, and heavy elements, respectively, and $X + Y + Z = 1$.

9.2.2 Solar fusion plasma

The solar fusion plasma is a weakly non-ideal gas, characterized by the plasma parameter

$$\Gamma = \frac{(Ze)^2}{akT} = \frac{\text{mean Coulomb potential energy}}{\text{mean kinetic (thermal) energy}}; \qquad (9.2.17)$$

$$a = n^{-1/3}$$

is order of average interparticle distance; n is average density.

$\Gamma \ll 1$ ideal plasma
$\Gamma < 1$ weakly non-ideal plasma
$\Gamma > 1$ high density / low temperature plasma.

9.2.3 Estimation of central temperature in the Sun

The basic condition for thermonuclear reactions between charged particles is that their thermal energy must be large enough to penetrate the Coulomb repulsion between them. Nuclear reactions are collision phenomena characterized by cross sections. The cross section σ of a reaction is defined as the probability that the reaction will occur if the incident flux consists of one particle and the target contains only one nucleus per unit area. The microscopic nature of the particles requires the quantum mechanical treatment of the collision problem. The number of reactions is directly proportional to the number density of the incident flux and the number density of the target. In the case of the nuclear fusion plasma within the Sun, thermal equilibrium is commonly assumed for the ensemble of nuclei. The distribution of the relative velocities among the nuclei is assumed to be Maxwell-Boltzmannian.

The thermonuclear raction rate is given by

$$r_{12} = n_1 n_2 < \sigma v >_{12}, \qquad (9.2.18)$$

where n_1 and n_2 denote the number densities of particles of type 1 and 2, respectively, and $< \sigma v >_{12}$ is the reaction probability in the unit volume per unit time. This definition of the reaction rate reveals immediately that the quantity

$$\tau_{12} = [n_2 < \sigma v >_{12}]^{-1}, \qquad (9.2.19)$$

has the dimension of time and can be considered to be the lifetime of particle 2 against reaction with particle 1. A suitable representation of the nuclear cross section contains two factors: A geometrical factor to which quantum mechanical interaction between two particles is always proportional, $\lambda^2 \sim (\mu v^2)^{-1}$ (where λ is the reduced de Broglie wave length, and μ is the reduced mass) and the probability for two particles of charge $Z_1 e$ and $Z_2 e$ to penetrate their electrostatic repulsion:

$$\sigma(v) = \frac{2S}{\mu v^2} \exp\left\{ -2\pi \frac{Z_1 Z_2 e^2}{\hbar v} \right\}. \tag{9.2.20}$$

The constant S is called astrophysical cross section factor and absorbs the intrinsical nuclear parts of the probability for the occurrence of a nuclear reaction. Then, the reaction probability is defined as the product of the cross section σ and the relative velocity v, averaged over the Maxwell-Boltzmann distribution of relative velocities of the reacting particles,

$$f(v)dv = \left(\frac{\mu}{2kT}\right)^{3/2} \exp\left\{ -\frac{\mu v^2}{2kT} \right\} 4\pi v^2 dv. \tag{9.2.21}$$

To investigate the competition between the exponential factors contained in the Maxwell-Boltzmann distribution function and the Gamov penetration factor the following order of magnitude estimation is pursued. For the number density of the particle gas we use the mean density of the Sun with mass $M_\odot$ and radius $R_\odot$ normalized to the mass of the proton, m_p,

$$n_2 = \frac{M_\odot}{R_\odot^3} \frac{1}{m_p}. \tag{9.2.22}$$

The velocity of the nuclei is assumed to be the root-mean-square velocity of the Maxwell-Boltzmann distribution,

$$v_{12} = \left(\frac{4kT}{m_p}\right)^{1/2}. \tag{9.2.23}$$

The nuclear energy generated in the Sun, which is lost by radiation, can be estimated in writing

$$E_{nuc} \approx X\Delta m M_\odot c^2, \tag{9.2.24}$$

where X is the fraction of mass the Sun can use for nuclear energy generation, $\Delta m M_\odot c^2$ is the fraction of mass of the Sun really converted into radiation energy. Thus, the nuclear lifetime of the Sun is of the order

$$\tau^{-1} \approx \frac{L_\odot}{E_{nuc}} \approx \frac{L_\odot}{X\Delta m M_\odot c^2}. \tag{9.2.25}$$

For the lifetime of particle 2 one has

$$\frac{1}{\tau_{12}} \approx \frac{L_\odot}{E_{nuc}} \approx n_2 \sigma_{12} v_{12}. \tag{9.2.26}$$

Thus,

$$\frac{L_\odot}{X\Delta m M_\odot c^2} \approx \frac{M_\odot}{R_\odot^3 m_p} \frac{2S}{m_p^{1/2}(kT)^{1/2}} \exp\left\{-\frac{2\pi e^2}{\hbar}\left(\frac{m_p}{4kT}\right)^{1/2}\right\}, \qquad (9.2.27)$$

and isolating the exponential term in this expression by setting it equal to unity and then taking the logarithm, one gets

$$2\pi\alpha_{el}\left(\frac{m_p c^2}{4kT}\right)^{1/2} \approx \ln\left\{\frac{2M_\odot^2 X\Delta m S c^2}{L_\odot R_\odot^3 m_p^{3/2}(kT)^{1/2}}\right\}. \qquad (9.2.28)$$

The numerical value of the logarithmic term on the right-hand-side in this equation is relatively insensitive to the values inserted for the various quantities in the brackets. Using solar values for the quantities, $M_\odot \approx 2 \times 10^{33} g, L_\odot \approx 4 \times 10^{33} ergs^{-1}, R_\odot \approx 7 \times 10^{10} cm, T_{c\odot} \approx 10^7 K, X = 0.1, \Delta m = 0.007, S_{pp} = 4 \times 10^{-22}$ keV barn, one obtains for the logarithmic term a numerical value of about 10. Then one obtains

$$kT = \left(\frac{(2\pi\alpha_{el})^2}{2^2 10^2}\right) m_p c^2 \approx 5 keV. \qquad (9.2.29)$$

This is the central temperature of the stationarily thermonuclear burning Sun. Actual central temperatures are about a factor 5 smaller or larger than this value due to the fact that the majority of nuclear reactions occur in the high-energy tail of the Maxwell-Boltzmann distribution function. *The Sun has to adjust this temperature through the competition between the distribution function of relative energies of the particles and the penetration factor of the reacting particles.*

Emden (polytropic gas spheres), Chandrasekhar (hydrostatic equilibrium), Bethe (nuclear energy generation), Fowler (thermonuclear reaction rates), Davis (solar neutrino detection).

9.3 Crucial Astrophysical Experiments: Data Analysis

9.3.1 The experiments

Davis: Detection of solar neutrinos (radiochemical: Homestake with 108 measurements 1970-1995, SAGE with 57 measurements 1990-2006, Gallex/GNO with 84 measurements 1996-2001; real time: SuperKamiokande with 184/358 measurements 1996-2001, SNO). The Sun is a gas sphere in hydrostatic equilibrium, slowly rotating, exhibiting magnetic fields, and oscillating. The solar neutrino problem, constituting the discrepancy between theoretically predicted and detected number of solar neutrinos, was (partially) solved by taking into account the Mikheyev-Smirnov-Wolfenstein effect (neutrino oscillations). Remaining question is whether the solar neutrino flux is varying over time, and if so, what is the physical mechanism that makes the flux varying?

Dicke: Sunspot cycle variations in the ~ 11 year half-cycle period all the way from 7.3 year to 17.1 yr. Random walk in the phase of the cycle? Superposition of different periodic cycles? Driver of the cycle (convective zone, tachocline, core)? Is there a chronometer hidden deep in the Sun?

Burlaga: The solar wind (Voyager) is a driven nonlinear nonequilibrium system consisting of a supersonic speed expanding fully ionized plasma that carries magnetic fields. The Sun injects matter, energy, momentum, magnetic fields. At distant heliosphere (approximately 90 Astronomical Units) the solar wind relaxes to a quasi-stationary, metastable state. The speed of the wind and the strength of the magnetic field show fluctuations over time and a fractal and multifractal scaling structure. How to describe possible deviations from thermodynamic equilibrium?

9.3.2 Analysis of the time series

Curve fitting: Attempts can be made to approximate (periodic) variations of measured physical quantities to different analytic functions. Three functions are preferred for this purpose: (i) gamma distribution depending on a power of the argument, (ii) lognormal function, and (iii) exponential distribution depending on a n-grade polynomial. Respective fitting parameters of these three functions can be calculated. Does such a function correspond to the solution of a reaction-diffusion equation governing the processes of reaction and diffusion (energy and mass transfer) of disturbances traveling from a source into an environment?

Fourier and wavelet analysis (time variation): Time series analysis is a rich field of mathematical and statistical analysis in which physical understanding of a time varying system can be gained through the analysis of time series measurements. Traditional methods of time series analysis are Fourier, wavelet, and autocorrelation analysis.

Fokker-Planck equation (deterministic and stochastic processes): Many natural phenomena are characterized by a degree of stochasticity. A long standing problem is the development of methods to model such phenomena. That is, given a set of data taken for a phenomenon, to develop an equation that can reproduce the data with an accuracy comparable to the measured data. If such a method is available, it can be utilized to reconstruct the original process with similar statistical properties; to understand the nature and properties of the stochastic process; and to predict the future behavior of the phenomenon, if it is time dependent, or its behavior over length scales, if it is length scale dependent. A preferred technique for this analysis, based on the Fokker-Planck equation (Langevin equation) is able to distinguish between deterministic and stochastic elements of a phenomenon by determining drift and diffusion coefficients.

We begin by describing the steps that lead to the development of a stochastic equation, based on the (stochastic) data set, which is then utilized to reconstruct the original data, as well as an equation that describes the phenomenon.

As the first step we check whether the data follow a Markov chain and, if so, estimate the Markov time (length) scale t_M. As is well-known, a given process with a degree of randomness or stochasticity may have a finite or an infinite Markov time (length) scale. The Markov time (length) scale is the minimum time interval over which the data can be considered as a Markov process. To determine the Markov scale t_M, we note that a complete characterization of the statistical properties of stochastic fluctuations of a quantity x in terms of a parameter t requires the evaluation of the joint probability distribution function (PDF) $P_n(x_1, t_1; \cdots ; x_n, t_n)$ for an arbitrary n, the number of the data points. If the phenomenon is a Markov process, an important simplification can be made, as the n-point joint PDF, P_n, is generated by the product of the conditional probabilities $p(x_{i+1}, t_{i+1}|x_i, t_i)$, for $i = 1, \cdots, n-1$. A necessary condition for a stochastic phenomenon to be a Markov process is that the Chapman-Kolmogorov (CK) equation,

$$p(x_2, t_2|x_1, t_1) = \int p(x_2, t_2|x_3, t_3)\, p(x_3, t_3|x_1, t_1)\mathrm{d}x_3 , \qquad (9.3.1)$$

should hold for any value of t_3 in the interval $t_2 < t_3 < t_1$. One should check the validity of the CK equation for different x_1 by comparing the directly-evaluated conditional probability functions $p(x_2, t_2|x_1, t_1)$ with the ones calculated according to right side of equation (9.3.1). The simplest way to determine t_M for stationary or homogeneous data is the numerical calculation of the quantity, $S = |p(x_2, t_2|x_1, t_1) - \int p(x_2, t_2|x_3, t_3)\, p(x_3, t_3|x_1, t_1)\mathrm{d}x_3|$, for given x_1 and x_2, in terms of, for example, $t_3 - t_1$ and considering the possible errors in estimating S. Then, $t_M = t_3 - t_1$ for that value of $t_3 - t_1$ for which S vanishes or is nearly zero (achieves a minimum).

Deriving an effective stochastic equation that describes the fluctuations of the quantity $x(t)$ constitutes the second step. The CK equation yields an evolution equation for the change of the distribution function $P(x, t)$ across the scales t. The CK equation, when formulated in differential form, yields a master equation which takes the form of a Fokker-Planck equation:

$$\frac{\mathrm{d}}{\mathrm{d}t}P(x, t) = \left[-\frac{\partial}{\partial x}D^{(1)}(x, t) + \frac{\partial^2}{\partial x^2}D^{(2)}(x, t) \right] P(x, t) . \qquad (9.3.2)$$

The drift and diffusion coefficients, $D^{(1)}(x, t)$ and $D^{(2)}(x, t)$, are estimated directly from the data and the moments $M^{(k)}$ of the conditional probability distributions:

$$D^{(k)}(x, t) = \frac{1}{k!}\lim_{\Delta t \to 0} M^{(k)},$$

$$M^{(k)} = \frac{1}{\Delta t}\int (x' - x)^k p(x', t + \Delta t|x, t)\mathrm{d}x'. \qquad (9.3.3)$$

We note that this Fokker-Planck equation is equivalent to the following Langevin equation:

$$\frac{d}{dt}x(t) = D^{(1)}(x) + \sqrt{D^{(2)}(x)} \, f(t) , \qquad (9.3.4)$$

where $f(t)$ is a random force with zero mean value and Gaussian statistics, δ-correlated in t, i.e., $\langle f(t)f(t')\rangle = 2\delta(t - t')$. Note that such a reconstruction of a stochastic process does *not* imply that the data do not contain any correlation, or that the above formulation ignores the correlations.

Regeneration of the stochastic process constitutes the third step. Eq. (9.3.4) enables us to regenerate a stochastic quantity which is similar to the original one *in the statistical sense*. The stochastic process is regenerated by iterating (9.3.4) which yields a series of data *without memory*. To compare the regenerated data with the original ones, we must take the spatial (or temporal) interval in the numerical discretization of (9.3.4) to be unity (or renormalize it to unity). However, the Markov length or time is typically greater than unity. Therefore, we should correlate the data over the Markov length or time scale. There are a number of methods to correlate the generated data in this interval. Here, we propose a new technique which we refer to as the *kernel* method, according to which one considers a kernel function $K(u)$ that satisfies the condition that,

$$\int_{-\infty}^{\infty} K(u)\mathrm{d}u = 1 , \qquad (9.3.5)$$

such that the data are determined by

$$x(t) = \frac{1}{nh} \sum_{i=1}^{n} x(t_i) K\left(\frac{t-t_i}{h}\right) , \qquad (9.3.6)$$

where h is the window width. For example, one of the most useful kernels is the standard normal density function, $K(u) = (2\pi)^{-1/2}\exp(-\frac{1}{2}u^2)$. In essence, the kernel method represents the data as a sum of 'bumps' placed at the observation points, with its function determining the shape of the bumps, and its window width h fixing their width. It is evident that, over the scale h, the kernel method correlates the data to each other.

Note 9.3.1: If a system exhibits a power law distribution, it can be described by a nonlinear Fokker-Planck equation. The establishment of states with power law distributions is regarded as a collective phenomenon. The power law distribution arises from the interactions between the subsystems of a many-body system. Alternatively, one can describe power law distribution by means of linear Fokker-Planck equations with state-dependent diffusion coefficients. In the context of linear Fokker-Planck equations, power law distributions describe a single system that is subjected to a multiplicative noise source or to some kind of temperature fluctuations. Currently, data analysis techniques are being developed that can be used to extract the model equations of systems described by Markov diffusion processes from experimental data.

9.4 Fundamental Equations for Nonequilibrium Processes

9.4.1 Chapman-Kolmogorov equation

In probability theory and in the theory of stochastic processes, the Chapman-Kolmogorov equation is an identity relating the joint probability distributions of different sets of coordinates on a stochastic process. Suppose that $\{f_i\}$ is an indexed collection of random variables, that is, a stochastic process. Let

$$p_{i_1,\ldots,i_n}(f_1,\ldots,f_n) \qquad (9.4.1)$$

be the joint probability density function of the random variables f_1 to f_n. Then the Chapman-Kolmogorov equation is

$$p_{i_1,\ldots,i_{n-1}}(f_1,\ldots,f_{n-1}) = \int_{-\infty}^{\infty} p_{i,1,\ldots,i_n}(f_1,\ldots,f_n)\mathrm{d}f_n \qquad (9.4.2)$$

i.e. a straightforward marginalization over the nuisance variable. When the stochastic process under consideration is Markovian, the Chapman-Kolmogorov equation is equivalent to an identity on transition densities. In the Markov chain setting, one assumes that $i_1 < \ldots < i_n$. Then, because of the Markov property,

$$p_{i_1,\ldots,i_n}(f_1,\ldots,f_n) = p_{i_1}(f_1)p_{i_2;i_1}(f_2|f_1)\cdots p_{i_n;i_{n-1}}(f_n|f_{n-1}), \qquad (9.4.3)$$

where the conditional probability $p_{i;j}(f_i|f_j)$ is the transition probability between the times $i > j$. So, the Chapman-Kolmogorov equation takes the form

$$p_{i_3;i_1}(f_3|f_1) = \int_{-\infty}^{\infty} p_{i_3;i_2}(f_3|f_2)p_{i_2;i_1}(f_2|f_1)\mathrm{d}f_2 . \qquad (9.4.4)$$

When the probability distribution on the state space of a Markov chain is discrete, the Chapman-Kolmogorov equation can be expressed in terms of (possibly infinite-dimensional) matrix multiplications, thus

$$T(t+s) = T(t)T(s) \qquad (9.4.5)$$

where $T(t)$ is the transition matrix, i.e., if X_t is the state of the process at time t, then for any two points i and j in the state space, one has

$$T_{ij}(t) = p(X_t = j|X_0 = i), \qquad (9.4.6)$$

that is, the probability that X_t is in state j given that X_0 was in state i.

9.4.2 Master equation

In physics, a master equation is a phenomenological first-order differential equation describing the time evolution of the probability of a system to occupy each one of a discrete set of states:

$$\frac{dp_k}{dt} = \sum_m T_{km} p_m,$$ (9.4.7)

where p_k is the probability for the system to be in the state k, while the matrix (T_{km}) is filled with a grid of transition rate constants T_{km}'s. In probability theory, this identifies the evolution as a continuous time Markov process with the integrated master equation obeying a Chapman-Kolmogorov equation. Note that

$$\sum_i T_{ik} = 0$$ (9.4.8)

(i.e. probability is conserved) so the equation may also be written as follows:

$$\frac{dp_k}{dt} = \sum_i (T_{ki} p_i - T_{ik} p_k).$$ (9.4.9)

If the matrix (T_{ik}) is symmetric, i.e. all the microscopic transition dynamics are state reversible so that

$$T_{ki} = T_{ik}$$ (9.4.10)

then this gives

$$\frac{dp_k}{dt} = \sum_i T_{ki} (p_i - p_k).$$ (9.4.11)

Many physical problems in classical and quantum mechanics can be reduced to the form of a master equation. One generalization of the master equation is the Fokker-Planck equation which describes the time evolution of a continuous probability distribution.

9.4.3 Fokker-Planck equation

The Fokker-Planck equation was used for the statistical description of Brownian motion of a particle in a fluid. Brownian motion follows the Langevin equation which can be solved for many different stochastic forcings with results being averaged (Monte Carlo method). However, instead of this computationally intensive approach, one can use the Fokker-Planck equation and consider $W(v,t)$, that is the probability density function of the particle having a velocity in the interval $(v, v+dv)$ when it starts its motion with v_0 at the time t_0. The general form of the Fokker-Planck equation for N variables is

$$\frac{\partial W}{\partial t} = \left[-\sum_{i=1}^N \frac{\partial}{\partial x_i} D_i^1(x_1,\ldots,x_N) + \sum_{i=1}^N \sum_{j=1}^N \frac{\partial^2}{\partial x_i \partial x_j} D_{ij}^2(x_1,\ldots,x_N) \right] W$$ (9.4.12)

where D^1 is the drift vector and D^2 the diffusion tensor, the latter of which results from the presence of the stochastic force.

Note 9.4.1: There are linear and nonlinear Fokker-Planck equations and there are generalizations of their standard representations: generalizations concerning the drift and diffusion coefficients, the transition probability densities related to the solutions of Fokker-Planck equations, and the Fokker-Planck operator as contained in the Fokker-Planck equation. Further to the Fokker-Planck equation, there are other types of evolution equations for probability distributions and density measures: Liouville equations, linear and nonlinear master equations, Boltzmann equations, fractional linear and nonlinear Fokker-Planck equations. Methods developed for Fokker-Planck equations determined by free energy measures can also be applied to nonlinear reaction-diffusion equations.

9.4.4 Langevin equation

In statistical physics, a Langevin equation is a stochastic differential equation describing Brownian motion in a potential. The first Langevin equation to be studied were those in which the potential is constant, so that the acceleration a of a Brownian particle of mass m is expressed as the sum of a viscous force which is proportional to the particle's velocity v (Stokes' law) and a noise term representing the effect of a continuous series of the collisions with the atoms of the underlying fluid:

$$ma = m\frac{dv}{dt} = -\beta v + \eta(t). \tag{9.4.13}$$

Often interesting results can be obtained, without solving the Langevin equation, from the fluctuation dissipation theorem. The main method of solution, if a solution is required, is by use of the Fokker-Planck equation, which provides a deterministic equation satisfied by the time dependent probability density. Alternatively, numerical solutions can be obtained by Monte Carlo simulation. Other techniques, such as path integration have also been used, drawing on the analogy between statistical physics and quantum mechanics (for example the Fokker-Planck equation can be transformed into a the Schroedinger equation by rescaling a few variables).

9.4.5 Reaction-diffusion equation

A specific form of the master equation is the reaction-diffusion equation. The simplest reaction-diffusion models are of the form

$$\frac{\partial \phi}{\partial t} = \xi \frac{\partial^2 \phi}{\partial x^2} + F(\phi) \tag{9.4.14}$$

where ξ is the diffusion constant and F is a nonlinear function representing the reaction kinetics. Examples of particular interest include the Fisher-Kolmogorov equation for which $F = \gamma\phi(1 - \phi^2)$ and the real Ginzburg-Landau equation for which

$F = \gamma\phi(1-\phi)$. The nontrivial dynamics of these systems arises from the competition between the reaction kinetics and diffusion.

Open macroscopic systems with reaction (transformation) and diffusion (transport): Evolution of a reaction-diffusion system involves three types of processes: (i) internal reaction (transformation), (ii) internal diffusion (transport), and (iii) interaction with the external environment. Of special interest are asymptotic states of reaction-diffusion systems that are reached after some time and wherein the system will remain unless internal or external disturbances bring the system out of this state. At one extreme, asymptotically the system may become a closed system with no interaction with the environment, relaxing to a state of internal thermodynamic equilibrium. Another extreme, when all internal transformations cease, the system reaches a state of transport equilibrium with the external environment. Both these asymptotic states are stationary. Starting from either of them and gradually switching on external transport or internal transformation, one obtains two basic branches (diffusion and reaction) of stationary asymptotic states. It may be the case that these two branches meet midway in such a manner that the stationary state remains unique and stable in the whole range of parameters. However, it may also occur that somewhere away from the two equilibrium limits both thermodynamic branches undergo some kind of bifurcation leading to their destabilization and to the emergence of a variety of other asymptotic states, not all of them being stationary, symmetric, or even ordered. Such phenomena are known as kinetic instabilities. The primary characteristic of a kinetic system is the kind of instabilities that may exhibit. Attempts to develop a unified theory of instabilities in nonequilibrium systems are contained in the works of Nicolis and Prigogine and Haken.

One of the best understood theoretical mechanism for pattern formation is the Turing instability of a homogeneous steady state in a two-species reaction-diffusion system. On its own, diffusion tends to smooth out irregularities; however, the differential diffusion of two distinct species coupled by nonlinear reaction terms may result in certain wavelengths becoming unstable so that pattern are produced.

The general form of a two-species reaction-diffusion model is

$$\frac{\partial n_1(x,t)}{\partial t} = \lambda f_1(n_1, n_2) + \nabla^2 n_1(x,t) \qquad (9.4.15)$$

$$\frac{\partial n_2(x,t)}{\partial t} = \lambda f_2(n_1, n_2) + d\nabla^2 n_2(x,t).$$

In these equations, $n_1(x,t)$ and $n_2(x,t)$ are the number densities for the two species. The functions f_1 and f_2 are generally nonlinear functions describing the reaction kinetics. The constant d is the ratio of the diffusion coefficients of species 2 to species 1, and $\lambda > 0$ is a scaling variable which can be interpreted as the characteristic size of the spatial domain or as the relative strengths of the reaction terms. The standard reaction-diffusion model is a diffusion-limited process in which the time for reactions to occur within a given reaction zone is considered to be much less than the time for reactants to diffuse between reaction zones. The reaction-diffusion model is also a mean-field model in which it is assumed that the reactions do not themselves introduce correlations between the diffusing species but are dependent only on local average concentrations. Thus microscopic fluctuations in $n(x,t)$ at the atomic level are ignored. If the concentration of species is spatially homogeneous, then the reaction-diffusion model reduces to the classical macroscopic rate equations from the law of mass action (ben Avraham, Havlin). The canonical model for Turing instability induced pattern formation is a reaction-diffusion equation with activator-inhibitor reaction kinetics: the above two equations with $\partial f_2/\partial n_1 > 0$ and $\partial f_1/\partial n_2 < 0$. In this case species 1 is an activator for production of species 2 and species 2 is an inhibitor for production of species 1. A linear stability analysis about the homogeneous steady-state solution, n_1^*, n_2^*, reveals that necessary conditions for Turing instability induced pattern formation are (Murray)

$$a_{11} + a_{22} < 0$$

$$a_{11}a_{22} - a_{12}a_{21} > 0$$

$$d > \left(\frac{1}{a_{11}}[(a_{11}a_{22} - a_{12}a_{21})^{1/2} + (-a_{12}a_{21})^{1/2}] \right)^2 \qquad (9.4.16)$$

where $a_{ij} = \partial f_i/\partial n_j$ is evaluated at the homogeneous steady-state solution. If the above conditions are met, then it can be shown that there is a range of wave numbers q defined by (Murray)

$$\frac{1}{2d}[(da_{11} + a_{22}) - ((da_{11} + a_{22})^2 - 4d(a_{11}a_{22} - a_{12}a_{21}))^{1/2}] \leq q^2$$

$$\leq \frac{1}{2d}[(da_{11} + a_{12}) + ((da_{11} + a_{22})^2 - 4d(a_{11}a_{22} - a_{12}a_{21}))^{1/2}] \qquad (9.4.17)$$

which will become excited and thus produce patterns. A necessary requirement for pattern formation, consistent with the above equations, is that the inhibitor diffuse faster than the activator ($d > 1$) in all activator-inhibitor systems.

Note 9.4.2: Recently, physical systems have been reported in which the diffusion rates of species cannot be characterized by a single parameter of the diffusion constant. Instead, the (anomalous) diffusion is characterized by a scaling parameter α as well as a diffusion constant D and the mean-square displacement of diffusing species $< r^2(t) >$ scales as a nonlinear power law in time $< r^2(t) > \sim t^\alpha$. The case $0 < \alpha < 1$ is called subdiffusion and, accordingly, the case $\alpha > 1$ is called superdiffusion. The problem of anomalous subdiffusion with reactions in terms of continuous-time random walks (CTRWs) with sources and sinks leads to a fractional activator-inhibitor model with a fractional order temporal derivative operating on the spatial Laplacian. The problem of anomalous superdiffusion with reactions has also been considered and in this case a fractional reaction-diffusion model has been proposed with the spatial Laplacian replaced by a spatial fractional differential operator.

9.5 Fractional Calculus

Mathematics of dynamical systems: There are three distinct paradigms for scientific understanding of dynamical systems. (i) In the Newtonian approach the system is modeled by a differential equation and subsequently solutions of the equations are obtained. (ii) In the approach through the geometric theory of differential equations (= qualitative theory) the system is also modeled by a differential equation but only qualitative information about the system is provided (Poincaré, Smale). (iii) Algorithmic modeling uses the computer, uses maps (discrete-time dynamical system) rather than differential equations (continuous-time dynamical system) that means to use algorithms instead of conventional formulas. This approach is a data driven modeling process.

Integer-order derivatives and their inverse operations (integer-order integrations) provide the language for formulating and analyzing many laws of physics. Integer calculus allows for geometrical interpretations of derivatives and integrals. The calculus of fractional derivatives and integrals does not have clear geometrical and physical interpretations. However the fractional calculus is almost as old as integer calculus. As early as 1695, Leibniz, in a reply to de l'Hospital, wrote "Thus it follows that $d^{1/2}x$ will be equal to $x\sqrt{dx:x},\ldots$ from which one day useful consequences will be drawn".

A first way to formally introduce fractional derivatives proceeds from the repeated differentiation of an integral power. (Formal definitions and a detailed discussion may be found in Chapter 2. Some definitions will be given here for the sake of completeness of the present discussion).

$$\frac{d^n}{dx^n}x^m = \frac{m!}{(m-n)!}x^{m-n}. \tag{9.5.1}$$

For an arbitrary power μ, repeated differentiation gives

$$\frac{d^n}{dx^n}x^\mu = \frac{\Gamma(\mu+1)}{\Gamma(\mu-n+1)}x^{\mu-n}. \tag{9.5.2}$$

with gamma functions replacing the factorials. The gamma functions allow for a generalization to differentiation of an arbitrary order α.

$$\frac{d^\alpha}{dx^\alpha}x^\mu = \frac{\Gamma(\mu+1)}{\Gamma(\mu-\alpha+1)}x^{\mu-\alpha}. \tag{9.5.3}$$

The extension defined by the latter equation corresponds to the Riemann-Liouville derivative. It is sufficient for handling functions that can be expanded in Taylor series. A second way to introduce fractional derivatives uses the fact that the nth derivative is an operation inverse to an n-fold repeated integration. Basic is the integral identity

$$\int_a^x \int_a^{y_1} \ldots \int_a^{y_{n-1}} dy_n \ldots dy_1 f(y_n) = \frac{1}{(n-1)!}\int_a^x dy f(y)(x-y)^{n-1}. \tag{9.5.4}$$

A generalization of the expression allows one to define a fractional integral of arbitrary order α via

$$_aD_x^{-\alpha}f(x) = \frac{1}{\Gamma(\alpha)} \int_a^x dy f(y)(x-y)^{\alpha-1} \quad (x \geq a). \tag{9.5.5}$$

A fractional derivative of an arbitrary order is defined through fractional integration and successive ordinary differentiation. The following is a causal convolution-type integral.

$$f(t) = \int_0^t d\tau h(\tau)g(t-\tau) \tag{9.5.6}$$

(transforms the input signal $h(t)$ into the output signal $f(t)$ via the memory function (the impulse response) $g(t)$. If $g(t)$ is the step function

$$g(t) = \begin{cases} 1 & \text{for } t \geq 0 \\ 0 & \text{for } t < 0 \end{cases} \tag{9.5.7}$$

then the latter expression is a first-order integral. And if $g(t) = \delta(t)$ is the Dirac delta-function, then the transformation represented by the former integral reproduces the input signal (this is the zeroth-order integral). It may be assumed that the fractional integration of order $v, (0 < v < 1)$,

$$f(t) = \frac{1}{\Gamma(v)} \int_0^t d\tau h(\tau)(t-\tau)^{v-1} \tag{9.5.8}$$

interpolates the memory function such that it lies between the delta-function (total absence of memory) and the step function (complete memory).

Stanislavsky developed a specific interpretation of fractional calculus: It was shown that there is a relation between stable probability distributions and the fractional integral. The time degree of freedom becomes stochastic. It is the sum of random time intervals and each of them is a random variable with a stable probability distribution. There exists a mathematically justified passage to the limit from discrete time steps (intervals) to a continuous limit. Corresponding processes have randomized operation time. The kinetic equations describing such processes are written in terms of time derivatives (or time integrals) of fractional order. The exponent of the fractional integral (derivative) is directly related to the parameter of the corresponding stable probability distribution. The occurrence of the fractional derivative (or integral) with respect to time in kinetic equations shows that these equations describe subordinate stochastic processes. Their directional process is directly related to a stochastic process with a stable probability distribution. This introduces a stochastic time arrow into the equations. In contrast to the traditional determinate time arrow with a "timer" counting equal time intervals, the stochastic "timer" has an irregular time step. This time step is a random variable with a stable probability distribution. This character of the probability distribution gives rise to long-term memory effects in the subordinate process, and the relaxation (reaction) in such a system has a power-law character. Although the above mentioned transformation of stochastic processes does not violate the laws of classical thermodynamics, it requires some modification of their macroscopic description. This manifests itself in the appearance of a generalized (fractional) operator with respect to time in the kinetic description of such anomalous systems. The order of this operator permits finding the parameter α corresponding to the stable distribution.

9.6 Nonextensive Statistical Mechanics

In 1865 Clausius introduced the concept of entropy in the context of classical thermodynamics without any reference to the microscopic world. Boltzmann discovered the fundamental description of the behavior of classical macroscopic bodies in equilibrium in terms of the properties of classical microscopic particles out of which they consist. He used both dynamical and statistical methods. Planck applied Boltzmann's method to radiation and discovered quantum mechanics. Einstein argued with Boltzmann and Planck that the statistical description of a physical system should be based on the dynamics (Newton's equation of motion) of the system. Boltzmann, in his research papers in 1868 and 1872, generalized the Maxwell-Boltzmann equilibrium velocity distribution for point particles in free space to the case of a number of material points that move under the influence of forces for which a potential function exists. It seems that he did not realize that he introduced probabilistic concepts in his "mechanical" considerations. Only in 1877 he proposed the relation between the second law of thermodynamics and probability theory with respect to the laws of thermal equilibrium and established the link between the thermodynamic entropy S and the probability W for the dynamical states of a physical system at a given total energy in phase space. According to Einstein, Boltzmann's statistical approach (without any reference to dynamics) only applies strictly to equilibrium. Later, Gibbs generalized Boltzmann's principle in μ-space ($S = k \ln W + \text{constant}$) to Γ-space ($S = -k \int f(\Gamma) \ln f(\Gamma) d\Gamma$) but the two approaches are equivalently valid only for equilibrium. Their generalization to nonequilibrium states remains an open problem.

The first classical non-Boltzmann-Gibbs statistics of physical systems was discovered by Tsallis. One of the properties within Clausius concept of entropy is the extensivity of the entropy, i.e., its proportionality to the number N of elements of the system. The Boltzmann-Gibbs entropy

$$S_{BG} = k \sum_{i=1}^{w} p_i \ln p_i \qquad (9.6.1)$$

(discrete version where w is the total number of microscopic states, with probabilities $\{p_i\}$). If the N elements (or subsystems) are probabilistically independent, then the joint probability is a product of the probabilities of individual events:

$$p_{i_1,i_2,\ldots,i_N} = p_{i_1} p_{i_2} \cdots p_{i_N}. \qquad (9.6.2)$$

It can be verified that

$$S_{BG}(N) \propto NS_{BG}(1). \tag{9.6.3}$$

If the correlations within the system are close to this ideal situation (e.g., local interactions), extensivity is still verified, in the sense that

$$S_{BG}(N) \propto N, \; N \to \infty. \tag{9.6.4}$$

However there are more complex situations for which S_{BG} is not extensive. For an important class of systems (e.g., asymptotically scale-invariant), a connection between S and W is known:

$$S_q = k \frac{1 - \sum_{j=1}^{W} p_i^q}{q-1} (q \in R; S_1 = S_{BG}). \tag{9.6.5}$$

This entropy was proposed by Tsallis as a possible basis for a generalization of Boltzmann-Gibbs statistics that is currently referred to as nonextensive statistical mechanics. In such a theory the energy is typically nonextensive whether or not the entropy is. The property

$$\frac{S_q(A+B)}{k} = \frac{S_q(A)}{k} + \frac{S_q(B)}{k} + (1-q)\frac{S_q(A)}{k}\frac{S_q(B)}{k} \tag{9.6.6}$$

which led to the term "nonextensive entropy", is valid only if the subsystems A and B are probabilistically independent. In some applications of q-statistics to physical problems, the entropic indices q can be computed from first principles when the precise dynamics is known. In other applications, when neither the microscopic nor the mesoscopic dynamics is accessible, only a phenomenological approach is possible, and q is determined by fitting.

9.7 Standard and Fractional Reaction

9.7.1 Boltzmann-Gibbs statistical mechanics

9.7.1.1 Differential equation

Which is the simplest ordinary differential equation? It is

$$\frac{dy}{dx} = 0, \tag{9.7.1}$$

whose solution (with $y(0) = 1$) is $y = 1$. What could be considered as the second in simplicity? It is

$$\frac{dy}{dx} = 1, \tag{9.7.2}$$

whose solution is $y = 1 + x$. And the next one? It is

$$\frac{dy}{dx} = y,\tag{9.7.3}$$

whose solution is $y = e^x$. Its inverse is $x = \ln y$, which coincides with the celebrated Boltzmann formula

$$S_{BG} = k \ln W,\tag{9.7.4}$$

where k is Boltzmann constant, and W is the measure of the space where the system is allowed to "live", taking into account total energy and similar constraints. If we have an isolated N-body Hamiltonian system (microcanonical ensemble in Gibbs notation), W is the dimensionless Euclidean *measure* (i.e., (hyper)volume) of the fixed-energy Riemann (hyper)surface in phase space (Gibbs' Γ-space) if the system microscopically follows *classical dynamics*, and it is the *dimension* of the associated Hilbert space if the system microscopically follows *quantum dynamics*. In what follows we indistinctively refer to classical or quantum systems. We shall nevertheless use, for simplicity, the wording "phase space" although we shall write down formulas where W is a natural number. If we introduce a natural scaling for x (i.e., if x carries physical dimensions) we must consider, instead of Eq. (9.7.3),

$$\frac{dy}{dx} = ay,\tag{9.7.5}$$

in such a way that ax is a dimensionless variable. The solution is now

$$y = e^{ax}.\tag{9.7.6}$$

This differential equation and its solution appear to admit at least three physical interpretations that are crucial in Boltzmann-Gibbs statistical mechanics. The *first* one is $(x, y, a) \rightarrow (t, \xi, \lambda)$, hence

$$\xi = e^{\lambda t},\tag{9.7.7}$$

where t is time,

$$\xi \equiv \lim_{\Delta x(0) \rightarrow 0} \frac{\Delta x(t)}{\Delta x(0)}$$

is the *sensitivity to initial conditions*, and λ is the (maximal) Lyapunov exponent associated with a typical phase-space variable x (the dynamically most unstable one, in fact). This sensitivity to initial conditions (with $\lambda > 0$) is of course the cause of the mixing in phase space which will guarantee *ergodicity*, the well known dynamical justification for the entropy in Eq. (9.7.4).

The *second* physical interpretation is given by $(x, y, a) \rightarrow (t, \Omega, -1/\tau)$, hence

$$\Omega = e^{-t/\tau},\tag{9.7.8}$$

where $\Omega \equiv \frac{\mathcal{O}(t) - \mathcal{O}(\infty)}{\mathcal{O}(0) - \mathcal{O}(\infty)}$, and τ is the characteristic time associated with the *relaxation* of a typical macroscopic observable $\mathcal{O}$ towards its value at the possible stationary state (*thermal equilibrium* for BG statistical mechanics). This relaxation occurs precisely because of the sensitivity to initial conditions, which guarantees strong chaos (essentially Boltzmann's 1872 *molecular chaos hypothesis*). It was Krylov the first to realize, over half a century ago, this deep connection. Indeed, τ typically scales like $1/\lambda$.

The *third* physical interpretation is given by $(x, y, a) \rightarrow (E_i, Z p_i, -\beta)$, hence

$$p_i = \frac{e^{-\beta E_i}}{Z}, \left(Z \equiv \sum_{j=1}^{W} e^{-\beta E_j} \right), \tag{9.7.9}$$

where E_i is the eigenvalue of the i-th quantum state of the Hamiltonian (with its associated boundary conditions), p_i is the probability of occurrence of the i-th state when the system is at its *macroscopic stationary state* in equilibrium with a thermostat whose temperature is $t \equiv 1/(k\beta)$ (canonical ensemble in Gibbs notation). It is a remarkable fact that the *exponential* functional form of the distribution which optimizes the Boltzmann-Gibbs generic entropy

$$S_{BG} = -k \sum_{i=1}^{W} p_i \ln p_i, \tag{9.7.10}$$

with the constraints

$$\sum_{i=1}^{W} p_i = 1, \tag{9.7.11}$$

and

$$\sum_{i=1}^{W} p_i E_i = U \quad (U \equiv \text{internal energy}), \tag{9.7.12}$$

precisely is the inverse functional form of the same entropy under the hypothesis of equal probabilities, i.e., $p_i = \frac{1}{W}$ for all i, hence the *logarithmic* Eq. (9.7.10). To the best of our knowledge, there is (yet) no clear generic mathematical linking for this fact, but it is nevertheless true. It might seem at first glance a quite bizarre thing to do that of connecting the standard Boltzmann-Gibbs exponential weight to the solution of a (linear) differential equation, in contrast with the familiar procedure consisting in extremizing an entropic functional (Eq. (9.7.10)) under appropriate constraints (Eqs. (9.7.11) and (9.7.12)). It might be helpful to remind to those readers who so think that it is precisely through a differential equation that Planck heuristically found the celebrated black-body radiation law in his October 1900 paper, considered by many as the beginning of the path that led to quantum mechanics.

In concluding the present remarks by saying that, when we stress that Eqs. (9.7.1), (9.7.2) and (9.7.3) naturally co-emerge within Boltzmann-Gibbs statistical mechanics, we only refer to the generic (or more typical) situations, *not to all* the situations. It is known, for example, that relaxation occurs through a power-law function of time at any typical second-order phase transition, whereas the Boltzmann-Gibbs weight remains exponential.

9.7.1.2 Mean value

The Boltzmann-Gibbs entropy in Eq. (9.7.10) can be rewritten as the following mean value:

$$S_{BG} = k \left\langle \ln \frac{1}{p_i} \right\rangle , \tag{9.7.13}$$

where $\langle \cdots \rangle \equiv \sum_{i=1}^{W} p_i(\cdots)$. The quantity $\ln(1/p_i)$ is some times called *surprise* or *unexpectedness*. We notice that the averaged quantity has the *same* functional form as that corresponding to the equal probability case Eq. (9.7.10), where $1/p_i$ plays the role of W.

9.7.1.3 Composition law for independent systems

Let us consider systems A and B as probabilistically independent, i.e., such that $p_{ij}^{A+B} = p_i^A p_j^B$ for all (i,j)). We can immediately prove that entropy (9.7.10) satisfies the following property

$$S_{BG}(A+B) = S_{BG}(A) + S_{BG}(B) , \tag{9.7.14}$$

referred from now on as *extensivity*. This property is sometimes referred to as *additivity*, reserving the word *extensivity* for the infinitely many body systems; we will for simplicity not make such a distinction here.

The *linear* property (9.7.14) of course encompasses the fact that, since $W_{A+B} = W_A W_B$, whenever we have equal probabilities, the logarithmic form (9.7.10) is absolutely fitted. For example, if we have N independent coins (or dices), it is $W = 2^N$ (or 6^N), hence $S_{BG} = Nk \ln 2$ (or $S_{BG} = Nk \ln 6$). If we have, as another example, a $d = 3$ regular lattice with ferromagnetic Heisenberg interactions between first neighbors at very high temperature, it is $W \sim A\rho^N$ (with $A > 0$, $\rho > 1$, and $N \to \infty$), hence $S_{BG} \sim Nk \ln \rho$. In all these cases, we have $S_{BG} \propto N$, which precisely fits the Clausius concept of thermodynamic entropy. It can be explored when the ubiquitous behavior $W(N) \sim \rho^N$ (with $N \gg 1$) is drastically violated, e.g., when $W \propto N^\gamma$, with $\gamma > 0$, which *also* appears to be ubiquitous in both natural and artificial systems.

9.7.2 Generalized Boltzmann-Gibbs statistical mechanics

There are several other properties than those discussed above, which also specifically characterize Boltzmann-Gibbs statistical mechanics, but we shall restrict the present analysis to only those, i.e., differential equations, mean value, entropy composition law. As we already mentioned, there is of course no logical-deductive manner to generalize a physical theory. Or, there is no generic or unique way of generalizing a logically consistent set of axioms into another one which also is logically consistent and which, by construction, recovers the original one as a particular case. It is therefore only metaphorically that we shall use, in what follows, the mathematical structure of Boltzmann-Gibbs statistical mechanics in order to generalize it.

9.7.2.1 Differential equations

Eqs. (9.7.1), (9.7.2) and (9.7.3) can be unified in a *single* differential equation through

$$\frac{dy}{dx} = a + by. \tag{9.7.15}$$

This can also be achieved with only one parameter through

$$\frac{dy}{dx} = y^q \quad (q \in \mathcal{R}) \tag{9.7.16}$$

Eqs. (9.7.1), (9.7.2) and (9.7.3) are respectively recovered for $q \to -\infty$, $q = 0$ and $q = 1$. The solution of Eq. (9.7.16) (with $y(0) = 1$) is given by

$$y = [1 + (1-q)x]^{1/(1-q)} \equiv e_q^x \quad (e_1^x = e^x). \tag{9.7.17}$$

The inverse function of the *q-exponential* is the *q-logarithm*, defined as follows

$$y = \frac{x^{1-q} - 1}{1-q} \equiv \ln_q x \quad (\ln_1 x = \ln x). \tag{9.7.18}$$

The Boltzmann principle, Eq. (9.6.1), can be generalized, for equal probabilities, as follows

$$S_q(p_i = 1/W, \forall i) = k \ln_q W = k \frac{W^{1-q} - 1}{1-q}. \tag{9.7.19}$$

As for the Boltzmann-Gibbs case, if x carries a physical dimension, we must consider, instead of Eq. (9.7.16),

$$\frac{dy}{dx} = a_q y^q \quad (a_1 = a), \tag{9.7.20}$$

hence

$$y = e_q^{a_q x}. \tag{9.7.21}$$

As for the Boltzmann-Gibbs case, we expect this solution to admit at least three different physical interpretations.

The first one corresponds to the sensitivity to initial conditions

$$\xi = e_q^{\lambda_q t}, \tag{9.7.22}$$

where λ_q generalizes the Lyapunov exponent or coefficient. This was conjectured in 1997, and, for unimodal maps, proved recently.

The second interpretation corresponds to relaxation, i.e.,

$$\Omega = e_q^{-t/\tau_q}. \tag{9.7.23}$$

There is (yet) no proof of this property, but there are several verifications (for instance, for a quantum chaotic system).

The third interpretation corresponds to the energy distribution at the stationary state, i.e.,

$$p_i = \frac{e_q^{-\beta_q E_i}}{Z_q}, \quad \left(Z_q \equiv \sum_{j=1}^{W} e_q^{-\beta_q E_j} \right). \tag{9.7.24}$$

This is precisely the form that comes out from the optimization of the generic entropy S_q under appropriate constraints. This form has been observed in a large variety of situations.

Before closing this subsection, let us stress that there is no reason for the values of q appearing in Eqs. (9.7.17), (9.7.18) and (9.7.19) be the same. Indeed, if we respectively denote them by q_{sen} (*sen* stands for *sensitivity*), q_{rel} (*rel* stands for *relaxation*) and q_{stat} (*stat* stands for *stationary state*), we typically (but not necessarily) have that $q_{sen} \leq 1$, $q_{rel} \geq 1$ and $q_{stat} \geq 1$. The possible connections between all these entropic indices are not (yet) known in general. However, for the edge of chaos of the z-logistic maps we do know some important properties. If we consider the multifractal $f(\alpha)$ function, the fractal or Hausdorff dimension d_f corresponds to the maximal height of $f(\alpha)$; also, we may denote by α_{min} and α_{max} the values of α at which $f(\alpha)$ vanishes (with $\alpha_{min} < \alpha_{max}$). It has been proved that

$$\frac{1}{1 - q_{sen}} = \frac{1}{\alpha_{mim}} - \frac{1}{\alpha_{max}}. \tag{9.7.25}$$

Moreover, there is some numerical evidence suggesting

$$\frac{1}{q_{rel} - 1} \propto (1 - d_f). \tag{9.7.26}$$

Unfortunately, we know not much about q_{stat}, but it would not be surprising if it was closely related to q_{rel}. They could even coincide, in fact.

9.7.2.2 Mean value

Since we have seen in the previous subsection that the logarithmic function naturally generalizes into the q-logarithmic one, let us define

$$S_q = k \left\langle \ln_q \frac{1}{p_i} \right\rangle, \tag{9.7.27}$$

where we may call $\ln_q(1/p_i)$ the q-surprise or q-unexpectedness. Then, it is straightforward to obtain

$$S_q = k\,\frac{1-\sum_{i=1}^{W}p_i^q}{q-1}\quad(S_1 = S_{BG}),\tag{9.7.28}$$

which is the entropy on which we shall base the present generalization of Boltzmann-Gibbs statistical mechanics.

9.7.2.3 Entropy composition law for independent systems

If we consider now the same two probabilistically independent systems A and B that we assumed before, we straightforwardly obtain

$$\frac{S_q(A+B)}{k} = \frac{S_q(A)}{k} + \frac{S_q(A)}{k} + (1-q)\frac{S_q(A)}{k}\,\frac{S_q(B)}{k}.\tag{9.7.29}$$

We re-obtain Eq. (9.7.10) in the limit $(1-q)/k \to 0$. Since S_q is always nonnegative, we see that, if $q < 1$ $(q > 1)$, we have that $S_q(A+B) > S_q(A) + S_q(B)$ $(S_q(A+B) < S_q(A) + S_q(B))$, which shall be referred as the superextensive (subextensive) case. It is from this property that the expression nonextensive statistical mechanics was coined (Tsallis).

9.7.3 Fractional reaction

In terms of Pochhammer's symbol

$$(\alpha)_n = \begin{cases} 1, n = 0, \alpha \neq 0 \\ \alpha(\alpha+1)\dots(\alpha+n-1), n \in N, \end{cases}\tag{9.7.30}$$

we can express the binomial series as

$$(1-x)^{-\alpha} = \sum_{r=0}^{\infty}\frac{(\alpha)_r x^r}{r!}, |x| < 1.\tag{9.7.31}$$

The Mittag-Leffler function is defined by

$$E_\alpha(x) = \sum_{n=0}^{\infty}\frac{z^n}{\Gamma(\alpha n+1)}.\tag{9.7.32}$$

This function was defined and studied by Mittag-Leffler. We note that this function is a direct generalization of an exponential series, since

$$E_1(z) = \exp(z).\tag{9.7.33}$$

It also includes the error functions and other related functions, for we have

$$E_{1/2}(\pm z^{1/2}) = e^z[1 + \mathrm{erf}(\pm z^{1/2})] = e^z \mathrm{erfc}(\mp z^{1/2}), \tag{9.7.34}$$

where

$$\mathrm{erf}(z) = \frac{2}{\pi^{1/2}} \int_0^z e^{-u^2} du, \mathrm{erfc}(z) = 1 - \mathrm{erf}(z), z \in C. \tag{9.7.35}$$

The equation

$$E_{\alpha,\beta}(z) = \sum_{n=0}^{\infty} \frac{z^n}{\Gamma(\alpha n + \beta)} \tag{9.7.36}$$

gives a generalization of the Mittag-Leffler function. When $\beta = 1$, Eq. (9.7.36) reduces to Eq. (9.7.32). Both the functions defined by Eqs. (9.7.32) and (9.7.36) are entire functions of order $1/\alpha$ and type 1. The Laplace transform of $E_{\alpha,\beta}(z)$ follows from the integral

$$\int_0^{\infty} e^{-pt} t^{\beta-1} E_{\alpha,\beta}(\lambda a t^{\alpha}) dt = p^{-\beta}(1 - ap^{-\alpha})^{-1}, \tag{9.7.37}$$

where $\Re(p) > |a|^{1/\alpha}, \Re(\beta) > 0$, which can be established by means of the Laplace integral

$$\int_0^{\infty} e^{-pt} t^{\rho-1} dt = \Gamma(\rho)/p^{\rho}, \tag{9.7.38}$$

where $\Re(p) > 0, \Re(\rho) > 0$. The Riemann-Liouville operator of fractional integration is defined as

$$_a D_t^{-v} f(t) = \frac{1}{\Gamma(v)} \int_a^t f(u)(t-u)^{v-1} du, v > 0, \tag{9.7.39}$$

with $_a D_t^0 f(t) = f(t)$. By integrating the standard kinetic equation

$$\frac{d}{dt} N_i(t) = -c_i N_i(t), (c_i > 0), \tag{9.7.40}$$

it is derived that

$$N_i(t) - N_0 = -c_i \; _0 D_t^{-1} N_i(t), \tag{9.7.41}$$

where $_0 D_t^{-1}$ is the standard Riemann integral operator. Here we recall that the number density of species $i, N_i = N_i(t)$, is a function of time and $N_i(t=0) = N_0$ is the number density of species i at time $t = 0$. By dropping the index i in Eq. (9.7.41), the solution of its generalized form

$$N(t) - N_0 = -c^v \; _0 D_t^{-v} N(t), \tag{9.7.42}$$

is obtained as

$$N(t) = N_0 \sum_{k=0}^{\infty} \frac{(-1)^k (ct)^{vk}}{\Gamma(vk+1)}. \tag{9.7.43}$$

By virtue of Eq. (9.7.36) we can rewrite Eq. (9.7.43) in terms of the Mittag-Leffler function in a compact form as

$$N(t) = N_0 E_v(-c^v t^v), v > 0. \tag{9.7.44}$$

Later we will investigate the solutions of three generalized forms of Eq. (9.7.40). The results are obtained in a compact form in terms of the generalized Mittag-Leffler function.

9.7.4 Thermonuclear reaction coefficient

Understanding the methods of evaluation of thermonuclear reaction rates is one of the most important goals of research in the field of stellar and cosmological nucleosynthesis. Practically all applications of fusion plasmas are controlled in some way or another by the theory of thermonuclear reaction rates under specific circumstances. After several decades of effort, a systematic and complete theory of thermonuclear reaction rates has been constructed. One of the basic ideas in this regard is that the motor of irreversibility and dissipation is the existence of reactions between individual nuclei. The latter produce a randomization of the energy and velocity distributions of particles. The effect of the reactions is balanced by the flow of the particles in a macroscopically inhomogeneous medium. As a result of this balance, the system reaches a quasi-stationary state close to equilibrium, in which steady fluxes of matter, energy, and momentum are present. The main ideas in the following are coming from statistical distribution theory and the theory of generalized special functions, mainly in the categories of Meijer's G-function and Fox's H-function of scalar, vector, and matrix arguments (Mathai, 1993; Aslam Chaudhry and Zubair, 2002). A fusion of mathematical and statistical techniques enabled us to evaluate thermonuclear reaction rate integrals in explicit closed forms. Some of the techniques which are used will be summarized here. In order to explain the ideas we will start with the evaluation of an integral over a real scalar variable first. Let

$$I(z;p,n,m) = p \int_0^\infty e^{-pt} t^{-n\rho} e^{-zt^{-n/m}} dt \qquad (9.7.45)$$

for $\Re(p) > 0, \Re(z) > 0$, n, m positive integers, where $\Re(.)$ denotes the real part of $(.)$. A particular case of this integral for $n\rho = -v, p = 1, n = 1, m = 2$,

$$I(z;1,1,2) = \int_0^\infty e^{-t} t^v e^{-zt^{-1/2}} dt \qquad (9.7.46)$$

is a thermonuclear function associated with equilibrium distributions in reaction rate theory under Maxwell-Boltzmannian approach. As can be seen from (9.7.45) that the usual mathematical techniques fail to obtain a closed-form representation of the basic integral in (9.7.45).

9.7.4.1 Statistical techniques

Certain special functions are related to particular probability laws governing products of independent exponential variables. Such laws can be related to the underlying physical processes. The integrand in (9.7.45) is a product of integrable positive functions and hence the integrand can be made into a product of statistical densities by normalizing them. Consider two statistically independent real scalar random variables x and y with the density functions $f_1(x) \geq 0, f_2(y) \geq 0$ for $0 < x < \infty, 0 < y < \infty$ and $f_1(x) = 0, f_2(y) = 0$ elsewhere. Let $u = xy$, the product of these random variables. Then from the transformation of the variables, $u = xy, v = x$, the density $g(u)$ of u is given by

$$g(u) = \int_0^\infty \frac{1}{v} f_1(v) f_2(u/v) dv. \tag{9.7.47}$$

The integral in (9.7.47) can be made equivalent to the integral in (9.7.45) by suitably selecting f_1 and f_2. Let

$$v = pt, u = pz^{m/n}, f_1(t) = t^{1-n\rho} e^{-t}, \text{ and } f_2(t) = e^{-t^{n/m}}, \tag{9.7.48}$$

excluding the normalizing constants. Then

$$\int_0^\infty \frac{1}{v} f_1(v) f_2(u/v) dv = \int_0^\infty \frac{1}{t} e^{-pt} (pt)^{1-n\rho} e^{-zt^{-n/m}} dt. \tag{9.7.49}$$

Hence

$$p^{n\rho} \int_0^\infty \frac{1}{t} e^{-pt} (pt)^{1-n\rho} e^{-zt^{n/m}} dt = p \int_0^\infty e^{-pt} t^{-n\rho} e^{-zt^{-n/m}} dt \tag{9.7.50}$$

which is exactly the integral to be evaluated in (9.7.45). We have identified the integral as the exact density of the product of two real scalar random variables, $u = xy$. Since this density is unique the idea is to evaluate this density through some other means. Notice that u is a product of positive variables and hence the method of moments can be exploited profitably. Consider the $(s-1)$th moment of u, denoted by expected value of u^{s-1}. That is, denoting the expectation operation by E,

$$E(u^{s-1}) = E(x^{s-1}) E(y^{s-1}) \tag{9.7.51}$$

due to statistical independence of x and y. Let

$$g_1(s) = E(x^{s-1}) \text{ and } g_2(s) = E(y^{s-1}). \tag{9.7.52}$$

Then $g_1(s)$ and $g_2(s)$ are the Mellin transforms of f_1 and f_2 respectively. Then from the inverse Mellin transform the unique density of u is available as

$$g(u) = \frac{1}{2\pi i} \int_L u^{-s} g_1(s) g_2(s) ds, i = \sqrt{-1}. \tag{9.7.53}$$

where L is a suitable contour. But, excluding the normalizing constants,

$$g_1(s) = \int_0^\infty t^{s-1} f_1(t) dt$$
$$= \int_0^\infty t^{1-n\rho+s-1} e^{-t} dt$$
$$= \Gamma(1 - n\rho + s) \text{ for } \Re(1 - n\rho + s) > 0, . \tag{9.7.54}$$

and

$$g_2(s) = \int_0^\infty t^{s-1} f_2(t) dt$$

$$= \int_0^\infty t^{s-1} e^{-t^{n/m}} dt$$

$$= \frac{m}{n} \Gamma(ms/n) \text{ for } \Re(s) > 0. \tag{9.7.55}$$

Then

$$\frac{1}{2\pi i} \int_L u^{-s} g_1(s) g_2(s) ds = \frac{1}{2\pi i} \int_L (m/n) \Gamma(ms/n) \Gamma(1 - n\rho + s) (pz^{m/n})^{-s} ds. \tag{9.7.56}$$

Therefore

$$p \int_0^\infty e^{-pt} t^{-n\rho} e^{-zt^{-n/m}} dt$$

$$= p^{n\rho} \int_0^\infty \frac{1}{t} e^{-pt} (pt)^{1-n\rho} e^{-zt^{-n/m}} dt$$

$$= p^{n\rho} (m/n) \frac{1}{2\pi i} \int_L \Gamma(ms/n) \Gamma(1 - n\rho + s) (pz^{m/n})^{-s} ds$$

$$= m p^{n\rho} \frac{1}{2\pi i} \int_{L_1} \Gamma(ms) \Gamma(1 - n\rho + ns) (z^m p^n)^{-s} ds \tag{9.7.57}$$

by replacing s/n by s. Our aim now is to evaluate the contour integral on the right side of (9.7.57) explicitly into computable forms. The contour integral in (9.7.57) can be written as an H-function which can then be reduced to a G-function since m and n are positive integers. G and H-functions are defined in Chapter 1 and hence they will not be discussed here.

Now, comparing (9.7.53) and (9.7.57) our starting integral is evaluated as follows

$$I(z; p, n, m) = m p^{n\rho} \frac{1}{2\pi i} \int_{L_1} \Gamma(ms) \Gamma(1 - n\rho + ns) (z^m p^n)^{-s} ds$$

$$= m p^{n\rho} H_{0,2}^{2,0} [(z^m p^n) |_{(0,m),(1-n\rho,n)}]. \tag{9.7.58}$$

The H-function in (9.7.58) can be reduced to a G-function which can again be reduced to computable series forms. For this purpose we expand the gammas in the integrand in (9.7.58) by using the multiplication formula for gamma functions, namely,

$$\Gamma(mz) = (2\pi)^{\frac{(1-m)}{2}} m^{mz-\frac{1}{2}} \Gamma(z) \Gamma\left(z + \frac{1}{m}\right) \cdots \Gamma\left(z + \frac{m-1}{m}\right) \tag{9.7.59}$$

$$m = 1, 2, \ldots \tag{9.7.60}$$

Expanding $\Gamma(ms)\Gamma(1-n\rho+ns)$ by using (9.7.59) we have

$$mp^{n\rho}\Gamma(ms)\Gamma(1-n\rho+ns) = (2\pi)^{\frac{1}{2}(2-m-n)} p^{n\rho} m^{\frac{1}{2}} n^{-n\rho+\frac{1}{2}} (m^m n^n)^s$$
$$\times \Gamma(s)\Gamma\left(s+\frac{1}{m}\right)\dots\Gamma\left(s+\frac{m-1}{m}\right)$$
$$\times \Gamma\left(s+\frac{1-n\rho}{n}\right)\dots\Gamma\left(s+\frac{n-n\rho}{n}\right).$$

Substituting these back and writing as a G-function we have

$$I(z;p,n,m) = p^{n\rho}(2\pi)^{\frac{1}{2}(2-m-n)} m^{\frac{1}{2}} n^{-n\rho+\frac{1}{2}}$$
$$\times G_{0,m+n}^{m+n,0}\left[\frac{z^m p^n}{m^m n^n}\bigg|_{0,\frac{1}{m},\dots,\frac{m-1}{m},\frac{1-n\rho}{n},\dots,\frac{n-n\rho}{n}}\right]. \tag{9.7.61}$$

9.7.4.2 Non-resonant thermonuclear reaction rate

The Maxwell-Boltzmannian form of the collision probability integral for non-resonant thermonuclear reactions is

$$I_1 = I(z;1,1,2) = \int_0^\infty y^v e^{-y} e^{-zy^{-1/2}} dy$$
$$= \pi^{-\frac{1}{2}} G_{0,3}^{3,0}\left[\frac{z^2}{4}\bigg|_{0,\frac{1}{2},1+v}\right]. \tag{9.7.62}$$

In stellar fusion plasmas the energies of the moving nuclei are assumed to be described by a Maxwell-Boltzmann distribution, $E\exp(-E/kt)$, where T is the local temperature, E is the energy and k the Boltzmann constant. Folding the cross section of a nuclear reaction, $\sigma(E)$, with this energy (or velocity) distribution leads to the nuclear reaction rate per pair of nuclei:

$$<\sigma v> = (8/\pi\mu)^{1/2}(kT)^{-3/2}\int_0^\infty \sigma(E)\exp(-E/kT)dE,$$

where v is the relative velocity of the pair of nuclei, E is the center-of-mass energy, and $\mu = m_1 m_2/(m_1+m_2)$ is the reduced mass of the entrance channel of the reaction. In order to cover the different evolution phases of the stars, i.e. from main sequence stars to supernovae, one must know the reaction rates over a wide range of temperatures, which in turn requires the availability of $\sigma(E)$ data over a wide range of energies. I is the challenge to the experimentalist to make precise $\sigma(E)$ measurements over a wide range of energies. For the class of charged-particle-induced reactions, there is a repulsive Coulomb barrier in the entrance channel of height $E_c = Z_1 Z_2 e^2/r$, where Z_1 and Z_2 are the integral nuclear charges of the interacting particles, e is the unit of electric charge, and r is the nuclear interaction radius. Due to the tunneling effect through the Coulomb barrier, $\sigma(E)$ drops nearly exponentially with decreasing energy:

$$\sigma(E) = S(E)E^{-1}\exp(-2\pi\eta),$$

where $\eta = Z_1 Z_2 e^2/hv$ is the Sommerfeld parameter, h is the Planck constant. The function $S(E)$ contains all the strictly nuclear effects, and is referred to as the astrophysical $S(E)$ factor. If the above equation for $\sigma(E)$ is inserted in the above equation for the nuclear reaction rate $<\sigma v>$, one obtains

$$<\sigma v> = (8/\pi\mu)^{1/2}(kT)^{-3/2}\int_0^\infty S(E)\exp(-E/kT - b/E^{1/2})\mathrm{d}E,$$

with $b = 2(2\mu)^{1/2}\pi^2 e^2 Z_1 Z_2/h$. Since for nonresonant reactions $S(E)$ varies slowly with energy, the steep energy dependence of the integrand in the equation for $<\sigma v>$ is governed by the exponential term, which is characterized by the peak near an energy E_0 that is usually much larger than kT, the mean thermal energy of the fusion plasma. The peak is referred to as the Gamow peak. For a constant $S(E)$ value over the energy region of the peak, one finds $E_0 = (bkT/2)^{2/3}$. This is the effective mean energy for a given reaction at a given temperature. If one approximates the peak by a Gaussian function, one finds an effective width $\delta = 4(E_0 kT)^{1/2}/3^{1/2}$. In the following, this approximation is not made and the respective integrals are analytically represented, beginning with

$$<\sigma v> = \left(\frac{8}{\pi\mu}\right)^{\frac{1}{2}}\sum_{v=0}^{2}\frac{1}{(kT)^{-\mu+\frac{1}{2}}}\frac{S^{(\mu)}(0)}{\mu!}$$
$$\times \int_0^\infty e^{-y}y^v e^{-zy^{-1/2}}\mathrm{d}y$$

where $S^{(\mu)}$ denotes the μ-th derivative. The G-function in (9.7.62) can be expressed as a computable power series as well as in closed-forms by using residue calculus. Writing the G-function in (9.7.62) as a Mellin-Barnes integral we have

$$G_{0,3}^{3,0}\left[\frac{z^2}{4}\Bigg|_{0,\frac{1}{2},1+v}\right] = \frac{1}{2\pi i}\int \Gamma(s)\Gamma(s+1/2)\Gamma(1+v+s)(z^2/4)^{-s}\mathrm{d}s. \quad (9.7.63)$$

Case (1): $v \neq \pm\frac{\lambda}{2}, \lambda = 0,1,2,\dots$ In this case the poles of the integrand are simple and the poles are at the points

$$s = 0,-1,-2,\dots; s = -\frac{1}{2},-\frac{1}{2}-1,\dots; s = -1-v,-2-v,\dots$$

The sum of the residues at $s = 0,-1,\dots$ is given by

$$\sum_{r=0}^{\infty}\frac{(-1)^r}{r!}\Gamma\left(\frac{1}{2}-r\right)\Gamma(1+v-r)\left(\frac{z^2}{4}\right)^r$$
$$= \Gamma\left(\frac{1}{2}\right)\Gamma(1+v)_0F_2\left(;\frac{1}{2},-v;-\frac{z^2}{4}\right) \quad (9.7.64)$$

where, in general, $_pF_q$ denotes a general hypergeometric function. The sum of the residues at $s = -\frac{1}{2}, -\frac{1}{2} - 1, \ldots$ is

$$\sum_{r=0}^{\infty} \frac{(-1)^r}{r!} \Gamma\left(-\frac{1}{2} - r\right) \Gamma\left(\frac{1}{2} + v - r\right) \left(\frac{z^2}{4}\right)^{\frac{1}{2}+r}$$

$$= \Gamma\left(-\frac{1}{2}\right) \Gamma\left(\frac{1}{2} + v\right) \left(\frac{z^2}{4}\right)^{\frac{1}{2}} {}_0F_2\left(;\frac{3}{2},\frac{1}{2} - v; -\frac{z^2}{4}\right). \quad (9.7.65)$$

The sum of the residues at $s = -1 - v, -1 - v - 1, \ldots$ is

$$\sum_{r=0}^{\infty} \frac{(-1)^r}{r!} \Gamma\left(-\frac{1}{2} - v - r\right) \Gamma(-1 - v - r) \left(\frac{z^2}{4}\right)^{1+v+r}$$

$$= \Gamma(-1 - v)\Gamma\left(-\frac{1}{2} - v\right) \left(\frac{z^2}{4}\right)^{1+v} {}_0F_2\left(;v + 2, v + \frac{3}{2}; -\frac{z^2}{4}\right)$$

$$(9.7.66)$$

Then from (9.7.64) to (9.7.66) we have

$$I(z;1,1,2) = \int_0^{\infty} y^v e^{-y} e^{-zy^{1/2}} dy$$

$$= \pi^{-\frac{1}{2}} G_{0,3}^{3,0}\left[\frac{z^2}{4} \Big|_{0,\frac{1}{2},1+v}\right]$$

$$= \Gamma(1 + v){}_0F_2\left(;\frac{1}{2}, -v; -\frac{z^2}{4}\right)$$

$$- 2\Gamma\left(\frac{1}{2} + v\right) \left(\frac{z^2}{4}\right)^{\frac{1}{2}} {}_0F_2\left(;\frac{3}{2};\frac{1}{2} - v; -\frac{z^2}{4}\right)$$

$$+ \frac{\Gamma(-1 - v)\Gamma\left(-\frac{1}{2} - v\right)}{\Gamma\left(\frac{1}{2}\right)} \left(\frac{z^2}{4}\right)^{1+v}$$

$$\times {}_0F_2\left(;v + 2, v + \frac{3}{2}; -\frac{z^2}{4}\right) \quad (9.7.67)$$

for $v \neq \pm\frac{\lambda}{2}, \lambda = 0,1,2,\ldots$. When v is a positive integer the poles at $s = -1 - v$, $-1 - v - 1, \ldots$ are of order 2 each. Hence the corresponding sum of residues can be written in terms of psi functions. Similarly when v is a negative integer, positive or negative half integer the corresponding sums will contain psi functions.

9.7.4.3 Modified non-resonant thermonuclear reaction rate: depletion

With deviations from the Maxwell-Boltzmann velocity distribution of nuclei in the fusion plasma, a modification which results in the depletion of the tail is introduced. In this case the collision probability integral will be of the following form:

$$I_2 = \int_0^\infty y^\nu e^{-y-y^\delta} e^{-zy^{-1/2}} dy. \tag{9.7.68}$$

We consider a general integral of this type. Let

$$I(z;\delta,a,b,m,n) = \int_0^\infty t^\rho e^{-at-bt^\delta-zt^{-n/m}} dt. \tag{9.7.69}$$

Expanding e^{-bt^δ} and then with the help of (9.7.67)) one can represent (9.7.69) in terms of a G-function as follows:

$$I(z;\delta,a,b,m,n) = \sum_{k=0}^\infty \frac{(-b)^k}{k!} a^{-(\rho+k\delta+1)} (2\pi)^{\frac{1}{2}(2-m-n)} m^{\frac{1}{2}} n^{\frac{1}{2}+\rho+k\delta}$$

$$\times G_{0,m+n}^{m+n,0} \left[\frac{z^m a^n}{m^m n^n} \Bigg|_{0,\frac{1}{m},\dots,\frac{m-1}{m},\frac{1+\rho+k\delta}{n},\dots,\frac{n+\rho+k\delta}{n}} \right] \tag{9.7.70}$$

for $\mathfrak{R}(z) > 0, \mathfrak{R}(a) > 0, \mathfrak{R}(b) > 0, m,n = 1,2,\dots$. the case in (9.7.68) is for $a = 1, b = 1, n = 1, m = 2$. With $\nu = \rho + k\delta$ the G-function in (9.7.68) corresponds to that in (9.7.70). When δ is irrational and ρ is rational, the poles of the integrand will be simple and the G-function is available in terms of hypergeometric functions. Other situations will involve psi functions. From the asymptotic behavior of the G-function, see for example Mathai (1993), one can write the integral in (9.7.68), for large values of z as follows:

$$I_2 \approx \pi^{\frac{1}{2}} (\beta/3)^{\frac{2\nu+1}{2}} e^{-\beta-(\beta/3)^\delta}, \beta = 3(z/2)^{2/3}. \tag{9.7.71}$$

9.7.4.4 Modified non-resonant thermonuclear reaction rate: cut-off

Another modification can be made by a cut-off of the high-energy tail of the Maxwell-Boltzmann distribution. In this case the collision probability integral to be evaluated is of the form

$$I_3 = \int_0^d t^{-\rho} e^{-at-zt^{-1/2}} dt, d < \infty. \tag{9.7.72}$$

We will consider a general integral of the form

$$I(z;d,a,\rho,n,m) = \int_0^d y^{-n\rho} e^{-ay-zy^{-n/m}} dy, d < \infty. \tag{9.7.73}$$

In order to evaluate (9.7.73) explicitly we will use statistical techniques as discussed earlier. Let x and y be two statistically independent real random variables having the densities $c_1 f_1(x), 0 < x < d$ and $c_2 f_2(y), 0 < y < \infty$ with $f_1(x)$ and $f_2(y)$ equal to zero elsewhere, where c_1 and c_2 are normalizing constants. Then taking

$$f_1(x) = x^{-n\rho+1} e^{-ax} \quad \text{and} \quad f_2(y) = e^{-y^n/m}$$

and proceeding as before one has the following result:

$$I(z;d,a,\rho,n,m) = \int_0^d t^{-n\rho} e^{-at-zt^{-n/m}} dt$$

$$= m^{\frac{1}{2}} n^{-1} (2\pi)^{(1-m)/2} d^{-n\rho+1} \sum_{r=0}^{\infty} \frac{(-ad)^r}{r!}$$

$$\times G_{n,m+n}^{m+n,0} \left[\frac{z^m}{d^n m^m} \Big|_{-\rho+\frac{r+1+j-1}{n}, j=1,...,n; \frac{j-1}{m}, j=1,...,m}^{-\rho+\frac{r+2+j-1}{n}, j=1,...,n} \right] \qquad (9.7.74)$$

for $\Re(z) > 0, > 0, \Re(a) > 0$. Then

$$I_3 = \int_0^d t^{-\rho} e^{-at-zt^{-1/2}} dt$$

$$= d^{-\rho+1} \pi^{-\frac{1}{2}} \sum_{r=0}^{\infty} \frac{(-ad)^r}{r!}$$

$$\times G_{1,3}^{3,0} \left[\frac{z^2}{4d} \Big|_{-\rho+r+1,0,\frac{1}{2}}^{-\rho+r+2} \right]. \qquad (9.7.75)$$

For large values of z the G-function behaves like $\pi^{\frac{1}{2}} x^{-\frac{1}{2}} e^{-2x^{\frac{1}{2}}}, x = \frac{z^2}{4d}$, see for example Mathai (1993). Then for large values of z,

$$I_3 \approx d^{-\rho+1} \left(\frac{z^2}{4d} \right)^{-\frac{1}{2}} e^{-ad-2(z^2/4d)^{\frac{1}{2}}}. \qquad (9.7.76)$$

Explicit series forms can be obtained for various values of the parameters with the help of residue calculus. For example, for $\Re(z) > 0, d > 0, \Re(a) > 0, -\rho+r+1 \neq \mu, \mu = 0,1,...$

$$I_3 = \int_0^d t^{-\rho} e^{-at-zt^{-1/2}} dt$$

$$= \pi^{-\frac{1}{2}} d^{-\rho+1} \sum_{r=0}^{\infty} \frac{(-ad)^r}{r!}$$

$$\times \left\{ \sum_{v=0, v \neq \rho}^{\infty} \frac{(-1)^v \Gamma\left(\frac{1}{2} - v\right)}{v!(-\rho+r+1-v)} \left(\frac{z^2}{4d} \right)^v \right.$$

$$+ \sum_{v=0}^{\infty} \frac{(-1)^v \Gamma\left(-\frac{1}{2} - v\right)}{v!\left(-\rho+r-v+\frac{1}{2}\right)} \left(\frac{z^2}{4d} \right)^{v+\frac{1}{2}}$$

$$+ \left(\frac{z^2}{4d} \right)^{\mu} \left[-\ln(\frac{z^2}{4d}) + A \right] B \right\}, \qquad (9.7.77)$$

where

$$A = \psi(\mu+1) + \psi\left(-\mu+\frac{1}{2}\right), \psi(z) = \frac{d}{dz} \ln \Gamma(z),$$

$$B = \frac{(-1)^{\mu}}{\mu!} \Gamma\left(-\mu+\frac{1}{2}\right).$$

9.7.4.5 Computations

For computational purposes we will consider the four basic integrals associated with the cases: non-resonant reactions, non-resonant "cut-off" reactions, non-resonant screened reactions, and non-resonant "depleted" reactions. Let

$$J_1(z,v) = \int_0^\infty y^v e^{-y-zy^{-1/2}} dy$$

$$J_2(z,d,v) = \int_0^d y^v e^{-y-zy^{-1/2}} dy$$

$$J_3(z,t,v) = \int_0^\infty y^v e^{-y-z(y+t)^{-\frac{1}{2}}} dy$$

$$J_4(z,\delta,b,v) = \int_0^\infty y^v e^{-y-by^\delta-zy^{-1/2}} dy. \tag{9.7.78}$$

The exact expressions for J_1 and J_2 are given in (9.7.67), (9.7.75) respectively. The symbolic evaluation of all these integrals cannot yet be achieved with Mathematica. Those integrals that involve no singularities are done by taking limits of the indefinite integrals. The definite versions of the integrals are done using the Marichev-Adamchik Mellin transform methods. The integration results are initially expressed in terms of Meijer's G-function, which are subsequently converted into hypergeometric functions using Slater's theorem. The notation for Meijer's G-function, belonging to the implemented special functions of Mathematica, is

$$\text{MeijerG}\left[\{\{a_1,...,a_n\},\{a_{n+1},...,a_p\}\},\{\{b_1,...,b_m\},\{b_{m+1},...,b_q\}\},z\right]. \tag{9.7.79}$$

Analytic expressions for the following Meijer's G-functions are available on Wolfram Research's Mathematical Functions web page:

$$G\{m,n,p,q\} = G\{3,0,0,3\} = \text{http://functions.wolfram.com/07.34.03.0948.01}, \tag{9.7.80}$$

and

$$G\{m,n,p,q\} = G\{3,0,1,3\} = \text{http://functions.wolfram.com/07.34.03.0955.01}. \tag{9.7.81}$$

Approximations for large values of z can be worked out with the help of the asymptotic behavior of G-functions, see for example Mathai (1993). These are the following for z very large:

$$J_1 \approx 2\left(\frac{\pi}{3}\right)^{\frac{1}{2}} \left(\frac{z^2}{4}\right)^{\frac{2v+1)}{6}} e^{-3(z^2/4)^{\frac{1}{3}}}$$

$$J_2 \approx d^{v+1} \left(\frac{z^2}{4d}\right)^{-\frac{1}{2}} e^{-d-2(z^2/4d)^{\frac{1}{2}}}$$

$$J_3 \approx 2 \left(\frac{\pi}{3}\right)^{\frac{1}{2}} \left(\frac{z^2}{4}\right)^{\frac{1}{6}} \left[\left(\frac{z^2}{4}\right)^{\frac{1}{3}} - t\right]^{\nu} e^{t - 3(z^2/4)^{\frac{1}{3}}}$$

$$J_4 \approx 2 \left(\frac{\pi}{3}\right)^{\frac{1}{2}} \left(\frac{z^2}{4}\right)^{\frac{2\nu+1}{6}} e^{-3(z^2/4)^{\frac{1}{3}} - b(z^2/4)^{\delta/3}}. \tag{9.7.82}$$

9.7.4.6 A generalization.

A mathematically interesting integral corresponding to (9.7.45) can be evaluated. Consider the integral

$$I = \int_0^\infty e^{-pt} t^{\rho-1} e^{-zt^{-\gamma}} dt. \tag{9.7.83}$$

Then take $f_1(x) = c_1 x^\rho e^{-px}, x > 0, f_2(y) = c_2 e^{-y^\gamma}, \gamma > 0, y > 0$ and $f_1(x) = 0, f_2(y) = 0$ elsewhere, where c_1 and c_2 are normalizing constants. Then $u = xy = z^{1/\gamma}$ and from (9.7.47) one has the integral in (9.7.83) evaluated as the following:

$$I = (\gamma p^\rho)^{-1} H_{0,2}^{2,0}[pz^{1/\gamma}|_{(p,1),(0,1/\gamma)}], 0 < z < \infty,$$

where $H(.)$ is the H-function defined in Chapter 1. When γ is rational the H-function can be rewritten in terms of a Meijer's G-function and then (9.7.83) can be evaluated in terms of the result given in (9.7.59). For specified values of γ and ρ one can obtain computable representations for the H-function.

9.8 Standard and Fractional Diffusion

9.8.1 Fick's first law of diffusion

- diffusion is known to be the equilibration of concentrations
- particle current has to flow against the concentration gradient
- in analogy with Ohm's law for the electric current and with Fourier's law for heat flow,

Fick assumed that the current j is proportional to the concentration gradient

$$j(r,t) = -D \frac{\partial c(r,t)}{\partial r} \tag{9.8.1}$$

D: diffusion coefficient; c: concentration if particles are neither created nor destroyed, then, according to the continuity equation

$$\frac{\partial c(r,t)}{\partial t} = -\frac{\partial j(r,t)}{\partial r}. \tag{9.8.2}$$

Combining Fick's first law with the continuity equation gives Fick's second law = diffusion equation

$$\frac{\partial c(r,t)}{\partial t} = D\frac{\partial^2 c(r,t)}{\partial r^2}, \quad [D] = \frac{L^2}{T}. \tag{9.8.3}$$

9.8.2 Einstein's approach to diffusion

- Fick's phenomenology missed the probabilistic point of view central to statistical mechanics
- in statistical mechanics particles move independently under the influence of thermal agitation
- the concentration of particles $c(r,t)$ at some point r is proportional to the probability $P(r,t)$ of finding a particle at r
- according to Einstein, the diffusion equation holds when probabilities are substituted for concentrations
- if a particle is initially placed at the origin of coordinates in d-dimensional space, then its evolution according to the diffusion equation is given by

$$P(r,t) = \frac{1}{(4\pi Dt)^{d/2}} \exp\left\{-\frac{r^2}{4Dt}\right\} \tag{9.8.4}$$

the mean squared displacement of the particle is thus

$$<r^2(t)> = \int r^2 P(r,t)\mathrm{d}^3 r = 2dDt. \tag{9.8.5}$$

$$<r^2(t)> \propto t$$

9.8.3 Fractional diffusion

In the following we derive the solution of the fractional diffusion equation using Eq. (9.8.3). The result is obtained in the form of the following: Consider the fractional diffusion equation

$$_0D_t^v N(x,t) - \frac{t^{-v}}{\Gamma(1-v)}\delta(x) = -c^v\frac{\partial^2}{\partial x^2}N(x,t), \tag{9.8.6}$$

with the initial condition

$$_0D_t^{v-k}N(x,t)|_{t=0} = 0, \quad (k=1,\ldots,n), \tag{9.8.7}$$

where $n = [\Re(v)]+1, c^v$ is a diffusion constant and $\delta(x)$ is Dirac's delta function. Then for the solution of (9.8.6) there exists the formula

$$N(x,t) = \frac{1}{(4\pi c^v t^v)^{1/2}} H_{1,2}^{2,0} \left[\frac{|x|^2}{4c^v t^v} \Big|_{(0,1),(1/2,1)}^{(1-\frac{v}{2},v)} \right] \tag{9.8.8}$$

In order to derive the solution, we introduce the Laplace-Fourier transform in the form

$$N^*(k,s) = \int_0^\infty \int_{-\infty}^\infty e^{-st+iks} N(x,t) dx \wedge dt. \tag{9.8.9}$$

Applying the Fourier transform with respect to the space variable x and Laplace transform with respect to the time variable t and using Eq. (9.8.6), we find that

$$s^v N^*(k,s) - s^{v-1} = -c^v k^2 N^*(k,s). \tag{9.8.10}$$

Solving for $N^*(k,s)$ gives

$$N^*(k,s) = \frac{s^{v-1}}{s^v + c^v k^2}. \tag{9.8.11}$$

To invert Eq. (9.8.11), it is convenient to first invert the Laplace transform and then the Fourier transform . Inverting the Laplace transform , we obtain

$$\tilde{N}(k,t) = E_v(-c^v k^2 t^v), \tag{9.8.12}$$

which can be expressed in terms of the H-function by using the definition of the generalized Mittag-Leffler functions in terms of a H-function as

$$\tilde{N}(k,t) = H_{1,2}^{1,1} \left[c^v k^2 t^v \Big|_{(0,1),(0,v)}^{(0,1)} \right]. \tag{9.8.13}$$

Using the integral

$$\frac{1}{2\pi} \int_{-\infty}^\infty e^{-ikx} f(k) dk = \frac{1}{\pi} \int_0^\infty f(k) \cos(kx) dk, \tag{9.8.14}$$

and the cosine transform of the H-function to invert the Fourier transform, we see that

$$\begin{aligned}
N(x,t) &= \frac{1}{k} \int_0^\infty \cos(kx) H_{1,2}^{1,1} \left[c^v k^2 t^v \Big|_{(0,1),(0,v)}^{(0,1)} \right] dk \\
&= \frac{1}{|x|} H_{3,3}^{2,1} \left[\frac{|x|^2}{c^v t^v} \Big|_{(1,2),(1,1),(1,1)}^{(1,1),(1,v),(1,1)} \right].
\end{aligned} \tag{9.8.15}$$

Applying a result of Mathai and Saxena (1978, p.4, eq. 1.2.1) the above expression becomes

$$N(x,t) = \frac{1}{|x|} H_{2,2}^{2,0} \left[\frac{|x|^2}{c^v t^v} \Big|_{(1,2),(1,1)}^{(1,v),(1,1)} \right]. \tag{9.8.16}$$

If we employ the formula (Mathai and Saxena, 1978,p. 4, eq. 1.2.4):

$$x^\sigma H_{p,q}^{m,n} \left[x \Big|_{(b_q,B_q)}^{(a_p,A_p)} \right] = H_{p,q}^{m,n} \left[x \Big|_{(b_q+\sigma B_q,B_q)}^{(a_p+\sigma A_p,A_p)} \right]. \tag{9.8.17}$$

Eq. (9.8.16) reduces to

$$N(x,t) = \frac{1}{(c^v t^v)^{1/2}} H_{2,2}^{2,0}\left[\frac{|x|^2}{c^v t^v} \Big|_{(0,2),(1/2,1)}^{(1-\frac{v}{2},v),(1/2,1)} \right]. \tag{9.8.18}$$

In view of the identity in Mathai and Saxena (1978, eq. 1.2.1), it yields

$$N(x,t) = \frac{1}{(c^v t^v)^{1/2}} H_{1,1}^{1,0}\left[\frac{|x|^2}{c^v t^v} \Big|_{(0,2)}^{(1-\frac{v}{2},v)} \right]. \tag{9.8.19}$$

Using the definition of the H-function, it is seen that

$$N(x,t) = \frac{1}{2\pi\omega(c^v t^v)^{1/2}} \int_\Omega \frac{\Gamma(-2\xi)}{\Gamma[1-\frac{v}{2}+v\xi]} \left[\frac{|x|^2}{c^v t^v} \right]^{-\xi} d\xi. \tag{9.8.20}$$

Applying the well-known duplication formula for the gamma function and interpreting the result thus obtained in terms of the H-function, we obtain the solution as

$$N(x,t) = \frac{1}{\sqrt{4\pi c^v t^v}} H_{1,2}^{2,0}\left[\frac{|x|^2}{4c^v t^v} \Big|_{(0,1),(1/2,1)}^{(1-\frac{v}{2},v)} \right]. \tag{9.8.21}$$

Finally the application of the result (Mathai and Saxena (1978, p.10, eq. 1.6.3) gives the asymptotic estimate

$$N(x,t) \sim O\left\{ \left[|x|^{\frac{v}{2-v}} \right] \left[\exp\left\{ -\frac{(2-v)(|x|^2 v^v)^{\frac{1}{2-v}}}{(4c^v t^v)^{\frac{1}{2-v}}} \right\} \right] \right\}, \quad (0 < v < 2). \tag{9.8.22}$$

9.8.4 Spatio-temporal pattern formation

Fractional reaction-diffusion equations provide models of diffusing and reacting species when the diffusion is anomalous sub-diffusion. These equations can be derived in the asymptotic long time limit from a mesoscopic description in terms of continuous time random walks (CTRW) with sources and sinks when the waiting time probability density corresponds to a heavy tailed distribution. The fractional reaction-diffusion system has fractional order temporal derivatives operating on the spatial Laplacian and reaction terms determined by the law of mass action. Turing instability induced pattern formation have been investigated in this model and in related models with fractional order temporal derivatives operating on both the spatial Laplacian and the reaction terms. Linear Turing instability analysis provides a reliable indicator of both the onset and the nature of the patterns that form. Anomalous diffusion with reactions can produce complex spatio-temporal patterns that do not occur in standard reaction-diffusion models.

References

A. Information theory and statistical distribution theory

Mathai, A.M. and Rathie, P.N. (1977): *Probability and Statistics*, Macmillan, London.

Mathai, A.M. and Rathie, P.N. (1975): *Basic Concepts in Information Theory and Statistics: Axiomatic Foundations and Applications*, Wiley Halsted, New York and Wiley Eastern, New Delhi.

Mathai, A.M. (1999): *An Introduction to Geometrical Probability: Distributional Aspects with Applications*, Gordon and Breach, Amsterdam.

B. Generalized special functions of mathematical physics

Mathai, A.M. and Saxena, R.K. (1973): *Generalized Hypergeometric Functions with Applications in Statistics and Physical Sciences*, Springer-Verlag, Heidelberg.

Mathai, A.M. and Saxena, R.K.(1978): *The H-function with Applications in Statistics and Other Disciplines*, Wiley Halsted, New York and Wiley Eastern, New Delhi.

Mathai, A.M. (1993): *A Handbook of Generalized Special Functions for Statistical and Physical Sciences*, Clarendon Press, Oxford.

C. Matrix transformations and functions of matrix argument

Mathai, A.M. and Provost S.B. (1992): *Quadratic Forms in Random Variables: Theory and Applications*, Marcel Dekker, New York.

Mathai, A.M., Provost, S.B., and Hayakawa, T. (1995): *Bilinear Forms and Zonal Polynomials*, Springer-Verlag, New York.

Mathai, A.M. (1997): *Jacobians of Matrix Transformations and Functions of Matrix Argument*, World Scientific, New York.

D. Fractional calculus

Srivastava, H.M. and Saxena, R.K. (2001): Operators of fractional integration and their applications. *Applied Mathematics and Computation*, **118**, 1-52.

E. Stable distributions

Jose, K.K. and Seetha Lekshmi, V. (2004): *Geometric Stable Distributions: Theory and Applications*, A SET Publication, Science Educational Trust, Palai.

F. Gamma functions

Chaudry, M.A. and Zubair, S.M. (2002): *On a Class of Incomplete Gamma Functions with Applications*, Chapman & Hall /CRC, New York.

Section 9.1

Boltzmann, L.: Entropie und Wahrscheinlichkeit (1872-1905). Ostwalds Klassiker der Exakten Wissenschaften, Band **286**, Verlag Harri Deutsch, Frankfurt am Main 2002.

Planck, M.: Die Ableitung der Strahlungsgesetze (1895-1900): Sieben Abhandlungen aus dem Gebiet der Elektrischen Strahlungstheorie. Ostwalds Klassiker der Exakten Wissenschaften, Band **206**, Verlag Harri Deutsch, Frankfurt am Main 2001.

Einstein, A. und von Smoluchowski, M.: Untersuchungen ueber die Theorie der Brownschen Bewegung; Abhandlung ueber die Brownsche Bewegung und verwandte Erscheinungen. Ostwalds Klassiker der Exakten Wissenschaften, Reprint der Baende **199** und **207**, Verlag Harri Deutsch, Frankfurt am Main 2001.

Pais, A. (1982): *Subtle is the Lord...: The Science and the Life of Albert Einstein*, Oxford University Press, Oxford.

Bach, A. (1990): Boltzmann's probability distribution of 1877. *Archive for History of Exact Sciences*, **41(1)**,1-40.

Nicolis, G. and Prigogine, I. (1977): *Self-Organization in Nonequilibrium Systems*, Wiley, New York.

Haken, H. (2000): *Information and Self-Organization: A Macroscopic Approach to Complex Systems*, Springer-Verlag, Berlin, Heidelberg.

Tsallis, C. and Gell-Mann, M. (Eds.) (2004): *Nonextensive Entropy: Interdisciplinary Applications*, Oxford University Press, New York.

Haubold, H.J., Mathai, A.M., and Saxena, R.K. (2004): Boltzmann-Gibbs entropy versus Tsallis entropy: Recent contributions to resolving the argument of Einstein concerning "Neither Herr Boltzmann nor Herr Planck has given a definition of W"? *Astrophysics and Space Science*, **290**, 241-245.

Masi, M.(2005): A step beyond Tsallis and Renyi entropies. *Physics Letters*, **A338**, 217-224.

Section 9.2

Emden, R. (1907): *Gaskugeln: Anwendungen der Mechanischen Waermetheorie auf Kosmologische und Meteorologische Probleme*, Verlag B.G. Teubner, Leipzig und Berlin.

Chandrasekhar, S. (1967): *An Introduction to the Study of Stellar Structure*, Dover, New York.

Stein, R.F. and Cameron, A.G.W. (Eds.) (1966): *Stellar Evolution*, Plenum Press, New York.

Kourganoff, V. (1973): *Introduction to the Physics of Stellar Interiors*, D. Reidel Publishing Company, Dordrecht.

Bethe, H.A. (1973): Energy production in stars. *Science*, **161**, 541-547.

Chandrasekhar, S. (1984): On stars, their evolution and their stability. *Reviews of Modern Physics*, **56**, 137-147.

Haubold, H.J. and Mathai, A.M. (1994): Solar nuclear energy generation and the chlorine solar neutrino experiment. in *Conference Proceedings No. 320: Basic Space Science*, American Institute of Physics, New York, pp. 102-116.

Haubold, H.J. and Mathai, A.M. (1995): Solar structure in terms of Gauss' hypergeometric function. *Astrophysics and Space Science*, **228**, 77-86.

Clayton, D.D. (1986): Solar structure without computers. *American Journal of Physics*, **54(4)**, 354-362.

Section 9.3

Davis Jr., R. (2003): A half-century with solar neutrinos. *Reviews of Modern Physics*, **75**, 985-994.

Davis Jr., R. 1996): A review of measurements of the solar neutrino flux and their variation. *Nuclear Physics*, **B48**, 284-298.

Smirnov, A.Yu. (2003): The MSW effect and solar neutrinos. In *Tenth International Workshop on Neutrino Telescopes, Proceedings*, ed. Milla Baldo Ceolin, Venezia, March 11-14, 2003, Instituto Veneto di Scienze, Lettere ed Arti, Campo Santo Stefano, edizionni papergraf, pp. 23-43.

Haubold, H.J. and Gerth, E. (1990): On the Fourier spectrum analysis of the solar neutrino capture rate. *Solar Physics*, **127**, 347-356.

Haubold, H.J. (1998): Wavelet analysis of the new solar neutrino capture rate data for the Homestake experiment. *Astrophysics and Space Science*, **258**, 201-218.

Dicke, R.H. (1978): Is there as chronometer hidden deep in the Sun? *Nature*, **276**, 676-680.

Kononovich, E.V. (2004): Mean variations of the solar activity cycles: analytical representations. In *Proceedings XXVII Seminar on Physics of Auroral Phenomena, Apatity*, Kola Science Center, Russian Academy of Science 2004, pp. 83-86.

Burlaga, L.F. and Vinas, A.F. (2005): Triangle for the entropic index q of non-extensive statistical mechanics observed by Voyager 1 in the distant heliosphere. *Physica*, **A356**, 375-384.

Siegert, S., Friedrich, R., and Peinke, J. (1998): Analysis of data sets of stochastic systems. *Physics Letters*, **A243**, 275-280.

Risken, H. (1996): *The Fokker-Planck Equation*, Springer-Verlag, Berlin Heidelberg.

Frank, T.D. (2005): *Nonlinear Fokker-Planck Equations*, Springer-Verlag, Berlin Heidelberg.

Section 9.4

Balescu, R. (2000): *Statistical Dynamics: Matter out of Equilibrium*, Imperial College Press, London.

Van Kampen, N.G. (2003): *Stochastic Processes in Physics and Chemistry*, Elsevier, Amsterdam.

Balescu, R. (2005): *Aspects of Anomalous Transport in Plasmas*, Institute of Physics Publishing, Bristol and Philadelphia.

Section 9.5

West, B.J., Bologna, M., and Grigolini, P.(2005): *Physics of Fractal Operators*, Springer-Verlag, New York.

Stanislavsky, A.A. (2004): Probability interpretation of the integral of fractional order. *Theoretical and Mathematical Physics*, **138**, 418-431.

Section 9.6

Cohen, E.G.D. (2005): Boltzmann and Einstein: statistics and dynamics - an unsolved problem. *Pramana Journal of Physics*, **64**, 635-643.

Boon, J.P. and Tsallis, C. (Eds.) (2005): Nonextensive Statistical Mechanics: New Trends, New Perspectives. *Europhysics News*, **36**, 183-231.

Tsallis, C. (2004): Dynamical scenario for nonextensive statistical mechanics. *Physica*, **A340**, 1-10.

Saxena, R.K., Mathai, A.M., and Haubold, H.J. (2004): Astrophysical thermonuclear functions for Boltzmann-Gibbs and Tsallis statistics. *Physica*, **A344**, 649-656.

Tsallis, C., Gell-Mann, M., and Sato, Y. (2005): Asymptotically scale-invariant occupancy of phase space makes the entropy S_q extensive. *Proceedings of The National Academy of Sciences of the USA*, **102**, 15377-15382.

Section 9.7

Ben-Avraham, D. and Havlin S. (2000): *Diffusion and Reactions in Fractals and Disordered Systems*, Cambridge University Press, Cambridge.

Fowler, W.A. (1984): Experimental and theoretical nuclear astrophysics: The quest for the origin of the elements. *Reviews of Modern Physics*, **56**, 149-179.

Haubold, H.J. and Mathai, A.M. (1995): A heuristic remark on the periodic variation in the number of solar neutrinos detected on Earth. *Astrophysics and Space Science*, **228**, 113-134.

Haubold, H.J. and Mathai, A.M. (1985): The Maxwell-Boltzmannian approach to the nuclear reaction rate theory. *Progress of Physics*, **33**, 623-644.

Anderson, W.J., Haubold, H.J., and Mathai, A.M. (1994): Astrophysical thermonuclear functions. *Astrophysics and Space Science*, **214**, 49-70.

Haubold, H.J. and Mathai, A.M. (2004): The fractional kinetic equation and thermonuclear functions. *Astrophysics and Space Science*, **273**, 53-63.

Tsallis, C. (2004): What should a statistical mechanics satisfy to reflect nature? *Physica*, **D193**, 3-34.

Section 9.8

Metzler, R. and Klafter, J. (2000): The Random Walk's Guide to Anomalous Diffusion: A Fractional Dynamics Approach. *Physics Reports*, **339**, 1-77.

Metzler, R. and Klafter, J. (2004): The restaurant at the end of the random walk: Recent developments in the description of anomalous transport by fractional dynamics. *Journal of Physics A: Math. Gen.*, **37**, R161-R208.

Saxena, R.K., Mathai, A.M., and Haubold, H.J. (2004): On generalized fractional kinetic equations. *Physica A*, **344**, 657-664.

Saxena, R.K., Mathai, A.M., and Haubold, H.J. (2004): Unified fractional kinetic equation and a fractional diffusion equation. *Astrophysics and Space Science*, **290**, 299-310.

Section 9.9

Haken, H. (2004): *Synergetics: Introduction and Advanced Topics*, Springer-Verlag, Berlin Heidelberg.

Wilhelmsson, H. and Lazzaro, E. (2001): *Reaction-Diffusion Problems in the Physics of Hot Plasmas*, Institute of Physics Publishing, Bristol and Philadelphia.

Murray, J.D. (2003): *Mathematical Biology. Volume I: An Introduction. Volume II: Spatial Models and Biomedical Applications*, Springer-Verlag, Berlin Heidelberg.

Adamatzky, A., De Lacy Costello, B., and Asai, T. (2005): *Reaction-Diffusion Computers*, Elsevier, Amsterdam.

Vlad, M.O. and Ross, J. (2002): Systematic derivation of reaction-diffusion equations with distributed delays and relations to fractional reaction-diffusion equations and hyperbolic transport equations: Applications to the theory of Neolithic transition. *Physical Review*, **E66**, 061908-1 - 061908-11.

Seki, K., Wojcik, M., and Tachiya, M. (2003): Fractional reaction-diffusion equations. *Journal of Chemical Physics*, **119**, 2165-2170.

Henry, B.I. and Wearne, S.L. (2000): Fractional reaction-diffusion. *Physica A*, **276**, 448-455.

Del-Castillo-Negrete, D., Carreras, B.A., and Lynch, V. (2003): Front dynamics in reaction-diffusion systems with Levy flights: A fractional diffusion approach. *Physical Review Letters*, **91**, 018302-1 - 018302-4.

Henry, B.I., Langlands, T.A.M., and Wearne, S.L. (2005): Turing pattern formation in fractional activator-inhibitor systems. *Physical Review*, **E72**, 026101-1 - 026101-14.

Chapter 10
An Introduction to Wavelet Analysis

[*This chapter is based on the lectures of Professor D.V. Pai, Department of Mathematics, Indian Institute of Technology Bombay, Powai, Mumbai - 400 076, India.*]

10.0 Introduction

During the last 20 years or so, the subject of "Wavelet analysis" has attracted a lot of attention from both mathematicians and engineers alike. Vaguely speaking the term "Wavelet" means a little wave, and it includes functions that are reasonably localized in the *time domain* as well as in the *frequency domain*. The idea seems to evolve from the limitation imposed by the *uncertainty principle* of Physics which puts a limit on simultaneous localization in both the time and the frequency domains.

From a historical perspective, although the idea of wavelet seems to originate with the work by Gabor and by Neumann in the late 1940s, this term seems to have been coined for the first time in the more recent seminal paper of Grossman and Morlet (1984). Nonethless, the techniques based on the use of translations and dilations are much older. This can be at least traced back to Calderón (1964) in his study of singular integral operators. Starting with the pioneering works reported in the early monographs contributed by Meyer (1992), Mallat (1989), Chui (1992), Daubechies (1992) and others, an ever increasing number of books, monographs and proceedings of international conferences which have appeared more recently in this field only point to its growing importance.

The main aim of these lectures is to attempt to present a quick introduction of this field to a beginner. We will mainly emphasize here the construction of orthonormal (o.n.) wavelets using the so-called *two-scale relation* . This will lead us to a natural classification of wavelets as well as to the classical multiresolution analysis. In particular, we will also attempt to highlight the spline wavelets of Chui and Wang (1993).

389

10.1 Fourier Analysis to Wavelet Analysis

Let $L^2(0,2\pi) = $ the space of all (equivalence classes) of 2π-periodic, Lebesgue measurable functions $f : \mathbb{R} \to \mathbb{C}$ such that $\int_0^{2\pi} |f(t)|^2 dt < \infty$. $L^2(0,2\pi)$ is a Hilbert space furnished with the inner product

$$(f,g) = \frac{1}{2\pi} \int_0^{2\pi} f(t)\overline{g(t)}dt, \quad f,g \in L^2(0,2\pi)$$

and the corresponding norm

$$\|f\|_2 = \left\{ \frac{1}{2\pi} \int_0^{2\pi} |f(t)|^2 dt \right\}^{1/2}.$$

Any f in $L^2(0,2\pi)$ has a *Fourier series representation*

$$f(t) = \sum_{k=-\infty}^{\infty} c_k e^{ikt}, \tag{10.1.1}$$

where the constants c_k, called the *Fourier coefficients* of f, are defined by

$$c_k = (f,w_k) = \frac{1}{2\pi} \int_0^{2\pi} f(t)\bar{e}^{ikt}dt, \quad w_k(t) = e^{ikt}. \tag{10.1.2}$$

This is a consequence of the important fact that $\{w_k(t) : k \in \mathbb{Z}\}$ is an *orthonormal basis* of $L^2(0,2\pi)$. Also recall that the Fourier series representation satisfies the so-called *Parseval identity*:

$$\|f\|_2^2 = \sum_{k=-\infty}^{\infty} |c_k|^2, \quad f \in L^2(0,2\pi).$$

Let us emphasize two interesting features in the Fourier series representation (10.1.1). Firstly, note that f is decomposed into an infinite sum of mutually orthogonal components $c_k w_k$. The second interesting feature to be noted of (10.1.1) is that the o.n. basis $\{w_k : k \in \mathbb{Z}\}$ is generated by "dilates" of a single function

$$w(t) := w_1(t) = e^{it};$$

that is, $w_k(t) = w(kt)$, $k \in \mathbb{Z}$, is, in fact, an *integral dilate* of $w(t)$. Let us reemphasize the following remarkable fact:

Every 2π-periodic square-integrable function is generated by a superposition of integral dilates of the single basic function $w(t) = e^{it}$.

The basic function $w(t) = \cos t + i\sin t$ is a sinusoidal wave. For any integer k with $|k|$ large, the wave $w_k(t) = w(kt)$ has high *frequency*, and for k in $\mathbb{Z}$ with $|k|$ small, the wave w_k has low frequency. Thus *every function in $L^2(0,2\pi)$ is composed of waves with various frequencies.*

Let $L^2(\mathbb{R}) := $ the space of all (equivalence classes) of complex measurable functions, defined on $\mathbb{R}$ for which

$$\int_{\mathbb{R}} |f(t)|^2 dt < \infty.$$

Note that the space $L^2(\mathbb{R})$ is a Hilbert space with the inner product

$$\langle f, g \rangle = \int_{\mathbb{R}} f(t)\overline{g(t)}dt \quad f, g \in L^2(\mathbb{R})$$

and the norm

$$\|f\|_2 = \left\{ \int_{\mathbb{R}} |f(t)|^2 dt \right\}^{1/2} \quad f \in L^2(\mathbb{R}).$$

Wavelet analysis also begins with a quest for a single function ψ in $L^2(\mathbb{R})$ to generate $L^2(\mathbb{R})$. Since any such function must decay to zero at $\pm\infty$, we must give up, as being too restrictive, the idea of using only linear combinations of dilates of ψ to recover $L^2(\mathbb{R})$. Instead, it is natural to consider both the *dilates* and the *translates*. The most convenient family of functions for this purpose is thus given by

$$\psi_{j,k}(t) = 2^{j/2}\psi(2^j t - k), \quad j, k \in \mathbb{Z}. \tag{10.1.3}$$

This involves a *binary dilation* (dilation by 2^j) and a *dyadic translation* (of $k/2^j$).

Lemma 10.1.1: Let ϕ, ψ be in $L^2(\mathbb{R})$. Then, for i, j, k, ℓ in $\mathbb{Z}$, we have:

(i) $\langle \psi_{j,k}, \phi_{j,\ell} \rangle = \langle \psi_{i,k}, \phi_{i,\ell} \rangle$;

(ii) $\|\psi_{j,k}\|_2 = \|\psi\|_2$.

Proof 10.1.1:

(i) We have

$$\langle \psi_{j,k}, \phi_{j,\ell} \rangle = \int_{\mathbb{R}} 2^{j/2}\psi(2^j t - k)2^{j/2}\overline{\phi(2^j t - \ell)}dt.$$

Put $t = 2^{i-j}x$ to get

$$\text{R.H.S.} = \int_{\mathbb{R}} 2^{i/2}\psi(2^i x - k)2^{i/2}\overline{\phi(2^i x - \ell)}dx$$
$$= \langle \psi_{i,k}, \phi_{i,\ell} \rangle.$$

Note that

$$\|\psi_{j,k}\|_2^2 = 2^j \int_{\mathbb{R}} |\psi(2^j t - k)|^2 dt$$
$$= \int_{\mathbb{R}} |\psi(x)|^2 dx = \|\psi\|_2^2. \text{ (We put } x = 2^j t - k.)$$

Remark 10.1.1: For $i, j \in \mathbb{Z}$, we have:
the set $\{\psi_{i,k} : k \in \mathbb{Z}\}$ is orthonormal
$\Leftrightarrow$ the set $\{\psi_{j,k} : k \in \mathbb{Z}\}$ is orthonormal.

Definition 10.1.1. A function $\psi \in L^2(\mathbb{R})$ is called an **orthonormal wavelet** (or an **o.n. wavelet**) if the family $\{\psi_{j,k}\}$, as defined in (10.1.3), is an orthonormal basis of $L^2(\mathbb{R})$; that is,

$$\langle \psi_{j,k}, \psi_{i,\ell} \rangle = \delta_{j,i}\delta_{k,\ell}, \quad j,k,i,\ell \in \mathbb{Z} \tag{10.1.4}$$

and every f in $L^2(\mathbb{R})$ has a representation

$$f(t) = \sum_{j,k=-\infty}^{\infty} c_{j,k}\psi_{j,k}(t), \tag{10.1.5}$$

where the convergence of the series in (10.1.5) is in $L^2(\mathbb{R})$:

$$\lim_{M_1,N_1,M_2,N_2 \to \infty} \left\| f - \sum_{j=-M_1}^{N_1} \sum_{k=-M_2}^{N_2} c_{j,k}\psi_{j,k} \right\|_2 = 0.$$

The series representation (10.1.5) of f is called a **wavelet series** and the coefficients $c_{j,k}$ given by

$$c_{j,k} = \langle f, \psi_{j,k} \rangle \tag{10.1.6}$$

are called the **wavelet coefficients**.

Example 10.1.1. Let us recall the definition of the **Haar function** $\psi^H(t)$ given below:

$$\psi^H(t) := \begin{cases} 1, & \text{if } 0 \le t < \frac{1}{2} \\ -1, & \text{if } \frac{1}{2} \le t < 1 \\ 0, & \text{otherwise.} \end{cases}$$

At this stage, the reader is urged as an exercise to verify that the family $\{\psi_{j,k}^H : j,k \in \mathbb{Z}\}$ is orthonormal in the space $L^2(\mathbb{R})$. We will come back again to this example in the next section. It will be shown there that ψ^H is, in fact, an o.n. wavelet.

Next, let us recall that the **Fourier transform** of a function f in $L^1(\mathbb{R})$ is the function $\hat{f}$ defined by

$$\hat{f}(w) := \int_{\mathbb{R}} f(t)\bar{e}^{iwt}\,dt, \quad w \in \mathbb{R}.$$

Definition 10.1.2. If a function $\psi \in L^2(\mathbb{R})$ satisfies the **admissibility condition:**

$$C_\psi := \int_{\mathbb{R}} \frac{|\hat{\psi}(w)|^2}{|w|}\,dw < \infty$$

then ψ is called a **"basic wavelet"**.

The definition is due to Grossman and Morlet (1984). It is related to the invertibility of the continuous wavelet transform as given by the next definition.

Definition 10.1.3. Relative to every basic wavelet ψ, consider the family of wavelets defined by

$$\psi_{a,b}(t) := |a|^{-1/2}\psi\left(\frac{t-b}{a}\right), a,b \in \mathbb{R}, \quad a \neq 0. \tag{10.1.7}$$

The **continuous wavelet transform** (CWT) corresponding to ψ is defined by

$$(W_\psi f)(a,b) = |a|^{-1/2}\int_{\mathbb{R}} f(t)\overline{\psi\left(\frac{t-b}{a}\right)}\,dt, \quad f \in L^2(\mathbb{R}) \tag{10.1.8}$$
$$= \langle f, \psi_{a,b}\rangle.$$

Let us note that the wavelet coefficients in (10.1.7) and (10.1.8) become

$$c_{j,k} = (W_\psi f)\left(\frac{1}{2^j}, \frac{k}{2^j}\right). \tag{10.1.9}$$

Thus wavelet series and the continuous wavelet transform are intimately related.

Let us also state the following *inversion theorem* for the continuous wavelet transform. The proof uses the Fourier transform of $\psi_{a,b}$, the Parseval identity and the fact that the Gaussian functions

$$g_\alpha(t) := \frac{1}{2\sqrt{\pi\alpha}}e^{-\frac{t^2}{4\alpha}}, \quad \alpha > 0$$

is an approximate identity in $L^1(\mathbb{R})$. Thus, for $f \in L^1(\mathbb{R})$, $\lim_{\alpha \to 0}(f \cdot g_\alpha)(t) = f(t)$ at every point t where f is continuous. The details of the proof are left to the reader as an exercise.

Theorem 10.1.1. *Let ψ in $L^2(\mathbb{R})$ be a basic wavelet which defines a continuous wavelet transform W_ψ. Then for any f in $L^2(\mathbb{R})$ and $t \in \mathbb{R}$ at which f is continuous,*

$$f(t) = \frac{1}{C_\psi}\int_{-\infty}^{\infty}\int_{-\infty}^{\infty}(W_\psi f)(a,b)\psi_{a,b}(t)\frac{da\,db}{a^2}, \tag{10.1.10}$$

where $\psi_{a,b}$ is defined by (10.1.7).

10.2 Construction of Orthonormal Wavelets

One of the first examples of an o.n. wavelet is due to Haar (1910). Let us recall (Example 10.1.1) that it is called the *Haar function* defined by

$$\psi^H(t) := \begin{cases} 1, & \text{if } 0 \leq t < \frac{1}{2} \\ -1, & \text{if } \frac{1}{2} \leq t < 1 \\ 0, & \text{otherwise.} \end{cases}$$

Most of the recent theories on wavelets are no doubt inspired by this example. However, as it turns out, its discontinuous nature is a serious drawback in many applications. Thus, in these lectures, one of our quests is to explore more examples. Let us begin with a real function ϕ in $L^2(\mathbb{R})$. As a first step let us assume that

(S_o) :   the family $\{\phi_{0,k}(t) = \phi(t-k) : k \in \mathbb{Z}\}$ is orthonormal.

Then it follows from Lemma 10.1.1, that the family $\{\phi_{j,k}(t) : k \in \mathbb{Z}\}$ is orthonormal, for each $j \in \mathbb{Z}$. Let us define

$$V_j = \overline{\text{span}}\{\phi_{j,k} : k \in \mathbb{Z}\}, \quad (j \in \mathbb{Z}),$$

the closure being taken in the topology of $L^2(\mathbb{R})$. It results from the next lemma that

$$V_j = \left\{ \sum_{k \in \mathbb{Z}} c_k \phi_{j,k} : c = \{c_k\} \in \ell^2(\mathbb{Z}) \right\}. \tag{10.2.1}$$

Lemma 10.2.1: Let $\{u_k : k \in \mathbb{Z}\}$ be an orthonormal bi-infinite sequence in a Hilbert space X. Then

$$\overline{\text{span}}\{u_k : k \in \mathbb{Z}\} = \left\{ \sum_{k=-\infty}^{\infty} c_k u_k : c = \{c_k\} \in \ell^2(\mathbb{Z}) \right\}.$$

Proof 10.2.1: Let V denote the L.H.S. set and U be the R.H.S. set. Note that for a sequence $\{c_k\} \in \ell^2(\mathbb{Z})$, the series $\sum_k c_k u_k$ converges, because its partial sums form a Cauchy sequence in X:

$$\left\| \sum_{k=-M}^{M} c_k u_k - \sum_{k=-N}^{N} c_k u_k \right\|^2 = \sum_{k=N+1}^{M} |c_k|^2 + \sum_{k=-M}^{-(N+1)} |c_k|^2 \to 0 \text{ as } M, N \to \infty.$$

Clearly, $U \subset V$. Also, U is a closed subspace being isometrically isomorphic to $\ell^2(\mathbb{Z})$. Since $u_k \in U$, we have $\text{span}\{u_k\} \subset U \Rightarrow \overline{\text{span}}\{u_k\} \subset \overline{U} = U$. Hence $V \subset U$

Let us assume, *in addition*, that $\phi \in V_1$. Then for a suitable $c \in \ell^2(\mathbb{Z})$, we will have

(S_1) $\phi(t) = \sum_{k \in \mathbb{Z}} c_k \phi(2t - k).$

This is called a **two-scale relation** or a **dilation equation**.

Lemma 10.2.2: Let ϕ in $L^2(\mathbb{R})$ satisfy (S_0) and (S_1). Then for all i, j in $\mathbb{Z}$,

$$\sum_k \left[c_{j-2k} c_{i-2k} + (-1)^{i+j} c_{1-j+2k} c_{1-i+2k} \right] = 2\delta_{j,i}. \tag{10.2.2}$$

Here the coefficients c_j's are as defined in (S_1).

Proof 10.2.2:

Case 1: $i+j$ is odd.

Then L.H.S. of (10.2.2) becomes

$$\sum_k c_{j-2k} c_{i-2k} - \sum_k c_{1-j+2k} c_{1-i+2k}.$$

Put $k = -r$ in the first sum and $k = r + \frac{i+j-1}{2}$ in the second sum to obtain

$$\sum_r c_{j+2r} c_{i+2r} - \sum_r c_{i+2r} c_{j+2r} = 0.$$

Case 2: $i+j$ is even.

Then L.H.S. of (10.2.2) becomes $\sum_k c_{j-2k} c_{i-2k} + \sum_k c_{1-j+2k} c_{1-i+2k}$. Put $k = -r$ in the first sum and $k = r + \frac{i+j}{2}$ in the second sum to obtain

$$\sum_r c_{j+2r} c_{i+2r} + \sum_r c_{i+2r+1} c_{j+2r+1} = \sum_k c_{j+k} c_{i+k} = \sum_\ell c_\ell c_{\ell+i-j}.$$

(We put $j+k = \ell$.)

Since the set $\{\phi_{1,k} : k \in \mathbb{Z}\}$ is orthonormal,

$$\sum_\ell c_\ell c_{\ell+i-j} = \langle \sum_\ell c_\ell \phi_{1,\ell}, \sum_\ell c_{\ell+i-j} \phi_{1\ell} \rangle.$$

Using (S_1), we have

$$\sum_\ell c_\ell \phi_{1\ell}(t) = \sum_\ell c_\ell 2^{1/2} \phi(2t - \ell) = 2^{1/2} \phi(t).$$

Likewise,

$$\sum_\ell c_{\ell+i-j} \phi_{1\ell}(t) = \sum_\ell c_{\ell+i-j} 2^{1/2} \phi(2t - \ell) \text{(put } \ell+i-j = m)$$

$$= \sum_m c_m 2^{1/2} \phi(2t + i - j - m)$$

$$= 2^{1/2} \sum_m c_m \phi(2(t - \frac{j-i}{2}) - m)$$

$$= 2^{1/2} \phi(t - \frac{j-i}{2}) = 2^{1/2} \phi_{0,\frac{j-i}{2}}.$$

Thus,

$$\sum_\ell c_\ell c_{\ell+i-j} = \langle 2^{1/2} \phi_{0,0}, 2^{1/2} \phi_{0,\frac{j-i}{2}} \rangle$$

$$= 2\delta_{j,i}, \text{ using } (S_0).$$

Lemma 10.2.3: Let ϕ in $L^2(\mathbb{R})$ satisfy (S_0) and (S_1). Let ψ in $L^2(\mathbb{R})$ be defined by

$$(W) \qquad \psi(t) = \sum_{k=-\infty}^{\infty} (-1)^k l c_{1-k} \phi(2t - k).$$

Then for $k \in \mathbb{Z}$,

$$\phi_{1,k} = 2^{-1/2} \sum_{m} \left[c_{k-2m} \phi_{0,m} + (-1)^k c_{1-k+2m} \psi_{0,m} \right]. \tag{10.2.3}$$

Here the coefficients c_j's are as defined in the two-scale relation (S_1).

Proof 10.2.3: By (S_1), $\qquad \phi(t) = \sum_i c_i \phi(2t - i)$

$$\Rightarrow \quad \phi(t - m) = \sum_i c_i \phi(2(t - m) - i) = \sum_i 2^{-1/2} c_i 2^{1/2} \phi(2t - 2m - i)$$

$$\Rightarrow \quad \phi_{0,m} = \sum_i 2^{-1/2} c_i \phi_{1,2m+i}.$$

Likewise,

$$\psi_{0,m} = \sum_i (-1)^i 2^{-1/2} c_{1-i} \phi_{1,2m+i}.$$

The (R.H.S.) of (10.2.3) can now be written as

$$2^{-1/2} \sum_{m} \left[c_{k-2m} \sum_i 2^{-1/2} c_i \phi_{1,2m+i} + (-1)^k c_{1-k+2m} \sum_i 2^{-1/2} (-1)^i c_{1-i} \phi_{1,2m+i} \right].$$

Changing the index i to r by the equation $r = 2m + i$, one obtains:

$$2^{-1} \sum_r \sum_m \left[c_{k-2m} c_{r-2m} + (-1)^{k+r} c_{1-k+2m} c_{1-r+2m} \right] \phi_{1,r} = 2^{-1} \sum_r 2\delta_{k,r} \phi_{1,r}.$$

The last expression equals $\phi_{1,k}$.

Lemma 10.2.4: Let U and V be closed subspaces of a Hilbert space X such that $U \perp V$. Then $U + V$ is closed.

Proof 10.2.4: Let $w_n = u_n + v_n, u_n \in U, v_n \in V$, be such that $w_n \to w$. Since

$$\|w_n - w_m\|^2 = \|u_n - u_m\|^2 + \|v_n - v_m\|^2$$

and $\{w_n\}$ is Cauchy, both the sequences $\{u_n\}, \{u_n\}$ are Cauchy. If $u = \lim u_n, v = \lim v_n$, then

$$w = u + v \in U + V.$$

Theorem 10.2.1. Let ϕ in $L^2(\mathbb{R})$ satisfy properties (S_0) and (S_1). Let ψ in $L^2(\mathbb{R})$ be defined by (W). Then the family $\{\psi_{j,k} : j, k \in \mathbb{Z}\}$ is orthonormal. Moreover, $V_1 = V_0 \oplus^{\perp} W_0$, where

$$W_0 = \overline{\text{span}}\{\psi_{0,k} : k \in \mathbb{Z}\}$$

Let us recall the property (W):

$$\psi(t) = \sum_{k=-\infty}^{\infty} (-1)^k c_{1-k} \phi(2t-k).$$

Proof 10.2.5: **Assertion 1:** For all n, m in $\mathbb{Z}$, $\phi_{0,n} \perp \psi_{0,m}$. Indeed,

$$\langle \phi_{0,n}, \psi_{0,m} \rangle = \int_{\mathbb{R}} \phi(t-n)\psi(t-m)dt$$

$$= \int_{\mathbb{R}} \left[\sum_k c_k \phi(2t-2n-k) \sum_\ell (-1)^\ell c_{1-\ell} \phi(2t-2m-\ell) \right] dt$$

$$= \sum_k \sum_\ell (=-1)^\ell c_k c_{1-\ell} \int_{\mathbb{R}} \phi(x-2n-k)\phi(x-2m-\ell)\frac{1}{2}dx$$

$$(\text{Put } 2t = x)$$

$$= \frac{1}{2} \sum_k \sum_\ell (-1)^\ell c_k c_{1-\ell} \delta_{2n+k,2m+\ell}$$

$$= \frac{1}{2} \sum_k (-1)^k c_k c_{p-k}$$

$$(2n+k = 2m+\ell \Rightarrow \ell = 2n-2m+k. \text{ Put } p = 2m-2n+1.)$$

$$= \frac{1}{4} \left[\sum_k (-1)^k c_k c_{p-k} + \sum_i (-1)^i c_i c_{p-i} \right] \quad (\text{Put } p-i = q.)$$

$$= \frac{1}{4} \left[\sum_k (-1)^k c_k c_{p-k} + \sum_q (-1)^{p-q} c_q c_{p-q} \right]$$

$$= \frac{1}{4} \left[\sum_k (-1)^k c_k c_{p-k} - \sum_q (-1)^q c_q c_{p-q} \right] = 0.$$

We have just established that $V_0 \perp W_0$.

Assertion 2: $\{\psi_{0,n} : n \in \mathbb{Z}\}$ is orthonormal. We have

$$\phi_{0,n}(t) = \phi(t-n) = \sum_k c_k \phi(2t-2n-k) = 2^{-1/2} \sum_k c_k \phi_{1,k+2n},$$

$$\psi_{0,n}(t) = \psi(t-n) = \sum_\ell (-1)^\ell c_{1-\ell} \phi(2t-2n-\ell) = 2^{-1/2} \sum_\ell (-1)^\ell c_{1-\ell} \phi_{1,\ell+2n}.$$

Hence

$$\langle \phi_{0,n}, \phi_{0,m} \rangle = 2^{-1} \langle \sum_k c_k \phi_{1,k+2n}, \sum_\ell c_\ell \phi_{1,\ell+2m} \rangle$$

$$= 2^{-1} \sum_k \sum_\ell c_k c_\ell \delta_{k+2n,\ell+2m} = 2^{-1} \sum_k c_k c_{k+2n-2m}.$$

A similar calculation gives

$$\langle \psi_{0,n}, \psi_{0,m} \rangle = 2^{-1} \sum_k (-1)^k (-1)^{k+2n-2m} c_{1-k} c_{1-2n+2m-k}$$

$$= 2^{-1} \sum_i C_i C_{i+2m-2n} = \delta_{m,n}.$$

Assertion 3:

$$V_1 = V_0 \overset{\perp}{\bigoplus} W_0.$$

From (S_1) and (W) we have $\phi \in V_1, \psi \in V_1$. The same is true for their integer shifts. Hence $V_0 + W_0 \subset V_1$. By Lemma 10.2.4, $V_0 + W_0$ is closed. Also, by Lemma 10.2.3, $\phi_{1,k} \in V_0 + W_0$. This implies $V_1 = \overline{\text{span}}\{\phi_{1,k}\} \subset V_0 + W_0$. Since $V_0 \perp W_0$, we conclude that $V_0 \overset{\perp}{\bigoplus} W_0 = V_1$. To complete the proof, let $W_j = \overline{\text{span}}\{\psi_{j,k} : k \in \mathbb{Z}\}$. On similar lines as before, one obtains

$$V_j = V_{j-1} \overset{\perp}{\bigoplus} W_{j-1}, \quad j \in \mathbb{Z}.$$

Assertion 4: $\langle \psi_{j,k}, \psi_{i,\ell} \rangle = \delta_{j,i} \delta_{k,\ell}.$

For $j = i$, this follows from Lemma 10.1.1. Assume $j \neq i$. Let $j < i$. Then $\psi_{j,k} \in W_j \subset V_{j+1} \subset V_i$ and $\psi_{i,\ell} \in W_i$. Hence $\psi_{j,k} \perp \psi_{i,\ell}$.

Remark 10.2.1: Assume $\phi \in L^2(\mathbb{R})$ satisfies properties (S_0) and (S_1). Then for $\ell \in \mathbb{Z}$,

$$\phi_{0,\ell}(t) = \phi(t-\ell) = \sum_k c_k \phi(2t - 2\ell - k)$$

$$= 2^{-1/2} \sum_k c_k 2^{1/2} \phi(2t - 2\ell - k) = 2^{-1/2} \sum_k c_k \phi_{1,2\ell+k}.$$

$$\phi_{j-1,k}(t) = 2^{\frac{j-1}{2}} \phi(2^{j-1}t - k)$$

$$= 2^{\frac{j-1}{2}} \sum_\ell c_\ell \phi(2^j t - 2k - \ell)$$

$$= 2^{-1/2} \sum_\ell c_\ell \phi_{j,2k+\ell}.$$

Thus

$$\phi_{j-1,k}(t) = 2^{-1/2} \sum_r c_{r-2k} \phi_{j,r}. \tag{10.2.4}$$

On the same lines, one sees that

$$\psi_{j-1,k} = 2^{-1/2} \sum_r (-1)^r c_{2k+1-r} \phi_{j,r}. \tag{10.2.5}$$

Theorem 10.2.2. *Let ϕ in $L^2(\mathbb{R})$ satisfy properties (S_0) and (S_1). Let ψ in $L^2(\mathbb{R})$ be defined by*

$$(W) \qquad \psi(t) = \sum_{k=-\infty}^{\infty} (-1)^k c_{1-k} \phi(2t - k)$$

(with coefficients c_j's as in (S_1)).
   In addition, assume that

$$\bigcap_{j \in \mathbb{Z}} V_j = \{0\} \text{ and } \overline{\bigcup_{j \in \mathbb{Z}} V_j} = L^2(\mathbb{R}).$$

Then $\{\psi_{j,k} : j, k \in \mathbb{Z}\}$ is an o.n. wavelet for $L^2(\mathbb{R})$.

Proof 10.2.6: In view of the previous theorem, we need only prove that the orthonormal set $\{\psi_{j,k} : j, k \in \mathbb{Z}\}$ is *complete*. To this end, we need only show that

$$f \in L^2(\mathbb{R}), f \perp \psi_{j,k}, j, k \in \mathbb{Z} \Rightarrow f = 0.$$

Assuming these hypotheses, we have $f \perp W_j$, for every $j \in \mathbb{Z}$. For each $j \in \mathbb{Z}$, let $P_j : L^2(\mathbb{R}) \to V_j$ denote the orthogonal projector onto V_j, and let $v_j := P_j f$. Thus $v_j \in V_j$ and $f - v_j \perp V_j$. Since

$$V_j = V_{j-1} \overset{\perp}{\bigoplus} W_{j-1},$$

we have

$$f - v_j \perp V_{j-1} \text{ and } f - v_j \perp W_{j-1} \Rightarrow v_j \in W_{j-1}^{\perp}.$$

Since $f \perp W_{j-1}$, we have $v_j \in V_{j-1}$. Also,

$$v_j \in V_{j-1} \text{ and } f - v_j \perp V_{j-1} \Rightarrow v_j = v_{j-1}.$$

Thus $\{v_j\}$ is a constant sequence. But the density of $\bigcup V_j$ and the nested property $V_j \subset V_{j+1}$, for all $j \in \mathbb{Z}$ of V_j's entail $v_j \to f$. Hence $v_j = f$ for all $j \in \mathbb{Z}$, from which one concludes that $f \in \bigcap_j V_j$. Thus $f = 0$.

Example 10.2.1. Perhaps, one simplest pair of functions illustrating the previous theorem is

$$\psi = \psi^H (\text{the Haar wavelet}), \quad \phi = \chi_{[0,1)}.$$

It is easy to verify that ϕ obeys the simple two-scale relation

$$\phi(t) = \phi(2t) + \phi(2t - 1).$$

Thus here, $c_0 = c_1 = 1$ and $c_k = 0$ for every $k \in \mathbb{Z} \setminus \{0, 1\}$, and ψ is given by

$$\psi(t) = \phi(2t) - \phi(2t - 1),$$

which is none other than the Haar wavelet.

It remains to check in the above example, the two properties

$$\bigcap_j V_j = \{0\} \text{ and } \overline{\bigcup V_j} = L^2(\mathbb{R}).$$

Here

$$\phi_{0,k} = \begin{cases} 1, & k \leq t < k+1 \\ 0, & \text{otherwise.} \end{cases}$$

Thus

$$V_0 := \{f \in L^2(\mathbb{R}) : f \text{ constant on } [k, k+1), \forall k \in \mathbb{Z}\}$$

and

$$V_j := \left\{ f \in L^2(\mathbb{R}) : f \text{ constant on } [\frac{k}{2^j}, \frac{k+1}{2^j}), \forall k \in \mathbb{Z} \right\}.$$

Clearly,

$$f \in V_0 \Rightarrow f \text{ constant on } [k, k+1), \forall k \in \mathbb{Z}$$
$$\Rightarrow f \text{ constant on } [\frac{k}{2}, \frac{k+1}{2}), \forall k \in \mathbb{Z}$$
$$\Rightarrow f \in V_1.$$

Thus $V_0 \subset V_1$. Likewise, $V_j \subset V_{j+1}, \forall j \in \mathbb{Z}$. Clearly, the space S of *step functions* is dense in $\overline{\bigcup_j V_j}$. It is well known that S is also dense in $L^2(\mathbb{R})$. Hence

$$\overline{\bigcup V_j} = L^2(\mathbb{R}).$$

Moreover,

$$f \in \bigcap_j V_j \Rightarrow f = \text{constant on } [0, \frac{1}{2^j}), \forall j \in \mathbb{Z}.$$

Letting $j \to -\infty$, we get $f = $ constant on $[0, \infty)$. Since $f \in L^2(\mathbb{R})$, this constant must be zero. It follows by a similar argument that f is identically 0 on $(-\infty, 0\}$. Thus,

$$\bigcap_j V_j = \{0\}.$$

Remark 10.2.2: We note that here it is easy to check directly that ϕ satisfies (S_0): the set $\{\phi_{0,k} : k \in \mathbb{Z}\}$ is orthonormal.

10.3 Classification of Wavelets and Multiresolution Analysis

Let us recall that if X is a separable Hilbert space, then a (Schauder) basis $\{x_n\}$ of X is said to be a **Riesz basis** of X if it is *equivalent* to an orthonormal basis $\{u_n\}$ of X, in the sense that, there exists a bounded invertible operator $T : X \to X$ such that $T(x_n) = u_n, \forall n \in \mathbb{N}$. From this definition, it is easy to prove the next proposition.

Proposition 10.3.1 Let X be a separable Hilbert space. Then the following statements are equivalent.

(a) $\{x_n\}$ is a Riesz basis for X.

(b) $\overline{\text{span}}\{x_n\} = X$ and for every $N \in \mathbb{N}$ and arbitrary constants $c_1, c_2, \ldots, c_N$, there are constants A, B with $0 < A \leq B < \infty$ such that

$$A \sum_{i=1}^{N} |c_i|^2 \leq \left\| \sum_{i=1}^{N} c_i x_i \right\|^2 \leq B \sum_{i=1}^{N} |c_i|^2.$$

Remark 10.3.1: Let $\{x_n\}$ be a Riesz basis in X. Then the series $\sum_{i=1}^{\infty} c_i x_i$ is convergent in X if and only if $c = \{c_i\} \in \ell^2$. As a result, each $x \in X$ has a unique representation

$$x = \sum_{i=1}^{\infty} c_i x_i, \quad c = \{c_i\} \in \ell^2.$$

The preceding discussion enables one to define a *Riesz basis* of X as follows.

Definition 10.3.1. A sequence $\{x_n\}$ in a Hilbert space X is said to constitute a **Riesz basis** of X if $\overline{\text{span}}\{x_n\}_{n \in \mathbb{N}} = X$ and there exists constants A, B with $0 < A \leq B < \infty$ such that

$$A \sum_{i=1}^{\infty} |c_j|^2 \leq \left\| \sum_{j=1}^{\infty} c_j x_j \right\|^2 \leq B \sum_{j=1}^{\infty} |c_j|^2$$

for every sequence $c = \{c_j\} \in \ell^2$.

We are now ready for the following definition.

Definition 10.3.2. A function ψ in $L^2(\mathbb{R})$ is called an $\mathcal{R}$-**function** if $\{\psi_{j,k} : j, k \in \mathbb{Z}\}$ as defined in (10.1.3) is a Riesz basis of $L^2(\mathbb{R})$ in the sense that

$$\overline{\text{span}}\{\psi_{j,k} : j, k \in \mathbb{Z}\} = L^2(\mathbb{R}),$$

and

$$A \|\{c_{j,k}\}\|_{\ell^2(\mathbb{Z})}^2 \leq \left\| \sum_{j,k \in \mathbb{Z}} c_{j,k} \psi_{j,k} \right\|^2 \leq B \|\{c_{j,k}\}\|_{\ell^2(\mathbb{Z})}^2$$

holds for all doubly bi-infinite sequences $\{c_{j,k}\} \in \ell^2(\mathbb{Z})$ and for suitable constants A, B such that $0 < A \leq B < \infty$.

Next suppose that ψ is an $\mathcal{R}$-function. By Hahn-Banach theorem, one can show that there exists a unique Riesz basis $\{\psi^{j,k}\}$ of $L^2(\mathbb{R})$ which is dual to the Riesz basis $\{\psi_{j,k}\}$:

$$\langle \psi_{j,k}, \psi^{l,m} \rangle = \delta_{j,l} \delta_{k,m}, \; j, k, l, m \in \mathbb{Z}. \tag{10.3.1}$$

Consequently, every function f in $L^2(\mathbb{R})$ admits the following(unique) series representation:

$$f(t) = \sum_{j,k=-\infty}^{\infty} \langle f, \psi_{j,k} \rangle \psi^{j,k}. \tag{10.3.2}$$

Note that, although the coefficients in this expansion are the values of CWT of f relative to ψ, the series (10.3.2) is, in general, *not a wavelet series*. In order that this be a wavelet series, there must exist some function $\tilde{\psi}$ in $L^2(\mathbb{R})$ such that

$$\psi^{j,k} = \tilde{\psi}_{j,k}, \quad j,k \in \mathbb{Z},$$

where $\tilde{\psi}_{j,k}$ is as defined in (10.1.3) from the function $\tilde{\psi}$.

Clearly, if $\{\psi_{j,k}\}$ is an o.n. basis of $L^2(\mathbb{R})$, then (10.3.1) holds with $\psi^{j,k} = \psi_{j,k}$, or $\tilde{\psi} = \psi$. In general, however, such a $\tilde{\psi}$ *does not exit*.

If ψ is chosen such that $\tilde{\psi}$ exists, then the pair $(\psi, \tilde{\psi})$ gives rise to the following convenient (dual) representation:

$$f(t) = \sum_{j,k=-\infty}^{\infty} \langle f, \psi_{j,k} \rangle \tilde{\psi}_{j,k}$$

$$= \sum_{j,k=-\infty}^{\infty} \langle f, \tilde{\psi}_{j,k} \rangle \psi_{j,k}$$

for any element f of $L^2(\mathbb{R})$.

Definition 10.3.3. A function ψ in $L^2(\mathbb{R})$ is called an $\mathcal{R}$-**wavelet** (or simply a **wavelet**) if it is an $\mathcal{R}$-function and there exists a function $\tilde{\psi}$ in $L^2(\mathbb{R})$, such that $\{\psi_{j,k}\}$ and $\{\tilde{\psi}_{j,k}\}$, as defined in (10.1.3), are dual bases of $L^2(\mathbb{R})$. If ψ is an $\mathcal{R}$-wavelet, then $\tilde{\psi}$ is called a **dual wavelet** corresponding to ψ.

Remark 10.3.2: A dual wavelet $\tilde{\psi}$ is unique and is itself an $\mathcal{R}$-wavelet. Moreover, ψ is the dual wavelet of $\tilde{\psi}$.

Remark 10.3.3: Every wavelet ψ, orthonormal or not, generates a "wavelet series" expansion of any f in $L^2(\mathbb{R})$:

$$f(t) = \sum_{j,k=-\infty}^{\infty} c_{j,k} \psi_{j,k}(t), \tag{10.3.3}$$

where each $c_{j,k}$ is the CWT of f relative to the dual $\tilde{\psi}$ of ψ evaluated at $(a,b) = \left(\frac{1}{2^j}, \frac{k}{2^j}\right)$.

We are now ready to look at an important decomposition of the space $L^2(\mathbb{R})$. Let ψ be any wavelet and consider the Riesz basis $\{\psi_{j,k}\}$ that it generates. For each $j \in \mathbb{Z}$, let

$$W_j = \overline{\text{span}}\{\psi_{j,k} : k \in \mathbb{Z}\}.$$

(10.3.3) suggests that $L^2(\mathbb{R})$ can be decomposed as a *direct sum* of the spaces W_j's:

$$L^2(\mathbb{R}) = \bigoplus_{j \in \mathbb{Z}} W_j \qquad (10.3.4)$$

in the sense that every f in $L^2(\mathbb{R})$ has a unique decomposition

$$f(t) = \ldots + g_{-1} + g_0 + g_1 + \ldots$$

where $g_j \in W_j, \forall j \in \mathbb{Z}$.

Moreover, if ψ is an o.n. wavelet, then in the above decomposition, the direct sum is, in fact, an *orthogonal direct sum*:

$$L^2(\mathbb{R}) = \bigoplus_{j \in \mathbb{Z}}^{\perp} W_j := \ldots \bigoplus^{\perp} W_{-1} \bigoplus^{\perp} W_0 \bigoplus^{\perp} W_1 \ldots \qquad (10.3.5)$$

Note that here, for $\forall j, l \in \mathbb{Z}, j \neq l$,

$$W_j \cap W_l = \{0\}, \quad W_j \perp W_l.$$

Definition 10.3.4. A wavelet ψ in $L^2(\mathbb{R})$ is called a **semi-orthogonal wavelet** (or **s.o. wavelet**) if the Riesz basis $\{\psi_{j,k}\}$ that it generates satisfies

$$\langle \psi_{j,k}, \psi_{l,m} \rangle = 0, \quad j \neq l \; \; j,k,l,m \in \mathbb{Z} \qquad (10.3.6)$$

Clearly, a semi-orthogonal wavelet also gives rise to an *orthogonal decomposition* (10.3.5) of $L^2(\mathbb{R})$.

We now come to the important concept of *multiresolution analysis* first introduced by Meyer(1986) and Mallat(1989). We saw that any wavelet ψ (semiorthogonal or not) generates a direct sum decomposition (10.3.4) of $L^2(\mathbb{R})$.

For each $j \in \mathbb{Z}$, let us consider the closed subspaces

$$V_j = \ldots \bigoplus W_{j-2} \bigoplus W_{j-1}$$

of $L^2(\mathbb{R})$. These subspaces satisfy the following properties:

(MR1) $\ldots \subset V_{-1} \subset V_0 \subset V_1 \subset \ldots$;
(MR2) $\overline{\bigcup_{j \in \mathbb{Z}} V_j} = L^2(\mathbb{R})$, the closure being taken in the topology of $L^2(\mathbb{R})$;
(MR3) $\bigcap_{j \in \mathbb{Z}} V_j = \{0\}$;
(MR4) $V_{j+1} = V_j \bigoplus W_j, \; j \in \mathbb{Z}$; and
(MR5) $f(t) \in V_j \Leftrightarrow f(2t) \in V_{j+1}, \; j \in \mathbb{Z}$.

If the initial subspace V_0 is generated by a single function ϕ in $L^2(\mathbb{R})$ in the sense that

$$V_0 = \overline{\text{span}}\{\phi_{0,k} : k \in \mathbb{Z}\}, \qquad (10.3.7)$$

then using (MR5) all the subspaces V_j are also generated by the same ϕ:

$$V_j = \overline{\text{span}}\{\phi_{j,k} : k \in \mathbb{Z}\}, \text{where } \phi_{j,k}(t) = 2^{\frac{j}{2}}\phi(2^j t - k). \qquad (10.3.8)$$

Definition 10.3.5. A function ϕ in $L^2(\mathbb{R})$ is said to generate a **multiresolution analysis (MRA)** if it generates a ladder of closed subspaces V_j that satisfy (MR1), (MR2), (MR3) and (MR5) in the sense of (10.3.8), and such that the following property holds.

(MR0) $\{\phi_{0,k} : k \in \mathbb{Z}\}$ forms a Riesz basis of V_0.

This means, there must exist constants A, B, with $0 < A \leq B < \infty$ such that

$$A \, \|\{c_k\}\|^2_{\ell^2(\mathbb{Z})} \leq \left\| \sum_{k \in \mathbb{Z}} c_k \phi_{0,k} \right\|^2_2 \leq B \, \|\{c_k\}\|^2_{\ell^2(\mathbb{Z})} \tag{10.3.9}$$

for all bi-infinite sequences $c = \{c_k\} \in \ell^2(\mathbb{Z})$.

In this case, ϕ is called a **scaling function**.

Using the Poisson's lemma (cf., e.g., [2], Lemma 2,24) and the Parseval's identity for Fourier transforms one shows that for any ϕ in $L^2(\mathbb{R})$, the following hold:

(A) The set $\{\phi(x-k) : k \in \mathbb{Z}\}$ is orthonormal.

$\Leftrightarrow$

The Fourier transform $\hat{\phi}$ of ϕ satisfies the identity

$$\sum_{-\infty}^{\infty} |\hat{\phi}(\omega + 2\pi k)|^2 = 1, \tag{10.3.10}$$

for almost all $\omega \in \mathbb{R}$.

(B) The family of functions $\{\phi(x-k) : k \in \mathbb{Z}\}$ satisfies the Riesz condition (10.3.9) with Riesz bounds A and B.

$\Leftrightarrow$

The Fourier transform $\hat{\phi}$ of ϕ satisfies

$$A \leq \sum_{-\infty}^{\infty} |\hat{\phi}(\omega + 2\pi k)|^2 \leq B, \text{a.e.} \tag{10.3.11}$$

Remark 10.3.4: The condition (MR5) implies

$$f(t) \in V_0 \Leftrightarrow f(2^j t) \in V_j.$$

Remark 10.3.5: The spaces V_j possess the following shift invariance property:

$$f(t) \in V_j \Leftrightarrow f\left(t + \frac{k}{2^j}\right) \in V_j, \forall k \in \mathbb{Z}.$$

The above remark follows from:

$$\phi_{j,\ell}\left(t + \frac{k}{2^j}\right) = 2^{\frac{j}{2}} \phi\left(2^j \left(t + \frac{k}{2^j}\right) - \ell\right)$$

$$= 2^{\frac{j}{2}} \phi(2^j t - (\ell - k)).$$

Next, we give a few examples of MRA of $L^2(\mathbb{R})$.

For $j, k \in \mathbb{Z}$, let us denote by $I_{j,k}$ the interval $[\frac{k}{2^j}, \frac{k+1}{2^j})$.

Example 10.3.1. For each $j \in \mathbb{Z}$, let V_j denote the space of piecewise constants:

$$V_j = \{ f \in L^2(\mathbb{R}) : f|_{I_{j,k}} \equiv \text{constant}, \forall k \in \mathbb{Z} \}.$$

Here V_0 is the closed linear span of the integer shifts of the characteristic function $\chi_{[0,1]}$, which is the scaling function ϕ. Here, it is easily verified that the set $\{ \phi_{0,k} : k \in \mathbb{Z} \}$ is orthonormal, and we have already checked that $\{ V_j : j \in \mathbb{Z} \}$ is a multiresolution. In this case, the wavelet is the *Haar wavelet*, which is, in fact, an o.n. wavelet.

Example 10.3.2. For each $j \in \mathbb{Z}$, let V_j be the $L^2(\mathbb{R})$-closure of the set S_j:

$$S_j = \{ f \in L^2(\mathbb{R}) \cap C(\mathbb{R}) : f|_{I_{j,k}} \text{ is linear}, \forall k \in \mathbb{Z} \}.$$

It is easy to check all the conditions of MRA except (MR0) similar to the previous example. Checking of (MR0) involves computation of Riesz bounds, which in this case, are $A = \frac{1}{3}, B = 1$. Here the scaling function ϕ can be taken to be the hat function:

$$\phi(t) = \begin{cases} t, & \text{if } 0 \le t \le 1 \\ 2 - t, & \text{if } 1 \le t \le 2 \\ 0, & \text{otherwise.} \end{cases}$$

Note that here

$$\phi_{0,k}(t) = \phi(t - k) = \begin{cases} t - k, & k \le t \le k+1 \\ k+2-t, & k+1 \le t \le k+2 \\ 0, & \text{otherwise.} \end{cases}$$

It is easy to see that the set $\{ \phi_{0,k} : k \in \mathbb{Z} \}$ is not orthonormal. Here using the two-scale relation and a variant of Theorem 10.2.2, we can show that the corresponding wavelet ψ is given by

$$\psi(t) = \phi(2t) - \frac{1}{2}\phi(2t - 1) - \frac{1}{2}\phi(2t + 1)$$

whose support is $[-\frac{1}{2}, \frac{3}{2}]$.

10.4 Spline Wavelets

For each positive integer m, we denote by $\mathcal{S}_m(2^{-j}\mathbb{Z}) =: \mathcal{S}_m^j$ the space of **cardinal splines** of order m and with the knot sequence $2^{-j}\mathbb{Z}$, for a fixed $j \in \mathbb{Z}$:

$$\mathcal{S}_m(2^{-j}\mathbb{Z}) = \{ f \in C^{m-2}(\mathbb{R}) : f|_{I_{j,k}} \in \mathcal{P}_m, \forall k \in \mathbb{Z} \}$$

(Here $\mathcal{P}_m$ denotes the class of polynomials of *order m*, *i.e.* of degree $\le m - 1$.) For each $m \in \mathbb{N}$, the mth **order cardinal B-spline** N_m is defined by

$$N_m = \chi_{[0,1)} * \cdots * \chi_{[0,1)} \ (m \text{ times convoluted}).$$

Put differently, N_m is defined recursively by:

$$N_m(t) = \int_{-\infty}^{\infty} N_{m-1}(t-s)N_1(s)\mathrm{d}s$$

$$= \int_0^1 N_{m-1}(t-s)\mathrm{d}s, \text{ with } N_1 := \chi_{[0,1)}.$$

The scaling functions in the two examples in the previous section are respectively the first order and the second order cardinal B-spline. It is well known that any $f \in S_m^j$ can be written as

$$f(t) = \sum_k c_k N_m(t-k). \tag{10.4.1}$$

Taking N_m as the scaling function, let us define

$$V_0^m = \overline{\mathrm{span}}S_m^0 = S_m(\mathbb{Z}) \tag{10.4.2}$$

Hence, a function f is in V_0^m if and only if it has a B-spline series representation (10.4.1) with the coefficient sequence $c = \{c_k\} \in \ell^2(\mathbb{Z})$. The other spaces V_j^m are defined by

$$f(t) \in V_j^m \Leftrightarrow f(2t) \in V_{j+1}^m, \quad j \in \mathbb{Z}.$$

In other words,

$$V_j^m = \overline{\mathrm{span}}S_m^j.$$

Clearly the subspaces $\{V_j^m : j \in \mathbb{Z}\}$ satisfy (MR1). The verification of (MR2) is immediate: The class of polynomials $\mathcal{P}$ is dense in $L^2(\mathbb{R})$ and $\mathcal{P} \subset V_j^m, \forall j \in \mathbb{Z}$. This implies

$$\overline{\bigcup_{j \in \mathbb{Z}} V_j^m} = L^2(\mathbb{R}).$$

The verification of (MR3) is exactly as in Example 10.3.1 The verification of (MR0) is carried out as in Example 10.3.2 with ϕ replaced by N_m. Here the smallest value of B is 1, and the largest value of A can be expressed in terms of the roots of the $Euler\text{-}Frobenius \ polynomial$

$$E_{2m-1}(z) = (2m-1)!z^{m-1}\sum_{k=-m+1}^{m-1} N_{2m}(m+k)z^k \tag{10.4.3}$$

From the nested sequence of spline spaces V_j^m, we have the orthogonal complementary subspaces W_j^m, given by

$$W_{j+1}^m = V_j^m \bigoplus W_j^m, \quad j \in \mathbb{Z}.$$

Just as the B-spline N_m is the minimally supported generator of $\{V_j^m\}$, we are interested in finding the minimally supported $\psi_m \in W_0$ that generates the mutually orthogonal subspaces W_j. Such compactly supported functions ψ_m are called the **B-wavelets** of order m. It turns out that

$$\text{support } N_m = [0,m]; \quad \text{support } \psi_m = [0, 2m-1], \forall m \in \mathbf{N}$$

We mention without working out further details that

$$\psi_m(t) = \sum_{k=0}^{3m-2} q_k N_m(2t-k), \qquad (10.4.4)$$

with

$$q_k = q_k^{(m)} = \frac{(-1)^k}{2^{m-1}} \sum_{l=0}^{m} \binom{m}{l} N_{2m}(k-l+1). \qquad (10.4.5)$$

For the relevant details, we refer the reader to Chui (1992, Chapter 6).

10.5 A Variant of Construction of Orthonormal Wavelets

Let us go back once again to Theorems 10.2.1 and 10.2.2. Suppose ϕ in $L^2(\mathbb{R})$ is such that ϕ does not satisfy (S_0), i.e., $\{\phi_{0,k} : k \in \mathbb{Z}\}$ is *not orthonormal*.

In this case, it seems natural to define Φ by requiring its Fourier transform to be using (10.3.10) and the Plencherel's thorem,

$$\hat{\Phi}(\omega) = \frac{\hat{\phi}(\omega)}{\{\sum_k |\hat{\phi}(\omega + 2\pi k)|^2\}^{1/2}}, \quad \omega \in \mathbb{R}. \qquad (10.5.1)$$

Theorem 10.5.1. *Let ϕ in $L^2(\mathbb{R})$ be such that it satisfies*

(MR0): $\qquad \{\phi_{0,k} : k \in \mathbb{Z}\}$ *is a Riesz basis of V_0.*

Define $\Phi \in L^2(\mathbb{R})$ by (10.5.1). Then $\{\Phi_{0,k} : k \in \mathbb{Z}\}$ is an orthonormal basis for the space V_0.

Theorem 10.5.2. *Let ϕ in $L^2(\mathbb{R})$ be such that $\{\phi_{0,k} : k \in \mathbb{Z}\}$ is a Riesz basis for V_0 and suppose $\phi \in V_1$. Then Φ as defined in (10.5.1) satisfies a two-scale relation*

$$\Phi(t) = \sum_{k=-\infty}^{\infty} a_k \Phi(2t-k), \quad (a = \{a_k\} \in \ell^2(\mathbb{Z})).$$

Let Ψ be defined by

$$\Psi(t) = \sum_k (-1)^k a_{1-k} \Phi(2t-k). \qquad (10.5.2)$$

Then the set $\{\Psi_{j,k} : j,k \in \mathbb{Z}\}$ is orthonormal.
Furthermore, if V_j's satisfy (MR2) and (MR3), then Ψ is an orthonormal wavelet.

The proofs of the above theorems follow from Theorems 10.2.1 and 10.2.2 by applying (10.3.10) and (10.3.11). The details are left to the reader as exercises.

Exercises 10.5

10.5.1. Let ψ^H be the Haar wavelet. Show that for integers $n < m$,

$$\int_n^m \psi^H(t)dt = 0.$$

10.5.2. Let $f, g \in L^2(\mathbb{R})$, and suppose that $f_{oj} \perp g_{0i}, \forall i, j \in \mathbb{Z}$. Show that $f_{nj} \perp g_{ni}, \forall n, i, j \in \mathbb{Z}$.

10.5.3. Let $\psi \in L^2(\mathbb{R}), n \in \mathbb{Z}, i \in \mathbb{Z}$. Define $\phi = \psi_{ni}$. Show that

$$\text{span}\{\phi_{kj} : k, j \in \mathbb{Z}\} = \text{span}\{\psi_{kj} : k, j \in \mathbb{Z}\}.$$

10.5.4. Let $\{u_n\}$ be an orthonormal sequence in a Hilbert space. Let α, β be in ℓ^2 such that $\alpha \perp \beta$. Define

$$w = \sum_k \alpha_k u_k, \quad v = \sum_k \beta_k u_k.$$

Show that $w \perp v$. Is the converse true?

10.5.5. Verify directly the orthonormality of the family of functions $\{\psi_{j,k}^H; j, k \in \mathbb{Z}\}$.

10.5.6. Give a proof of Theorem 10.5.1.

10.5.7. Give a proof of Theorem 10.5.2.

References

Calderòn, A. (1964). Intermediate spaces and interpolation, the complex method, *Studia Math.* **24**, 113–190.

Chui, C.K. (1992). *"An introduction to wavelets"*, Academic Press, San Diego.

Chui, C.K. and Wang, J.Z. (1993) An analysis of cardinal-spline wavelets, *J. Approx. Theory,* **72(1)**, 54–68.

Daubechies, I. (1992). *"Ten lectures on wavelets"*, SIAM, Philadelphia.

Daubechies, I. (1998). Orthonormal bases of compactly supported wavelets, *Comm. Pure Appl.Math.* **41**, 909–996.

Mallat, S. (1998). *"A wavelet tour of signal processing"*, Academic Press, New York.

Mallat, S. (1989). Multiresolution approximations and wavelet orthonormal bases of $L^2(\mathbb{R})$, *Trans. Amer. Math. Soc.* **315**, 69–87.

Meyer, Y. (1992). *"Wavelets and Operators"* Cambridge University Press.

Grossman, A. and Morlet, J. (1984). Decomposition of Hardy functions into square integrable wavelets of constant shape, *SIAM J. Math. Anal.* **15**, 723–736.

Chapter 11
Jacobians of Matrix Transformations

[*This Chapter is based on the lectures of Professor A.M. Mathai of McGill University, Canada (Director of the SERC Schools).*]

11.0 Introduction

Real scalar functions of matrix argument, when the matrices are real, will be dealt with. It is difficult to develop a theory of functions of matrix argument for general matrices. Let $X = (x_{ij}), i = 1 \cdots, m$ and $j = 1, \cdots, n$ be an $m \times n$ matrix where the x_{ij}'s are real elements. It is assumed that the readers have the basic knowledge of matrices and determinants. The following standard notations will be used here. A prime denotes the transpose, $X' = $ transpose of $X, |(.)|$ denotes the determinant of the square matrix, $m \times m$ matrix $(\cdot)$. The same notation will be used for the absolute value also. $\mathrm{tr}(X)$ denotes the trace of a square matrix X, $\mathrm{tr}(X) = $ sum of the eigenvalues of $X = $ sum of the leading diagonal elements in X. A real symmetric positive definite X (definiteness is defined only for symmetric matrices when real) will be denoted by $X = X' > 0$. Then $0 < X = X' < I \Rightarrow X = X' > 0$ and $I - X > 0$. Further, $\mathrm{d}X$ will denote the wedge product or skew symmetric product of the differentials $\mathrm{d}x_{ij}$'s.

That is, when $X = (x_{ij})$, an $m \times n$ matrix

$$\mathrm{d}X = \mathrm{d}x_{11} \wedge \cdots \wedge \mathrm{d}x_{1n} \wedge \mathrm{d}x_{21} \wedge \cdots \mathrm{d}x_{2n} \wedge \cdots \wedge \mathrm{d}x_{mn}. \qquad (11.0.1)$$

If $X = X'$, that is symmetric, and $p \times p$, then

$$\mathrm{d}X = \mathrm{d}x_{11} \wedge \mathrm{d}x_{21} \wedge \mathrm{d}x_{22} \wedge \mathrm{d}x_{31} \wedge \cdots \wedge \mathrm{d}x_{pp} \qquad (11.0.2)$$

a wedge product of $1 + 2 + \cdots + p = p(p+1)/2$ differentials.

A wedge product or skew symmetric product is defined in Chapter 1.

409

11.1 Jacobians of Linear Matrix Transformations

Some standard Jacobians, that we will need later, will be illustrated here. For more on Jacobians see Mathai (1997). First we consider a very basic linear transformation involving a vector of real variables going to a vector of real variables.

Theorem 11.1.1. *Let X and Y be $p \times 1$ vectors of real scalar variables, functionally independent (no element in X is a function of the other elements in X and similarly no element in Y is a function of the other elements in Y), and let $Y = AX$, $|A| \neq 0$, where $A = (a_{ij})$ is a nonsingular $p \times p$ matrix of constants (A is free of the elements in X and Y; each element in Y is a linear function of the elements in X, and vice versa). Then*

$$Y = AX, \ |A| \neq 0 \Rightarrow dY = |A| dX. \tag{11.1.1}$$

Proof 11.1.1:

$$Y = \begin{bmatrix} y_1 \\ \vdots \\ y_p \end{bmatrix} = AX = \begin{bmatrix} a_{11} & a_{12} & \cdots & a_{1p} \\ a_{21} & a_{22} & \cdots & a_{2p} \\ \vdots & \vdots & \cdots & \vdots \\ a_{p1} & a_{p2} & \cdots & a_{pp} \end{bmatrix} \begin{bmatrix} x_1 \\ x_2 \\ \vdots \\ x_p \end{bmatrix} \Rightarrow$$

$$y_i = a_{i1} x_1 + \ldots + a_{ip} x_p, i = 1, \ldots, p.$$

$$\frac{\partial y_i}{\partial x_j} = a_{ij} \Rightarrow \left(\frac{\partial y_i}{\partial x_j} \right) = (a_{ij}) = A \Rightarrow J = |A|.$$

Hence,

$$dY = |A| dX.$$

That is, Y and X, $p \times 1$, A is $p \times p$, $|A| \neq 0$, A is a constant matrix, then

$$Y = AX, \ |A| \neq 0 \Rightarrow dY = |A| dX.$$

Example 11.1.1. Consider the transformation $Y = AX$ where $Y' = (y_1, y_2, y_3)$ and $X' = (x_1, x_2, x_3)$ and let the transformation be

$$y_1 = x_1 + x_2 + x_3$$
$$y_2 = 3x_2 + x_3$$
$$y_3 = 5x_3.$$

Then write dY in terms of dX.

Solution 11.1.1: From the above equations, by taking the differentials we have

$$dy_1 = dx_1 + dx_2 + dx_3$$
$$dy_2 = 3dx_2 + dx_3$$
$$dy_3 = 5dx_3.$$

Then taking the product of the differentials we have

$$dy_1 \wedge dy_2 \wedge dy_3 = [dx_1 + dx_2 + dx_3] \wedge [3dx_2 + dx_3] \wedge [5dx_3].$$

Taking the product directly and then using the fact that $dx_2 \wedge dx_2 = 0$ and $dx_3 \wedge dx_3 = 0$ we have

$$dY = dy_1 \wedge dy_2 \wedge dy_3$$
$$= 15dx_1 \wedge dx_2 \wedge dx_3 = 15dX$$
$$= |A|dX$$

where

$$A = \begin{bmatrix} 1 & 1 & 1 \\ 0 & 3 & 1 \\ 0 & 0 & 5 \end{bmatrix}.$$

This verifies Theorem 11.1.1 also. This theorem is the standard result that is seen in elementary textbooks. Now we will investigate more elaborate linear transformations.

Theorem 11.1.2. *Let X and Y be $m \times n$ matrices of functionally independent real variables and let $A, m \times m$ be a nonsingular constant matrix. Then*

$$Y = AX \Rightarrow dY = |A|^n dX. \tag{11.1.2}$$

Proof 11.1.2: Let $Y = AX = (AX^{(1)}, AX^{(2)}, \ldots, AX^{(n)})$ where $X^{(1)}, \ldots, X^{(n)}$ are the columns of X. Then the Jacobian matrix for X going to Y is of the form

$$\begin{bmatrix} A & O & \ldots & O \\ O & A & \ldots & O \\ \vdots & \vdots & \ldots & \vdots \\ O & O & \ldots & A \end{bmatrix} \Rightarrow \begin{vmatrix} A & O & \ldots & O \\ \vdots & \vdots & \ldots & \vdots \\ O & O & \ldots & A \end{vmatrix} = |A|^n = J \tag{11.1.3}$$

where O denotes a null matrix and J is the Jacobian for the transformation of X going to Y or $dY = |A|^n dX$.

In the above linear transformation the matrix X was pre-multiplied by a nonsingular constant matrix A. Now let us consider the transformation of the form $Y = XB$ where X is post-multiplied by a nonsingular constant matrix B.

Theorem 11.1.3. *Let X be a $m \times n$ matrix of functionally independent real variables and let B be an $n \times n$ nonsingular matrix of constants. Then*

$$Y = XB, \ |B| \neq 0 \Rightarrow dY = |B|^m dX, \tag{11.1.4}$$

Proof 11.1.3:

$$Y = XB = \begin{bmatrix} X^{(1)}B \\ \vdots \\ X^{(m)}B \end{bmatrix}$$

where $X^{(1)}, \ldots, X^{(m)}$ are the rows of X. The Jacobian matrix is of the form,

$$\begin{bmatrix} B & 0 & \cdots & 0 \\ \vdots & \vdots & \cdots & \vdots \\ 0 & 0 & \cdots & B \end{bmatrix} \Rightarrow \begin{vmatrix} B & 0 & \cdots & 0 \\ \vdots & \vdots & & \vdots \\ & & \cdots & B \end{vmatrix} = |B|^m \Rightarrow dY = |B|^m dX.$$

Then combining the above two theorems we have the Jacobian for the most general linear transformation.

Theorem 11.1.4. *Let X and Y be $m \times n$ matrices of functionally independent real variables. Let A be $m \times m$ and B be $n \times n$ nonsingular matrices of constants. Then*

$$Y = AXB, |A| \neq 0, |B| \neq 0, Y, m \times n, X, m \times n, \Rightarrow dY = |A|^n |B|^m dX. \qquad (11.1.5)$$

Proof 11.1.4: For proving this result first consider the transformation $Z = AX$ and then the transformation $Y = ZB$, and make use of Theorems 11.1.2 and 11.1.3.

In Theorems 11.1.2 to 11.1.4 the matrix X was rectangular. Now we will examine a situation where the matrix X is square and symmetric. If X is $p \times p$ and symmetric then there are only $1 + 2 + \cdots + p = p(p+1)/2$ functionally independent elements in X because, here $x_{ij} = x_{ji}$ for all i and j. Let $Y = Y' = AXA', X = X', |A| \neq 0$. Then we can obtain the following result:

Theorem 11.1.5. *Let $X = X'$ be a $p \times p$ real symmetric matrix of $p(p+1)/2$ functionally independent real elements and let A be a $p \times p$ nonsingular constant matrix. Then*

$$Y = AXA', X = X', |A| \neq 0, \Rightarrow dY = |A|^{p+1} dX. \qquad (11.1.6)$$

Proof 11.1.5: This result can be proved by using the fact that a nonsingular matrix such as A can be written as a product of elementary matrices in the form

$$A = E_1 E_2 \cdots E_k$$

where $E_1, \cdots, E_k$ are elementary matrices. Then

$$Y = AXA' \Rightarrow E_1 E_2 \cdots E_k X E_k' \cdots E_1'.$$

where E_j' is the transpose of E_j. Let $Y_k = E_k X E_k', Y_{k-1} = E_{k-1} Y_k E_{k-1}'$, and so on, and finally $Y = Y_1 = E_1 Y_2 E_1'$. Evaluate the Jacobians in these transformations to obtain the result, observing the following facts. If, for example, the elementary matrix E_k is formed by multiplying the i-th row of an identity matrix by the nonzero scalar

c then taking the wedge product of differentials we have $dY_k = c^{p+1}dX$. Similarly, for example, if the elementary matrix E_{k-1} is formed by adding the i-th row of an identity matrix to its j-th row then the determinant remains the same as 1 and hence $dY_{k-1} = dY_k$. Since these are the only two types of basic elementary matrices, systematic evaluation of successive Jacobians gives the final result as $|A|^{p+1}$.

Note 11.1.1: From the above theorems the following properties are evident: If X is a $p \times q$ matrix of functionally independent real variables and if c is a scalar quantity and B is a $p \times q$ constant matrix then

$$Y = cX \Rightarrow dY = c^{pq}dX \qquad (11.1.7)$$

$$Y = cX + B \Rightarrow dY = c^{pq}dX. \qquad (11.1.8)$$

Note 11.1.2: If X is a $p \times p$ symmetric matrix of functionally independent real variables, a is a scalar quantity and B is a $p \times p$ symmetric constant matrix then

$$Y = aX + B, X = X', B = B' \Rightarrow dY = a^{p(p+1)/2}dX. \qquad (11.1.9)$$

Note 11.1.3: For any $p \times p$ lower triangular (or upper triangular) matrix of $p(p+1)/2$ functionally independent real variables, $Y = X + X'$ is a symmetric matrix, where X' denoting the transpose of $X = (x_{ij})$, then observing that the diagonal elements in $Y = (y_{ij})$ are multiplied by 2, that is, $y_{ii} = 2x_{ii}$, $i = 1, \ldots, p$, we have

$$Y = X + X' \Rightarrow dY = 2^p dX. \qquad (11.1.10)$$

Example 11.1.2. Let X be a $p \times q$ matrix of pq functionally independent random variables having a matrix-variate Gaussian distribution with the density given by

$$f(X) = c \exp\{-\mathrm{tr}[A(Z-M)B(X-M)']\}$$

where, A is a $p \times p$ positive definite constant matrix, B is a $q \times q$ positive definite constant matrix, M is $p \times q$ constant matrix, $\mathrm{tr}(\cdot)$ denotes the trace of the matrix $(\cdot)$ and c is the normalizing constant, then evaluate c.

Solution 11.1.2: Since A and B are positive definite matrices we can write $A = A_1 A_1'$ and $B = B_1 B_1'$ where A_1 and B_1 are nonsingular matrices, that is, $|A_1| \neq 0$, $|B_1| \neq 0$. Also we know that for any two matrices P and Q, $\mathrm{tr}(PQ) = \mathrm{tr}(QP)$ as long as PQ and QP are defined, PQ need not be equal to QP. Then

$$\mathrm{tr}[A(X-M)B(X-M)'] = \mathrm{tr}[A_1 A_1'(X-M)B_1 B_1'(X-M)']$$
$$= \mathrm{tr}[A_1'(X-M)B_1 B_1'(X-m)'A_1]$$
$$= \mathrm{tr}(YY')$$

where

$$Y = A_1'(X-M)B_1,$$

But from Theorem 11.1.4

$$dY = |A_1'|^q |B_1|^p d(X - M) = |A_1|^q |B_1|^p d(X - M)$$

since $|A_1|' = |A_1|$

$$= |A|^{q/2} |B|^{p/2} dX$$

since $|A| = |A_1|^2$, $|B| = |B_1|^2$, $d(X - M) = d(X)$, M being a constant matrix. If $f(X)$ is a density then the total integral is unity, that is,

$$1 = \int_X f(X) dX$$
$$= c \int_X \exp\{-\text{tr}[A(X - M)B(X - M)']\} dX$$
$$= c \int_Y \exp\{-\text{tr}[YY']\}$$

where, for example, $\int_X$ denotes the integral over all elements in X. Note that for any real matrix P, trace of PP' is the sum of squares of all the elements in P. Hence

$$\int_Y \exp\{-\text{tr}[YY']\} = \int_Y \exp\{-\sum_{i,j} y_{ij}^2\} dY$$
$$= \prod_{i,j} \int_{-\infty}^{\infty} e^{-y_{ij}^2} dy_{ij}.$$

But

$$\int_{-\infty}^{\infty} e^{-u^2} du = \sqrt{\pi}$$

and therefore

$$1 = c \, |A|^{q/2} |B|^{p/2} \sqrt{\pi}^{pq} \Rightarrow$$
$$c = (|A|^{q/2} |B|^{p/2} \pi^{pq/2})^{-1}.$$

Note 11.1.4: What happens in the transformation $Y = X + X'$ where both X and Y are $p \times p$ matrices of functionally independent real elements. When $X = X'$, then $Y = 2X$ and this case is already covered before. If $X \ne X'$ then Y has become symmetric with $p(p+1)/2$ variables whereas in X there are p^2 variables and hence this is not a one-to-one transformation.

Example 11.1.3. Consider the transformation

$$y_{11} = x_{11} + x_{21}, \quad y_{12} = x_{11} + x_{21} + 2x_{12} + 2x_{22},$$
$$y_{13} = x_{11} + x_{21} + 2x_{13} + x_{23}, \quad y_{21} = x_{11} + 3x_{21},$$
$$y_{22} = x_{11} + 3x_{21} + 2x_{12} + 6x_{22}, \quad y_{23} = x_{11} + 3x_{21} + 2x_{13} + 6x_{23}.$$

Write this transformation in the form $Y = AXB$ and then evaluate the Jacobian in this transformation.

Solution 11.1.3: Writing the transformation in the form $Y = AXB$ we have

$$Y = \begin{bmatrix} y_{11} & y_{12} & y_{13} \\ y_{21} & y_{22} & y_{23} \end{bmatrix}, \quad X = \begin{bmatrix} x_{11} & x_{12} & x_{13} \\ x_{21} & x_{22} & x_{23} \end{bmatrix},$$

$$A = \begin{bmatrix} 1 & 1 \\ 1 & 3 \end{bmatrix}, \quad B = \begin{bmatrix} 1 & 1 & 1 \\ 0 & 2 & 0 \\ 0 & 0 & 2 \end{bmatrix}.$$

Hence the Jacobian is

$$J = |A|^3 |B|^2 = (2^3)(4^2) = 128.$$

This can also be verified by taking the differentials in the starting explicit forms and then taking the wedge products. This verification is left to the reader.

Exercises 11.1.

11.1.1. If X and A are $p \times p$ lower triangular matrices where $A = (a_{ij})$ is a constant matrix with $a_{jj} > 0$, $j = 1, \ldots, p, X = (x_{ij})$ and x_{ij}'s, $i \geq j$ are functionally independent real variables then show that

$$Y = XA \Rightarrow \mathrm{d}Y = \left\{ \prod_{j=1}^{p} a_{jj}^{p-j+1} \right\} \mathrm{d}X,$$

$$Y = AX \Rightarrow \mathrm{d}Y = \left\{ \prod_{j=1}^{p} a_{jj}^{j} \right\} \mathrm{d}X,$$

and

$$Y = aX \Rightarrow \mathrm{d}Y = a^{p(p+1)/2} \mathrm{d}X \tag{11.1.11}$$

where a is a scalar quantity.

11.1.2. Let X and B be upper triangular $p \times p$ matrices where $B = (b_{ij})$ is a constant matrix with $b_{jj} > 0, j = 1, \ldots, p$, $X = (x_{ij})$ where the x_{ij}'s, $i \leq j$ be functionally independent real variables and b be a scalar quantity, then show that

$$Y = XB \Rightarrow \mathrm{d}Y = \left\{ \prod_{j=1}^{p} b_{jj}^{j} \right\} \mathrm{d}X,$$

$$Y = BX \Rightarrow \mathrm{d}Y = \left\{ \prod_{j=1}^{p} b_{jj}^{p+1-j} \right\} \mathrm{d}X,$$

and

$$Y = bX \Rightarrow \mathrm{d}Y = b^{p(p+1)/2} \mathrm{d}X. \tag{11.1.12}$$

11.1.3. Let X, A, B be $p \times p$ lower triangular matrices where $A = (a_{ij})$ and $B = (b_{ij})$ be constant matrices with $a_{jj} > 0, b_{jj} > 0, \ j = 1, \ldots, p$ and $X = (x_{ij})$ with x_{ij}'s, $i \geq j$ be functionally independent real variables. Then show that

$$Y = AXB \Rightarrow dY = \left\{ \prod_{j=1}^{p} a_{jj}^j b_{jj}^{p+1-j} \right\} dX,$$

and

$$Z = A'X'B' \Rightarrow dZ = \left\{ \prod_{j=1}^{p} b_{jj}^j a_{jj}^{p+1-j} \right\} dX. \tag{11.1.13}$$

11.1.4. Let $X = -X'$ be a $p \times p$ skew symmetric matrix of functionally independent $p(p-1)/2$ real variables and let $A, |A| \neq 0$, be a $p \times p$ constant matrix. Then prove that

$$Y = AXA', X' = -X, |A| \neq 0 \Rightarrow dY = |A|^{p-1} dX. \tag{11.1.14}$$

11.1.5. Let X be a lower triangular $p \times p$ matrix of functionally independent real variables and $A = (a_{ij})$ be a lower triangular matrix of constants with $a_{jj} > 0, \ j = 1, \ldots, p$. Then show that

$$Y = XA + A'X' \Rightarrow dY = 2^p \left\{ \prod_{j=1}^{p} a_{jj}^{p+1-j} \right\} dX, \tag{11.1.15}$$

and

$$Y = AX + X'A' \Rightarrow dY = 2^p \left\{ \prod_{j=1}^{p} a_{jj}^j \right\} dX. \tag{11.1.16}$$

11.1.6. Let X and A be as defined in Exercise 11.1.5. Then show that

$$Y = A'X + X'A \rightarrow dY = 2^p \left\{ \prod_{j=1}^{p} a_{jj}^j \right\} dX, \tag{11.1.17}$$

and

$$Y = AX' + XA' \rightarrow dY = 2^p \left\{ \prod_{j=1}^{p} a_{jj}^{p+1-j} \right\} dX. \tag{11.1.18}$$

11.1.7. Consider the transformation $Y = AX$ where

$$Y = \begin{bmatrix} y_{11} & y_{12} \\ 0 & y_{22} \end{bmatrix}, \quad X = \begin{bmatrix} x_{11} & x_{12} \\ 0 & x_{22} \end{bmatrix}, \quad A = \begin{bmatrix} 2 & 1 \\ 0 & 3 \end{bmatrix}.$$

Writing AX explicitly and then taking the differentials and wedge products show that $dY = 12dX$ and verify that $J = \prod_{j=1}^{p} a_{jj}^{p+1-j} = (2^2)(3^1) = 12$.

11.1.8. Let Y and X be as in Exercise 11.1.7 and consider the transformation $Y = XB$ where, $B = A$ in Exercise 11.1.7. Then writing XB explicitly, taking differentials and then the wedge products show that $dY = 18dX$ and verify the result that $J = \prod_{j=1}^{p} b_{jj}^j = (2^1)(3^2) = 18$.

11.1.9. Let Y, X, A be as in Exercise 11.1.7. Consider the transformation $Y = AX + X'A'$. Evaluate the Jacobian from first principles of taking differentials and wedge products and then verify the result that the Jacobian is $2^p = 2^2 = 4$ times the Jacobian in Exercise 11.1.7.

11.1.10. Let Y, X, B be as in Exercise 11.1.8. Consider the transformation $Y = XB + B'X'$. Evaluate the Jacobian from first principles and then verify that the Jacobian is $2^p = 2^2 = 4$ times the Jacobian in Exercise 11.1.8.

11.2 Jacobians in Some Nonlinear Transformations

Some basic nonlinear transformations will be considered in this section and some more results will be given in the exercises at the end of this section. The most popular nonlinear transformation is when a positive definite (naturally real symmetric also) matrix is decomposed into a triangular matrix and its transpose. This will be discussed first.

Example 11.2.1. Let X be $p \times p$, symmetric positive definite and let $T = (t_{ij})$ be a lower triangular matrix. Consider the transformation $Z = TT'$. Obtain the conditions for this transformation to be one-to-one and then evaluate the Jacobian.

Solution 11.2.1:

$$X = (x_{ij}) = \begin{bmatrix} x_{11} & x_{12} & \cdots & x_{1p} \\ \vdots & \vdots & \cdots & \vdots \\ x_{p1} & x_{p2} & \cdots & x_{pp} \end{bmatrix}$$

with $x_{ij} = x_{ji}$ for all i and j, $X = X' > 0$. When X is positive definite, that is, $X > 0$ then $x_{jj} > 0, j = 1, \ldots, p$ also.

$$TT' = \begin{bmatrix} t_{11} & 0 & \cdots & 0 \\ t_{21} & t_{22} & \cdots & 0 \\ \vdots & \vdots & \cdots & \vdots \\ t_{p1} & t_{p2} & \cdots & t_{pp} \end{bmatrix} \begin{bmatrix} t_{11} & t_{21} & \cdots & t_{p1} \\ 0 & t_{22} & \cdots & t_{p2} \\ \vdots & \vdots & \cdots & \vdots \\ 0 & 0 & \cdots & t_{pp} \end{bmatrix} = X \Rightarrow$$

$x_{11} = t_{11}^2 \Rightarrow t_{11} = \pm\sqrt{x_{11}}$. This can be made unique if we impose the condition $t_{11} > 0$. Note that $x_{12} = t_{11}t_{21}$ and this means that t_{21} is unique if $t_{11} > 0$. Continuing like this, we see that for the transformation to be unique it is sufficient that $t_{jj} > 0$, $j = 1, \ldots, p$. Now, observe that,

$$x_{11} = t_{11}^2, x_{22} = t_{21}^2 + t_{22}^2, \ldots, x_{pp} = t_{p1}^2 + \ldots + t_{pp}^2$$

and $x_{12} = t_{11}t_{21}, \ldots, x_{1p} = t_{11}t_{p1}$, and so on.

$$\frac{\partial x_{11}}{\partial t_{11}} = 2t_{11}, \frac{\partial x_{11}}{\partial t_{21}} = 0, \ldots, \frac{\partial x_{11}}{\partial t_{p1}} = 0,$$

$$\frac{\partial x_{12}}{\partial t_{21}} = t_{11}, \dots, \frac{\partial x_{1p}}{\partial t_{p1}} = t_{11},$$

$$\frac{\partial x_{22}}{\partial t_{22}} = 2t_{22}, \frac{\partial x_{22}}{\partial t_{31}} = 0, \cdots, \frac{\partial x_{22}}{\partial t_{p1}} = 0,$$

and so on. Taking the x_{ij}'s in the order $x_{11}, x_{12}, \cdots, x_{1p}, x_{22}, \cdots, x_{2p}, \cdots, x_{pp}$ and the t_{ij}'s in the order $t_{11}, t_{21}, t_{22}, \cdots, t_{pp}$ we have the Jacobian matrix a triangular matrix with the diagonal elements as follows: t_{11} is repeated p times, t_{22} is repeated $p - 1$ times and so on, and finally t_{pp} appearing once. The number 2 is appearing a total of p times. Hence the determinant is the product of the diagonal elements, giving,

$$2^p t_{11}^p t_{22}^{p-1} \cdots t_{pp}.$$

Therefore, for $X = X' > 0, T = (t_{ij}), t_{ij} = 0, i < j, t_{jj} > 0, j = 1, \cdots p$ we have

Theorem 11.2.1. *Let $X = X' > 0$ be a $p \times p$ real symmetric positive definite matrix and let $X = TT'$ where T is lower triangular with positive diagonal elements, $t_{jj} > 0, j = 1, \dots, p$. Then*

$$X = TT' \Rightarrow dX = 2^p \left\{ \prod_{j=1}^p t_{jj}^{p+1-j} \right\} dT. \tag{11.2.1}$$

Example 11.2.2. If X is $p \times p$ real symmetric positive definite then evaluate the following integral, we will call it *matrix-variate real gamma* , denoted by $\Gamma_p(\alpha)$:

$$\Gamma_p(\alpha) = \int_X |X|^{\alpha - \frac{p+1}{2}} e^{-\mathrm{tr}(X)} dX \tag{11.2.2}$$

and show that

$$\Gamma_p(\alpha) = \pi^{\frac{p(p-1)}{4}} \Gamma(\alpha) \Gamma\left(\alpha - \frac{1}{2}\right) \cdots \Gamma\left(\alpha - \frac{p-1}{2}\right) \tag{11.2.3}$$

for $\Re(\alpha) > \frac{p-1}{2}$.

Solution 11.2.2: Make the transformation $X = TT'$ where T is lower triangular with positive diagonal elements. Then

$$|TT'| = \prod_{j=1}^p t_{jj}^2, \quad dX = 2^p \left\{ \prod_{j=1}^p t_{jj}^{p+1-j} \right\} dT$$

and

$$\mathrm{tr}(X) = t_{11}^2 + (t_{21}^2 + t_{22}^2) + \dots + (t_{p1}^2 + \dots + t_{pp}^2).$$

Then substituting these, the integral over X reduces to the following:

$$\int_X |X|^{\alpha - \frac{p+1}{2}} e^{-\mathrm{tr}(X)} dX = \int_T \left\{ \prod_{j=1}^p \int_0^\infty 2t_{jj}^{\alpha - \frac{j}{2}} e^{-t_{jj}^2} dt_{jj} \right\} \prod_{i>j} \int_{-\infty}^\infty e^{-t_{ij}^2} dt_{ij}.$$

Observe that

$$2 \int_0^\infty t_{jj}^{\alpha-\frac{j}{2}} e^{-t_{jj}^2} dt_{jj} = \Gamma\left(\alpha - \frac{j-1}{2}\right), \ \Re(\alpha) > \frac{j-1}{2}$$
$$\int_{-\infty}^\infty e^{-t_{ij}^2} dt_{ij} = \sqrt{\pi}$$

and there are $p(p-1)/2$ factors in $\prod_{i>j}$ and hence

$$\Gamma_p(\alpha) = \pi^{\frac{p(p-1)}{4}} \Gamma(\alpha)\Gamma\left(\alpha - \frac{1}{2}\right) \cdots \Gamma\left(\alpha - \frac{p-1}{2}\right)$$

and the condition $\Re(\alpha - \frac{j-1}{2})$, $j = 1, \cdots, p \Rightarrow \Re(\alpha) > \frac{p-1}{2}$. This establishes the result.

Notation 11.2.1.

$$\Gamma_p(\alpha): \quad \textbf{Real matrix-variate gamma}$$

Definition 11.2.1. Real matrix-variate gamma $\Gamma_p(\alpha)$: It is defined by equations (11.2.2) and (11.2.3) where (11.2.2) gives the integral representation and (11.2.3) gives the explicit form.

Remark 11.2.1: If we try to evaluate the integral $\int_X |X|^{\alpha-\frac{p+1}{2}} e^{-\text{tr}(X)} dX$ from first principles, as a multiple integral, notice that even for $p = 3$ the integral is practically impossible to evaluate. For $p = 2$ one can evaluate after going through several stages.

The transformation in (11.2.1) is a nonlinear transformation whereas in (11.1.4) it is a general linear transformation involving mn functionally independent real x_{ij}'s. When X is a square and nonsingular matrix its regular inverse X^{-1} exists and the transformation $Y = X^{-1}$ is a one-to-one nonlinear transformation. What will be the Jacobian in this case?

Theorem 11.2.2. *For* $Y = X^{-1}$ *where* X *is a* $p \times p$ *nonsingular matrix we have*

$$Y = X^{-1}, |X| \neq 0 \Rightarrow dY = |X|^{-2p} \text{ for a general } X$$
$$= |X|^{-(p+1)} \text{ for } X = X'. \tag{11.2.4}$$

Proof 11.2.1: This can be proved by observing the following: When X is nonsingular, $XX^{-1} = I$ where I denotes the identity matrix. Taking differentials on both sides we have

$$(dX)X^{-1} + X(dX^{-1}) = O \Rightarrow$$
$$(dX^{-1}) = -X^{-1}(dX)X^{-1} \tag{11.2.5}$$

where (dX) means the matrix of differentials. Now we can apply Theorem 11.1.4 treating X^{-1} as a constant matrix because it is free of the differentials since we are taking only the wedge product of differentials on the left side.

Note 11.2.1: If the square matrix X is nonsingular and skew symmetric then proceeding as above it follows that

$$Y = X^{-1}, \ |X| \neq 0, \ X' = -X \Rightarrow dY = |X|^{-(p-1)} dX. \qquad (11.2.6)$$

Note 11.2.2: If X is nonsingular and lower or upper triangular then, proceeding as before we have

$$Y = X^{-1} \Rightarrow dY = |X|^{-(p+1)} \qquad (11.2.7)$$

where $|X| \neq 0$, X is lower or upper triangular.

Theorem 11.2.3. Let $X = (x_{ij})$ be $p \times p$ symmetric positive definite matrix of functionally independent real variables with $x_{jj} = 1$, $j = 1, \ldots, p$. Let $T = (t_{ij})$ be a lower triangular matrix of functionally independent real variables with $t_{jj} > 0$, $j = 1, \ldots, p$. Then

$$X = TT', \text{ with } \sum_{j=1}^{i} t_{ij}^2 = 1, \ i = 1, \ldots, p \Rightarrow$$

$$dX = \left\{ \prod_{j=2}^{p} t_{jj}^{p-j} \right\} dT, \qquad (11.2.8)$$

and

$$X = T'T, \text{ with } \sum_{i=j}^{p} t_{ij}^2 = 1, \ j = 1, \ldots, p \Rightarrow$$

$$dX = \left\{ \prod_{j=1}^{p-1} t_{jj}^{j-1} \right\} dT. \qquad (11.2.9)$$

Proof 11.2.2: Since X is symmetric with $x_{jj} = 1$, $j = 1, \ldots, p$ there are only $p(p-1)/2$ variables in X. When $X = TT'$ take the x_{ij}'s in the order $x_{21}, \ldots, x_{p1}$, $x_{32}, \ldots, x_{p2}, \ldots, x_{pp-1}$ and the t_{ij}'s also in the same order and form the matrix of partial derivatives. We obtain a triangular format and the product of the diagonal elements gives the required Jacobian.

Example 11.2.3. Let $R = (r_{ij})$ be a $p \times p$ real symmetric positive definite matrix such that $r_{jj} = 1$, $j = 1, \ldots, p$, $-1 < r_{ij} = r_{ji} < 1$, $i \neq j$. (This is known as the correlation matrix in statistical theory). Then show that

$$f(R) = \frac{[\Gamma(\alpha)]^p}{\Gamma_p(\alpha)} |R|^{\alpha - \frac{p+1}{2}}$$

is a density function for $\Re(\alpha) > \frac{p-1}{2}$.

Solution 11.2.3: Since R is positive definite $f(R) \geq 0$ for all R. Let us check the total integral. Let T be a lower triangular matrix as defined in the above theorem and let $R = TT'$. Then

$$\int_R |R|^{\alpha - \frac{p+1}{2}} dR = \int_T \left\{ \prod_{j=2}^{p} (t_{jj}^2)^{\alpha - \frac{j+1}{2}} \right\} dT.$$

Observe that

$$t_{jj}^2 = 1 - t_{j1}^2 - \ldots - t_{jj-1}^2$$

where $-1 < t_{ij} < 1$, $i > j$. Then let

$$B = \int_R |R|^{\alpha - \frac{P+1}{2}} dR = \prod_{i>j} \Delta_j$$

where

$$\Delta_j = \int_{w_j} (1 - t_{j1}^2 - \ldots - t_{jj-1}^2)^{\alpha - \frac{j+1}{2}} dt_{j1} \cdots dt_{jj-1}$$

where $w_j = (t_{jk})$, $-1 < t_{jk} < 1$, $k = 1, \ldots, j-1$, $\sum_{k=1}^{j-1} t_{jk}^2 < 1$.

Evaluating the integral with the help of Dirichlet integral of Chapter 1 and then taking the product we have the final result showing that $f(R)$ is a density.

Exercises 11.2.

11.2.1. Let $X = X' > 0$ be $p \times p$. Let $T = (t_{ij})$ be an upper triangular matrix with positive diagonal elements. Then show that

$$X = TT' \Rightarrow dX = 2^p \left\{ \prod_{j=1}^{p} t_{jj}^j \right\} dT. \tag{11.2.10}$$

11.2.2. Let $x_1, \cdots, x_p$ be real scalar variables. Let $y_1 = x_1 + \cdots + x_p$, $y_2 = x_1 x_2 + x_1 x_3 + \cdots x_{p-1} x_p$ (sum of products taken two at a time), $\cdots$, $y_k = x_1 \cdots x_k$. Then for $x_j > 0$, $j = 1, \cdots, k$ show that

$$dy_1 \wedge \cdots \wedge dy_k = \left\{ \prod_{i=1}^{p-1} \prod_{j=i+1}^{p} |x_i - x_j| \right\} dx_1 \wedge \cdots \wedge dx_p. \tag{11.2.11}$$

11.2.3. Let $x_1, \cdots, x_p$ be real scalar variables. Let

$$x_1 = r \sin \theta_1$$
$$x_j = r \cos \theta_1 \cos \theta_2 \cdots \cos \theta_{j-1} \sin \theta_j, \quad j = 2, 3, \cdots, p-1$$
$$x_p = r \cos \theta_1 \cos \theta_2 \cdots \cos \theta_{p-1}$$

for $r > 0, -\frac{\pi}{2} < \theta_j \leq \frac{\pi}{2}, j = 1, \cdots, p-2, -\pi < \theta_{p-1} \leq \pi$. Then show that

$$dx_1 \wedge \cdots \wedge dx_p = r^{p-1} \left\{ \prod_{j=1}^{p-1} |\cos \theta_j|^{p-j-1} \right\} dr \wedge d\theta_1 \wedge \cdots \wedge d\theta_{p-1}. \tag{11.2.12}$$

11.2.4. Let $X = \frac{T}{|T|}$ where X and T are $p \times p$ lower triangular or upper triangular matrices of functionally independent real variables with positive diagonal elements. Then show that

$$dX = (p-1)|T|^{-p(p+1)/2}dT. \tag{11.2.13}$$

11.2.5. For real symmetric positive definite matrices X and Y show that

$$\lim_{t \to \infty} \left| I + \frac{XY}{t} \right|^{-t} = e^{-\text{tr}(XY)} = \lim_{t \to \infty} \left| I - \frac{XY}{t} \right|^{t}. \tag{11.2.14}$$

11.2.6. Let $X = (x_{ij}), W = (w_{ij})$ be lower triangular $p \times p$ matrices of distinct real variables with $x_{jj} > 0, w_{jj} > 0, j = 1, \cdots, p, \sum_{k=1}^{j} w_{jk}^2 = 1, j = 1, \cdots, p$. Let $D = \text{diag}(\lambda_1, \cdots, \lambda_p), \lambda_j > 0, j = 1, \cdots, p$, real and distinct where $\text{diag}(\lambda_1, \cdots, \lambda_p)$ denotes a diagonal matrix with diagonal elements $\lambda_1, \cdots, \lambda_p$. Show that

$$X = DW \Rightarrow dX = \left\{ \prod_{j=1}^{p} \lambda_j^{j-1} w_{jj}^{-1} \right\} dD \wedge dW. \tag{11.2.15}$$

11.2.7. Let X, A, B be $p \times p$ nonsingular matrices where A and B are constant matrices and X is a matrix of functionally independent real variables. Then, ignoring the sign, show that

$$Y = AX^{-1}B \Rightarrow dY = |AB|^p |X|^{-2p}dX \text{ for a general } X, \tag{11.2.16}$$

$$= |AX^{-1}|^{p+1}dX \text{ for } X = X', B = A', \tag{11.2.17}$$

$$= |AX^{-1}|^{p-1} \text{ for } X' = -X, B = A'. \tag{11.2.18}$$

11.2.8. Let X and A be $p \times p$ matrices where A is a nonsingular constant matrix and X is a matrix of functionally independent real variables such that $A + X$ is nonsingular. Then, ignoring the sign, show that

$$Y = (A+X)^{-1}(A-X) \text{ or } (A-X)(A+X)^{-1} \Rightarrow$$

$$dY = 2^{p^2} |A|^p |A+X|^{-2p}dX \text{ for a general } X, \tag{11.2.19}$$

$$= 2^{\frac{p(p+1)}{2}} |I+X|^{-(p+1)}dX \text{ for } A = I, X = X'. \tag{11.2.20}$$

11.2.9. Let X and A be real $p \times p$ lower triangular matrices where A is a constant matrix and X is a matrix of functionally independent real variables such that A and $A + X$ are nonsingular. Then, ignoring the sign, show that

$$Y = (A+X)^{-1}(A-X) \Rightarrow$$

$$dY = 2^{\frac{p(p+1)}{2}} |A+X|_{+}^{-(p+1)} \left\{ \prod_{j=1}^{p} |a_{jj}|^{p+1-j} \right\} dX, \tag{11.2.21}$$

and

$$Y = (A - X)(A + X)^{-1} \Rightarrow$$

$$dY = 2^{\frac{p(p+1)}{2}} |A + X|_+^{-(p+1)} \left\{ \prod_{j=1}^{p} |a_{jj}|^j \right\} dX \qquad (11.2.22)$$

where $|\cdot|_+$ denotes that the absolute value is taken.

11.2.10. State and prove the corresponding results in Exercise 11.2.9 for upper triangular matrices.

11.3 Transformations Involving Orthonormal Matrices

Here we will consider a few matrix transformations involving orthonormal and semiorthonormal matrices. Some basic results that we need later on are discussed here. For more on these types and various other types of transformations see Mathai (1997). Since the proofs in many of the results are too involved and beyond the scope of this School we will not go into the details of the proofs.

Theorem 11.3.1. *Let X be a $p \times n, n \geq p$, matrix of rank p and of functionally independent real variables, and let $X = TU_1'$ where T is $p \times p$ lower triangular with distinct nonzero diagonal elements and U_1' a unique $n \times p$ semiorthonormal matrix, $U_1'U_1 = I_p$, all are of functionally independent real variables. Then*

$$X = TU_1' \Rightarrow dX = \left\{ \prod_{j=1}^{p} |t_{jj}|^{n-j} \right\} dT \wedge U_1'(dU_1) \qquad (11.3.1)$$

where

$$\int \wedge U_1'(dU_1) = \frac{2^p \pi^{\frac{pn}{2}}}{\Gamma_p \left(\frac{n}{2} \right)}. \qquad (11.3.2)$$

(see equation (11.2.3) for $\Gamma_p(\cdot)$).

Proof 11.3.1: For proving the main part of the theorem take the differentials on both sides of $X = TU_1'$ and then take the wedge product of the differentials systematically. Since it involves many steps the proof of the main part is not given here. The second part can be proved without much difficulty. Consider X the $p \times n, n \geq p$ real matrix. Observe that

$$\mathrm{tr}(XX') = \sum_{ij} x_{ij}^2,$$

that is, the sum of squares of all elements in $X = (x_{ij})$ and there are np terms in $\sum_{ij} x_{ij}^2$. Now consider the integral

$$\int_X e^{-\text{tr}(X)} dX = \prod_{ij} \int_{-\infty}^{\infty} e^{-x_{ij}^2} dx_{ij}$$
$$= \pi^{np/2}$$

since each integral over x_{ij} gives $\sqrt{\pi}$. Now let us evaluate the same integral by using Theorem 11.3.1. Consider the same transformation as in Theorem 11.3.1, $X = TU_1'$. Then

$$\pi^{np/2} = \int_X e^{-\text{tr}(X)} dX$$
$$= \int_T \left\{ \prod_{j=1}^{p} |t_{jj}|^{n-j} \right\} e^{-(\Sigma_{i \geq j} t_{ij}^2)} dT \int \wedge U_1'(dU_1).$$

But for $0 < t_{jj} < \infty, -\infty < t_{ij} < \infty, i > j$ and U_1 unrestricted semiorthonormal, we have

$$\int_T \left\{ \prod_{j=1}^{p} |t_{jj}|^{n-j} \right\} e^{-(\Sigma_{i \geq j} t_{ij}^2)} dT = 2^{-p} \Gamma_p \left(\frac{n}{2} \right) \tag{11.3.3}$$

observing that for $j = 1, \ldots, p$ the p integrals

$$\int_0^{\infty} |t_{jj}|^{n-j} e^{-t_{jj}^2} dt_{jj} = 2^{-1} \Gamma \left(\frac{n}{2} - \frac{j-1}{2} \right), \; n > j - 1, \tag{11.3.4}$$

and each of the $p(p-1)/2$ integrals

$$\int_{-\infty}^{\infty} e^{-t_{ij}^2} dt_{ij} = \sqrt{\pi}, \; i > j. \tag{11.3.5}$$

Now, substituting these the result in (11.3.2) is established.

Remark 11.3.1: For the transformation $X = TU_1'$ to be unique, either one can take T with the diagonal elements $t_{jj} > 0, \; j = 1, \ldots, p$ and U_1 unrestricted semiorthonormal matrix or $-\infty < t_{jj} < \infty$ and U_1 a unique semiorthonormal matrix.

From the outline of the proof of Theorem 11.3.1 we have the following result:

$$\int_{V_{p,n}} \wedge U_1'(dU_1) = \frac{2^p \pi^{pn/2}}{\Gamma_p(\frac{n}{2})} \tag{11.3.6}$$

where $V_{p,n}$ is the *Stiefel manifold*, or the set of semiorthonormal matrices of the type $U_1, n \times p$, such that $U_1'U_1 = I_p$ where I_p is an identity matrix of order p. For $n = p$ the Stiefel manifold becomes the *full orthogonal group*, denoted by O_p. Then we have for, $n = p$,

$$\int_{O_p} \wedge U_1'(dU_1) = \frac{2^p \pi^{p^2}}{\Gamma_p(\frac{n}{2})}. \tag{11.3.7}$$

Following through the same steps as in Theorem 11.3.1 we can have the following theorem involving an upper triangular matrix T_1.

Theorem 11.3.2. *Let X_1 be an $n \times p$, $n \geq p$, matrix of rank p and of functionally independent real variables and let $X_1 = U_1 T_1$ where T_1 is a real $p \times p$ upper triangular matrix with distinct nonzero diagonal elements and U_1 is a unique real $n \times p$ semiorthonormal matrix, that is, $U_1'U_1 = I_p$. Then, ignoring the sign,*

$$X_1 = U_1 T_1 \Rightarrow dX_1 = \left\{ \prod_{j=1}^{p} |t_{jj}|^{n-j} \right\} dT_1 \wedge U_1'(dU_1) \qquad (11.3.8)$$

As a corollary to Theorem 11.3.1 or independently we can prove the following result:

Corollary 11.3.1: *Let X_1, T, U_1 be as defined in Theorem 11.3.1 with the diagonal elements of T positive, that is, $t_{jj} > 0$, $j = 1, \ldots, p$ and U_1 an arbitrary semiorthonormal matrix, and let $A = XX'$, which implies, $A = TT'$ also. Then*

$$A = XX' \Rightarrow \qquad (11.3.9)$$

$$dA = 2^p \left\{ \prod_{j=1}^{p} t_{jj}^{p+1-j} \right\} dT \qquad (11.3.10)$$

$$\Rightarrow$$

$$dT = 2^{-p} \left\{ \prod_{j=1}^{p} t_{jj}^{-p-1-j} \right\} dA. \qquad (11.3.11)$$

In practical applications we would like to have dX in terms of dA or vice versa after integrating out $\wedge U_1'(dU_1)$ over the Stiefel manifold $V_{p,n}$. Hence we have the following corollary.

Corollary 11.3.2: *Let X_1, T, U_1 be as defined in Theorem 11.3.1 with $t_{jj} > 0$, $j = 1, \ldots, p$ and let $S = XX' = TT'$. Then, after integrating out $\wedge U_1'(dU_1)$ we have*

$$X = TU_1' \text{ and } S = XX' = TT' \Rightarrow \qquad (11.3.12)$$

$$dX = \left\{ \prod_{j=1}^{p} (t_{jj}^2)^{\frac{n}{2} - \frac{j}{2}} \right\} dT \wedge U_1'(dU_1) \qquad (11.3.13)$$

$$\int_{V_{p,n}} \wedge U_1'(dU_1) = \frac{2^p \pi^{np/2}}{\Gamma_p(\frac{n}{2})} \qquad (11.3.14)$$

$$dS = 2^p \left\{ \prod_{j=1}^{p} (t_{jj}^2)^{\frac{p+1}{2} - \frac{j}{2}} \right\} dT \qquad (11.3.15)$$

$$|S| = \prod_{j=1}^{p} t_{jj}^2 \qquad (11.3.16)$$

and, finally,

$$dX = |S|^{\frac{n}{2} - \frac{p+1}{2}} dS. \qquad (11.3.17)$$

Example 11.3.1. Let X be a $p \times n, n \geq p$ random matrix having an np-variate real Gaussian density

$$f(X) = \frac{e^{-\frac{1}{2}\text{tr}(V^{-1}XX')}}{(2\pi)^{\frac{np}{2}}|V|^{\frac{n}{2}}}, \ V = V' > 0.$$

Evaluate the density of $S = XX'$.

Solution 11.3.1: Consider the transformation as in Theorem 11.3.1, $X = TU_1'$ where T is a $p \times p$ lower triangular matrix with positive diagonal elements and U_1 is a arbitrary $n \times p$ semiorthonormal matrix, $U_1'U_1 = I_p$. Then

$$\text{d}X = \left\{ \prod_{j=1}^{p} t_{jj}^{n-j} \right\} \text{d}T \wedge U_1(\text{d}U_1).$$

Integrating out $\wedge U_1(\text{d}U_1)$ we have the marginal density of T, denoted by $g(T)$. That is,

$$g(T)\text{d}T = \frac{2^p e^{-\frac{1}{2}\text{tr}(V^{-1}TT')}}{2^{np/2}|V|^{\frac{n}{2}}} \left\{ \prod_{j=1}^{p} t_{jj}^{n-j} \right\} \text{d}T.$$

Now substituting from Corollary 11.3.2, S and $\text{d}S$ in terms of T and $\text{d}T$ we have the density of S, denoted by, $h(S)$, given by

$$h(S) = C_1|S|^{\frac{n}{2} - \frac{p+1}{2}} e^{-\frac{1}{2}V^{-1}S}, \ S = S' > 0.$$

Since the total integral, $\int_S h(S)\text{d}S = 1$, we have

$$C_1 = [2^{np/2}\Gamma_p\left(\frac{n}{2}\right)|V|^{n/2}]^{-1}.$$

Exercises 11.3.

11.3.1. Let $X = (x_{ij})$, $W = (w_{ij})$ be $p \times p$ lower triangular matrices of distinct real variables such that $x_{jj} > 0$, $w_{jj} > 0$ and $\sum_{k=1}^{j} w_{jk}^2 = 1$, $j = 1, \ldots, p$. Let $D = \text{diag}(\lambda_1, \ldots, \lambda_p)$, $\lambda_j > 0$, $j = 1, \ldots, p$ be real positive and distinct. Let $D^{\frac{1}{2}} = \text{diag}(\lambda_1^{\frac{1}{2}}, \ldots, \lambda_p^{\frac{1}{2}})$. Then show that

$$X = DW \Rightarrow \text{d}X = \left\{ \prod_{j=1}^{p} \lambda_j^{j-1} w_{jj}^{-1} \right\} \text{d}D \wedge \text{d}W. \tag{11.3.18}$$

11.3.2. Let X, D, W be as defined in Exercise 11.3.1 then show that

$$X = D^{\frac{1}{2}}W \Rightarrow \text{d}X = \left\{ 2^{-p} \prod_{j=1}^{p} (\lambda_j^{\frac{1}{2}})^{j-2} w_{jj}^{-1} \right\} \text{d}D \wedge \text{d}W. \tag{11.3.19}$$

11.3.3. Let X, D, W be as defined in Exercise 11.3.1 then show that

$$X = D^{\frac{1}{2}}WW'D^{\frac{1}{2}} \Rightarrow dX = \left\{ \prod_{j=1}^{p} \lambda_j^{\frac{p-1}{2}} w_{jj}^{p-j} \right\} dD \wedge dW, \tag{11.3.20}$$

and

$$X = W'DW \Rightarrow dX = \left\{ \prod_{j=1}^{p} (\lambda_j w_{jj})^{j-1} \right\} dD \wedge dW. \tag{11.3.21}$$

11.3.4. Let X, D, W be as defined in Exercise 11.3.1 then show that

$$X = WD \Rightarrow dX = \left\{ \prod_{j=1}^{p} \lambda_j^{p-j} w_{jj}^{-1} \right\} dD \wedge dW, \tag{11.3.22}$$

$$X = WD^{\frac{1}{2}} \Rightarrow dX = \left\{ 2^{-p} \prod_{j=1}^{p} (\lambda_j^{\frac{1}{2}})^{p-j-1} w_{jj}^{-1} \right\} dD \wedge dW, \tag{11.3.23}$$

$$X = WDW' \Rightarrow dX = \left\{ \prod_{j=1}^{p} (\lambda_j w_{jj})^{p-j} \right\} dD \wedge dW, \tag{11.3.24}$$

and

$$X = D^{\frac{1}{2}}W'WD^{\frac{1}{2}} \Rightarrow dX = \left\{ \prod_{j=1}^{p} \lambda_j^{\frac{p-1}{2}} w_{jj}^{j-1} \right\} dD \wedge dW. \tag{11.3.25}$$

11.3.5. Let X, T, U be $p \times p$ matrices of functionally independent real variables where all the principal minors of X are nonzero, T is lower triangular and U is lower triangular with unit diagonal elements. Then, ignoring the sign, show that

$$X = TU' \Rightarrow dX = \left\{ \prod_{j=1}^{p} |t_{jj}|^{p-j} \right\} dT \wedge dU, \tag{11.3.26}$$

and

$$X = T'U \Rightarrow dX = \left\{ \prod_{j=1}^{p} |t_{jj}|^{j-1} \right\} dT \wedge dU. \tag{11.3.27}$$

11.3.6. Let X be a $p \times p$ symmetric matrix of functionally independent real variables and with distinct and nonzero eigenvalues $\lambda_1 > \lambda_2 > \ldots > \lambda_p$ and let $D = \mathrm{diag}(\lambda_1, \ldots, \lambda_p)$, $\lambda_j \neq 0$, $j = 1, \ldots, p$. Let U be a unique $p \times p$ orthonormal matrix $U'U = I = UU'$ such that $X = UDU'$. Then, ignoring the sign, show that

$$dX = \left\{ \prod_{i=1}^{p-1} \prod_{j=i+1}^{p} |\lambda_i - \lambda_j| \right\} dD \wedge U'(dU). \tag{11.3.28}$$

11.3.7. For a 3×3 matrix X such that $X = X' > 0$ and $I - X > 0$ show that

$$\int_X dX = \frac{\pi^2}{90}.$$

11.3.8. For a $p \times p$ matrix X of p^2 functionally independent real variables with positive eigenvalues, show that

$$\int_Y |Y'Y|^{\alpha - \frac{p+1}{2}} e^{-\text{tr}(Y'Y)} dY = 2^{-p} \frac{\Gamma_p(\alpha - \frac{1}{2})\pi^{\frac{p^2}{2}}}{\Gamma_p(\frac{p}{2})}.$$

11.3.9. Let X be a $p \times p$ matrix of p^2 functionally independent real variables. Let $D = \text{diag}(\mu_1, \ldots, \mu_p), \mu_1 > \mu_2 > \ldots > \mu_p$, μ_j real for $j = 1, \ldots, p$. Let U and V be orthonormal matrices such that

$$X = UDV' \Rightarrow$$

$$dX = \left\{ \prod_{i<j} |\mu_i^2 - \mu_j^2| \right\} dD \wedge dG \wedge dH \qquad (11.3.29)$$

where $(dG) = U'(dU)$ and $(dH) = (dV')V$, and the μ_j's are known as the singular values of X.

11.3.10. Let $\lambda_1 > \ldots > \lambda_p > 0$ be real variables and $D = \text{diag}(\lambda_1, \ldots, \lambda_p)$. Show that

$$\int_D e^{-\text{tr}(D^2)} \left\{ \prod_{i<j} |\lambda_i^2 - \lambda_j^2| \right\} dD = \frac{[\Gamma_p(\frac{p}{2})]^2}{2^p \pi^{\frac{p^2}{2}}}.$$

References

Mathai, A.M. (1978). Some results on functions of matrix argument, *Math. Nachr.* **84**, 171-177.

Mathai, A.M. (1993). *A Handbook of Generalized Special Functions for Statistical and Physical Sciences*, Oxford University Press, Oxford.

Mathai, A.M. (1997). *Jacobians of Matrix Transformations and Functions of Matrix Argument*, World Scientific Publishing, New York.

Mathai, A.M. (2004). *Modules 1,2,3*, Centre for Mathematical Sciences, India.

Mathai, A.M., Provost, S.B. and Hayakawa, T. (1995). *Bilinear Forms and Zonal Polynomials*, Springer-Verlag Lecture Notes in Statistics, **102**, New York.

Chapter 12
Special Functions of Matrix Argument

[*This Chapter is based on the lectures of Professor A.M. Mathai of McGill University, Canada (Director of the SERC Schools).*]

12.0 Introduction

Real scalar functions of matrix argument, when the matrices are real, will be dealt with in this chapter. It is difficult to develop a theory of functions of matrix argument for general matrices. The notations that we have used in Chapter 11 will be used in the present chapter also. A discussion of scalar functions of matrix argument when the elements of the matrices are in the complex domain may be seen from Mathai (1997).

12.1 Real Matrix-Variate Scalar Functions

When dealing with matrices it is often difficult to define uniquely fractional powers such as square roots even when the matrices are real square or even symmetric. For example

$$A_1 = \begin{bmatrix} 1 & 0 \\ 0 & 1 \end{bmatrix}, A_2 = \begin{bmatrix} -1 & 0 \\ 0 & 1 \end{bmatrix}, A_3 = \begin{bmatrix} 1 & 0 \\ 0 & -1 \end{bmatrix}, A_4 = \begin{bmatrix} -1 & 0 \\ 0 & -1 \end{bmatrix}$$

all give $A_1^2 = A_2^2 = A_3^2 = A_4^2 = I_2$ where I_2 is a 2×2 identity matrix. Thus, even for I_2, which is a nice, square, symmetric, positive definite matrix there are many matrices which qualify to be square roots of I_2. But if we confine to the class of positive definite matrices, when real, then for the square root of I_2 there is only one candidate, namely, $A_1 = I_2$ itself. Hence the development of the theory of scalar functions of matrix argument is confined to positive definite matrices, when real, and hermitian positive definite matrices when in the complex domain.

429

12.1.1 Real matrix-variate gamma

In the real scalar case the integral representation for a gamma function is the following:

$$\Gamma(\alpha) = \int_0^\infty x^{\alpha-1} e^{-x} dx, \ \Re(\alpha) > 0. \tag{12.1.1}$$

Let X be a $p \times p$ real symmetric positive definite matrix and consider the integral

$$\Gamma_p(\alpha) = \int_{X=X'>0} |X|^{\alpha - \frac{p+1}{2}} e^{-\text{tr}(X)} dX \tag{12.1.2}$$

where, when $p = 1$ the equation (12.1.2) reduces to (12.1.1). We have already evaluated this integral in Chapter 11 as an exercise to the basic nonlinear matrix transformation $X = TT'$ where T is a lower triangular matrix with positive diagonal elements. Hence the derivation will not be repeated here. We will call $\Gamma_p(\alpha)$ the real matrix-variate gamma. Observe that for $p = 1$, $\Gamma_p(\alpha)$ reduces to $\Gamma(\alpha)$.

12.1.2 Real matrix-variate gamma density

With the help of (12.1.2) we can create the real matrix-variate gamma density as follows, where X is a $p \times p$ real symmetric positive definite matrix:

$$f(X) = \begin{cases} \frac{1 \times 1^{\alpha - \frac{p+1}{2}}}{\Gamma_p(\alpha)} e^{-\text{tr}(X)}, X = X' > 0, \ \Re(\alpha) > \frac{p-1}{2} \\ 0, \text{ elsewhere .} \end{cases} \tag{12.1.3}$$

If another parameter matrix is to be introduced then we obtain a gamma density with parameters $(\alpha, B), B = B' > 0$, as follows:

$$f_1(X) = \begin{cases} \frac{|B|^\alpha}{\Gamma_p(\alpha)} |X|^{\alpha - \frac{p+1}{2}} e^{-\text{tr}(BX)}, X = X' > 0, B = B' > 0, \Re(\alpha) > \frac{p-1}{2} \\ 0, \text{elsewhere.} \end{cases}$$
$$\tag{12.1.4}$$

Remark 12.1.1: In $f_1(X)$ if B is replaced by $\frac{1}{2}V^{-1}, V = V' > 0$ and α is replaced by $\frac{n}{2}, n = p, p+1, \dots$ then we have the most important density in multivariate statistical analysis known as the nonsingular Wishart density.

As in the scalar case, two matrix random variables X and Y are said to be independently distributed if the joint density of X and Y is the product of their marginal densities. We will examine the densities of some functions of independently distributed matrix random variables. To this end we will introduce a few more functions.

Definition 12.1.1. A real matrix-variate beta function, denoted by $B_p(\alpha,\beta)$, is defined as

$$B_p(\alpha,\beta) = \frac{\Gamma_p(\alpha)\Gamma_p(\beta)}{\Gamma_p(\alpha+\beta)}, \, \Re(\alpha) > \frac{p-1}{2}, \Re(\beta) > \frac{p-1}{2}. \tag{12.1.5}$$

The quantity in (12.1.5), analogous to the scalar case $(p=1)$, is the real matrix-variate beta function. Let us try to obtain an integral representation for the real matrix-variate beta function of (12.1.5). Consider

$$\Gamma_p(\alpha)\Gamma_p(\beta) = \left[\int_{X=X'>0} |X|^{\alpha-\frac{p+1}{2}} e^{-\text{tr}(X)} dX\right]$$
$$\times \left[\int_{Y=Y'>0} |Y|^{\beta-\frac{p+1}{2}} e^{-\text{tr}(Y)} dY\right]$$

where both X and Y are $p \times p$ matrices.

$$= \int\int |X|^{\alpha-\frac{p+1}{2}} |Y|^{\beta-\frac{p+1}{2}} e^{-\text{tr}(X+Y)} dX \wedge dY.$$

Put $U = X+Y$ for a fixed X. Then

$$Y = U - X \Rightarrow |Y| = |U-X| = |U||I - U^{-\frac{1}{2}} X U^{-\frac{1}{2}}|$$

where, for convenience, $U^{\frac{1}{2}}$ is the symmetric positive definite square root of U. Observe that when two matrices A and B are nonsingular where AB and BA are defined, even if they do not commute,

$$|I - AB| = |I - BA|$$

and if $A = A' > 0$ and $B = B' > 0$ then

$$|I - AB| = |I - A^{\frac{1}{2}} BA^{\frac{1}{2}}| = |I - B^{\frac{1}{2}} AB^{\frac{1}{2}}|.$$

Now,

$$\Gamma_p(\alpha)\Gamma_p(\beta) = \int_U \int_X |U|^{\beta-\frac{p+1}{2}} |X|^{\alpha-\frac{p+1}{2}} |I - U^{-\frac{1}{2}} X U^{-\frac{1}{2}}|^{\beta-\frac{p+1}{2}} e^{-\text{tr}(U)} dU \wedge dX.$$

Let $Z = U^{-\frac{1}{2}} X U^{-\frac{1}{2}}$ for fixed U. Then $dX = |U|^{\frac{p+1}{2}} dZ$ by using Theorem 11.1.5. Now,

$$\Gamma_p(\alpha)\Gamma_p(\beta) = \int_Z |Z|^{\alpha-\frac{p+1}{2}} |I-Z|^{\beta-\frac{p+1}{2}} dZ \int_{U=U'>0} |U|^{\alpha+\beta-\frac{p+1}{2}} e^{-\text{tr}(U)} dU.$$

Evaluation of the U-integral by using (12.1.2) yields $\Gamma_p(\alpha+\beta)$. Then we have

$$B_p(\alpha,\beta) = \frac{\Gamma_p(\alpha)\Gamma_p(\beta)}{\Gamma_p(\alpha+\beta)} = \int_Z |Z|^{\alpha-\frac{p+1}{2}} |I-Z|^{\beta-\frac{p+1}{2}} dZ.$$

Since the integral has to remain non-negative we have $Z = Z' > 0, I - Z > 0$. Therefore, one representation of a real matrix-variate beta function is the following, which is also called the type-1 beta integral.

$$B_p(\alpha, \beta) = \int_{0<Z=Z'<I} |Z|^{\alpha - \frac{p+1}{2}} |I - Z|^{\beta - \frac{p+1}{2}} dZ, \Re(\alpha) > \frac{p-1}{2}, \Re(\beta) > \frac{p-1}{2}. \tag{12.1.6}$$

By making the transformation $V = I - Z$ note that α and β can be interchanged in the integral which also shows that $B_p(\alpha, \beta) = B_p(\beta, \alpha)$ in the integral representation also.

Let us make the following transformation in (12.1.6).

$$W = (I - Z)^{-\frac{1}{2}} Z (I - Z)^{-\frac{1}{2}} \Rightarrow W = (Z^{-1} - I)^{-1} \Rightarrow W^{-1} = Z^{-1} - I$$

$$\Rightarrow |W|^{-(p+1)} dW = |Z|^{-(p+1)} dZ \Rightarrow dZ = |I + W|^{-(p+1)} dW.$$

Under this transformation the integral in (12.1.6) becomes the following: Observe that

$$|Z| = |W||I + W|^{-1}, |I - Z| = |I + W|^{-1}.$$

$$B_p(\alpha, \beta) = \int_{W=W'>0} |W|^{\alpha - \frac{p+1}{2}} |I + W|^{-(\alpha+\beta)} dW, \tag{12.1.7}$$

for $\Re(\alpha) > \frac{p-1}{2}, \Re(\beta) > \frac{p-1}{2}$.

The representation in (12.1.7) is known as the *type-2 integral* for a real matrix-variate beta function. With the transformation $V = W^{-1}$ the parameters α and β in (12.1.7) will be interchanged. With the help of the type-1 and type-2 integral representations one can define the type-1 and type-2 beta densities in the real matrix-variate case.

Definition 12.1.2. **Real matrix-variate type-1 beta density for the $p \times p$ real symmetric positive definite matrix X such that $X = X' > 0$, $I - X > 0$.**

$$f_2(X) = \begin{cases} \frac{1}{B_p(\alpha,\beta)} |X|^{\alpha - \frac{p+1}{2}} |I - X|^{\beta - \frac{p+1}{2}} = 0 < X = X' < I, \Re(\alpha) > \frac{p-1}{2}, \Re(\beta) > \frac{p-1}{2}, \\ 0, \text{ elsewhere.} \end{cases}$$

$$\tag{12.1.8}$$

Definition 12.1.3. **Real matrix-variate type-2 beta density for the $p \times p$ real symmetric positive definite matrix X.**

$$f_3(X) = \begin{cases} \frac{\Gamma_p(\alpha+\beta)}{\Gamma_p(\alpha)\Gamma_p(\beta)} |X|^{\alpha - \frac{p+1}{2}} |I + X|^{-(\alpha+\beta)}, X = X' > 0, \\ \qquad\qquad \Re(\alpha) > \frac{p-1}{2}, \Re(\beta) > \frac{p-1}{2} \\ 0, \text{ elsewhere.} \end{cases} \tag{12.1.9}$$

Example 12.1.1. Let X_1, X_2 be $p \times p$ matrix random variables, independently distributed as (12.1.3) with parameters α_1 and α_2 respectively. Let $U = X_1 + X_2, V = (X_1 + X_2)^{-\frac{1}{2}} X_1 (X_1 + X_2)^{-\frac{1}{2}}, W = X_2^{-\frac{1}{2}} X_1 X_2^{-\frac{1}{2}}$. Evaluate the densities of U, V and W.

Solutions 12.1.1: The joint density of X_1 and X_2, denoted by $f(X_1, X_2)$, is available as the product of the marginal densities due to independence. That is,

$$f(X_1, X_2) = \frac{|X_1|^{\alpha_1 - \frac{p+1}{2}} |X_2|^{\alpha_2 - \frac{p+1}{2}} e^{-\text{tr}(X_1 + X_2)}}{\Gamma_p(\alpha_1) \Gamma_p(\alpha_2)}, \quad X_1 = X_1' > 0, \, X_2 = X_2' > 0,$$

$$\Re(\alpha_1) > \frac{p-1}{2}, \quad \Re(\alpha_2) > \frac{p-1}{2}. \tag{12.1.10}$$

$$U = X_1 + X_2 \Rightarrow |X_2| = |U - X_1| = |U||I - U^{-\frac{1}{2}} X_1 U^{-\frac{1}{2}}|.$$

Then the joint density of $(U, U_1) = (X_1 + X_2, X_1)$, the Jacobian is unity, is available as

$$f_1(U, U_1) = \frac{1}{\Gamma_p(\alpha_1) \Gamma_p(\alpha_2)} |U_1|^{\alpha_1 - \frac{p+1}{2}} |U|^{\alpha_2 - \frac{p+1}{2}} |I - U^{-\frac{1}{2}} U_1 U^{-\frac{1}{2}}|^{\alpha_2 - \frac{p+1}{2}} e^{-\text{tr}(U)}.$$

Put $U_2 = U^{-\frac{1}{2}} U_1 U^{-\frac{1}{2}} \Rightarrow dU_1 = |U|^{\frac{p+1}{2}} dU_2$ for fixed U. Then the joint density of U and $U_2 = V$ is available as the following:

$$f_2(U, V) = \frac{1}{\Gamma_p(\alpha_1) \Gamma_p(\alpha_2)} |U|^{\alpha_1 + \alpha_2 - \frac{p+1}{2}} e^{-\text{tr}(U)} |V|^{\alpha_1 - \frac{p+1}{2}} |I - V|^{\alpha_2 - \frac{p+1}{2}}.$$

Since $f_2(U, V)$ is a product of two functions of U and V, $U = U' > 0, V = V' > 0$, $I - V > 0$ we see that U and V are independently distributed. The densities of U and V, denoted by $g_1(U)$, $g_2(V)$ are the following:

$$g_1(U) = c_1 |U|^{\alpha_1 + \alpha_2 - \frac{p+1}{2}} e^{-\text{tr}(U)}, U = U' > 0$$

and

$$g_2(V) = c_2 |V|^{\alpha_1 - \frac{p+1}{2}} |I - V|^{\alpha_2 - \frac{p+1}{2}}, \, V = V' > 0, \, I - V > 0,$$

where c_1 and c_2 are the normalizing constants. But from the gamma density and type-1 beta density note that

$$c_1 = \frac{1}{\Gamma_p(\alpha_1 + \alpha_2)}, \, c_2 = \frac{\Gamma_p(\alpha_1 + \alpha_2)}{\Gamma_p(\alpha_1) \Gamma_p(\alpha_2)}, \quad \Re(\alpha_1) > 0, \, \Re(\alpha_2) > 0.$$

Hence U is gamma distributed with the parameter $(\alpha_1 + \alpha_2)$ and V is type-1 beta distributed with the parameters α_1 and α_2 and further that U and V are independently distributed. For obtaining the density of $W = X_2^{-\frac{1}{2}} X_1 X_2^{-\frac{1}{2}}$ start with (12.1.10). Change (X_1, X_2) to (X_1, W) for fixed X_2. Then $dX_1 = |X_2|^{\frac{p+1}{2}} dW$. The joint density of X_2 and W, denoted by $f_{w, x_2}(W, X_2)$, is the following, observing that

$$\text{tr}(X_1 + X_2) = \text{tr}[X_2^{\frac{1}{2}} (I + X_2^{-\frac{1}{2}} X_1 X_2^{-\frac{1}{2}}) X_2^{\frac{1}{2}}]$$

$$= \text{tr}[X_2^{\frac{1}{2}} (I + W) X_2^{\frac{1}{2}}] = \text{tr}[(I + W) X_2]$$

$$= \text{tr}[(I + W)^{\frac{1}{2}} X_2 (I + W)^{\frac{1}{2}}]$$

by using the fact that $\text{tr}(AB) = \text{tr}(BA)$ for any two matrices where AB and BA are defined.

$$f_{w,x_2}(W,X_2) = \frac{1}{\Gamma_p(\alpha_1)\Gamma_p(\alpha_2)}|W|^{\alpha_1-\frac{p+1}{2}}|X_2|^{\alpha_1+\alpha_2-\frac{p+1}{2}}e^{-\text{tr}[(I+W)^{\frac{1}{2}}X_2(I+W)^{\frac{1}{2}}]}.$$

Hence the marginal density of W, denoted by $g_w(W)$, is available by integrating out X_2 from $f_{w,x_2}(W,X_2)$. That is,

$$g_w(W)=\frac{1}{\Gamma_p(\alpha_1)\Gamma_p(\alpha_2)}|W|^{\alpha_1-\frac{p+1}{2}}\int_{X_2=X_2'>0}|X_2|^{\alpha_1+\alpha_2-\frac{p+1}{2}}e^{-\text{tr}[(I+W)^{\frac{1}{2}}X_2(I+W)^{\frac{1}{2}}]}dX_2.$$

Put $X_3 = (I+W)^{\frac{1}{2}}X_2(I+W)^{\frac{1}{2}}$ for fixed W, then $dX_3 = |I+W|^{\frac{p+1}{2}}dX_2$. Then the integral becomes

$$\int_{X_2=X_2'>0}|X_2|^{\alpha_1+\alpha_2-\frac{p+1}{2}}e^{-\text{tr}[(I+W)^{\frac{1}{2}}X_2(I+W)^{\frac{1}{2}}]}dX_2$$

$$= \Gamma_p(\alpha_1 + \alpha_2)|I+W|^{-(\alpha_1+\alpha_2)}.$$

Hence,

$$g_w(W) = \begin{cases} \frac{\Gamma_p(\alpha_1+\alpha_2)}{\Gamma_p(\alpha_1)\Gamma_p(\alpha_2)}|W|^{\alpha_1-\frac{p+1}{2}}|I+W|^{-(\alpha_1+\alpha_2)}, W = W' > 0, \\ \qquad \Re(\alpha_1) > \frac{p-1}{2}, \ \Re(\alpha_2) > \frac{p-1}{2} \\ 0, \text{elsewhere}, \end{cases}$$

which is a type-2 beta density with the parameters α_1 and α_2. Thus, W is real matrix-variate type-2 beta distributed.

Exercises 12.1.

12.1.1. For a real $p \times p$ matrix X such that $X = X' > 0, 0 < X < I$ show that

$$\int_X dX = \frac{[\Gamma_p(\frac{p+1}{2})]^2}{\Gamma_p(p+1)}.$$

12.1.2 Let X be a 2×2 real symmetric positive definite matrix with eigenvalues in $(0,1)$. Then show that

$$\int_{0<X<I}|X|^\alpha dX = \frac{\pi}{(\alpha+1)(\alpha+2)(2\alpha+3)}.$$

12.1.3 For a 2×2 real positive definite matrix X show that

$$\int_X |I+X|^{-3}dX = \frac{\pi}{6}.$$

12.1.4 For a 4×4 real positive definite matrix X such that $0 < X < I$, show that

$$\int_X dX = \frac{2\pi^4}{7!5}.$$

12.1.5 If the $p \times p$ real positive definite matrix random variable X is distributed as a real matrix-variate type-1 beta (having a type-1 beta density), evaluate the density of $Y = A^{\frac{1}{2}}XA^{\frac{1}{2}}$ where the constant matrix $A = A' > 0$.

12.2 The Laplace Transform in the Matrix Case

If $f(x_1, \cdots, x_k)$ is a scalar function of the real scalar variables $x_1, \cdots, x_k$ then the Laplace transform of f, with the parameters $t_1, \cdots, t_k$, is given by

$$L_f(t_1, \cdots, t_k) = \int_0^\infty \cdots \int_0^\infty e^{-t_1 x_1 - \cdots - t_k x_k} f(x_1, \cdots, x_k) dx_1 \wedge \cdots \wedge dx_k. \quad (12.2.1)$$

If $f(X)$ is a real scalar function of the $p \times p$ real symmetric positive definite matrix X then the Laplace transform of $f(X)$ should be consistent with (12.2.1). When $X = X'$ there are only $p(p+1)/2$ distinct elements, either $x_{ij}'s, i \le j$ or $x_{ij}'s, i \ge j$. Hence what is needed is a linear function of all these variables. That is, in the exponent we should have the linear function $t_{11}x_{11} + (t_{21}x_{21} + t_{22}x_{22}) + \cdots + (t_{p1}x_{p1} + \cdots + t_{pp}x_{pp})$. Even if we take a symmetric matrix $T = (t_{ij}) = T'$ then the trace of TX,

$$\text{tr}(TX) = t_{11}x_{11} + \cdots + t_{pp}x_{pp} + 2 \sum_{i<j=1}^p t_{ij}x_{ij}.$$

Hence if we take a symmetric matrix of parameters t_{ij}'s such that

$$T^* = \begin{bmatrix} t_{11} & \frac{1}{2}t_{12} & \cdots & \frac{1}{2}t_{1p} \\ \frac{1}{2}t_{21} & t_{22} & \cdots & \frac{1}{2}t_{2p} \\ \vdots & & & \\ \frac{1}{2}t_{p1} & \frac{1}{2}t_{p2} & \cdots & t_{pp} \end{bmatrix}, \quad T^* = (t_{ij}^*) \Rightarrow t_{jj}^* = t_{jj}, t_{ij}^* = \frac{1}{2}t_{ij}, \ i \ne j$$

then

$$\text{tr}(T^*X) = t_{11}x_{11} + \cdots + t_{pp}x_{pp} + \sum_{i=1}^p \sum_{j=1, i>j}^p t_{ij}x_{ij}.$$

Hence the Laplace transform in the matrix case, for real symmetric positive definite matrix X, is defined with the parameter matrix T^*.

Definition 12.2.1. Laplace transform in the matrix case.

$$L_f(T^*) = \int_{X=X'>0} e^{-\text{tr}(T^*X)} f(X) dX, \quad (12.2.2)$$

whenever the integral is convergent.

Example 12.2.1. Evaluate the Laplace transform for the two-parameter gamma density in (12.1.4).

Solution 12.2.1: Here,

$$f(X) = \frac{|B|^\alpha}{\Gamma_p(\alpha)} |X|^{\alpha - \frac{p+1}{2}} e^{-\text{tr}(BX)}, X = X' > 0, B = B' > 0, \ \Re(\alpha) > \frac{p-1}{2}. \quad (12.2.3)$$

Hence the Laplace transform of f is the following:

$$L_f(T^*) = \frac{|B|^\alpha}{\Gamma_p(\alpha)} \int_{X=X'>0} |X|^{\alpha-\frac{p+1}{2}} e^{-\text{tr}(T^*X)} e^{-\text{tr}(BX)} dX.$$

Note that since T^*, B and X are $p \times p$,

$$\text{tr}(T^*X) + \text{tr}(BX) = \text{tr}[(B+T^*)X].$$

Thus for the integral to converge the exponent has to remain positive definite. Then the condition $B + T^* > 0$ is sufficient. Let $(B+T^*)^{\frac{1}{2}}$ be the symmetric positive definite square root of $B + T^*$. Then

$$\text{tr}[(B+T^*)X] = \text{tr}[(B+T^*)^{\frac{1}{2}}X(B+T^*)^{\frac{1}{2}}],$$

$$(B+T^*)^{\frac{1}{2}}X(B+T^*)^{\frac{1}{2}} = Y \Rightarrow dX = |B+T^*|^{-\frac{p+1}{2}} dY$$

and

$$|X|^{\alpha-\frac{p+1}{2}} dX = |B+T^*|^{-\alpha}|Y|^{\alpha-\frac{p+1}{2}} dY.$$

Hence,

$$L_f(T^*) = \frac{|B|^\alpha}{\Gamma_p(\alpha)} \int_{Y=Y'>0} |B+T^*|^{-\alpha}|Y|^{\alpha-\frac{p+1}{2}} e^{-\text{tr}(Y)} dY$$
$$= |B|^\alpha |B+T^*|^{-\alpha} = |I+B^{-1}T^*|^{-\alpha}. \tag{12.2.4}$$

Thus for known B and arbitrary T^*, (12.1.3) will uniquely determine (12.1.2) through the uniqueness of the inverse Laplace transform. The conditions for the uniqueness will not be discussed here. For some results in this direction see Mathai (1993, 1997) and the references therein.

12.2.1 A convolution property for Laplace transforms

Let $f_1(X)$ and $f_2(X)$ be two real scalar functions of the real symmetric positive definite matrix X and let $g_1(T^*)$ and $g_2(T^*)$ be their Laplace transforms. Let

$$f_3(X) = \int_{0<S=S'<X} f_1(X-S)f_2(S)dS. \tag{12.2.5}$$

Then g_1g_2 is the Laplace transform of $f_3(X)$.
   This result can be established from the definition itself.

$$L_{f_3}(T^*) = \int_{X=X'>0} e^{-\text{tr}(T^*X)} f_3(X)dX$$
$$= \int_{x>0}\int_{S<X} e^{-\text{tr}(T^*X)} f_1(X-S)f_2(S)dS \wedge dX.$$

Note that $\{S < X, X > 0\}$ is also equivalent to $\{X > S, S > 0\}$. Hence we may interchange the integrals. Then

$$L_{f_3}(T^*) = \int_{S>0} f_2(S) \left[\int_{X>S} e^{-\mathrm{tr}(T^*X)} f_1(X-S) dX \right] \wedge dS.$$

Put $X - S = Y \Rightarrow X = Y + S$ and then

$$L_{f_3}(T^*) = \int_{S>0} e^{-\mathrm{tr}(T^*S)} f_2(S) \left[\int_{Y>0} e^{-\mathrm{tr}(T^*Y)} f_1(Y) dY \right] \wedge dS$$
$$= g_2(T^*) g_1(T^*).$$

Example 12.2.2. Using the convolution property for the Laplace transform and an integral representation for the real matrix-variate beta function show that

$$B_p(\alpha, \beta) = \Gamma_p(\alpha) \Gamma_p(\beta) / \Gamma_p(\alpha + \beta).$$

Solution 12.2.2: Let us start with the integral representation

$$B_p(\alpha, \beta) = \int_{0<X<I} |X|^{\alpha - \frac{p+1}{2}} |I - X|^{\beta - \frac{p+1}{2}} dX,$$

$$\Re(\alpha) > \frac{p-1}{2}, \Re(\beta) > \frac{p-1}{2}.$$

Consider the integral

$$\int_{0<U<X} |U|^{\alpha - \frac{p+1}{2}} |X - U|^{\beta - \frac{p+1}{2}} dU = |X|^{\beta - \frac{p+1}{2}} \int_{0<U<X} |U|^{\alpha - \frac{p+1}{2}}$$
$$\times |I - X^{-\frac{1}{2}} U X^{-\frac{1}{2}}|^{\beta - \frac{p+1}{2}} dU$$
$$= |X|^{\alpha + \beta - \frac{p+1}{2}} \int_{0<Y<I} |Y|^{\alpha - \frac{p+1}{2}} |I - Y|^{\beta - \frac{p+1}{2}} dY,$$
$$Y = X^{-\frac{1}{2}} U X^{-\frac{1}{2}}.$$

Then

$$B_p(\alpha, \beta) |X|^{\alpha + \beta - \frac{p+1}{2}} = \int_{0<U<X} |U|^{\alpha - \frac{p+1}{2}} |X - U|^{\beta - \frac{p+1}{2}} dU. \qquad (12.2.6)$$

Take the Laplace transform on both sides to obtain the following:
On the left side,

$$B_p(\alpha, \beta) \int_{X>0} |X|^{\alpha + \beta - \frac{p+1}{2}} e^{-\mathrm{tr}(T^*X)} dX = B_p(\alpha, \beta) |T^*|^{-(\alpha + \beta)} \Gamma_p(\alpha + \beta).$$

On the right side we get,

$$\int_{X>0} e^{-\mathrm{tr}(T^*X)} \left[\int_{0<U<X} |U|^{\alpha - \frac{p+1}{2}} |X - U|^{\beta - \frac{p+1}{2}} dU \right] dX$$
$$= \Gamma_p(\alpha) \Gamma_p(\beta) |T^*|^{-(\alpha + \beta)} \quad \text{(by the convolution property in (12.2.5).)}$$

Hence

$$B_p(\alpha, \beta) = \Gamma_p(\alpha)\Gamma_p(\beta)/\Gamma_p(\alpha + \beta).$$

Example 12.2.3. Let $h(T^*)$ be the Laplace transform of $f(X)$, that is, $h(T^*) = L_f(T^*)$. Then show that the Laplace transform of $|X|^{-\frac{p+1}{2}}\Gamma_p(\frac{p+1}{2})f(X)$ is equivalent to $\int_{U>T^*} h(U)\mathrm{d}U$.

Solution: From (12.2.3) observe that for symmetric positive definite constant matrix B the following is an identity.

$$|B|^{-\alpha} = \frac{1}{\Gamma_p(\alpha)} \int_{X>0} |X|^{\alpha - \frac{p+1}{2}} e^{-\mathrm{tr}(BX)} \mathrm{d}X, \Re(\alpha) > \frac{p-1}{2}. \tag{12.2.7}$$

Then we can replace $|X|^{-\frac{p+1}{2}}\Gamma_p(\frac{p+1}{2})$ by an equivalent integral.

$$|X|^{-\frac{p+1}{2}}\Gamma_p\left(\frac{p+1}{2}\right) \equiv \int_{Y>0} |Y|^{\frac{p+1}{2} - \frac{p+1}{2}} e^{-\mathrm{tr}(XY)}\mathrm{d}Y = \int_{Y>0} e^{-\mathrm{tr}(XY)}\mathrm{d}Y.$$

Then the Laplace transform of $|X|^{-\frac{p+1}{2}}\Gamma_p\left(\frac{p+1}{2}\right)f(X)$ is given by,

$$\int_{X>0} e^{-\mathrm{tr}(T^*X)} f(X)[\int_{Y>0} e^{-\mathrm{tr}(YX)}\mathrm{d}Y] \wedge \mathrm{d}X$$

$$= \int_{X>0}\int_{Y>0} e^{-\mathrm{tr}[(T^*+Y)X]} f(X)\mathrm{d}Y \wedge \mathrm{d}X. \ (\text{Put } T^* + Y = U \Rightarrow U > T^*)$$

$$= \int_{Y>0} h(T^* + Y)\mathrm{d}Y = \int_{U>T^*} h(U)\mathrm{d}U.$$

Example 12.2.4. For $X > B, B = B' > 0$ and $v > -1$ show that the Laplace transform of $|X - B|^v$ is $|T|^{-(v+\frac{p+1}{2})}e^{-\mathrm{tr}(T^*B)}\Gamma_p(v + \frac{p+1}{2})$.

Solution 12.2.3: Laplace transform of $|X - B|^v$ with parameter matrix T^* is given by,

$$\int_{X>B} |X - B|^v e^{-\mathrm{tr}(T^*X)}\mathrm{d}X = e^{-\mathrm{tr}(BT^*)} \int_{Y>0} |Y|^v e^{-\mathrm{tr}(T^*Y)}\mathrm{d}Y, Y = X - B$$

$$= e^{-\mathrm{tr}(BT^*)} \Gamma_p\left(v + \frac{p+1}{2}\right) |T^*|^{-(v+\frac{p+1}{2})}$$

(by writing $v = v + \frac{p+1}{2} - \frac{p+1}{2}$) for $v + \frac{p+1}{2} > \frac{p-1}{2} \Rightarrow v > -1$.

Exercises 12.2.

12.2.1. By using the process in Example 12.2.3, or otherwise, show that the Laplace transform of $[\Gamma_p(\frac{p+1}{2})|X|^{-\frac{p+1}{2}}]^n f(X)$ can be written as

$$\int_{W_1 > T^*} \int_{W_2 > W_1} \cdots \int_{W_n > W_{n-1}} h(W_n) dW_1 \wedge \cdots \wedge dW_n$$

where $h(T^*)$ is the Laplace transform of $f(X)$.

12.2.2. Show that the Laplace transform of $|X|^n$ is $|T^*|^{-n-\frac{p+1}{2}}\Gamma_p(n+\frac{p+1}{2})$ for $n > -1$.

12.2.3. If the $p \times p$ real matrix random variable X has a type-1 beta density with parameters (α_1, α_2) then show that

(i) $U = (I - X)^{-\frac{1}{2}} X (I - X)^{-\frac{1}{2}} \sim$ type-2 beta (α_1, α_2)

(ii) $V = X^{-1} - I \sim$ type-2 beta (α_2, α_1)

where " $\sim$ " indicates "distributed as", and the parameters are given in the brackets.

12.2.4. If the $p \times p$ real symmetric positive definite matrix random variable X has a type-2 beta density with parameters α_1 and α_2 then show that

(i) $U = X^{-1} \sim$ type-2 beta (α_2, α_1)

(ii) $V = (I + X)^{-1} \sim$ type-1 beta (α_2, α_1)

(iii) $W = (I + X)^{-\frac{1}{2}} X (I + X)^{-\frac{1}{2}} \sim$ type-1 beta (α_1, α_2).

12.2.5. If the Laplace transform of $f(X)$ is $g(T^*) = L_{T^*}(f(X))$, where X is real symmetric positive definite and $p \times p$ then show that

$$\Delta^n g(T^*) = L_{T^*}(|X|^n f(X)), \quad \Delta = (-1)^p \left| \frac{\partial}{\partial T^*} \right|$$

where $|\frac{\partial}{\partial T^*}|$ means that first the partial derivatives with respect to t_{ij}'s for all i and j are taken, then written in the matrix form and then the determinant is taken, where $T^* = (t_{ij}^*)$.

12.3 Hypergeometric Functions with Matrix Argument

There are essentially three approaches available in the literature for defining a hypergeometric function of matrix argument. One approach due to Bochner (1952) and Herz (1955) is through Laplace and inverse Laplace transforms. Under this approach, a hypergeometric function is defined as the function satisfying a pair of integral equations, and explicit forms are available for $_0F_0$ and $_1F_0$. Another approach is available from James (1961) and Constantine (1963) through a series form involving zonal polynomials. Theoretically, explicit forms are available for general parameters or for a general $_pF_q$ but due to the difficulty in computing higher order zonal polynomials, computations are feasible only for small values of p and q. For a detailed discussion of zonal polynomials see Mathai, Provost and Hayakawa (1995). The third approach is due to Mathai (1978, 1993) with the help of a generalized matrix transform or M-transform. Through this definition a hypergeometric function is defined as a class of functions satisfying a certain integral equation. This definition is the one most suited for studying various properties of hypergeometric functions. The series form is least suited for this purpose. All these definitions are introduced for symmetric functions in the sense that

$$f(X) = f(X') = f(QQ'X) = f(Q'XQ) = f(D), \ D = \text{diag}(\lambda_1, ..., \lambda_p).$$

If X is $p \times p$ and real symmetric then there exists an orthonormal matrix Q, that is, $QQ' = I, Q'Q = I$ such that $Q'XQ = \text{diag}(\lambda_1, ..., \lambda_p)$ where $\lambda_1, ..., \lambda_p$ are the eigenvalues of X. Thus, $f(X)$, a scalar function of the $p(p+1)/2$ functionally independent elements in X, is essentially a function of the p variables $\lambda_1, ..., \lambda_p$ when the function $f(X)$ is symmetric in the above sense.

12.3.1 Hypergeometric function through Laplace transform

Let $_rF_s(a_1, ..., a_r; b_1, ..., b_s; Z)$ be the hypergeometric function of the matrix argument Z to be defined, $Z = Z'$. Consider the following pair of Laplace and inverse Laplace transforms.

$$_{r+1}F_s(a_1, ..., a_r, c; b_1, ..., b_s; -\wedge^{-1})| \wedge |^{-c}$$

$$= \frac{1}{\Gamma_p(c)} \int_{U=U'>0} e^{-\text{tr}(\wedge U)} {}_rF_s(a_1, ..., a_r; b_1, ..., b_s; -U)|U|^{c-\frac{p+1}{2}} dU \quad (12.3.1)$$

and

$$_rF_{s+1}(a_1, ..., a_r; b_1, ..., b_r, c; -\wedge)| \wedge |^{c-\frac{p+1}{2}}$$

$$= \frac{\Gamma_p(c)}{(2\pi i)^{p(p+1)/2}} \int_{\Re(Z)=X>X_0} e^{\text{tr}(\wedge Z)} {}_rF_s(a_1, ..., a_r; b_1, ..., b_s; -Z^{-1})|Z|^{-c} dZ$$

$$(12.3.2)$$

where $Z = X + iY$, $i = \sqrt{-1}$, $X = X' > 0$, and X and Y belong to the class of symmetric matrices with the non-diagonal elements weighted by $\frac{1}{2}$. The function $_rF_s$ satisfying (12.3.1) and (12.3.2) can be shown to be unique under certain conditions and that function is defined as the hypergeometric function of matrix argument $\wedge$, according to this definition.

Then by taking $_0F_0(;;-\wedge) = e^{-\text{tr}(\wedge)}$ and by using the convolution property of the Laplace transform and equations (12.3.1) and (12.3.2) one can systematically build up. The Bessel function $_0F_1$ for matrix argument is defined by Herz (1955). Thus we can go from $_0F_0$ to $_1F_0$ to $_0F_1$ to $_1F_1$ to $_2F_1$ and so on to a general $_pF_q$.

Example 12.3.1. Obtain an explicit form for $_1F_0$ from the above definition by using $_0F_0\,(;;-U) = e^{-\text{tr}(U)}$.

Solution 12.3.1: From (12.3.1)

$$\frac{1}{\Gamma_p(c)} \int_{U=U'>0} |U|^{c-\frac{p+1}{2}} e^{-\text{tr}(\wedge U)} {}_0F_0(;;-U) dU$$

$$= \frac{1}{\Gamma_p(c)} \int_{U>0} |U|^{c-\frac{p+1}{2}} e^{-\text{tr}[(I+\wedge)U]} dU = |I+\wedge|^{-c},$$

since

$$_0F_0(;;-U) = e^{-\text{tr}(U)}.$$

But

$$|I+\wedge|^{-c} = |\wedge|^{-c} |I+\wedge^{-1}|^{-c}.$$

Then from (12.3.1)

$$_1F_0(c;;-\wedge^{-1}) = |I+\wedge^{-1}|^{-c}$$

which is an explicit representation.

12.3.2 Hypergeometric function through zonal polynomials

Zonal polynomials are certain symmetric functions in the eigenvalues of the $p \times p$ matrix Z. They are denoted by $C_K(Z)$ where K represents the partition of the positive integer k, $K = (k_1,...,k_p)$ with $k_1 + \cdots + k_p = k$. When Z is 1×1 then $C_K(z) = z^k$. Thus, $C_K(Z)$ can be looked upon as a generalization of z^k in the scalar case. For details see Mathai, Provost and Hayakawa (1995). In terms of $C_K(Z)$ we have the representation for a

$$_0F_0(;;Z) = e^{\text{tr}(Z)} = \sum_{k=0}^{\infty} \frac{(\text{tr}(Z))^k}{k!} = \sum_{k=0}^{\infty} \sum_K \frac{C_K(Z)}{k!}. \qquad (12.3.3)$$

The binomial expansion will be the following:

$$_1F_0(\alpha;;Z) = \sum_{k=0}^{\infty} \sum_K \frac{(\alpha)_K C_K(Z)}{k!} = |I - Z|^{-\alpha}, \qquad (12.3.4)$$

for $0 < Z < I$, where,

$$(\alpha)_K = \prod_{j=1}^{p} \left(\alpha - \frac{j-1}{2} \right)_{k_j} , K = (k_1, ..., k_p), k_1 + \cdots + k_p = k. \qquad (12.3.5)$$

In terms of zonal polynomials a hypergeometric series is defined as follows:

$$_pF_q(a_1, ..., a_p; b_1, ..., b_q; Z) = \sum_{k=0}^{\infty} \sum_{K} \frac{(a_1)_K \cdots (a_p)_K}{(b_1)_K \cdots (b_q)_K} \frac{C_K(Z)}{k!}. \qquad (12.3.6)$$

For (12.3.6) to be defined, none of the denominator factors is equal to zero, $q \geq p$, or $q = p + 1$ and $0 < Z < I$. For other details see Constantine (1963). In order to study properties of a hypergeometric function with the help of (12.3.6) one needs the Laplace and inverse Laplace transforms of zonal polynomials. These are the following:

$$\int_{X=X'>0} |X|^{\alpha - \frac{p+1}{2}} e^{-\text{tr}(XZ)} C_K(XT) dX = |Z|^{-\alpha} C_K(TZ^{-1}) \Gamma_p(\alpha, K) \qquad (12.3.7)$$

where

$$\Gamma_p(\alpha, K) = \pi^{p(p-1)/4} \prod_{j=1}^{p} \Gamma \left[\alpha + k_j - \frac{j-1}{2} \right] = \Gamma_p(\alpha)(\alpha)_K. \qquad (12.3.8)$$

$$\frac{1}{(2\pi i)^{p(p+1)/2}} \int_{\Re(Z)=X>X_0} e^{\text{tr}(SZ)} |Z|^{-\alpha} C_K(Z) dZ$$

$$= \frac{1}{\Gamma_p(\alpha, K)} |S|^{\alpha - \frac{p+1}{2}} C_K(S), i = \sqrt{-1} \qquad (12.3.9)$$

for $Z = X + iY$, $X = X' > 0$, X and Y are symmetric and the nondiagonal elements are weighted by $\frac{1}{2}$. If the non-diagonal elements are not weighted then the left side in (12.3.9) is to be multiplied by $2^{p(p-1)/2}$. Further,

$$\int_{0<X<I} |X|^{\alpha - \frac{p+1}{2}} |I - X|^{\beta - \frac{p+1}{2}} C_K(TX) dX = \frac{\Gamma_p(\alpha, K) \Gamma_p(\beta)}{\Gamma_p(\alpha + \beta, K)} C_K(T)$$

$$\Re(\alpha) > \frac{p-1}{2}, \Re(\beta) > \frac{p-1}{2}. \qquad (12.3.10)$$

Example 12.3.2. By using zonal polynomials establish the following results:

$$_2F_1(a, b; c; X) = \frac{\Gamma_p(c)}{\Gamma_p(a)\Gamma_p(c-a)}$$

$$\times \int_{0<\wedge<I} |\wedge|^{a - \frac{p+1}{2}} |I - \wedge|^{c-a-\frac{p+1}{2}} |I - \wedge X|^{-b} d\wedge \qquad (12.3.11)$$

for $\Re(a) > \frac{p-1}{2}$, $\Re(c-a) > \frac{p-1}{2}$.

Solution 12.3.2: Expanding $|I - \wedge X|^{-b}$ in terms of zonal polynomials and then integrating term by term the right side reduces to the following:

$$|I - \wedge X|^{-b} = \sum_{k=0}^{\infty} \sum_{K} (b)_K \frac{C_K(\wedge X)}{k!} \quad \text{for } 0 < \wedge X < I$$

and

$$\int_{0 < \wedge < I} |\wedge|^{a - \frac{p+1}{2}} |I - \wedge|^{c - a - \frac{p+1}{2}} C_K(\wedge X) \mathrm{d}\wedge = \frac{\Gamma_p(a, K)\Gamma_p(c - a)}{\Gamma_p(c, K)} C_K(X)$$

by using (12.3.10). But

$$\frac{\Gamma_p(a, K)\Gamma_p(c - a)}{\Gamma_p(c, K)} = \frac{\Gamma_p(a)\Gamma_p(c - a)}{\Gamma_p(c)} \frac{(a)_K}{(c)_K}.$$

Substituting these back, the right side becomes

$$\sum_{k=0}^{\infty} \sum_{K} \frac{(a)_K (b)_K}{(c)_K} \frac{C_K(X)}{k!} = {}_2F_1(a, b; c; X).$$

This establishes the result.

Example 12.3.3. Establish the result

$${}_2F_1(a, b; c; I) = \frac{\Gamma_p(c)\Gamma_p(c - a - b)}{\Gamma_p(c - a)\Gamma_p(c - b)} \tag{12.3.12}$$

for $\Re(c - a - b) > \frac{p-1}{2}$, $\Re(c - a) > \frac{p-1}{2}$, $\Re(c - b) > \frac{p-1}{2}$.

Solution 12.3.3: In (12.3.11) put $X = I$, combine the last factor on the right with the previous factor and integrate out with the help of a matrix-variate type-1 beta integral.

Uniqueness of the ${}_pF_q$ through zonal polynomials, as given in (12.3.6), is established by appealing to the uniqueness of the function defined through the Laplace and inverse Laplace transform pair in (12.3.1) and (12.3.2), and by showing that (12.3.6) satisfies (12.3.1) and (12.3.2).

The next definition, introduced by Mathai in a series of papers is through a special case of Weyl's fractional integral.

12.3.3 Hypergeometric functions through M-transforms

Consider the class of $p \times p$ real symmetric definite matrices and the null matrix O. Any member of this class will be either positive definite or negative definite or null. Let α be a complex parameter such that $\Re(\alpha) > \frac{p-1}{2}$. Let $f(S)$ be a scalar symmetric function in the sense $f(AB) = f(BA)$ for all A and B when AB and BA are defined. Then the M-transform of $f(S)$, denoted by $M_\alpha(f)$, is defined as

$$M_\alpha(f) = \int_{U=U'>0} |U|^{\alpha - \frac{p+1}{2}} f(U) \mathrm{d}U. \tag{12.3.13}$$

Some examples of symmetric functions are $e^{\pm \mathrm{tr}(S)}$, $|I \pm S|^\beta$ for nonsingular $p \times p$ matrices A and B such that,

$$e^{\pm \mathrm{tr}(AB)} = e^{\pm \mathrm{tr}(BA)}; \; |I \pm AB|^\beta = |I \pm BA|^\beta.$$

Is it possible to recover $f(U)$, a function of $p(p+1)/2$ elements in $U = (u_{ij})$ or a function of p eigenvalues of U, that is a function of p variables, from $M_\alpha(t)$ which is a function of one parameter α? In a normal course the answer is in the negative. But due to the properties that are seen, it is clear that there exists a set of sufficient conditions by which $M_\alpha(f)$ will uniquely determine $f(U)$. It is easy to note that the class of functions defined through (12.3.13) satisfy the pair of integral equations (12.3.1) and (12.3.2) defining the unique hypergeometric function.

A hypergeometric function through M-transform is defined as a class of functions $_rF_s^*$ satisfying the following equation:

$$\int_{X=X'>0} |X|^{\alpha - \frac{p+1}{2}} {}_rF_s^*(a_1, ..., a_p; b_1, ..., b_q; -X) \mathrm{d}X$$

$$= \frac{\{\prod_{j=1}^s \Gamma_p(b_j)\}}{\{\prod_{j=1}^r \Gamma_p(a_j)\}} \frac{\{\prod_{j=1}^r \Gamma_p(a_j - \rho)\}}{\{\prod_{j=1}^s \Gamma_p(b_j - \rho)\}} \Gamma_p(\rho) \tag{12.3.14}$$

where ρ is an arbitrary parameter such that the gammas exist.

Example 12.3.4. Re-establish the result

$$L_T(|X - B|^\nu) = \Gamma_p\left(\nu + \frac{p+1}{2}\right) |T|^{-(\nu + \frac{p+1}{2})} e^{-\mathrm{tr}(TB)} \tag{12.3.15}$$

by using M-transforms.

Solution 12.3.4: We will show that the M-transforms on both sides of (12.3.15) are one and the same. Taking the M-transform of the left-side, with respect to the parameter ρ, we have,

$$\int_{T>0} |T|^{\rho - \frac{p+1}{2}} \{L_T(|X-B|^\nu)\} \mathrm{d}T = \int_{T>0} |T|^{\rho - \frac{p+1}{2}} \left[\int_{X>B} |X-B|^\nu e^{-\mathrm{tr}(TX)} \mathrm{d}X\right] \mathrm{d}T$$

$$= \int_{T>0} |T|^{\rho - \frac{p+1}{2}} e^{-\mathrm{tr}(TB)} \left[\int_{Y>0} |Y|^\nu e^{-\mathrm{tr}(TY)} \mathrm{d}Y\right] \mathrm{d}T.$$

Noting that $\nu = \nu + \frac{p+1}{2} - \frac{p+1}{2}$ the Y-integral gives $|T|^{-\nu - \frac{p+1}{2}} \Gamma_p(\nu + \frac{p+1}{2})$. Then the T-integral gives

$$M_\rho(\text{left-side}) = \Gamma_p\left(v + \frac{p+1}{2}\right)\Gamma_p\left(\rho - v - \frac{p+1}{2}\right) |B|^{-\rho+v+\frac{p+1}{2}}.$$

M-transform of the right side gives,

$$M_\rho(\text{right-side}) = \int_{T>0} |T|^{\rho-\frac{p+1}{2}} \left\{ \Gamma_p\left(v + \frac{p+1}{2}\right) |T|^{-(v+\frac{p+1}{2})} e^{-\text{tr}(TB)} \right\} dT$$

$$= \Gamma_p\left(v + \frac{p+1}{2}\right)\Gamma_p\left(\rho - v - \frac{p+1}{2}\right) |B|^{-\rho+v+\frac{p+1}{2}}.$$

The two sides have the same M-transform.

Starting with $_0F_0(;;X) = e^{\text{tr}(X)}$, we can build up a general $_pF_q$ by using the M-transform and the convolution form for M-transforms, which will be stated next.

12.3.4 A convolution theorem for M-transforms

Let $f_1(U)$ and $f_2(U)$ be two symmetric scalar functions of the $p \times p$ real symmetric positive definite matrix U, with M-transforms $M_\rho(f_1) = g_1(\rho)$ and $M_\rho(f_2) = g_2(\rho)$ respectively. Let

$$f_3(S) = \int_{U>0} |U|^\beta f_1(U^{\frac{1}{2}} S U^{\frac{1}{2}}) f_2(U) dU \qquad (12.3.16)$$

then the M-transform of f_3 is given by,

$$M_\rho(f_3) = g_1(\rho) g_2\left(\beta - \rho + \frac{p+1}{2}\right). \qquad (12.3.17)$$

The result can be easily established from the definition itself by interchanging the integrals.

Example 12.3.5. Show that

$$_1F_1(a;c;-\wedge) = \frac{\Gamma_p(c)}{\Gamma_p(a)\Gamma_p(c-a)} \int_{0<U<I} |U|^{a-\frac{p+1}{2}} |I-U|^{c-a-\frac{p+1}{2}} e^{-\text{tr}(\wedge U)} dU.$$

$$(12.3.18)$$

Solution 12.3.5: We will establish this by showing that both sides have the same M-transforms. From the definition in (12.3.14) the M-transform of the left side with respect to the parameter ρ is given by the following:

$$M_\rho(\text{left-side}) = \int_{\wedge=\wedge'>0} |\wedge|^{\rho-\frac{p+1}{2}} \, _1F_1(a;c;-\wedge) d\wedge$$

$$= \left[\frac{\Gamma_p(a-\rho)}{\Gamma_p(c-\rho)}\Gamma_p(\rho)\right] \frac{\Gamma_p(c)}{\Gamma_p(a)}.$$

$$M_\rho(\text{right-side}) = \int_{\wedge > 0} |\wedge|^{\rho - \frac{p+1}{2}} \left\{ \frac{\Gamma_p(c)}{\Gamma_p(a)\Gamma_p(c-a)} \right.$$
$$\left. \times \int_{0 < U < I} |U|^{a - \frac{p+1}{2}} |I - U|^{c-a-\frac{p+1}{2}} e^{-\text{tr}(\wedge U)} dU \right\} d\wedge.$$

Take,

$$f_1(U) = e^{-\text{tr}(U)} \text{ and } f_2(U) = |U|^{a-\frac{p+1}{2}} |I - U|^{c-a-\frac{p+1}{2}}.$$

Then

$$M_\rho(f_1) = g_1(\rho) = \int_{U > 0} |U|^{\rho - \frac{p+1}{2}} e^{-\text{tr}(U)} dU = \Gamma_p(\rho), \Re(\rho) > \frac{p-1}{2}.$$

$$M_\rho(f_2) = g_2(\rho) = \int_{U > 0} |U|^{\rho - \frac{p+1}{2}} |U|^{a-\frac{p+1}{2}} |I - U|^{c-a-\frac{p+1}{2}} dU$$

$$= \frac{\Gamma_p(a+\rho-\frac{p+1}{2})\Gamma_p(c-a)}{\Gamma_p(c+\rho-\frac{p+1}{2})}, \Re(c-a) > \frac{p-1}{2},$$
$$\Re(a+\rho) > p, \Re(c+\rho) > p.$$

Taking f_3 in (12.3.16) as the second integral on the right above we have

$$M_\rho(\text{right-side}) = \left\{ \frac{\Gamma_p(c)}{\Gamma_p(a)} \right\} \Gamma_p(\rho) \frac{\Gamma_p(a-\rho)}{\Gamma_p(c-\rho)} = M_\rho(\text{left-side}).$$

Hence the result.

Almost all properties, analogous to the ones in the scalar case for hypergeometric functions, can be established by using the M-transform technique very easily. These can then be shown to be unique, if necessary, through the uniqueness of Laplace and inverse Laplace transform pair. Theories for functions of several matrix arguments, Dirichlet integrals, Dirichlet densities, their extensions, Appell's functions, Lauricella functions, and the like, are available. Then all these real cases are also extended to complex cases as well. For details see Mathai (1997). Problems involving scalar functions of matrix argument, real and complex cases, are still being worked out and applied in many areas such as statistical distribution theory, econometrics, quantum mechanics and engineering areas. Since the aim in this brief note is only to introduce the subject matter, more details will not be given here.

Exercises 12.3.

12.3.1. Show that for $\wedge = \wedge' > 0$ and $p \times p$,
$$_1F_1(a;c;-\wedge) = e^{-\text{tr}(\wedge)} {}_1F_1(c-a;c;\wedge).$$

12.3.2. For $p \times p$ real symmetric positive definite matrices $\wedge$ and $\vee$ show that
$$_1F_1(a;c;-\wedge) = \frac{\Gamma_p(c)}{\Gamma_p(a)\Gamma_p(c-a)} |\wedge|^{-(c-\frac{p+1}{2})} \int_{0 < \vee < \wedge} e^{-\text{tr}(\vee)}$$
$$\times |\vee|^{a-\frac{p+1}{2}} |\wedge - \vee|^{c-a-\frac{p+1}{2}} d\vee.$$

12.3.3. Show that for ε a scalar and A a $p \times p$ matrix with p finite

$$\lim_{\varepsilon \to 0} |I + \varepsilon A|^{-\frac{1}{\varepsilon}} = \lim_{\varepsilon \to \infty} |I + \frac{A}{\varepsilon}|^{-\varepsilon} = e^{-\text{tr}(A)}.$$

12.3.4. Show that

$$\lim_{a \to \infty} {}_1F_1(a;c;-\frac{Z}{a}) = \lim_{\varepsilon \to 0} {}_1F_1\left(\frac{1}{\varepsilon};c;-\varepsilon Z\right)$$
$$= {}_0F_1(;c;-Z).$$

12.3.5. Show that

$$_1F_1(a;c;-\wedge) = \frac{\Gamma_p(c)}{(2\pi i)^{\frac{p(p+1)}{2}}} \int_{\Re(Z)=X>X_0} e^{\text{tr}(Z)}|Z|^{-c}|I + \wedge Z^{-1}|^{-a}dZ.$$

12.3.6. Show that

$$_2F_1(a,b;c;X) = |I - X|^{-\beta} {}_2F_1(c - a,b;c;-X(I - X)^{-1}).$$

12.3.7. For $\Re(s) > \frac{p-1}{2}, \Re(b-s) > \frac{p-1}{2}, \Re(c-a-s) > \frac{p-1}{2}$, show that

$$\int_{0<X<I} |X|^{s-\frac{p+1}{2}} |I - X|^{b-s-\frac{p+1}{2}} {}_2F_1(a,b;c;X)dX$$

$$= \frac{\Gamma_p(c)\Gamma_p(s)\Gamma_p(b-s)\Gamma_p(c-a-s)}{\Gamma_p(b)\Gamma_p(c-a)\Gamma_p(c-s)}.$$

12.3.8. Defining the Bessel function $A_r(S)$ with $p \times p$ real symmetric positive definite matrix argument S, as

$$A_r(S) = \frac{1}{\Gamma_p(r+\frac{p+1}{2})} {}_0F_1(;r+\frac{p+1}{2};-S), \qquad (12.3.19)$$

show that

$$\int_{S>0} |S|^{\delta-\frac{p+1}{2}} A_r(S)e^{-\text{tr}(\wedge S)}dS = \frac{\Gamma_p(\delta)}{\Gamma_p(r+\frac{p+1}{2})}|\wedge|^{-\delta} {}_1F_1\left(\delta;r+\frac{p+1}{2};-\wedge^{-1}\right).$$

12.3.9. If

$$M(\alpha,\beta;A) = \int_{X=X'>0} |X|^{\alpha-\frac{p+1}{2}}|I + X|^{\beta-\frac{p+1}{2}}e^{-\text{tr}(AX)}dX,$$
$$\Re(\alpha) > \frac{p-1}{2}, A = A' > 0$$

then show that

$$\int_{X>0} |X + A|^{\nu}e^{-\text{tr}(TX)}dX = |A|^{\nu+\frac{p+1}{2}}M\left(\frac{p+1}{2},\nu+\frac{p+1}{2};A^{\frac{1}{2}}TA^{\frac{1}{2}}\right).$$

12.3.10. If Whittaker function W is defined as

$$\int_{Z>0} |Z|^{\mu-\frac{p+1}{2}} |I+Z|^{\nu-\frac{p+1}{2}} e^{-\mathrm{tr}(AZ)} dZ$$
$$= |A|^{-\frac{\mu+\nu}{2}} \Gamma_p(\mu) e^{\frac{1}{2}\mathrm{tr}(A)} W_{\frac{1}{2}(\nu-\mu),\frac{1}{2}(\nu+\mu-\frac{(p+1)}{2})}(A)$$

then show that

$$\int_{X>U} |X+B|^{2\alpha-\frac{p+1}{2}} |X-U|^{2q-\frac{p+1}{2}} e^{-\mathrm{tr}(MX)} dX$$
$$= |U+B|^{\alpha+q-\frac{p+1}{2}} |M|^{-(\alpha+q)} e^{\frac{1}{2}\mathrm{tr}[(B-U)M]} \Gamma_p(2q)$$
$$\times W_{(\alpha-q),(\alpha+q-\frac{(p+1)}{4})}(S), S = (U+B)^{\frac{1}{2}} M(U+B)^{\frac{1}{2}}.$$

12.4 A Pathway Model

As an application of a real scalar function of matrix argument, we will introduce a general real matrix-variate probability model, which covers almost all real matrix-variate densities used in multivariate statistical analysis. Through the density introduced here, a pathway is created to go from one functional form to another, to go from matrix-variate type-1 beta to matrix-variate type-2 beta to matrix-variate gamma to matrix-variate Gaussian densities.

12.4.1 The pathway density

Let $X = (x_{ij}), i = 1, ..., p, j = 1, ..., r, r \geq p$ be of rank p and of real scalar variables x_{ij}'s for all i and j, and having the density $f(X)$, where

$$f(X) = c|A^{\frac{1}{2}} X B X' A^{\frac{1}{2}}|^{\alpha} |I - a(1-q) A^{\frac{1}{2}} X B X' A^{\frac{1}{2}}|^{\frac{\beta}{1-q}} \qquad (12.4.1)$$

for $A = A' > 0$ and $p \times p, B = B' > 0$ and $r \times r$ with $I - a(1-q) A^{\frac{1}{2}} X B X' A^{\frac{1}{2}} > 0$, A and B are free of the elements in X, a, β, q are scalar constants with $a > 0, \beta > 0$, and c is the normalizing constant. $A^{\frac{1}{2}}$ and $B^{\frac{1}{2}}$ denote the real positive definite square roots of A and B respectively.

For evaluating the normalizing constant c one can go through the following procedure: Let

$$Y = A^{\frac{1}{2}} X B^{\frac{1}{2}} \Rightarrow dY = A^{\frac{r}{2}} |B|^{\frac{p}{2}} dX$$

by using Theorem 11.1.5. Let

$$U = YY' \Rightarrow dY = \frac{\pi^{\frac{rp}{2}}}{\Gamma_p\left(\frac{r}{2}\right)} |U|^{\frac{r}{2}-\frac{p+1}{2}} dU$$

by using Theorem 11.2, where $\Gamma_P(\cdot)$ is the real matrix-variate gamma function. Let, for $q < 1$,

$$V = a(1-q)U \Rightarrow dV = [a(1-q)]^{\frac{p(p+1)}{2}} dU$$

from the same Theorem 11.1.5. If $f(X)$ is a density then the total integral is 1 and therefore,

$$1 = \int_X f(X) dX = \frac{c}{|A|^{\frac{r}{2}} |B|^{\frac{p}{2}}} \int_Y |YY'|^{\alpha} |I - a(1-q)YY'|^{\frac{\beta}{1-q}} dY \tag{12.4.2}$$

$$= \frac{\pi^{\frac{rp}{2}}}{\Gamma_p\left(\frac{r}{2}\right) |A|^{\frac{r}{2}} |B|^{\frac{p}{2}}} \int_U |U|^{\alpha + \frac{r}{2} - \frac{p+1}{2}} |I - a(1-q)U|^{\frac{\beta}{1-q}} dU. \tag{12.4.3}$$

Note 12.4.1: Note that from (12.4.2) and (12.4.3) we can also infer the densities of Y and U respectively.

At this stage we need to consider three cases.

Case (1): $q, < 1$. Then $a(1-q) > 0$. Make the transformation $V = a(1-q)U$, then

$$c^{-1} = \frac{\pi^{\frac{rp}{2}}}{\Gamma_p\left(\frac{r}{2}\right) |A|^{\frac{r}{2}} |B|^{\frac{p}{2}} [a(1-q)]^{p(\alpha + \frac{r}{2})}} \int_V |V|^{\alpha + \frac{r}{2} - \frac{p+1}{2}} |I - V|^{\frac{\beta}{1-q}} dV. \tag{12.4.4}$$

The integral in (12.4.4) can be evaluated by using a real matrix-variate type-1 beta integral. Then we have

$$c^{-1} = \frac{\pi^{\frac{rp}{2}}}{\Gamma_p\left(\frac{r}{2}\right) |A|^{\frac{r}{2}} |B|^{\frac{p}{2}} [a(1-q)]^{p(\alpha + \frac{r}{2})}} \frac{\Gamma_p\left(\alpha + \frac{r}{2}\right) \Gamma_p\left(\frac{\beta}{1-q} + \frac{p+1}{2}\right)}{\Gamma_p\left(\alpha + \frac{r}{2} + \frac{\beta}{1-q} + \frac{p+1}{2}\right)} \tag{12.4.5}$$

for $\alpha + \frac{r}{2} > \frac{p-1}{2}$.

Note 12.4.2: In statistical problems usually the parameters are real and hence we will assume the parameters to be real here as well as in the discussions to follow. If α is in the complex domain then the condition will reduce to $\Re(\alpha) + \frac{r}{2} > \frac{p-1}{2}$.

Case (ii): $q > 1$.

In this case $1 - q = -(q-1)$ where $q - 1 > 0$. Then in (12.4.3) one factor in the integrand becomes

$$|I - a(1-q)U|^{\frac{\beta}{1-q}} = |I + a(q-1)U|^{-\frac{\beta}{q-1}} \tag{12.4.6}$$

and then making the substitution $V = a(q-1)U$ and then evaluating the integral by using a type-2 beta integral we have

$$c^{-1} = \frac{\pi^{\frac{rp}{2}}}{\Gamma_p\left(\frac{r}{2}\right) |A|^{\frac{r}{2}} |B|^{\frac{p}{2}} [a(q-1)]^{p(\alpha + \frac{r}{2})}} \int_V |V|^{\alpha + \frac{r}{2} - \frac{p+1}{2}} |I + V|^{-\frac{\beta}{q-1}} dV.$$

Evaluate the integral by using a real matrix-variate type-2 beta integral. We have the following:

$$c^{-1} = \frac{\pi^{\frac{rp}{2}}}{\Gamma_p\left(\frac{r}{2}\right)|A|^{\frac{r}{2}}|B|^{\frac{p}{2}}[a(q-1)]^{p(\alpha+\frac{r}{2})}} \frac{\Gamma_p\left(\alpha+\frac{r}{2}\right)\Gamma_p\left(\frac{\beta}{q-1}-\alpha-\frac{r}{2}\right)}{\Gamma_p\left(\frac{\beta}{q-1}\right)} \qquad (12.4.7)$$

for $\alpha+\frac{r}{2} > \frac{p-1}{2}$, $\frac{\beta}{q-1}-\alpha-\frac{r}{2} > \frac{p-1}{2}$.

Case (iii): $q = 1$.

When q approaches 1 from the left or from the right it can be shown that the determinant containing q in (12.4.3) and (12.4.6) approaches an exponential form, which will be stated as a lemma:

Lemma 12.4.1:

$$\lim_{q\to1}|I - a(1-q)U|^{\frac{\beta}{1-q}} = e^{-a\beta \, \text{tr}(U)}. \qquad (12.4.8)$$

This lemma can be proved easily by observing that for any real symmetric matrix U there exists an orthonormal matrix Q such that $QQ' = I = Q'Q$, $Q'UQ = \text{diag}(\lambda_1,...,\lambda_p)$ where λ_j's are the eigenvalues of U. Then

$$|I - a(1-q)U| = |I - a(1-q)QQ'UQQ'|$$
$$= |I - a(1-q)Q'UQ|$$
$$= |I - a(1-q)\text{diag}(\lambda_1,...,\lambda_p)|$$
$$= \prod_{j=1}^{p}(1 - a(1-q)\lambda_j).$$

But

$$\lim_{q\to1}(1 - a(1-q)\lambda_j)^{\frac{\beta}{1-q}} = e^{-a\beta\lambda_j}.$$

Then

$$\lim_{q\to1}|I - a(1-q)U|^{\frac{\beta}{1-q}} = e^{-a\beta\text{tr}(U)}.$$

Hence in case (iii), for $q \to 1$, we have

$$c^{-1} = \frac{\pi^{\frac{rp}{2}}}{\Gamma_p\left(\frac{r}{2}\right)|A|^{\frac{r}{2}}|B|^{\frac{p}{2}}} \int_U |U|^{\alpha+\frac{r}{2}-\frac{p+1}{2}} e^{-a\beta\text{tr}(U)} dU$$

$$= \frac{\pi^{\frac{rp}{2}}}{\Gamma_p\left(\frac{r}{2}\right)|A|^{\frac{r}{2}}|B|^{\frac{p}{2}}} \frac{\Gamma_p\left(\alpha+\frac{r}{2}\right)}{(a\beta)^{p(\alpha+\frac{r}{2})}}, \quad \alpha+\frac{r}{2} > \frac{p-1}{2} \qquad (12.4.9)$$

by evaluating the integral with the help of a real matrix-variate gamma integral.

12.4.2 A general density

For X, A, B, a, β, q as defined in (12.4.1) the density $f(X)$ there has three different forms for three different situations of q. That is,

$$f(X) = c_1 |A^{\frac{1}{2}} X B X' A^{\frac{1}{2}}|^{\alpha} |I - a(1-q) A^{\frac{1}{2}} X B X' A^{\frac{1}{2}}|^{\frac{\beta}{1-q}}, \text{ for } q < 1 \qquad (12.4.10)$$

$$= c_2 |A^{\frac{1}{2}} X B X' A^{\frac{1}{2}}|^{\alpha} |I + a(q-1) A^{\frac{1}{2}} X B X' A^{\frac{1}{2}}|^{-\frac{\beta}{q-1}}, \text{ for } q > 1 \qquad (12.4.11)$$

$$= c_3 |A^{\frac{1}{2}} X B X' A^{\frac{1}{2}}|^{\alpha} \exp\left\{-a\beta \text{tr}[A^{\frac{1}{2}} X B X' A^{\frac{1}{2}}]\right\}, \text{ for } q = 1 \qquad (12.4.12)$$

where $c_1 = c$ for $q < 1, c_2 = c$ for $q > 1$ and $c_3 = c$ for $q = 1$, given in (12.4.5), (12.4.6) and (12.4.9) respectively.

Note 12.4.3: Observe that $f(X)$ maintains a generalized real matrix-variate type-1 beta form for $-\infty < q < 1$, $f(X)$ maintains a generalized real matrix-variate type-2 beta form for $1 < q < \infty$ and $f(X)$ keeps a generalized real matrix-variate gamma form when $q \to 1$.

Note 12.4.4: If a location parameter matrix is to be introduced then in $f(X)$, replace X by $X - M$ where M is a $p \times r$ constant matrix. All properties and derivations remain the same except that now X is located at M instead of at the origin O.

Remark 12.4.1: The parameter q in the density $f(X)$ can be taken as a pathway parameter. It defines a pathway from a generalized type-1 beta form to a type-2 beta form to a gamma form. Thus a wide variety of probability models are available from $f(X)$. If the experimenter needs a model with a thicker tail or thinner tail or the right and left tails cut off, all such models are available from $f(X)$ for various values of q. For $\alpha = 0$ one has the matrix-variate Gaussian form coming from $f(X)$.

12.4.3 Arbitrary moments

Arbitrary moments of the determinant $|A^{\frac{1}{2}} X B X' A^{\frac{1}{2}}|$ is available from the normalizing constant itself for various values of q. That is, denoting the expected values by E,

$$E|A^{\frac{1}{2}} X B X' A^{\frac{1}{2}}|^h = \frac{1}{[a(1-q)]^{ph}} \frac{\Gamma_p\left(\alpha + h + \frac{r}{2}\right)}{\Gamma_p\left(\alpha + \frac{r}{2}\right)} \frac{\Gamma_p\left(\alpha + \frac{r}{2} + \frac{\beta}{1-q} + \frac{p+1}{2}\right)}{\Gamma_p\left(\alpha + h + \frac{r}{2} + \frac{\beta}{1-q} + \frac{p+1}{2}\right)}$$

$$(12.4.13)$$

$$\text{for } q < 1, \ \alpha + h + \frac{r}{2} > \frac{p-1}{2}$$

$$= \frac{1}{[a(q-1)]^{ph}} \frac{\Gamma_p\left(\alpha + h + \frac{r}{2}\right)}{\Gamma_p\left(\alpha + \frac{r}{2}\right)} \frac{\Gamma_p\left(\frac{\beta}{q-1} - \alpha - h - \frac{r}{2}\right)}{\Gamma_p\left(\frac{\beta}{q-1} - \alpha - \frac{r}{2}\right)} \qquad (12.4.14)$$

$$\text{for } q > 1, \frac{\beta}{q-1} - \alpha - h - \frac{r}{2} > \frac{p-1}{2}, \ \alpha + h + \frac{r}{2} > \frac{p-1}{2}$$

$$= \frac{1}{(a\beta)^{ph}} \frac{\Gamma_p\left(\alpha + h + \frac{r}{2}\right)}{\Gamma_p\left(\alpha + \frac{r}{2}\right)} \text{ for } q = 1, \alpha + h + \frac{r}{2} > \frac{p-1}{2}. \tag{12.4.15}$$

12.4.4 Quadratic forms

The current theory in statistical literature is based on a Gaussian or normal population and quadratic and bilinear forms in a simple random sample coming from such a normal population, or a quadratic form and bilinear forms in normal variables, independently distributed or jointly normally distributed. But from the structure of the density $f(X)$ in (12.4.1) it is evident that we can extend the theory to a much wider class. For $p = 1, r > p$ the constant matrix A is a scalar quantity. For convenience let us take it as 1. Then we have

$$A^{\frac{1}{2}}XBX'A^{\frac{1}{2}} = (x_1,...,x_r)B \begin{bmatrix} x_1 \\ x_2 \\ \vdots \\ x_r \end{bmatrix} = u(\text{say}). \tag{12.4.16}$$

Here u is a real positive definite quadratic form in the first row of X, and this row is denoted by $(x_1,..,x_r)$. Now observe that the density of u is available as a special case in $f(X)$, from (12.4.10) for $q < 1$, from (12.4.11) for $q > 1$ and from (12.4.12) for $q = 1$. [Write down the exact density in the three cases as an exercise].

12.4.5 Generalized quadratic form

For a general p, $U = A^{\frac{1}{2}}XBX'A^{\frac{1}{2}}$ is the generalized quadratic form in X where X has the density $f(X)$ in (12.4.1). The density of U is available from (12.4.3) in the following form, denoting it by $f_1(U)$. Then

$$f_1(U) = c^*|U|^{\alpha + \frac{r}{2} - \frac{p+1}{2}}|I - a(1-q)U|^{\frac{\beta}{1-q}} \tag{12.4.17}$$

where

$$c^* = \frac{[a(1-q)]^{p(\alpha + \frac{r}{2})}\Gamma_p\left(\alpha + \frac{r}{2} + \frac{\beta}{1-q} + \frac{p+1}{2}\right)}{\Gamma_p\left(\alpha + \frac{r}{2}\right)\Gamma_p\left(\frac{\beta}{1-q} + \frac{p+1}{2}\right)} \tag{12.4.18}$$

for $q < 1, \alpha + \frac{r}{2} > \frac{p-1}{2}$,

$$= \frac{[a(q-1)]^{p(\alpha+\frac{r}{2})}\Gamma_p\left(\frac{\beta}{q-1}\right)}{\Gamma_p\left(\alpha+\frac{r}{2}\right)\Gamma_p\left(\frac{\beta}{q-1}-\alpha-\frac{r}{2}\right)}, \qquad (12.4.19)$$

for $q > 1, \alpha + \frac{r}{2} > \frac{p-1}{2}, \frac{\beta}{q-1} - \alpha - \frac{r}{2} > \frac{p-1}{2},$

$$= \frac{(a\beta)^{p(\alpha+\frac{r}{2})}}{\Gamma_p\left(\alpha+\frac{r}{2}\right)}, \qquad (12.4.20)$$

for $q = 1, \alpha + \frac{r}{2} > \frac{p-1}{2}.$

$$(12.4.21)$$

12.4.6 Applications to random volumes

Another connection to geometrical probability problems is established in Mathai (2005). This is coming from the fact that the rows of the $p \times r, r \geq p$ matrix of rank p can be considered to be p linearly independent points in a r-dimensional Euclidean space. Then the determinant of XX' represents the square of the volume content of the p-parallelotope generated by the convex hull of the p linearly independent points represented by X. If the points are random points in some sense, see for example a discussion of random points and random geometrical configurations from Mathai (1999), then we are dealing with a random volume in $|XX'|^{\frac{1}{2}}$. The distribution of this random volume is of interest in geometrical probability problems when the points have specified distributions. For problems of this type see Mathai (1999). Then the distributions of such random volumes will be based on the distribution of X where X has the very general density given in (12.4.1). Thus the existing theory in this area is extended to a very general class of basic distributions covered by the $f(X)$ of (12.4.1).

Exercises 12.4.

12.4.1. By using Stirling's approximation for gamma functions, namely,

$$\Gamma(z+a) \approx \sqrt{2\pi}\, z^{z+a-\frac{1}{2}}\, e^{-z} \qquad (12.4.22)$$

for $|z| \to \infty$ and a a bounded quantity, show that the moment expressions in (12.4.13) and (12.4.14) reduce to the moment expression in (12.4.15).

12.4.2. By opening up $\Gamma_p(\cdot)$ in terms of gamma functions and by examining the structure of the gamma products in (12.4.13) show that for $q < 1$ we can write

$$E[|a(1-q)A^{\frac{1}{2}}XBX'A^{\frac{1}{2}}|^h] = \prod_{j=1}^{p} E(x_j^h) \qquad (12.4.23)$$

where x_j is a real scalar type-1 beta with the parameters

$$\left(\alpha + \frac{r}{2} - \frac{j-1}{2}, \frac{\beta}{1-q} + \frac{p+1}{2}\right), \; j = 1,...,p.$$

12.4.3. By going through the procedure in Exercise 12.4.2 show that, for $q > 1$,

$$E[|a(q-1)A^{\frac{1}{2}}XBX'A^{\frac{1}{2}}|^h] = \prod_{j=1}^{p} E(y_j^h) \tag{12.4.24}$$

where y_j is a real scalar type-2 beta random variable.

12.4.4. Let $q < 1, a(1-q) = 1, Y = XBX', Z = A^{\frac{1}{2}}XBX'A^{\frac{1}{2}}$ where X has the density $f(X)$ of (12.4.1). Then show that Y has the non-standard real matrix-variate type-1 beta density and Z has standard type-1 beta density.

12.4.5. Let $q < 1, a(1-q) = 1, \alpha + \frac{r}{2} = \frac{p+1}{2}, \beta = 0, Z = A^{\frac{1}{2}}XBX'A^{\frac{1}{2}}$ where X has the density $f(X)$ of (12.4.1). Then show that Z has a standard uniform density.

12.4.6. Let $q < 1, \alpha = 0, a(1-q) = 1, \frac{\beta}{1-q} = \frac{1}{2}(m-p-r-1)$. Then show that the $f(X)$ of (12.4.1) reduces to the inverted T density of Dickey.

12.4.7. Let $q > 1, a(q-1) = 1, Y = XBX', Z = A^{\frac{1}{2}}XBX'A^{\frac{1}{2}}$. Then when X has the density in (12.4.1) show that Y has the non-standard matrix-variate type-2 beta density and Z has the standard type-2 beta density.

12.4.8. Let $q = 1, a = 1, \beta = 1, \alpha + \frac{r}{2} = \frac{n}{2}, Y = XBX', A = \frac{1}{2}V^{-1}$. Then show that Y has a Wishart density when X has the density in (12.4.1).

12.4.9. Let $q = 1, a = 1, \beta = 1, \alpha = 0$ in $f(X)$ of (12.4.1). Then show that $f(X)$ reduces to the real matrix-variate Gaussian density.

12.4.10. Let $q > 1, a(q-1) = 1, \alpha + \frac{r}{2} = \frac{p+1}{2}, \frac{\beta}{q-1} = 1, Y = A^{\frac{1}{2}}XBX'A^{\frac{1}{2}}$. Then if X has the density in (12.4.1) show that Y has a standard real matrix-variate Cauchy density.

References

Bochner, S. (1952). Bessel functions and modular relations of higher type and hyperbolic differential equations, *Com. Sem. Math. de l'Univ. de Lund, Tome supplémentaire dédié à Marcel Riesz*, 12-20

Constantine, A.G. (1963). Some noncentral distribution problems in multivariate analysis, *Annals of Mathematical Statistics*, **34**, 1270-1285.

Herz, C.S. (1955). Bessel functions of matrix argument, *Annals of Mathematics*, **61** (3), 474-523.

James, A.T. (1961). Zonal polynomials of the real positive definite matrices, *Annals of Mathematics*, **74**, 456-469.

Mathai, A.M. (1978). Some results on functions of matrix argument, *Math. Nachr.* **84**, 171-177.

Mathai, A.M. (1993). *A Handbook of Generalized Special Functions for Statistical and Physical Sciences*, Oxford University Press, Oxford.

Mathai, A.M. (1997). *Jacobians of Matrix Transformations and Functions of Matrix Argument*, World Scientific Publishing, New York.

Mathai, A.M. (1999). *An Introduction to Geometrical Probability: Distributional Aspects with Applications*, Gordon and Breach, New York.

Mathai, A.M., Provost, S.B. and Hayakawa, T. (1995). *Bilinear Forms and Zonal Polynomials*, Springer-Verlag Lecture Notes in Statistics, **102**, New York.

Mathai, A.M. (2005). A pathway to matrix-variate gamma and normal densities, *Linear Algebra and Its Applications*, **396**, 317-328.

Author Index

Subject Index